Textbook of Biochemistry

Textbook of Biochemistry

Edited by **Artie Weissberg**

Callisto Reference
New York

Published by Callisto Reference,
106 Park Avenue, Suite 200,
New York, NY 10016, USA
www.callistoreference.com

Textbook of Biochemistry
Edited by Artie Weissberg

International Standard Book Number: 978-1-63239-705-8 (Hardback)

Printed in China.

Contents

Preface

This book has been an outcome of determined endeavour from a group of educationists in the field. The primary objective was to involve a broad spectrum of professionals from diverse cultural background involved in the field for developing new researches. The book not only targets students but also scholars pursuing higher research for further enhancement of the theoretical and practical applications of the subject.

Biochemistry is the branch of science which deals with studying the chemical processes taking place in or related to living organisms. It incorporates studies regarding cells, examining components like proteins, lipids, organelles, etc. The scientists working in this field also study the interactions between cells and examine how it takes place, while fighting illness or during growth. This book outlines the processes and applications related to this area. It discusses the fundamentals as well as modern approaches of this field. It presents researches and studies performed by experts across the globe. Scientists and students actively engaged in this field will find this book full of crucial and unexplored concepts. It will serve as beneficial source of reference for readers.

It was an honour to edit such a profound book and also a challenging task to compile and examine all the relevant data for accuracy and originality. I wish to acknowledge the efforts of the contributors for submitting such brilliant and diverse chapters in the field and for endlessly working for the completion of the book. Last, but not the least; I thank my family for being a constant source of support in all my research endeavours.

Editor

1

Molecular Cloning and Characterization of a Candidate Plant Growth-Related and Time-Keeping Constitutive Cell Surface Hydroquinone (NADH) Oxidase (ENOX1) from *Arabidopsis lyrata*

Xiaoyu Tang, Laura M. C. Ades, Dorothy M. Morré, D. James Morré*

MorNuCo, Inc., Purdue Research Park, West Lafayette, IN, USA
Email: *Dj_morre@yahoo.com

Abstract

ENOX (ECTO-NOX) proteins are proteins of the external surface of the plasma membrane that catalyze oxidation of both NADH and hydroquinones as well as carry out protein disulfide-thiol interchange. They exhibit both prion-like and time-keeping (clock) properties. The oxidative and interchange activities alternate to generate a regular period of 24 min in length. Here we report the cloning, expression, and characterization of a plant candidate constitutive ENOX (CNOX or ENOX1) protein from *Arabidopsis lyrata*. The gene encoding the 335 (165) amino acid protein is found in accession XP-002882467. Functional motifs characteristics of ENOX proteins previously identified by site-directed mutagenesis and present in the candidate ENOX1 protein from plants include adenine nucleotide and copper binding motifs along with essential cysteines. However, the drug binding motif (EEMTE) sequence of human ENOX2 is absent. The activities of the recombinant protein expressed in *E. coli* were unaffected by capsaicin, EGCg, and other ENOX2-inhibiting substances. Periodic oxidative activity was exhibited both with NAD(P)H and reduced coenzyme Q as substrate. Bound copper was necessary for activity and activity was inhibited by the ENOX1-specific inhibitor simalikalactone D. Addition of melatonin phased the 24-min period such that the next complete period began 24 min after the melatonin addition as appeared to be characteristic of ENOX1 activities in general. Periodic protein disulfide-thiol interchange activity also was demonstrated along with the 2 oxidative plus 3 interchange activity pattern characteristics of the 24-min ENOX1 protein period. Concentrated solutions of the purified plant ENOX1 protein formed insoluble aggregates, devoid of enzymatic activity, resembling amyloid. Activity was restored to

*Corresponding author.

aggregate preparations by isoelectric focusing.

Keywords

ENOX1, ECTO-NOX Proteins, Plant Growth, Hydroquinone (NADH) Oxidation, Plasma Membrane Electron Transport, *Arabadopsis lyrata*

1. Introduction

One or more hormone-responsive external plasma membrane hydroquinone oxidases capable of catalyzing protein disulfide interchange and that oxidize NAD(P)H as an alternate substrate (External NADH oxidase = ENOX) have been exhibited by all animal and plant species, thus far investigated, as well as by yeast [1]-[3]. Based on activity characteristics, three related ENOX proteins have emerged. One (CNOX or ENOX1) is both constitutive and widely distributed [4]. A second ENOX activity designated tNOX or ENOX2 is tumor or cancer-associated [5] [6]. ENOX2 proteins are inhibited by a series of quinone site inhibitors all with anticancer activity including the vanilloid capsaicin [7] [8], the antitumor sulfonylurea LY181984 [9] [10], the isoflavone phenoxodiol [11] and the quassinoid glaucarubolone [12] whereas ENOX1 proteins are resistant [11] to these substances but inhibited specifically by the quassinoid simalikalactone D which has no effect on ENOX2 activity [13]. Shed forms of both activities are found in culture media conditioned by the growth of mammalian [10] or plant cells. A 33.5-kDa protein with capsaicin-inhibited NADH oxidase activity was purified from spent media of cultured HeLa cells [14] and from sera of cancer patients [5]. The NADH oxidase activities from sera of healthy volunteers or spent media of non-cancer cells were identified as ENOX1 [15]. A third group of proteins with ENOX-like activity designated arNOX/ENOX3 [16] is age-related [17] and evident only in aged individuals, late passage cultured cells, or senescing [19] short-lived plant parts [17] [18].

The ENOX1 and ENOX 2 proteins function in the enlargement phase of cell growth [19]-[21]. When their activity is inhibited, cells are unable to enlarge [2] [7] [13] [22]. Cancer cells unable to enlarge to a size sufficient to allow division following inhibition of ENOX2 undergo apoptosis. *Arabidopsis* seedlings treated with the quassinoid ENOX1 inhibitor simalikalactone D ceased growth until the drug was metabolized but did not die [13].

A distinguishing characteristic of ENOX proteins that permits their unequivocal identification is that the proteins exhibit two activities that alternate [1]-[3]. The first activity is that of a hydroquinone oxidase (NADH serves as an alternate substrate for this activity) [23]. The second is that of a protein disulfide-thiol interchange measured either from the restoration of activity to inactive (scrambled) RNase [24] or from the cleavage of dithiodipyridine substrates [25]. Each activity generates a distinct oscillatory activity with a period length of 24 min for ENOX1. The strictly periodic activity of the NOX proteins (ENOX1 and ENOX2 as well as arNOX) distinguishes their activity from all other oxidase or protein disulfide isomerase forms [6] and imparts to the ENOX1 protein a potential role as an ultradian (with period lengths of less than 24 h) oscillator of the cellular biological clock [26].

In a previous report [15], purification of a protease-resistant protein fragment with drug-resistant ENOX1 activity and a period length of 24 min from human serum protein, was described. The protein fragment was blocked to direct sequencing and was resistant to further protease digestion. Polyclonal antibodies raised to the fragment partially blocked total NOX activity of human sera from the surface of human non-cancer cells. They did not react with recombinant ENOX2 or with molecular species identified as ENOX2 from sera of cancer patients or from cancer cell lines grown in culture. The human ENOX1 was the first candidate ENOX1 to be cloned and expressed as a recombinant protein [4].

ENOX proteins lack iron or iron-sulfur clusters but still reduce oxygen. The previously cloned and sequenced ENOX2 protein contains a copper site conserved with the copper site of the enzyme Mg^{2+} superoxide dismutase [6]. Site-directed mutagenesis of this copper site resulted in the loss of enzymatic activity when the mutant protein was expressed in *Escherichia coli* [6] [27]. These findings were extended to the candidate plant ENOX1 protein with the discovery of a second copper site also required for activity [4]. These findings raise the possibility that the concerted four electron transfers required to reduce molecular oxygen to water are carried out by dimers of the dicopper-containing ENOX monomers. In this report, a candidate plant ENOX1 protein with functional motifs consistent with this activity model was cloned and characterized from *Arabidopsis lyrata.*

2. Methods

2.1. Search for Candidate Constitutive ENOX1 (ENOX1) from Plants

Protein BLAST (Basic Local Alignment Search Tool) with either ENOX1 or ENOX2 sequences as a query was used for similarity searches in different databases (non-redundant protein sequences, UniProt, EST, and others) [28] with no plant proteins having significant similarity being found. However, sequence of a cloned ENOX1 from *Saccharomyces cerevisiae* [29] did reveal significant homology. The homologous protein from *Arabidopsis lyrata* was selected for evaluation as a candidate for the constitutive ENOX1 from plants.

2.2. Plasmids Construction

Plasmids carrying the *Arabidopsis* ENOX1 (M458 to V580 of hypothetical protein RALYDRAFT_477943 [*Arabidopsis lyrata* aubsp. *Lyrata*] XP_002882467) sequence were prepared by inserting the pET11a vector (between NheI and BamHI sites) within the *Arabidopsis* ENOX1 sequence. The *Arabidopsis* ENOX1 sequence was synthesized by GenScript USA Inc. (Piscataway, NJ). DNA sequences of the ligation products (pET11a-AraENOX1) were confirmed by DNA sequencing.

2.3. Expression of Recombinant *Arabidopsis* ENOX1

The pET11a-AraENOX1 was transformed to BL21 (DE3) competent cells. A single colony was picked and inoculated into the 5 ml LB + ampicillin (LB/AMP) medium. The overnight culture (1 ml) was diluted into 100 ml LB/AMP media (1:100 dilution). The cells were grown with vigorous shaking (250 rpm) at 37°C to an OD_{600} of 0.4 - 0.6 and IPTG (0.5 mM) was added for induction. Cultures were collected after 5 h incubation with shaking (250 rpm) at 37°C.

Cells were centrifuged at 5000 g for 6 min. Pellets were then resuspended in 20 mM Tris-HCl, pH 8.0, containing 0.5 mM PMSF, 1 mM benzamidine and 1 mM 6-aminocaproic and lysed by three passages through a French pressure cell (SLM Aminco) at 20,000 psi. Expression of the recombinant *Arabidopsis* ENOX1 of about 14 kDa was confirmed by SDS-PAGE with silver staining. Transformed cells were stored at −80°C in a standard glycerol stock solution. The *Arabidopsis* ENOX1 proteins were further purified on Criterion IEF gels (Bio-Rad, Hercules, CA). The IEF gel was cut into seven equal segments. The pH represented by each slice was based on IEF standards (Bio-Rad). The slices were soaked in 15 mM Tris-Mes buffer, pH 7, at 4°C for overnight with shaking. The gel-free extracts were assayed for ENOX1 activity.

2.4. Protein Determination

Protein concentrations were determined by the bicinchoninic acid (BCA) method [30] (BCA Protein Assay Kit, Thermo Scientific, Rockford, IL, USA) with bovine serum albumin as the standard.

2.5. Enzyme Activity Assays

Oxidation of NADH was determined spectrophotometrically from the disappearance of NADH measured at 340 nm in a reaction mixture containing 25 mM Tris-MES (pH 7.2), 100 μM GSH, 1 mM KCN to inhibit mitochondrial oxidase activity, 150 μM NADH and the enzyme at 37°C with temperature control (±0.5°C) and stirring. Prior to assay, 1 μM reduced glutathione was added to reduce the protein in the presence of substrate. After 10 min, 0.03% hydrogen peroxide was added to reoxidize the protein under renaturing conditions and in the presence of substrate to start the reaction. Activities were measured using paired Hitachi U3210 or paired SLM Aminco 2000 spectrophotometers both with continuous recording. Assays were for 1 min and were repeated on the same sample at intervals of 1.5 min for the times indicated. An extinction coefficient of 6.22 cm^{-1} mM^{-1} was used to determine specific activity.

Oxidation of reduced coenzyme Q_{10} ($CoQ_{10}H_2$) was measured as the disappearance of $CoQ_{10}H_2$ at both 290 nM and 410 nM [23]. The reaction was started with the addition of 40 μl of 5 mM $Q_{10}H_2$ (Tischcon Corp., Westbury, NY). An extinction coefficient of 0.805 mM^{-1} cm^{-1} was used to calculate the rate of $Q_{10}H_2$ oxidation.

Protein disulfide-thiol interchange was determined spectrophotometrically from the increase in absorbance at 340 nm resulting from the cleavage of dithiodipyridine (DTDP) [25]. DTDP cleavage was buffered 50 mM Tris-MES, pH 7. The assay was preincubated (1 h at room temperature) with 0.5 μmoles DTDP in 5 μl of dimethyl sulfoxide (DMSO) to react with endogenous reductants present with the plasma membranes. After 10 min, a further 3.5 μmoles DTDP were added in 35 μl DMSO to start the reaction. The final reaction volume was 2.5 ml. The reaction was monitored from the increase in absorbance at 340 nm. Specific activities were calculated using a milimolar absorption coefficient of 6.21.

2.6. Removal of Copper (II) from ENOX1

IEF purified ENOX1 was concentrated to 0.7 mg/ml by using a Centricon concentrator (Millipore Corporation, Danvers, MA) fitted with a 10,000 nominal molecular weight limit ultracel YM membrane. Samples (50 μl) were combined with 1 μl of trifluoroacetic acid (TFA) in the presence or absence of 15 μl 10 mM bathocuproine. After 2 h of incubation at room temperature, the samples were dialyzed (Spectra/Pro Dialysis membrane, molecular weight cut-off 6000 - 8000, Spectrum Laboratories (Rancho Dominquez, CA) against 20 mM Tris-HCl, pH 8, at 4°C overnight.

2.7. Site-Directed Mutagenesis

Amino acids indicated were replaced by alanines by site directed mutagenesis according to Braman *et al.* [31]. Numbered amino acids and nucleotide positions of splice variant products refer to numbers assigned to amino acids of the full length transcript.

2.8. Statistical Analyses

Means and standard deviations were analyzed for statistical significance using a two-tailed test.

3. Results

The identification of the candidate plant the ENOX1 from *Arabidopsis lyrata* was based on a homology (BLAST) search by comparison with the ENOX1 (YML117W) from *Saccharomyces cerevisiae* [29]. The 14 kDa amino acid sequence selected (**Figure 1**) had 37% identity and 58% similarity between amino acids 84 and 126 of XP002882467 from *Arabidopsis* and amino acids 932-968 of EDN64277 (YML117W) from yeast.

Potential functional motifs within the 14 kDa transcript included a potential NADH binding site at G570XGXXL which aligned with G958XGXXV in YML117W. Potential protein disulfide sites were located at M458XXXXCC and M527XXXXXXC along with C534. Potential copper sites were at H466PY, Y531LY (which over laps M527XXXXXXC) and Y479XXXXH.

Expression of the recombinant *Arabidopsis* ENOX1 with a molecular weight of about 14 kDa was confirmed by SDS-PAGE with silver staining (**Figure 2**).

3.1. Protein Characterization

A continuous trace of an IEF-purified preparation of the recombinant *Arabidopsis* ENOX1 protein illustrates the characteristic oscillatory activity (**Figure 3**). Intervals of rapid activity (arrows) were interspersed with intervals of less activity. The period length was 24 min. No oscillations were observed with NADH alone or with the plant ENOX1 in the absence of NADH.

For more detailed evaluations, rates averaged over 1 min every 1.5 min with recombinant plant ENOX1 expressed in *E. coli* exhibited more clearly the oscillatory pattern of oxidation of exogenously supplied NADH characteristic of ENOX1 proteins (**Figure 4**). The repeating pattern was that of five maxima, two of which were separated by six min (arrows) and the remainder separated by 4.5 min [$6 + (4 \times 4.5) = 24$ min]. As is characteristic of ENOX1 proteins from other sources, the oscillatory pattern could be phased by the addition of 1 μM melatonin (**Figure 5**). A new maximum was observed exactly 24 min after melatonin addition and continued thereafter as phased by the melatonin addition.

As is characteristic of ENOX proteins in general, the proteins also exhibited protein disulfide-thiol interchange (protein disulfide isomerase) activity illustrated by the time-dependent cleavage of a dithiodipyridyl sub-

a. NCBI Reference Sequence: XP_002882467.1

1 MARNSNSDEA FSSEEEEERV KDNEEEDEEE LEAVARSSGS DDDEVVAADE SPVSDGEAAP
61 VEDDYEDEED EEKAEISKRE KARLKEMQKL KKQKIQEMLE SQNASIDADM NNKGKGRLKY
121 LLQQTELFAH FAKGDASSSQ KKAKGRGRHA SKITEEEEDE EYLKEEEDGL TGSGNTRLLT
181 QPSCIQGKMR DYQLAGLNWL IRLYENGING ILADEMGLGK TLQTISLLAY LHEYRGINGP
241 HMVVAPKSTL GNWMNEIRRF CPVLRAVKFL GNPEERRHIR DDLLVAGKFD ICVTSFEMAI
301 KEKTALRRFS WRYIIIDEAH RIKNENSLLS KTMRLFSTNY RLLITGTPLQ NNLHELWALL
361 NFLLPEIFSS AETFDEWFQI SGENDQQEVV QQLHKVLRPF LLRRLKSDVE KGLPPKKETI
421 LKVGMSQMQK QYYKALLQKD LEAVNAGGER KRLLNIA**MQL** **RKCCNHPYLF** **QGAEPGPPYT**
481 **TGDHLITNAG** **KMVLLDKLLP** **KLKERDSRVL** **IFSQMTRLLD** **ILEDYLMYRG** **YLYCRIDGNT**
541 **GGDERDASIE** **AYNKPGSEKF** **VFLLSTRAGG** **LGINLATADV** VILYDSDWNP QVDLQAQDRA
601 HRIGQKKEVQ VFRFCTESAI EEKVIERAYK KLALDALVIQ QGRLAEQKTV NKDELLQMVR
661 YGAEMVFSSK DSTITDEDID RIIAKGEEAT AELDAKMKKF TEDAIQFKMD DSADFYDFDD
721 DNKDENKLDF KKIVSDNWND PPKRERKRNY SESEYFKQTL RQGAPAKPKE PRIPRMPQLH
781 DFQFFNIQRL TELYEKEVRY LMQTHQKNQL KDTIDVEEPE GGDPLTAEEV EEKELLLEEG
841 FSTWSRRDFN TFLRACEKYG RNDIKSIASE MEGKTEEEVE RYAKVFKERY KELNDYDRII
901 KNIERGEARI SRKDEIMKAI GKKLDRYRNP WLELKIQYGQ NKGKLYNEEC DRFMICMIHK
961 LGYGNWDELK AAFRTSPLFR FDWFVKSRTS QELARRCDTL IRLIEKENQE FDERERQARK
1021 EKKLAKSATP SKRPLGRQAS ESPSSTKKRK HLSMR

b. Alignment of YML117W and *Arabidopsis* ENOX1

```
1 1      ----------------------------MASMQLRKCCNHPYLFQGAEPGPPYTTGDHLITNAGKMVLLDK--------- 43
2 801    CCWINFMNISSAISLVEEMNKESTVQNESGEVTLKRATEEK--FGGRYKGLLINYGKDRCGNINKNLIAGKNSRFYKKVK 878

1 44     ------LLPKLKER------DSRVLIFSQMTRL--LDILEDYLMYRGYLYCRIDGNTGGDERDASIEAYNKPGSEKFVFL 109
2 879    RPSYNIRLSKLEEKRRQNEIDKKEKAFDKPLNLESLGISLDAHKDNGGGETGTANNTG-HENESELEAENENGNETGSF- 956

1 110    LSTRAGGLGINLATADV
2 957    -----GGLGLAVASSDV
```

Figure 1. Alignment of *Saccharomyces cerevisiae* YML117W and *Arabidopsis* ENOX1. Sequence 1 is the recombinant *Arabidopsis* ENOX1. Sequence 2 is the YML117W. There is 37% (16/43) identity and 58% (25/43) similarity between recombinant *Arabidopsis* ENOX1 amino acids 84 to 126 and YML117W amino acids 932-968.

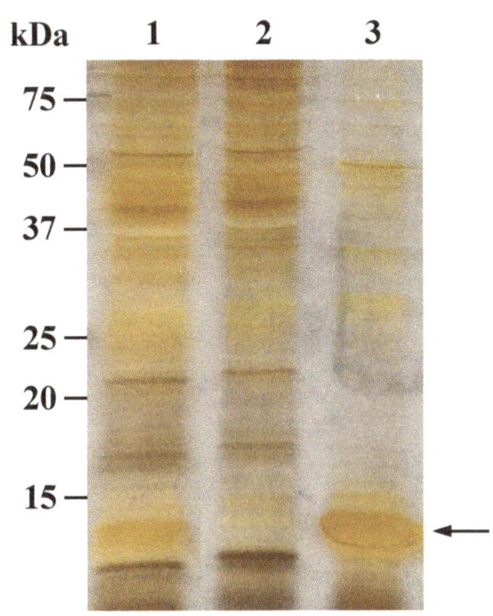

Figure 2. Expression of 14 kD recombinant *Arabidopsis* ENOX1 shown on 15% SDS-PAGE with silver staining. Lanes 1 and 2: whole cells of pET11a-AraENOX1 transformed *E. coli* (2 µl); lanes 3 and 4: supernatant of French pressed pET11a-AraENOX1 transformed *E. coli* (2 µl); lanes 5 and 6: pellet of French pressed pET11a-AraENOX1 transformed *E. coli* (2 µl). The expressed recombinant *Arabidopsis* ENOX1 was found in the pellet of French pressed *E. coli.*

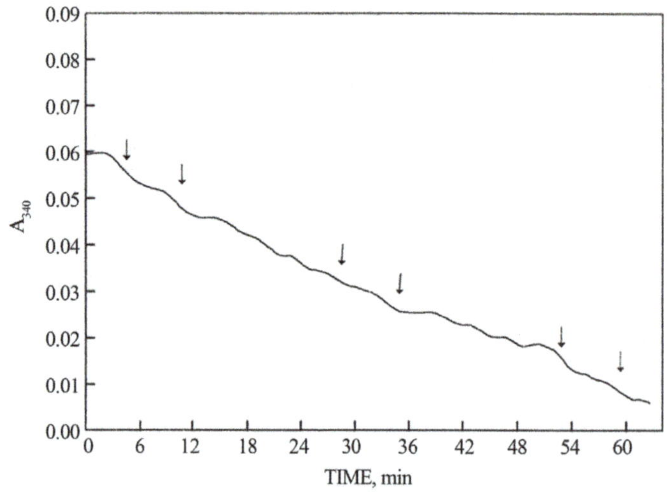

Figure 3. Continuous trace showing the decrease in A_{340} as a measure of consumption of NADH over 12 min for a fraction of IEF purified *Arabidopsis* ENOX1. The assay conditions were as described ([24], Materials and Methods) except that the NADH concentration was 0.75 mM and the data were collected automatically and stored using a SPECTRA max 340PC microplate reader. The mixture contained ca. 20 μg ENOX1 in a total volume of 200 μl.

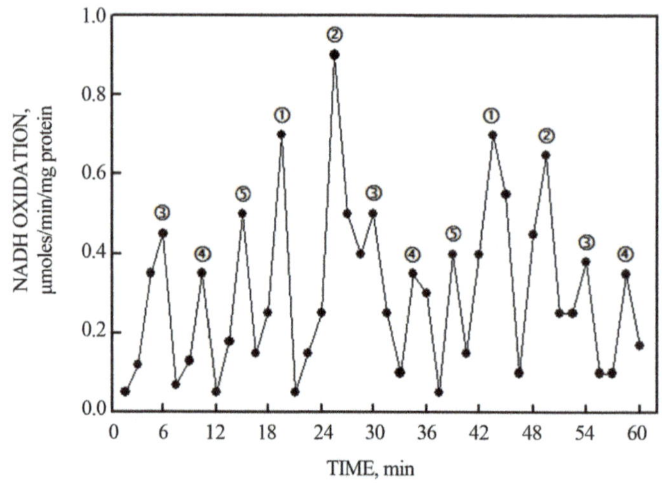

Figure 4. NADH oxidase activity of IEF purified recombinant ENOX1 of *Arabidopsis*. Illustrated is the oscillatory pattern of 5 maxima. The major maxima separated by 6 min, are labeled ①and ②. The three minor maxima that follow are separated from the major maxima and each other by 4.5 min creating the 24 min period [6 + (4.5 × 4) = 24].

strate (**Figure 6**). An oscillatory pattern similar to that for NADH oxidation was observed with a period length of 24 min (arrows). As reported previously [2], the principal maxima of the two activities, NADH oxidation and protein disulfide interchange, alternated.

The recombinant ENOX1 oxidizes reduced coenzyme Q in a standard assay (**Figure 7**) with activity measured either at A_{410} (**Figure 7(upper)**) or at A_{290} (**Figure 7(lower)**). As with NADH oxidation (**Figure 4**) and dithiodipyri-dine cleavage (**Figure 6**), the characteristic pattern of oscillations with a 24 min period was reproduced (**Figure 7**). Hydroquinones of the plasma membrane (reduced coenzyme Q for animals/reduced coenzyme Q or phylloquinone for plants) are the physiological substrates for ENOX proteins.

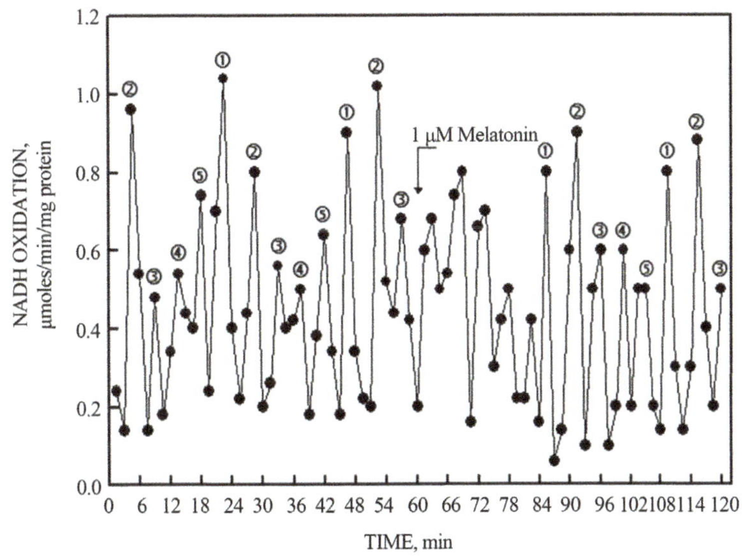

Figure 5. The NADH oxidase activity of IEF-purified recombinant ENOX1 of *Arabidopsis* and response to 1 μM melatonin. After addition of melatonin, new maxima appear 24 min following melatonin addition (arrow), an ENOX1 characteristic.

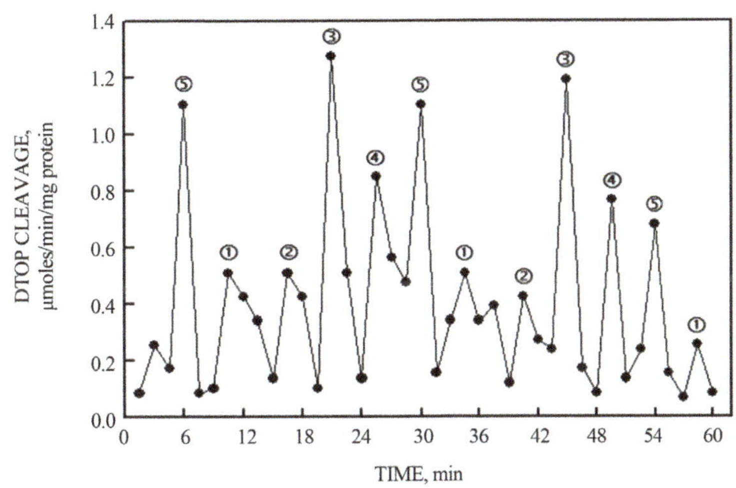

Figure 6. Protein disulfide-thiol interchange activity of IEF-purified recombinant *Arabidopsis* ENOX1 measured from the cleavage of a dithiodipyridine (DTDP) substrate. An oscillatory activity was observed with the activities were most strongly associated with the three maxima separated by 4.5 min rather than with the two maxima separated by 6 min.

Primarily through reduction of the aggregation of the recombinant proteins, further purification by isoelectric focusing was required to achieve the reported specific activities. Highest specific activities were achieved at a focusing pH of about 6.9 (**Figure 8**) which approximates the calculated isoelectric point of the recombinant protein.

The ENOX activity eluted from the IEF gel was further identified as ENOX1 by its resistance to various ENOX2 inhibitors including cis-platinum, phenoxodiol, EGCg and capsaicin all tested at concentrations sufficient to inhibit ENOX2 activity completely (**Table 1**). Also, without effect, were tyrosol and gallic acid, specific inhibitors of arNOX (ENOX 3) and 2,4-dichlorophenoxyacetic acid (2,4-D) which serves as a specific activator of plant arNOX (ENOX 5) (**Table 1**). With the recombinant *Arabidopsis* ENOX1 protein eluted from the IEF gels, no inhibition was observed. Activity was inhibited by the ENOX1-specific quassinoid inhibitor simalika

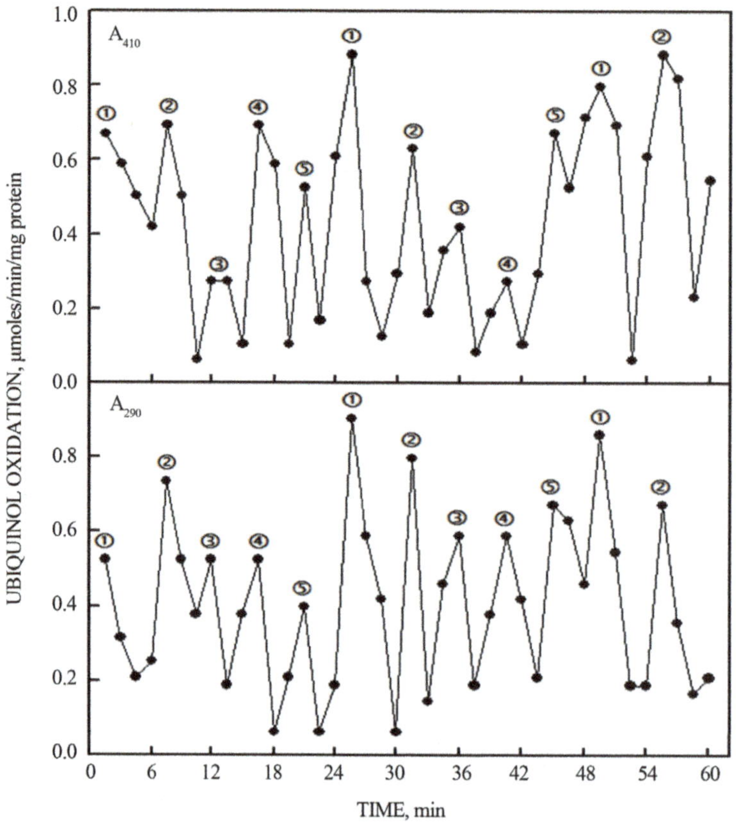

Figure 7. Ability of recombinant *Arabidopsis* ENOX1 to oxidize hydroquinone (reduced coenzyme Q) measured either by an increase in A_{410} (upper) or a decrease in A_{290} (lower). As with NADH oxidation of **Figure 3**, the activity oscillates with prominent maxima separated by 6 min, ① and ②, to create a 24 min period containing 3 additional maxima separated by 4.5 min (total of 5 maxima).

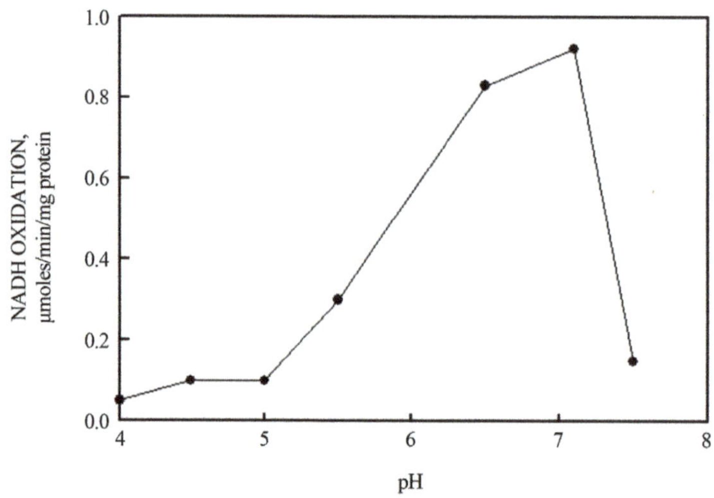

Figure 8. Purification and activation of recombinant *Arabidopsis* ENOX1 by isoelectric focusing.

lactone D (**Figure 9**) along with the growth regulating herbicides mefluidide and sulfosulfuron (**Table 1**). The

Table 1. NADH oxidase activity of IEF-purified ENOX1 recombinant Arabidopsis and response to ENOX2 inhibitors, 2,4-D and the ENOX1 inhibitor simalikalactone D. Average of 3 determinations ± standard deviations.

Inhibitor	μmoles/min/mg protein
None	3.7 ± 0.4
Cis-platinum (100 μM)	3.5 ± 0.02
Phenoxodiol (10 μM)	3.7 ± 0.05
EGCg (500 μM)	3.9 ± 0.05
Capsaicin (1 μM)	3.7 ± 0.1
Tyrosol (10 μM)	3.4 ± 0.2
Gallic acid (100 μM)	3.0 ± 0.5
Simalikalactone D (1 μM)	0.9 ± 0.1**
2,4-Dichlorophenoxyacetic acid (1 μM)	3.7 ± 0.1
Mefluidide (*N*-[2,4-dimethyl-5-[(trifluoromethyl)sulfonyl]amino]phenyl]acetamide) (100 μM)	1.8 ± 0.15*
Sulfosulfuron sulfonylurea herbicide (Trade Name: Outrider) (100 μM)	1.4 ± 0.5*

The concentration of cis-platinum, phenoxodiol, EGCg and capsaicin used results in >90% inhibition of recombinant human ENOX2 assayed in parallel. The concentrations of tyrosol and gallic acid, specific inhibitors of arNOX (ENOX3), which, in separate experiments, resulted in >90% inhibition of ENOX3 and 2,4-D at 1 μM, which specifically activates the dNOX of plants were without effect. Simalikalactone D is a general ENOX 1 inhibitor. *Very significant (p = 0.005 - 0.0034). **Extremely significant p = 0.0003).

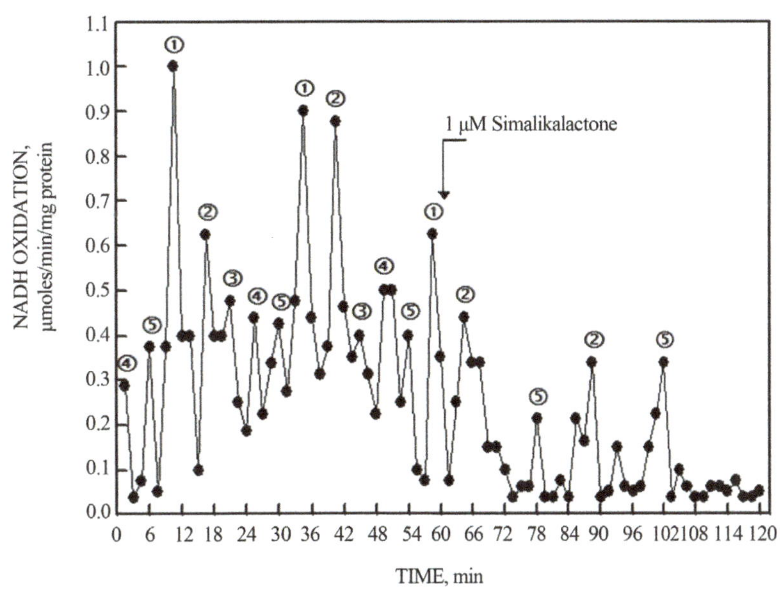

Figure 9. Inhibition of recombinant *Arabidopsis* ENOX1 by the specific ENOX1 quassinoid inhibitor simalikalactone D.

auxin herbicide 2,4-dichlorophenoxyacetic acid (2,4-D) which stimulates the dNOX (=ENOX5) activity of soybean plasma membranes approximately two-fold at 1 μM, was without effect (**Table 1**).

3.2. ENOX1 Activity Requires the Presence of Copper

Copper presence was necessary for ENOX1 activity (**Figure 10**). The IEF-purified ENOX1 when unfolded in the presence of trifluoroacetic acid retained activity after dialysis and at physiological pH. However, if the

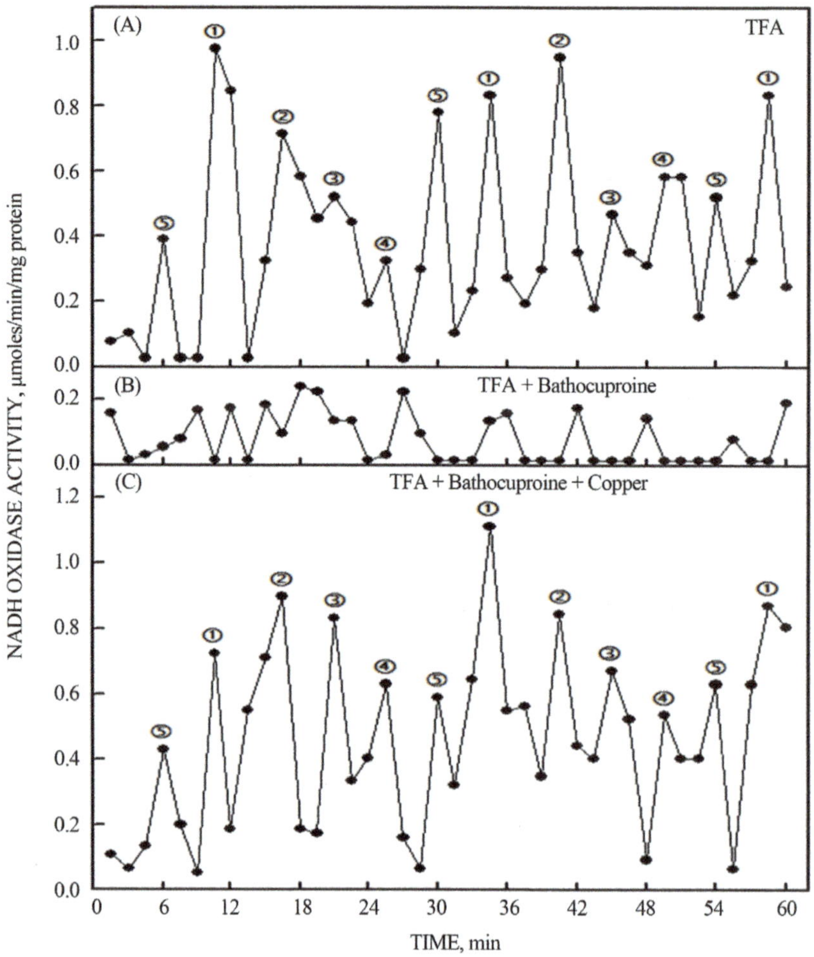

Figure 10. NADH oxidase activity of recombinant *Arabidopsis* ENOX1 diminished with TFA + bathocuproine. (A) In the presence of TFA, the 24 min period was unaffected; (B) When assayed with TFA and bathocuproine, the 24 min period was much reduced; (C) Removal of bathocuproine by dialysis and readdition of copper restored full activity.

ENOX1 was unfolded with trifluoroacetic acid in the presence of the copper chelator bathocuproine, activity was lost (**Figure 10(B)**). Activity was subsequently restored by refolding in the presence of copper at physiological pH (**Figure 10(C)**).

3.3. Period Length in Deuterium Oxide

ENOX1 activity when assayed in heavy water yielded a pattern of activities with the period length increased from 24 min to about 30 min (**Figure 11**).

3.4. Site-Directed Mutagenesis

Confirmation of functional assignments of motifs common to ENOX1 protein in general is provided for the specific functional motifs of the *Arabidopsis* ENOX1 by site-directed mutagenesis (**Table 2**). The G570A replacement in the putative adenine nucleotide binding motif largely eliminated NADH oxidation and reduced protein disulfide-thiol interchange by 75%. Reduced activities were seen with both C463A and C534A replacements. A response to the C463A replacement was especially evident with protein disulfide-thiol interchange. Putative copper site replacements, Y468A and Y531A reduced activities as well. Activities were reduced to near background levels with the Y531A replacement.

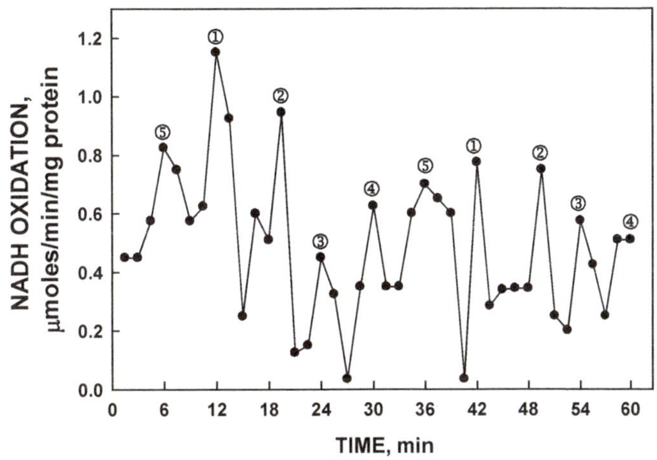

Figure 11. NADH oxidase activity of *Arabidopsis* ENOX1 when assayed in D_2O exhibited an increase in period length from 24 min to 30 min. The effect of heavy water to increase period length is one of the hallmarks of the biological clock.

Table 2. Confirmation of functional motifs of *Arabidopsis* ENOX1 by site-directed mutagenesis. Average of 3 determinations ± standard deviations.

Mutation	NADH oxidation	Disulfide-thiol interchange
	µmoles/min/mg protein	
Wild type	0.81 ± 0.04	1.00 ± 0.16
G570A	0.09 ± 0.05*	0.25 ± 0.02*
C463A	0.12 ± 0.06**	0.01 ± 0.01**
C534A	0.33 ± 0.07**	0.19 ± 0.06**
Y468A	0.18 ± 0.05**	0.25 ± 0.02*
Y479A	0.80 ± 0.01	0.95 ± 0.16
Y531A	0.02 ± 0.01**	0.07 ± 0.01**

*Very significant (p = 0.001 - 0.005), **Extremely significant (p = 0.0001 - 0.0004).

4. Discussion

The sequence of the putative plant ENOX1 protein appeared to be highly similar to that of ENOX1 of the yeast *Saccharomyces cerevisiae* [27]. Except for putative functional motifs, no similarity of the plant ENOX1 was found for human ENOX2 [6] or ENOX1 [4] or the yeast ENOX1. A putative NADH-binding site, a disulfide-thiol interchange site, and copper-binding sites of both the plant and yeast ENOX1, were also found in the human ENOX1 and ENOX2 proteins while the putative anti-cancer drug binding site of ENOX2 was not.

Based on the high level of its similarity to ENOX1 of yeast, the 14 kDa constitutive ECTO-NOX protein of plants is designated as an ENOX1 (see Hugo Gene Nomenclature Committee Designation http://www.gene.ucl.ac.uk/nomenclature/ for ENOX terminology). The cloned protein expressed in *E. coli* is listed as NCBI Accession No. BK008839.

When expressed in bacteria, the plant ENOX1 from *Arabidopsis lyrata* exhibited activity characteristics similar to those of ENOX1 proteins from mammalian [1] [15] [32] or other plant [1] [33]-[35] sources. Activity measured either as NADH oxidation (artificial substrate) or oxidation of reduced coenzyme Q (native substrate) oscillated with 5 activity maxima within each 24 min period. Period length was determined from two major activity maxima separated by 6 min and indicated by as ① and ② in the figures. Other maxima were separated from each other and from the two major activity maxima by intervals of 4.5 min (6 min + 4 × 4.5 min = 24 min).

The 24 min period length is distinct from the 22 min period length of mammalian ENOX2 [6] [26] and the 26 min period length of mammalian, plant, or yeast arNOX [16]-[18].

The period of ENOX1 is phased by addition of melatonin as has been reported for ENOX1 proteins from other sources [36]. Upon addition of melatonin, a new period occurs exactly 24 min after melatonin addition and continues with the newly established period thereafter. This is a characteristic not shared either with ENOX2 or arNOX (ENOX3). Additionally, the *Arabidopsis* ENOX1 was unaffected by the ENOX3 inhibitors tyrosol and gallic acid or by the activator of the plant ENOX5, 2,4-D (**Table 1**). Also identifying the *Arabidopsis* protein as an ENOX1 protein is its inability to be inhibited by the ENOX2 inhibitor capsaicin and other ENOX2 inhibitors including EGCg, phenoxodiol and cis-platinum. This is consistent with the absence from its amino acid sequence of the putative drug binding site of ENOX2 [6].

We are unable to exclude the possibility that additional ENOX1 proteins may be discovered from plants with sequences different from the ENOX1 reported here. Thus far, proteins from a number of plant sources have been purified to homogeneity but have failed to yield N-terminal amino acid sequence due to being blocked to N-terminal sequencing and being resistant to proteolytic degradation including proteinase K proteolysis. Like ENOX2 [37] [38], the ENOX1 proteins share characteristics in common with prions [39] [40], Alzheimer's Aβ protein [41], and aggregated tNOX proteins [42] especially those properties of resistance to protease degration and the ability to aggregate. The aggregated preparations of ENOX1 lacked enzymatic activity and were refractory to resolubilization. Size exclusion chromatography revealed approximately 10% dimers, <2.5% monomers and low molecular weight fragments with the remainder consisting of aggregates in the range of 250,000 to >1,000,000 kDa. The *Arabidopsis* ENOX1 amyloid, while insoluble even in SDS was completely disaggregated by boiling in SDS not unlike amyloid-β [1]-[42] for example [43]. For reasons possibly related to dimerization rather than multimerization, ENOX1 proteins prepared by isoelectric focusing or eluted from isoelectric focusing gels did not aggregate, retained enzymatic activity and were stable in solution for several weeks or months when stored frozen or at 4˚C.

An important consideration from the studies reported here is that the human ENOX1 protein bound approximately 2 moles of copper/mole of protein [4]. As the prevailing soluble form of ENOX proteins may be dimers based on size exclusion chromatography, the presence of two coppers per monomer permits the ENOX1 to be viewed as a dimeric protein containing 4 coppers/dimer that is capable of carrying out 4 electron transfers from NADH or reduced coenzyme Q directly to molecular oxygen as required to form water. ENOX1 proteins do reduce molecular oxygen to water [44] despite the fact that they lack flavin and/or electron carriers other than copper [6].

This protein disulfide isomerase-like activity measured either as restoration of activity to denatured and inactive ribonuclease A or by cleavage of dithiodipyridine (DTDP) substrate as in the present study, also exhibits the oscillatory activity patterns characteristic of ENOX proteins in general [4]. For ENOX1, the period length is again 24 min due to the asymmetry imparted by two of the 5 maxima per period being separated by 6 min rather than the usual 4.5 min. The protein disulfide-thiol interchange activity of the ENOX protein has been implicated as obligatory to the cell enlargement phase of cell growth for both animal [19] and plant [20] [21] cells. However, the mechanism whereby disulfide bonds are formed and broken through the action of NOX proteins may differ substantially from that of the classical flavoprotein disulfide isomerases. The portion of the *Arabidopsis* ENOX1 protein involved in protein disulfide-thiol interchange is unknown. A marked loss of activity resulted from both the C463A and the C534A replacements (**Table 2**). A source of protons and electrons is not required for the interchange activity as exemplified by the ability of ENOX1 alone to catalyze the restoration of activity to inactive and denatured ribonuclease A. Neither glutathione nor dithiothreitol support the protein disulfide-thiol interchange activity.

The high capacity of proliferating mammalian cells to transfer electrons from cytosolic NADH to extracellular acceptors like oxygen via plasma membrane electron transport [45] [46] is now generally regarded as being via a plasma membrane electron transport chain involving reduced quinones as the transmembrane carrier of electrons and protons [3] [47] [48]. The identification and cloning of the constitutive ENOX1 protein of plants whose proposed function is as a terminal oxidase for the plasma membrane electron transport chain [47] rather than superoxide generation, and as a driver of both cell enlargement and of the biological clock represents an important advance in our understanding of these enzyme proteins of the plasma membrane which have, until recently, been largely ill defined [48].

5. Conclusion

A gene encoding a candidate constitutive ECTO-NADH oxidase (ENOX1) was identified based on significant sequence homology with ENOX1 of *Saccharomyces cerevisiae*. Functional motifs conserved with human ENOX1 were identified by site directed mutagenesis. The *Arabidopsis* ENOX1 has catalytic properties of previously characterized ENOX proteins, requires bound copper for activity and exhibits time-keeping activity oscillations with a 24-min period phased by melatonin. As the prevailing soluble form is that of a dimer, the *Arabidopsis* ENOX1 emerges as dimeric protein containing 4 copper atoms per dimer capable of 4 electron transfers from NADH or reduced coenzyme Q directly to molecular oxygen to form water.

Acknowledgements

We thank Aya Ryuzoji for preparation of the figures, Peggy Runck for manuscript preparation and Ping Gao for assistance with site-directed mutagenesis.

Nucleotide Sequence Data

Nucleotide sequence data are available in The Third Party Annotation Section of the DOB/EMBL/GenBank databases under the accession number TPA: BK008839.

References

[1] Morré, D.J. (1998) NADH Oxidase: A Multifunctional Ectoproteins of the Eukaryotic Cell Surface. In: Asard, H., Bérczi, A. and Caubergs, R., Eds., *Plasma Membrane Redox Systems and Their Role in Biological Stress and Disease*, Kluwer Academic Publishers, Dordrecht, 121-156. http://dx.doi.org/10.1007/978-94-017-2695-5_5

[2] Morré, D.J. and Morré, D.M. (2003) Cell Surface NADH Oxidases (ECTO-NOX Proteins) with Roles in Cancer, Cellular Time-Keeping, Growth, Aging and Neurodegenerative Disease. *Free Radical Research*, **37**, 795-808. http://dx.doi.org/10.1080/1071576031000083107

[3] Morré, D.J. and Morré, D.M. (2013) ECTO-NOX Proteins. Springer, New York, 507 p. http://dx.doi.org/10.1007/978-1-4614-3958-5

[4] Jiang, Z., Gorenstein, N.M., Morré, D.M. and Morré, D.J. (2008) Molecular Cloning and Characterization of a Candidate Human Growth-Related and Time-Keeping Constitutive Cell Surface Hydroquinone (NADH) Oxidase. *Biochemistry*, **47**, 14028-14038. http://dx.doi.org/10.1021/bi801073p

[5] Chueh, P.-J., Morré, D.J., Wilkinson, F.E., Gibson, J. and Morré, D.M. (1997) A 33.5 kDa Heat- and Protease-Resistant NADH Oxidase Inhibited by Capsaicin from Sera of Cancer Patients. *Archives of Biochemistry and Biophysics*, **342**, 38-47. http://dx.doi.org/10.1006/abbi.1997.9992

[6] Chueh, P.-J., Kim, C., Cho, N., Morré, D.M. and Morré, D.J. (2002) Molecular Cloning and Characterization of a Tumor-Associated, Growth-Related and Time-Keeping Hydroquinone (NADH) Oxidase (NOX) of the HeLa Cell Surface. *Biochemistry*, **41**, 3732-3741. http://dx.doi.org/10.1021/bi012041t

[7] Morré, D.J., Chueh, P.-J. and Morré, D.M. (1995) Capsaicin Inhibits Preferentially the NADH Oxidase and Growth of Transformed Cells in Culture. *Proceedings of the National Academy of Sciences USA*, **92**, 1831-1835. http://dx.doi.org/10.1073/pnas.92.6.1831

[8] Morré, D.J., Sun, E., Geilen, C., Wu, L.-Y., De Cabo, R., Krasagakis, K., Orfanos, C.E. and Morré, D.M. (1996) Capsaicin Inhibits Plasma Membrane NADH Oxidase and Growth of Human and Mouse Melanoma Lines. *European Journal of Cancer*, **32A**, 1995-2003. http://dx.doi.org/10.1016/0959-8049(96)00234-1

[9] Morré, D.J., Wu, L.-Y. and Morré, D.M. (1995) The Antitumor Sulfonylurea N-(4-methylphenylsulfonyl)-N(-(chlorophenyl)urea (LY181984) Inhibits NADH Oxidase Activity of HeLa Plasma Membranes. *Biochimica et Biophysica Acta*, **1240**, 11-17. http://dx.doi.org/10.1016/0005-2736(95)00164-7

[10] Morré, D.J., Wilkinson, F.E., Kim, C., Cho, N., Lawrence, J., Morré, D.M. and McClure, D. (1996) Antitumor Sulfonylurea-Inhibited NADH Oxidase of Cultured HeLa Cells Shed into Media. *Biochimica et Biophysica Acta*, **1280**, 197-206. http://dx.doi.org/10.1016/0005-2736(95)00290-1

[11] Morré, D.J., Chueh, P.-J., Yagiz, K., Balicki, A., Kim, C. and Morré, D.M. (2007) ECTO-NOX Target for the Anticancer Isoflavene Phenoxodiol. *Oncology Research*, **16**, 299-312.

[12] Morré, D.J., Grieco, P.A. and Morré, D.M. (1998) Mode of Action of the Anticancer Quassinoids-Inhibition of the Plasma Membrane NADH Oxidase. *Life Science*, **63**, 595-604. http://dx.doi.org/10.1016/S0024-3205(98)00310-5

[13] Morré, D.J. and Greico, P.A. (1999) Glaucarubolone and Simalikalactone D, Respectively, Preferentially Inhibit Aux-

in-Induced and Constitutive Components of Plant Cell Enlargement and the Plasma Membrane NADH Oxidase. *International Journal of Plant Sciences*, **160**, 291-297. http://dx.doi.org/10.1086/314133

[14] Wilkinson, F.E., Kim, C., Cho, N., Chueh, P.-J., Leslie, S., Moya-Camarena, S., Wu, L.-Y., Morré, D.M. and Morré, D.J. (1996) Isolation and Identification of a Protein with Capsaicin-Inhibited NADH Oxidase Activity from Culture Media Conditioned by Growth of HeLa Cells. *Archives of Biochemistry and Biophysics*, **336**, 275-282. http://dx.doi.org/10.1006/abbi.1996.0558

[15] Sedlak, D., Morré, D.M. and Morré, D.J. (2001) A Drug-Unresponsive and Protease-Resistant CNOX Protein from Human Sera. *Archives of Biochemistry and Biophysics*, **386**, 106-116. http://dx.doi.org/10.1006/abbi.2000.2180

[16] Morré, D.M. and Morré, D.J. (2003) Specificity of Coenzyme Q Inhibition of an Aging-Related Cell Surface NADH Oxidase (ECTO-NOX) That Generates Superoxide. *BioFactors*, **18**, 33-43. http://dx.doi.org/10.1002/biof.5520180205

[17] Dick, S., Phung, C., Morré, D.M. and Morré, D.J. (2013) Molecular Cloning and Characterization of an ECTO-NOX3 (ENOX3) of *Saccharomyces cerevisiae*. *Advances in Biological Chemistry*, **3**, 505-511. http://dx.doi.org/10.4236/abc.2013.35055

[18] Morré, D.M., Guo, F. and Morré, D.J. (2003) An Aging-Related Cell Surface NADH Oxidase (arNOX) Generates Superoxide and Is Inhibited by Coenzyme Q. *Molecular and Cellular Biochemistry*, **254**, 101-109. http://dx.doi.org/10.1023/A:1027301405614

[19] Pogue, R., Morré, D.M. and Morré, D.J. (2000) CHO Cell Enlargement Oscillates with a Temperature-Compensated Period of 24 Minutes. *Biochimica et Biophysica Acta*, **1498**, 44-51. http://dx.doi.org/10.1016/S0167-4889(00)00076-8

[20] Morré, D.J., Pogue, R. and Morré, D.M. (2001) Soybean Cell Enlargement Oscillates with a Temperature-Compensated Period Length of ca. 24 Min. *In Vitro Cellular & Developmental Biology-Plant*, **37**, 19-23. http://dx.doi.org/10.1007/s11627-001-0004-3

[21] Morré, D.J., Ternes, P. and Morré, D.M. (2002) Cell Enlargement of Plant Tissue Explants Oscillates with a Temperature-Compensated Period Length of ca. 24 Min. *In Vitro Cellular & Developmental Biology-Plant*, **38**, 18-28. http://dx.doi.org/10.1079/IVP2001249

[22] Morré, D.J., Brightman, A.O., Hidalgo, A. and Navas, P. (1995) Selective Inhibition of Auxin-Stimulated NADH Oxidase Activity and Elongation Growth of Soybean Hypocotyls by Thiol Reagents. *Plant Physiology*, **107**, 1285-1291.

[23] Kishi, T., Morré, D.M. and Morré, D.J. (1999) The Plasma Membrane NADH Oxidase of HeLa Cells Has Hydroquinone Oxidase Activity. *Biochimica et Biophysica Acta*, **1412**, 66-77. http://dx.doi.org/10.1016/S0005-2728(99)00049-3

[24] Morré, D.J., De Cabo, R., Jacobs, E. and Morré, D.M. (1995) Auxin-Modulated Protein Disulfide-Thiol Interchange Activity from Soybean Plasma Membranes. *Plant Physiology*, **109**, 573-578.

[25] Morré, D.J., Gomez-Rey, M.L., Schramke, C., Em, O., Lawler, J., Hobeck, J. and Morré, D.M. (1999) Use of Dipyridyl-Dithio Substrates to Measure Directly the Protein Disulfide-Thiol Interchange Activity of the Auxin Stimulated NADH: Protein Disulfide Reductase (NADH Oxidase) of Soybean Plasma Membranes. *Molecular and Cellular Biochemistry*, **200**, 7-13. http://dx.doi.org/10.1023/A:1006916116297

[26] Morré, D.J., Chueh, P.-J., Pletcher, J., Tang, X.Y., Wu, L.-Y. and Morré, D.M. (2002) Biochemical Basis for the Biological Clock. *Biochemistry*, **41**, 11941-11945. http://dx.doi.org/10.1021/bi020392h

[27] Chueh, P.-J., Morré, D.M. and Morré, D.J. (2002) A Site-Directed Mutagenesis Analysis of tNOX Functional Domains. *Biochimica et Biophysica Acta*, **1594**, 74-83. http://dx.doi.org/10.1016/S0167-4838(01)00286-2

[28] Altschul, S.F., Madden, T.L., Schäffer, A.A., Zhang, J., Zhang, Z., Miller, W. and Lipman, D.J. (1997) Gapped BLAST and PSI-BLAST: A New Generation of Protein Database Search Programs. *Nucleic Acids Research*, **25**, 3389-3402. http://dx.doi.org/10.1093/nar/25.17.3389

[29] Dick, S.S., Ryuzoji, A., Morré, D.M. and Morré, D.J. (2013) Identification of the Constitutive Ultradian Oscillator of the Circadian Clock (ENOX1) in *Saccharomyces cerevisiae*. *Advances in Biological Chemistry*, **3**, 320-328. http://dx.doi.org/10.4236/abc.2013.33036

[30] Smith, P.K., Krohn, R.I., Hermanson, G.T., Mailia, A.K., Gartner, F.F., Provenzano, M.D., Fujimoto, E.K., Groeke, N.M., Olson, B.J. and Klenk, D.C. (1985) Measurement of Protein Using Bicinchoninic Acid. *Analytical Biochemistry*, **150**, 76-85. http://dx.doi.org/10.1016/0003-2697(85)90442-7

[31] Braman, J., Papworth, C. and Greener, A. (1996) Site-Directed Mutagenesis Using Double-Stranded Plasmid DNA Templates. *Methods Molecular Biology*, **57**, 31-44.

[32] Brightman, A.O., Wang, J., Miu, R.K., Sun, I.L., Barr, R., Crane, F.L. and Morré, D.J. (1992) A Growth Factor- and Hormone-Stimulated NADH Oxidase from Rat Liver Plasma Membrane. *Biochimica et Biophysica Acta*, **1105**, 109-117. http://dx.doi.org/10.1016/0005-2736(92)90168-L

[33] Morré, D.J. and Brightman, A.O. (1991) NADH Oxidase of Plasma Membranes. *Journal of Bioenergetics and Biomembranes*, **23**, 469-489. http://dx.doi.org/10.1007/BF00771015

[34] Brightman, A.O., Barr, R., Crane, F.L. and Morré, D.J. (1988) Auxin-Stimulated NADH Oxidase Purified from Plasma Membrane of Soybean. *Plant Physiology*, **86**, 1264-1269. http://dx.doi.org/10.1104/pp.86.4.1264

[35] Morré, D.J. and Morré, D.M. (1998) NADH Oxidase Activity of Soybean Plasma Membrane Oscillates with a Temperature Compensated Period of 24 Min. *The Plant Journal*, **16**, 279-284. http://dx.doi.org/10.1046/j.1365-313x.1998.00293.x

[36] Morré, D.J., Penel, C., Greppin, H. and Morré, D.M. (2002) The Plasma Membrane-Associated NADH Oxidase of Spinach Leaves Responds to Blue Light. *International Journal of Plant Sciences*, **613**, 543-547. http://dx.doi.org/10.1086/340543

[37] Kelker, M., Kim, C., Chueh, P.J., Guimont, R., Morré, D.M. and Morré, D.J. (2001) Cancer Isoform of a Tumor-Associated Cell Surface NADH Oxidase (tNOX) Has Properties of a Prion. *Biochemistry*, **40**, 7351-7354. http://dx.doi.org/10.1021/bi010596i

[38] Kim, C. and Morré, D.J. (2004) Prion Proteins and ECTO-NOX Proteins Exhibit Similar Oscillating Redox Activities. *Biochemical and Biophysical Research Communications*, **315**, 1140-1146. http://dx.doi.org/10.1016/j.bbrc.2004.02.007

[39] Griffith, J.S. (1964) Self-Replication and Scrapie. *Nature*, **315**, 1043-1044.

[40] Prusiner, S.B. (1994) Biology and Genetics of Prion Diseases. *Annual Review of Microbiology*, **48**, 655-686. http://dx.doi.org/10.1146/annurev.mi.48.100194.003255

[41] Multhaup, G. (1997) Amyloid Precursor Protein, Copper and Alzheimer's Disease. *Biomedicine & Pharmacotherapy*, **51**, 105-111. http://dx.doi.org/10.1016/S0753-3322(97)86907-7

[42] del Castillo-Olivares, A., Yantiri, F., Chueh, P.-J., Wang, S., Sweeting, M., Sedlak, D., Morré, D.M., Burgess, J. and Morré, D.J. (1998) A Drug-Responsive and Protease-Resistant Peripheral NADH Oxidase Complex from the Surface of HeLa S Cells. *Archives of Biochemistry and Biophysics*, **358**, 125-140. http://dx.doi.org/10.1006/abbi.1998.0823

[43] Rangachari, V., Morré, B.D., Reed, D.K., Sonoda, L.K., Bridges, A.W., Conboy, E., Hartigan, D. and Rosenberry, T.L. (2007) Amyloid-β (1-42) Rapidly Forms Protofibrils and Oligomers by Distinct Pathways in Low Concentrations of Sodium Dodecylsulfate. *Biochemistry*, **46**, 12451-12462. http://dx.doi.org/10.1021/bi701213s

[44] Orczyk, J., Morré, D.M. and Morré, D.J. (2005) Periodic Fluctuations in Oxygen Consumption Comparing HeLa (Cancer) and CHO (Non-Cancer) Cells and Response to External $NAD(P)^+/NAD(P)H$. *Molecular and Cellular Biochemistry*, **273**, 161-167. http://dx.doi.org/10.1007/s11010-005-0326-2

[45] Berridge, M.V. and Tan, A.S. (2000) High-Capacity Redox Control at the Plasma Membrane of Mammalian Cells: Trans-Membrane, Cell Surface, and Serum NADH-Oxidases. *Antioxidants & Redox Signaling*, **2**, 231-242. http://dx.doi.org/10.1089/ars.2000.2.2-231

[46] Hyun, D.-H., Hunt, N.D., Emerson, S.S., Hernandez, J.O., Mattson, M.P. and de Cabo, R. (2007) Up-Regulation of Plasma Membrane-Associated Redox Activities in Neuronal Cells Lacking Functional Mitochondria. *Journal of Neurochemistry*, **100**, 1364-1374. http://dx.doi.org/10.1111/j.1471-4159.2006.04411.x

[47] Crane, F.L. and Low, H. (2008) Reactive Oxygen Species Generation at the Plasma Membrane for Antibody Control. *Autoimmunity Reviews*, **7**, 518-522. http://dx.doi.org/10.1016/j.autrev.2008.04.004

[48] Baker, M.A. and Lawen, A. (2000) Plasma Membrane NADH-Oxidoreductase System: A Critical Review of the Structural and Functional Data. *Antioxidants & Redox Signaling*, **2**, 197-212. http://dx.doi.org/10.1089/ars.2000.2.2-197

Abbreviations

arNOX, age-related cell surface NADH oxidase = ENOX; CNOX, constitutive and drug-unresponsive cell surface NADH oxidase (ENOX1); DTDP, dithiodipyridine dNOX, 2,4-D (auxin-activated cel surface NADH oxidase (ENOX5); ENOX, ECTO-NOX; tNOX, tumor (cancer)-associated and drug-(capsaicin) inhibited cell surface NADH oxidase (ENOX2); IEF, isoelectric focusing; PMSF, ρ-chloromercurisuifunic acid; SDS-PAGE, sodium dodecylsulfate polyacrylamide gel electrophoresis; TFA, tritluroacetic acid; tNOX, tumor (cancer)-associated and drug (capsaicin) inhibited cell surface NADH oxidase (ENOX2).

2

Gene Regulation through mRNA Expression

Haruko Inose*, **Kotomi Mukai, Misa Ito, Seiji Masuda***

Division of Integrated Life Science, Graduate School of Biostudies, Kyoto University, Kyoto, Japan
Email: *inose.haruko.24n@st.kyoto-u.ac.jp, *masuda@kais.kyoto-u.ac.jp

Abstract

In eukaryotes, gene expression is achieved by four steps: transcription, processing, nuclear export, and translation. Each step requires multiple factors, and frequently two or more pathways are used by a single gene, enabling strictly regulated gene expression. Importantly, eukaryotes, taking advantage of the separated structures of the nucleus and the cytoplasm, have evolved complex and organized mRNA processing mechanisms that permit sophisticated biological activity. The processes are much more complicated than those found in prokaryotes, in which transcription and translation occur linearly in time and place. Here, we review gene expression, focusing on mRNA processing in the nucleus and the gene regulatory systems found at each step. Combination of gene regulation shows the typical phenotype in each cell. Further understanding of the uncertain mechanisms will uncover the gene regulation through mRNA expression.

Keywords

Capping, Splicing, Polyadenylation, Export, Half Life

1. Introduction

Gene expression in eukaryotes is characterized by the spatially separated processes of transcription and translation. Despite having an identical genome, varying levels of gene expression within differing types of cells permit different biological activities in response to both the internal (developmental, cell cycle-related) and external environments. This provides the range of functions necessary for higher eukaryotes.

Gene regulation is found at each of the six steps of gene expression: 1) transcriptional regulation; 2) mRNA processing control; 3) mRNA transport and localization; 4) translational regulation; 5) protein activity; and (6) mRNA half-life. In the last two decades, it has been shown that mRNA metabolism plays important roles in gene regulation.

In the first section, we provide an overview of gene expression, mostly in higher eukaryotes, focusing on

*Corresponding authors.

mRNA processing: capping, splicing, polyadenylation, and mRNA export. In the second section, we summarize gene regulation, focusing on mRNA processing control, mRNA transport, and mRNA half-life.

2. Gene Expression in Eukaryotes

2.1. Capping

Nascent mRNA, when about 20 bases have been synthesized by RNA polymerase II (polII), is first modified by "capping" [1]. Capping is a hallmark of transcription initiation that functions to protect mRNA from degradation by 5'-3' exonuclease activity (**Figure 1**). Three enzymatic reactions are required for capping: hydrolysis of the 5' terminal phosphate by RNA triphosphatase, addition of a guanine residue to the trimmed diphosphate terminus by RNA guanyltransferase, and methylation of the guanine residue by RNA methyltransferase. After capping, nascent mRNA contains an N7-guanosine cap "m7GpppN" at its 5' terminus via an unusual 5'-5' phosphodiester linkage. The cap structure is required not only for mRNA nuclear export but also for translation: the cap-binding complex (CBC) formed by CBP80 and CBP20 binds to the cap in the nucleus, whereas eukaryotic elongation factor (eIF) 4E binds to the cap in the cytoplasm.

2.2. Splicing

In eukaryotes, most genes consist of protein-coding regions called exons, which are interspersed by untranslated regions called introns. The removal of introns by splicing is required to generate contiguous coding regions in the nucleus. The exact position of the 5'splice site (at the 5' terminus of the intron) and the 3' splice site (at the 3' terminus of the intron) is determined by the RNA sequence as well as by trans-acting factors and uridine-rich

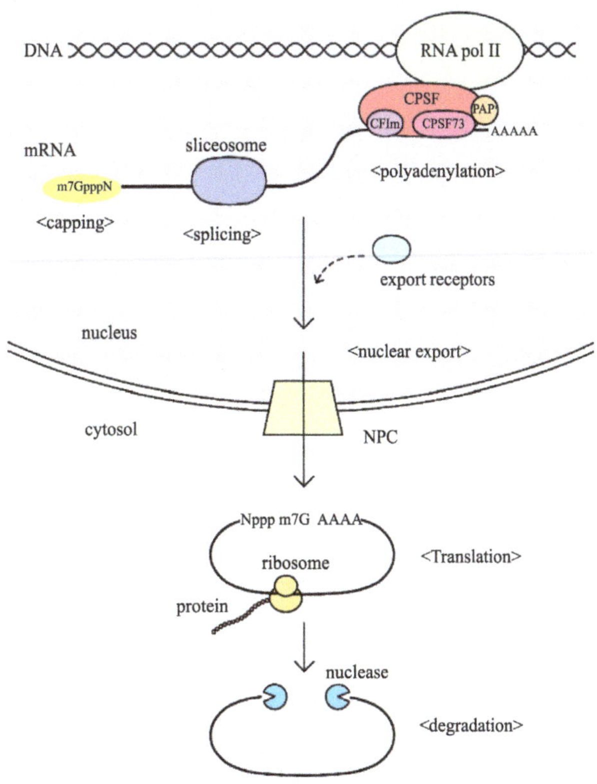

Figure 1. mRNA biogenesis. Pol II transcribes the protein coding genes in the nucleus. Transcribed pre-mRNA undergoes the various processing steps, capping, splicing and polyadenylation. The properly-processed mRNA is exported from the nucleus to the cytoplasm. In the cytoplasm, mRNA serves as a template for protein synthesis.

small nuclear ribonucleoprotein particles (U snRNPs). Typically, the 5' splice site has a consensus "GU" dinucleotide, and the 3' splice site has "AG". In addition, a branch point site and polypyrimidine (Py) tract are also required. The branch point site is an adenine residue that attacks the 5' splice site and dissociates the exon from the intron; it is usually found several tens of bases upstream of the 3' splice site. Py is a sequence rich in adenine and uracil found between the branch point site and the 3' splice site.

Splicing is an energy-consuming process that is catalyzed by the spliceosome [2] [3]. The spliceosome is a large complex of about 150 proteins and five U snRNAs. Intriguingly, uridine-rich small nuclear ribonucleic acids (U snRNAs), which are small RNAs with complementarity to pre-mRNA, form the active site of the spliceosome. As indicated in **Figure 2**, the 5' splice site is first recognized by U1 snRNP via base-pairing. However, the Py and 3' splice site are recognized by U2AF, which results in the recruitment of branch point-binding protein (BBP). BBP is then replaced by U2 snRNP, which makes the branch point site protrude from the RNA chain. Next, U4, U5, and U6 snRNPs undergo complex interactions with each other. U1 snRNP then dissociates from the RNA, to be replaced by U6 snRNP. After U4 snRNP dissociates, U2 snRNP and U6 snRNP bind each other via base-pairing. Importantly, U2 snRNP positions itself at the branch point site while U6 snRNP is positioned at the 5' splice site at this point, thus enclosing 5' splice site and branch point site in proximal region. As a result, the branch point site attacks the 5' splice site phosphate and the intron is spliced out. Finally, the 3' terminus of the upstream exon attacks the downstream exon 5' terminus that is dissociated from the 3' splice site, connecting them together. This complex procedure for splicing, in which the components of the spliceosome are sequentially replaced to ultimately generate a catalytically active site, enables cells to avoid aberrant splicing.

2.3. Polyadenylation

Most mRNA has a poly-A tail at its 3' terminus, which is not encoded by the genome. When RNA polII reaches the end of a gene, two polyadenylation factors, cleavage and polyadenylation specificity factor (CPSF) and cleavage stimulation factor (CstF), are recruited onto its C-terminal domain (CTD). CPSF, which recognizes the

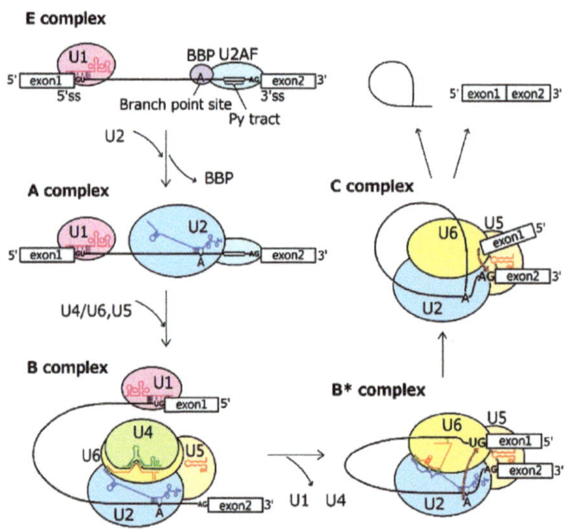

Figure 2. Dynamics of mRNA splicing. Formation of spliceosome is a stepwise event, starting from E complex followed by forming A, B, B' and C complexes in this order. In E complex, U1 snRNP is associated at 5' splice site, and Branch point binding protein (BBP) and U2AF are associated at brunch site and 3' splice site, respectively. In A complex, U2 snRNP is recruited to 3' splice site. In B complex, U4/U6 and U5 snRNPs are recruited to U2 snRNP. In B* complex, U1 and U4 snRNPs are dissociated from pre-mRNA, and U2 and U6 snRNP form duplex to enable 5' and 3' splice sites in a proximal region. In C complex, splicing occurs in two-step reaction and intron is removed from mRNA.

poly-A signal "AAUAAA", and CstF, which recognizes a GU-rich element downstream of "AAUAAA", are translocated onto the mRNA when RNA polII transcribes these sequences. The cleavage factors I and II cleave nascent mRNA with the support of CPSF and CstF, followed by the recruitment of poly-A polymerase (PAP) to the end of the transcript. PAP then adds approximately 200 adenines to the 3' terminus of the nascent mRNA [4].

The poly-A tail plays important roles in two phases. First, it is recognized by poly-A-binding protein (PABP) [5]. In the cytoplasm, PABP interacts with the 5' cap via interaction with eIF4G, a protein bound to eIF4E. Interaction between the poly-A tail and the 5' cap results in circularization of the mRNA, enabling efficient ribosomal cycling and translation. Second, PABP bound on the poly-A tail protects the mRNA from 3'-5' nuclease attack, extending its half-life. The poly-A tail is gradually shortened over time. When it becomes too short to bind PABP, the exosome recognizes and degrades mRNA in the 3' to 5' direction.

2.4. Nuclear Export

Macromolecules usually require receptors for their nuclear trafficking through the nuclear pore complex (NPC); this also applies to mRNA export. Two proteins, TAP (also called NXF1) and CRM1 (also called XPO1) have been identified as nuclear export receptors for different classes of mRNA export.

TAP is responsible for bulk mRNA export [6] [7]. This export pathway is highly coupled to splicing in humans. In contrast, Mex67, its homolog in *Saccharomyces cerevisiae*, is recruited onto primary transcripts because only approximately 5% of genes in this species contain introns [8] [9]. Intron-containing nascent transcripts are recognized and bound by the spliceosome via their introns, which results in the recruitment of TREX during splicing [10]-[13]. TREX ultimately recruits TAP—the recruitment of TAP to mRNA therefore depends on splicing. So, how does TAP export mRNA from the nucleus to the cytoplasm? TAP is known to bind the FG-repeat in nucleoporins, the major components of the NPC. This binding ability promotes mRNA export to the cytoplasm [14]-[17]. In addition, on the cytoplasmic side of the NPC, DBP5, a DEAD-box helicase connected to NPC fibrils, remodels mRNPs in concert with Gle1 and IP6 and releases mRNPs into the cytoplasm by dissociating several proteins including TAP [18]-[20]. It is considered that the TAP-mediated mRNA export pathway, by virtue of its splicing dependence, prevents the export of incompletely spliced mRNA.

However, several kinds of mRNA use CRM1 as their export receptor [21]. The CRM1-mediated export pathway seems to be independent of splicing. While TAP is recruited onto mRNAs via splicing in humans, CRM1 is recruited onto mRNAs via specific adapter proteins that bind to a cis-RNA element in the 3' untranslated region (UTR). Moreover, the CRM1-mediated pathway is different from the TAP-mediated pathway in the way it remodels the exported mRNPs at the cytoplasmic face of the NPC. When an mRNP is released into the cytoplasm, Ran plays a key role in the CRM1-mediated pathway (the equivalent of DBP5 in the TAP pathway) [22] [23]. Ran is a small GTPase that is present as Ran-GTP in the nucleus, where GTP exists at a high concentration by the catalytic activity of the chromatin-bound guanine nucleotide exchange factor RCC1 [24]. In contrast, Ran exists in a GDP-bound form in the cytoplasm, where GTP is hydrolyzed by the catalytic activity of the Ran GTPase-activating protein RanGAP. Ran works like a switch: CRM1 bound to Ran-GTP in the nucleus binds to the adapter protein with specific RNA elements including mRNA, while CRM1 dissociates from Ran-GDP in the cytoplasm to release the adapter protein.

3. Gene Regulation

3.1. Gene Regulation via Capping

Historically, it was considered that capping did not actively affect gene regulation because it was a universal process. However, this view has now been challenged. The complete cap structure is required for recognition by CBC, and thus an accelerated capping reaction may enhance gene expression. Several mechanisms have been identified that enhance methylation, the final step in cap generation (**Figure 3**). In yeast, S-adenosylmethionine (SAM), which serves as a methyl donor in capping, and its related factors, have been shown to affect cap methylation. SAM synthase activates Abd1, a yeast RNA (guanine-7-) methyltransferase (RNMT) that catalyzes SAM to produce a methylated cap and S-adenosyl-L-homocystein (SAH) [25]. SAH, in turn, inhibits the activity of Abd1 until it is hydrolyzed by S-adenosyl-L-homocysteine hydrolase (SAHH) [26].

In humans, importin α has been identified as a binding partner of CBC that improves the efficiency of cap

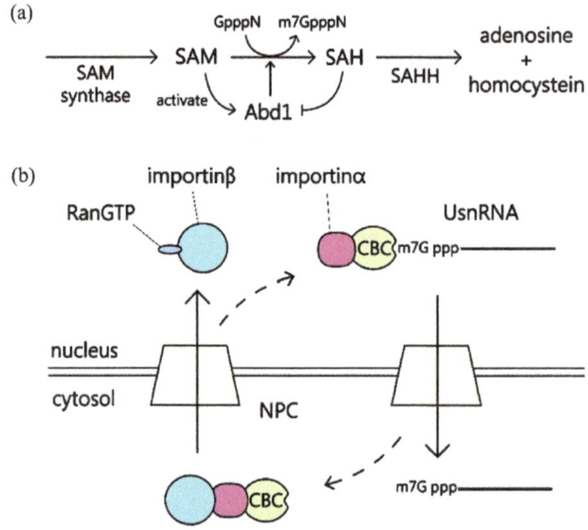

Figure 3. Regulation of cap methylation. (a) Feedback regulation of the cap methylation. Abd1 is activated by SAM followed by catalyzing cap methylation. SAH, by-product of cap methylation, inhibits Abd1; (b) Regulation of the cap methylation by importin α and β, and Ran-GTP.

methylation. In yeast, importin α is normally bound to CBC, which means that CBC in complex with importin α, not CBC alone, recognizes capped U snRNA [27]. However, the binding of importin β to importin α triggers the dissociation of the importin α-CBC complex from capped U snRNA [27]-[29]. Thus, in the nucleus, where importin β is bound to Ran-GTP, the importin α-CBC complex binds to capped U snRNA, while in the cytoplasm, where importin β is free from Ran, the importin α-CBC complex dissociates from capped U snRNA. Importantly, the activity of Ran is known to be enhanced by growth factors, suggesting that external signals promote the formation of the importin α-CBC-cap complex, resulting in the promotion of cap methylation [29].

m7GpppN is not the only nucleotide to be methylated in the capping reaction. The ribose adjacent to the cap, namely, the first transcribed ribose, and following riboses are often methylated to produce methylated-ribose1 (cap1), cap2, and so on. Cap0, the GpppN that is added to the 5' terminus of mRNA, and cap1, are found in all human mRNA, while mcap2 is found in only half of human mRNA [30]. The observation that the 5' mRNA terminus can have different methylation levels implies the existence of something similar to the "histone code" to regulate gene expression. Cap4, a hypermethylated structure found in *Trypanosoma brucei* mRNA, supports this idea by the fact that the cap4 structure is required for maximum translation [31]. It is as yet unknown how a difference in methylation levels can affect gene expression, and similarly what determines the methylation levels, although recent studies have attempted to identify the consensus sequence for cap1 and cap2, and investigate the structure of hMTR1 and hMTR2, which are the human methylases for cap1 and cap2, respectively [32]-[34].

Emerging data have shown that decapping also functions as a gene regulatory step. DCP2 is a canonical decapping enzyme in humans [35] [36]. When the poly-A tail becomes too short to bind PABP, the circular formation between the 5' cap and the poly-A tail is no longer maintained, and DCP2 hydrolyzes the cap structure, resulting in rapid degradation of mRNA by Xrn1, a 5'-3' exonuclease that recognizes the 5' monophosphate terminus. Decapping is upregulated in conditions of environmental stress such as starvation or heat shock. Importantly, long non-coding (lnc) RNA, as well as mRNA, is a target of decapping by DCP2. Some lncRNAs repress the transcription of stress-inducible genes by forming an R-loop (DNA:RNA hybrid) with the corresponding genomic region. The reduction in the amount of relevant lncRNA is followed by increased expression of stress-inducible genes. Thus, cells can modulate their metabolism by reducing canonical mRNA, meanwhile enabling the expression of stress-inducible genes.

In addition, a non-canonical decapping enzyme has also been identified, which is suggested to function in the capping quality control mechanism. In yeast, Rai1 (DOM3Z in humans) has been shown to degrade mRNA with a methylation-defective cap [37]. In Rai1-deficient yeast, the accumulation of mRNA with an aberrantly methylated cap can be observed under glucose or amino acid starvation, which indicates that Rai1 controls mRNA translation by degrading aberrantly capped mRNA.

3.2. Alternative Splicing

Most exons are constitutively spliced so as to be retained in the mature mRNA. Meanwhile, some exons are spliced in different patterns; this process is called "alternative splicing". Alternative splicing plays important roles in the versatility of proteins derived from a single gene in different cell types or developmental stages. When alternative splicing is disrupted, it impairs a wide range of developmental and physiological functions, often with pathological consequences. Here, we describe the classification and mechanisms of alternative splicing, although the details of the so-called "splicing code" remain to be understood.

The major forms of alternative splicing are categorized by five patterns [38]. They are: 1) exon skipping; 2) alternative 3' splice site usage; 3) alternative 5' splice site usage; 4) intron retention; and 5) mutually exclusive exons. Exon skipping is the most common form. Complicated patterns of splicing arising from combinations of two mechanisms are also observed.

Alternative splicing is typically controlled by cis-acting splicing regulatory elements (SRE), which can be observed in both exons and introns, and function as either splicing enhancers or silencers [39] [40]. Splice site selection is regulated by SREs and trans-acting splicing factors bound to SREs. According to their locations and activities, SREs are classified into four types: exonic splicing enhancers and intronic splicing enhancers (that stimulate splicing), and exonic splicing silencers and intronic splicing silencers (that inhibit splicing). Accessibility to splice sites or SREs is determined by the structures of the pre-mRNA and RNA-binding proteins [3] [41]. Serine/arginine-rich (SR) protein and heterogeneous nuclear ribonucleoprotein (hnRNP) are important trans-acting factors. SR protein promotes splicing when it is bound to exons, whereas it inhibits splicing when it is bound to introns [42] [43]. hnRNP also controls splicing positively and/or negatively according to the particular pre-mRNA it is associated with [43] [44].

SRSF1 (also celled SF2/ASF), a member of the SR family, affects the alternative splicing of various targets, including the proto-oncogene *RON*. SRSF1 promotes the skipping of *RON* exon 11, which results in cell motility and invasion [45]. SRSF1 promotes the expression of anti-apoptotic isoforms which cannot interact with pro-apoptotic factors or inhibit the action of pro-apoptotic factors such as MYC [46]. BIM belongs to the BCL2 family and promotes apoptosis by inhibiting anti-apoptotic BCL2 family members [47]. BIN1 interacts with MYC and promotes apoptosis. Overexpression of SRSF1 promotes the expression of two *BIM* isoforms, called $\gamma 1$ and $\gamma 2$, lacking exon 2 and 3, and *BIN1* lacking exon 13 [46]. An increase in BIM $\gamma 1$, which lacks pro-apoptotic activity, and a decrease in exon 13-containing BIN1, which has lost its interaction with MYC, contribute to the SRSF1-induced phenotype. In short, SRSF1 regulates apoptosis by promoting the expression of BIN1 and BIM isoforms lacking pro-apoptotic activity.

hnRNP L, a member of the hnRNP family, regulates splicing in a position-dependent manner. hnRNP L was initially discovered as a splicing activator of the human *eNOS*. It binds to the CA-repeat region of the intron upstream of exon 13 [48]. However, two reports have shown that hnRNP L represses alternative splicing by competing with U2AF for intron binding close to the 3' splice site, as demonstrated for the *TJP1* [49] [50]. Thus, hnRNP L represses the splicing of alternative exons when it is bound to intronic regions upstream of these exons, and activates splicing when it is bound to intronic regions downstream of alternative exons [51].

In the forskolin/protein kinase A pathway, hnRNP K binds to the KARRE element upstream of the 3' splice site. This binding functions to repress exon 5a of *Snap*25 and competes with the binding of U2AF to the 3' splice site. hnRNP K also controls splicing of genes related to neurological diseases, such as *Runx1* [52].

In humans, more than 90% of transcripts undergo alternative splicing. It is precisely controlled, according to the required tissue and organ function. Mutations in SREs affecting alternative splicing cause human genetic diseases and cancers [53]. Overall, 15% - 50% of genetic diseases are caused by aberrant splicing [54]. One example is Alzheimer's disease. HMGA1a, originally described as an oncogene product, has a high affinity for presenilin2 (*PS*2) pre-mRNA, but its binding results in aberrant splicing that produces mRNA lacking exon 5 (*PS*2*V*). PS2V, the protein product of this aberrant splicing, accumulates in the brain of people with Alzheimer's disease, and can cause cell death [55] [56].

In addition to cis- and trans-acting factors, chromatin modification and transcription rate have also been suggested to regulate alternative splicing [53]. The precise regulation of alternative splicing is still not fully understood; advances in our understanding will inform on the mechanisms of gene regulation and help to develop therapeutic strategies for human disease.

3.3. Alternative Polyadenylation

Alternative polyadenylation (APA) is a phenomenon that produces multiple isoforms of mRNA because of the existence of multiple polyadenylation sites in one gene. APA often controls the fate of mRNA by diversifying the length of the mRNA 3' UTR. A longer 3' UTR is considered to have more cis-elements and therefore have a chance for regulating the mRNA stability (see chapter 3-5 in detail).

Recent genome-wide analyses have revealed that APA is observed widely throughout the genome, and that its pattern varies according to developmental stage and tissue organization [57] [58]. In *Drosophila*, mRNAs in the central nervous system tend to have longer 3' UTRs, whereas those in the testis have shorter 3' UTRs [59]. An extended 3' UTR is common among transcription factors and RNA-binding proteins, suggesting a relationship with auto- and post-transcriptional regulation. Ongoing studies are attempting to identify consensus sequences for APA and reveal the mechanistic details.

3.4. Gene Regulation via mRNA Export

Gene regulation by mRNA export is frequently observed in retroviruses such as human immunodeficiency virus type 1 (HIV-1). HIV-1 possesses a single promoter in its genome and produces just one primary transcript. However, it produces more than 40 multiple mRNA isoforms by alternative splicing. In the early phase, the HIV-1 primary transcript, encoding the Rev protein, is exported into the cytoplasm in a splicing-dependent manner (**Figure 4**). The Rev protein is translocated into the nucleus, where it is translated in the early phase [60] [61]. The Rev protein binds to the Rev response element (RRE) in the primary transcript [62]. Then, CRM1, an RNA export receptor, recognizes the nuclear export signal in Rev and binds to the primary transcript via Rev proteins [63]. This results in the export of the HIV-1 primary transcript to the cytoplasm via the CRM1-dependent mRNA export pathway. When Rev is bound to the RRE, Rev also interacts with CBP, thus suppressing the recruitment of the TREX components and the TAP-dependent mRNA export pathway [64].

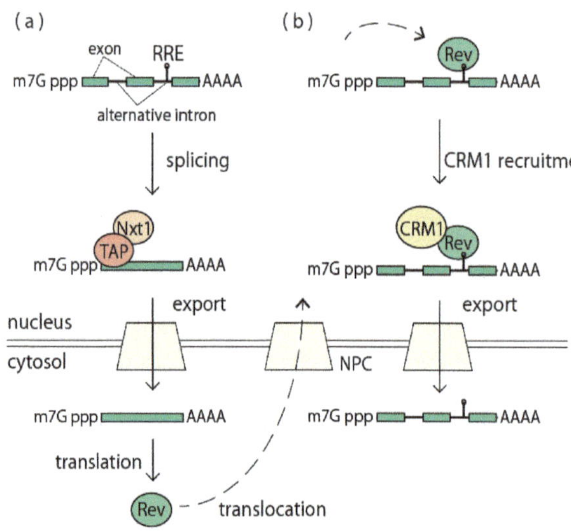

Figure 4. Strategy to export alternatively spliced mRNAs of HIV-1. (a) In the early phase, transcripts of HIV-1 (**Figure 4** shows representative isoforms) are exported into the cytoplasm in a splicing-dependent manner. mRNAs exported in this way encode proteins including Rev. Rev is translocated in the nucleus; (b) In the late phase, Rev binds to RRE which is present in an intron of primary transcript of HIV-1. The nuclear export receptor CRM1 is recruited to Rev. As a result, intron-containing mRNAs of HIV-1 are exported to the cytoplasm to propagate.

TAP controls its expression level by regulating its own mRNA export by a process similar to the "CTE system" in Mason-Pfizer Monkey Virus (MPMV) [65]. MPMV is a retrovirus whose gene expression is regulated by its RNA element, the constitutive transport element (CTE). The CTE is directly recognized and bound by TAP, so mRNA containing a CTE is efficiently exported into the cytoplasm [66], even in an intron-containing form [67]-[70]. Li *et al.* found that intron 10 of TAP contained an element similar to the CTE found in MPMV and its relatives, which was directly recognized by TAP [65]. When excessive TAP is expressed by chance, saturated TAP directly recognizes and binds the CTE, which results in export of *TAP* mRNA in an intron-containing form. TAP mRNA in an intron-containing form is likely to be degraded by nonsense-mediated decay, providing a mechanism for auto-regulation.

Recent research has revealed that gene regulation via mRNA export may be more important than previously recognized. GANP is known to promote TAP-dependent mRNA export [71]. Now it has been shown that GANP promotes mRNAs that are highly expressed, short-lived, and highly enriched in central components of the gene expression machineries [72]. It is suggested that GANP enables rapid changes in gene expression by facilitating nuclear export of specific classes of mRNA, by bridging target genes and the nuclear pore.

3.5. Half-Life of mRNA

Some genes regulate their expression level by controlling their mRNA half-life. mRNA with a shorter half-life results in lower protein expression level because of reduced translation; this system is frequently found in factors related to the cell cycle, development, and immune responses. Such mRNAs usually have cis-elements in their 3' UTR, including an AU-rich element (ARE) and/or target sites for micro RNAs (miRNAs).

mRNAs containing an ARE tend to be rapidly degraded because the poly-A tail is quickly shortened. AREs are classified into three types according to their sequence, which determines the degradation pathway taken [73] [74]: class I has one to three repeats of "AUUUA"; class II has multiple repeats of "AUUUA", and class III has an A-and-U cluster without any repeats of "AUUUA". ARE-binding proteins (AUBPs) also determine mRNA fate. There are many AUBPs, such as tristetraprolin (TTP), BRF1, AUF1, the Hu protein family, and ataxin 2. TTP, BRF1, and AUF1 destabilize ARE-containing mRNAs, while the Hu protein family and ataxin 2 stabilize them [75]-[80].

Class II AREs are recognized and bound by TTP. Mice deficient in TTP develop complicated immunological disease soon after birth because stabilization of mRNAs containing class II AREs results in excess expression of TNFα, a major regulator of the immune system, whose mRNA contains a class II ARE [76]. TTP is known to be required for deadenylation, and details of its destabilization mechanism are now emerging. TTP contains a zinc finger CCCH domain and directly binds to mRNAs via their ARE [81] [82]. However, TTP also binds to the CCR4-NOT multi-functional complex via its subunit, NOT1 [83]. NOT1 recruits the deadenylases Caf1 and CCR4, which results in deadenylation of mRNA containing a class II ARE [84].

miRNAs are small RNAs approximately 22 nt long that regulate gene expression post-transcriptionally via RNA silencing. Most of miRNAs are transcribed as primary miRNA (pri-miRNA) by polII in the nucleus, and has a 5' cap, poly-A tail, and local hairpin structure [85]-[87]. The nuclear RNAse III Drosha and its essential cofactor DGCR8 crop unnecessary stem-loops and single-stranded RNA to form pre-miRNA. After processing, the pre-miRNA is exported to the cytoplasm by exportin 5 with RanGTP, where it is processed by Dicer, and loaded onto Argonaute (AGO) protein to form the RNA-induced silencing complex (RISC). miRNA induces translational repression and/or mRNA degradation by the binding to 3' UTRs. The RISC recognizes the target mRNA by being guided by a mostly complementary miRNA sequence. More than 1000 miRNAs have been identified, implying that the expression of more than 5% of mRNAs is affected by miRNAs [88]. In fact, microarray data have suggested that almost one third of mRNAs are potential targets for miRNAs, suggesting that this represents a global function rather than a "fine-tuning" mechanism. There appear to be at least two mechanisms for miRNA-mediated gene regulation—initiation-dependent and initiation-independent [87]—but the exact mechanisms are still to be elucidated.

It has also been reported that miRNAs participate in the degradation of ARE-containing mRNAs. Qing *et al.* showed that components of the RISC such as AGO and miR16 are required for destabilization of ARE-containing mRNA [89]. Interestingly, TTP is also required for this pathway. Because of its apparent importance, uncovering these mechanisms will lead to advances in the understanding of human diseases and developmental processes.

4. Conclusion

Gene regulation is a complex process. The regulation steps include transcription, mRNA processing, mRNA transport and translocation, translation, protein activation and mRNA half-life control. In capping, methylation step is regulated by feedback loop. The meaning of methylation at first and second nucleotide, however, is unclear. In splicing, the selection of each exon is an important step for the different types of protein expression. SR proteins and hnRNPs play important roles for the splice site selection. In APA, selection of APA site determines the mRNA half-life. Specific sequence located mostly at 3' UTR is the target site for AUBPs as well as miRNAs and regulates the stability of mRNA. Output of multiple gene regulations results in specific expression pattern and determines the cell specific phenotype.

Acknowledgements

This work was supported by grants-in-aid from the Ministry of Education, Culture, Sports, Science and Technology of Japan, the Skylark Food Science Institute, The TojuroIijima Foundation for Food Science and Technology, The Foundation for Dietary Scientific Research, the Noda Institute for Scientific Research and Technology, and The Fuji Foundation for Protein Research to S. M.

References

[1] Hirose, Y. and Manley, J.L. (2000) RNA Polymerase II and the Integration of Nuclear Events. *Genes & Development*, **14**, 1415-1429.

[2] Staley, J.P. and Guthrie, C. (1998) Mechanical Devices of the Spliceosome: Motors, Clocks, Springs, and Things. *Cell*, **92**, 315-326. http://dx.doi.org/10.1016/S0092-8674(00)80925-3

[3] Matera, A.G. and Wang, Z. (2014) A Day in the Life of the Spliceosome. *Nature Reviews Molecular Cell Biology*, **15**, 108-121. http://dx.doi.org/10.1038/nrm3742

[4] Colgan, D.F. and Manley, J.L. (1997) Mechanism and Regulation of mRNA Polyadenylation. *Genes & Development*, **11**, 2755-2766. http://dx.doi.org/10.1101/gad.11.21.2755

[5] Mangus, D.A., Evans, M.C. and Jacobson, A. (2003) Poly(A)-Binding Proteins: Multifunctional Scaffolds for the Post-Transcriptional Control of Gene Expression. *Genome Biology*, **4**, 223. http://dx.doi.org/10.1186/gb-2003-4-7-223

[6] Culjkovic-Kraljacic, B. and Borden, K.L. (2013) Aiding and Abetting Cancer: mRNA Export and the Nuclear Pore. *Trends in Cell Biology*, **23**, 328-335. http://dx.doi.org/10.1016/j.tcb.2013.03.004

[7] Natalizio, B.J. and Wente, S.R. (2013) Postage for the Messenger: Designating Routes for Nuclear mRNA Export. *Trends in Cell Biology*, **23**, 365-373. http://dx.doi.org/10.1016/j.tcb.2013.03.006

[8] Strasser, K., Masuda, S., Mason, P., Pfannstiel, J., Oppizzi, M., Rodriguez-Navarro, S., Rondon, A.G., Aguilera, A., Struhl, K., Reed, R. and Hurt, E. (2002) TREX Is a Conserved Complex Coupling Transcription with Messenger RNA Export. *Nature*, **417**, 304-308. http://dx.doi.org/10.1038/nature746

[9] Gwizdek, C., Iglesias, N., Rodriguez, M.S., Ossareh-Nazari, B., Hobeika, M., Divita, G., Stutz, F. and Dargemont, C. (2006) Ubiquitin-Associated Domain of Mex67 Synchronizes Recruitment of the mRNA Export Machinery with Transcription. *Proceedings of the National Academy of Sciences of the United States of America*, **103**, 16376-16381. http://dx.doi.org/10.1073/pnas.0607941103

[10] Zhou, Z., Luo, M.J., Straesser, K., Katahira, J., Hurt, E. and Reed, R. (2000) The Protein Aly Links Pre-Messenger-RNA Splicing to Nuclear Export in Metazoans. *Nature*, **407**, 401-405. http://dx.doi.org/10.1038/35030160

[11] Zhou, Z., Licklider, L.J., Gygi, S.P. and Reed, R. (2002) Comprehensive Proteomic Analysis of the Human Spliceosome. *Nature*, **419**, 182-185.

[12] Masuda, S., Das, R., Cheng, H., Hurt, E., Dorman, N. and Reed, R. (2005) Recruitment of the Human TREX Complex to mRNA during Splicing. *Genes & Development*, **19**, 1512-1517. http://dx.doi.org/10.1101/gad.1302205

[13] Yamazaki, T., Fujiwara, N., Yukinaga, H., Ebisuya, M., Shiki, T., Kurihara, T., Kioka, N., Kambe, T., Nagao, M., Nishida, E. and Masuda, S. (2010) The Closely Related RNA Helicases, UAP56 and URH49, Preferentially form Distinct mRNA Export Machineries and Coordinately Regulate Mitotic Progression. *Molecular Biology of the Cell*, **21**, 2953-2965. http://dx.doi.org/10.1091/mbc.E09-10-0913

[14] Rout, M.P., Aitchison, J.D., Suprapto, A., Hjertaas, K., Zhao, Y. and Chait, B.T. (2000) The Yeast Nuclear Pore Complex: Composition, Architecture, and Transport Mechanism. *Journal of Cell Biology*, **148**, 635-651. http://dx.doi.org/10.1083/jcb.148.4.635

[15] Fribourg, S., Braun, I.C., Izaurralde, E. and Conti, E. (2001) Structural Basis for the Recognition of a Nucleoporin FG

Repeat by the NTF2-Like Domain of the TAP/p15 mRNA Nuclear Export Factor. *Molecular Cell*, **8**, 645-656. http://dx.doi.org/10.1016/S1097-2765(01)00348-3

[16] Ribbeck, K. and Gorlich, D. (2001) Kinetic Analysis of Translocation through Nuclear Pore Complexes. *The EMBO Journal*, **20**, 1320-1330. http://dx.doi.org/10.1093/emboj/20.6.1320

[17] Grant, R.P., Hurt, E., Neuhaus, D. and Stewart, M. (2002) Structure of the C-Terminal FG-Nucleoporin Binding Domain of Tap/NXF1. *Nature Structural Biology*, **9**, 247-251. http://dx.doi.org/10.1038/nsb773

[18] Alcazar-Roman, A.R., Tran, E.J., Guo, S. and Wente, S.R. (2006) Inositol Hexakisphosphate and Gle1 Activate the DEAD-Box Protein Dbp5 for Nuclear mRNA Export. *Nature Cell Biology*, **8**, 711-716. http://dx.doi.org/10.1038/ncb1427

[19] Weirich, C.S., Erzberger, J.P., Flick, J.S., Berger, J.M., Thorner, J. and Weis, K. (2006) Activation of the DExD/H-Box Protein Dbp5 by the Nuclear-Pore Protein Gle1 and Its Coactivator $InsP_6$ Is Required for mRNA Export. *Nature Cell Biology*, **8**, 668-676. http://dx.doi.org/10.1038/ncb1424

[20] Tran, E.J., Zhou, Y., Corbett, A.H. and Wente, S.R. (2007) The DEAD-Box Protein Dbp5 Controls mRNA Export by Triggering Specific RNA:Protein Remodeling Events. *Molecular Cell*, **28**, 850-859. http://dx.doi.org/10.1016/j.molcel.2007.09.019

[21] Hutten, S. and Kehlenbach, R.H. (2007) CRM1-Mediated Nuclear Export: To the Pore and Beyond. *Trends in Cell Biology*, **17**, 193-201. http://dx.doi.org/10.1016/j.tcb.2007.02.003

[22] Fornerod, M., Ohno, M., Yoshida, M. and Mattaj, I.W. (1997) CRM1 Is an Export Receptor for Leucine-Rich Nuclear Export Signals. *Cell*, **90**, 1051-1060. http://dx.doi.org/10.1016/S0092-8674(00)80371-2

[23] Kehlenbach, R.H., Dickmanns, A., Kehlenbach, A., Guan, T. and Gerace, L. (1999) A Role for RanBP1 in the Release of CRM1 from the Nuclear Pore Complex in a Terminal Step of Nuclear Export. *Journal of Cell Biology*, **145**, 645-657. http://dx.doi.org/10.1083/jcb.145.4.645

[24] Bischoff, F.R. and Ponstingl, H. (1991) Catalysis of Guanine Nucleotide Exchange on Ran by the Mitotic Regulator RCC1. *Nature*, **354**, 80-82.

[25] Schwer, B., Saha, N., Mao, X., Chen, H.W. and Shuman, S. (2000) Structure-Function Analysis of Yeast mRNA Cap Methyltransferase and High-Copy Suppression of Conditional Mutants by AdoMet Synthase and the Ubiquitin Conjugating Enzyme Cdc34p. *Genetics*, **155**, 1561-1576.

[26] Chiang, P.K., Gordon, R.K., Tal, J., Zeng, G.C., Doctor, B.P., Pardhasaradhi, K. and McCann, P.P. (1996) S-Adenosylmethionine and Methylation. *The FASEB Journal*, **10**, 471-480.

[27] Gorlich, D., Kraft, R., Kostka, S., Vogel, F., Hartmann, E., Laskey, R.A., Mattaj, I.W. and Izaurralde, E. (1996) Importin Provides a Link between Nuclear Protein Import and U snRNA Export. *Cell*, **87**, 21-32. http://dx.doi.org/10.1016/S0092-8674(00)81319-7

[28] Wen, Y. and Shatkin, A.J. (2000) Cap Methyltransferase Selective Binding and Methylation of GpppG-RNA Are Stimulated by Importin-*α*. *Genes & Development*, **14**, 2944-2949. http://dx.doi.org/10.1101/gad.848200

[29] Dias, S.M., Wilson, K.F., Rojas, K.S., Ambrosio, A.L. and Cerione, R.A. (2009) The Molecular Basis for the Regulation of the Cap-Binding Complex by the Importins. *Nature Structural & Molecular Biology*, **16**, 930-937. http://dx.doi.org/10.1038/nsmb.1649

[30] Furuichi, Y., Morgan, M., Shatkin, A.J., Jelinek, W., Salditt-Georgieff, M. and Darnell, J.E. (1975) Methylated, Blocked 5 Termini in HeLa Cell mRNA. *Proceedings of the National Academy of Sciences of the United States of America*, **72**, 1904-1908. http://dx.doi.org/10.1073/pnas.72.5.1904

[31] Zamudio, J.R., Mittra, B., Campbell, D.A. and Sturm, N.R. (2009) Hypermethylated Cap 4 Maximizes Trypanosoma Brucei Translation. *Molecular Microbiology*, **72**, 1100-1110. http://dx.doi.org/10.1111/j.1365-2958.2009.06696.x

[32] Belanger, F., Stepinski, J., Darzynkiewicz, E. and Pelletier, J. (2010) Characterization of hMTr1, a Human Cap1 2'-O-Ribose Methyltransferase. *The Journal of Biological Chemistry*, **285**, 33037-33044. http://dx.doi.org/10.1074/jbc.M110.155283

[33] Werner, M., Purta, E., Kaminska, K.H., Cymerman, I.A., Campbell, D.A., Mittra, B., Zamudio, J.R., Sturm, N.R., Jaworski, J. and Bujnicki, J.M. (2011) 2'-O-Ribose Methylation of Cap2 in Human: Function and Evolution in a Horizontally Mobile Family. *Nucleic Acids Research*, **39**, 4756-4768. http://dx.doi.org/10.1093/nar/gkr038

[34] Smietanski, M., Werner, M., Purta, E., Kaminska, K.H., Stepinski, J., Darzynkiewicz, E., Nowotny, M. and Bujnicki, J.M. (2014) Structural Analysis of Human 2'-O-Ribose Methyltransferases Involved in mRNA Cap Structure Formation. *Nature Communications*, **5**, Article Number: 3004.

[35] Van Dijk, E., Cougot, N., Meyer, S., Babajko, S., Wahle, E. and Seraphin, B. (2002) Human Dcp2: A Catalytically Active mRNA Decapping Enzyme Located in Specific Cytoplasmic Structures. *The EMBO Journal*, **21**, 6915-6924. http://dx.doi.org/10.1093/emboj/cdf678

[36] Piccirillo, C., Khanna, R. and Kiledjian, M. (2003) Functional Characterization of the Mammalian mRNA Decapping Enzyme hDcp2. *RNA*, **9**, 1138-1147.

[37] Jiao, X., Xiang, S., Oh, C., Martin, C.E., Tong, L. and Kiledjian, M. (2010) Identification of a Quality-Control Mechanism for mRNA 5'-End Capping. *Nature*, **467**, 608-611. http://dx.doi.org/10.1038/nature09338

[38] Cartegni, L., Chew, S.L. and Krainer, A.R. (2002) Listening to Silence and Understanding Nonsense: Exonic Mutations that Affect Splicing. *Nature Reviews Genetics*, **3**, 285-298. http://dx.doi.org/10.1038/nrg775

[39] Fu, X.D. and Ares Jr., M. (2014) Context-Dependent Control of Alternative Splicing by RNA-Binding Proteins. *Nature Reviews Genetics*, **15**, 689-701. http://dx.doi.org/10.1038/nrg3778

[40] Wang, Z. and Burge, C.B. (2008) Splicing Regulation: From a Parts List of Regulatory Elements to an Integrated Splicing Code. *RNA*, **14**, 802-813. http://dx.doi.org/10.1261/rna.876308

[41] Warf, M.B. and Berglund, J.A. (2010) Role of RNA Structure in Regulating Pre-mRNA Splicing. *Trends in Biochemical Sciences*, **35**, 169-178. http://dx.doi.org/10.1016/j.tibs.2009.10.004

[42] Wang, Y., Xiao, X., Zhang, J., Choudhury, R., Robertson, A., Li, K., Ma, M., Burge, C.B. and Wang, Z. (2013) A Complex Network of Factors with Overlapping Affinities Represses Splicing through Intronic Elements. *Nature Structural & Molecular Biology*, **20**, 36-45. http://dx.doi.org/10.1038/nsmb.2459

[43] Erkelenz, S., Mueller, W.F., Evans, M.S., Busch, A., Schoneweis, K., Hertel, K.J. and Schaal, H. (2013) Position-Dependent Splicing Activation and Repression by SR and hnRNP Proteins Rely on Common Mechanisms. *RNA*, **19**, 96-102. http://dx.doi.org/10.1261/rna.037044.112

[44] Wang, Y., Ma, M., Xiao, X. and Wang, Z. (2012) Intronic Splicing Enhancers, Cognate Splicing Factors and Context-Dependent Regulation Rules. *Nature Structural & Molecular Biology*, **19**, 1044-1052. http://dx.doi.org/10.1038/nsmb.2377

[45] Ghigna, C., Giordano, S., Shen, H., Benvenuto, F., Castiglioni, F., Comoglio, P.M., Green, M.R., Riva, S. and Biamonti, G. (2005) Cell Motility Is Controlled by SF2/ASF through Alternative Splicing of the *Ron* Protooncogene. *Molecular Cell*, **20**, 881-890. http://dx.doi.org/10.1016/j.molcel.2005.10.026

[46] Anczukow, O., Rosenberg, A.Z., Akerman, M., Das, S., Zhan, L., Karni, R., Muthuswamy, S.K. and Krainer, A.R. (2012) The Splicing Factor SRSF1 Regulates Apoptosis and Proliferation to Promote Mammary Epithelial Cell Transformation. *Nature Structural & Molecular Biology*, **19**, 220-228. http://dx.doi.org/10.1038/nsmb.2207

[47] O'Connor, L., Strasser, A., O'Reilly, L.A., Hausmann, G., Adams, J.M., Cory, S. and Huang, D.C. (1998) Bim: A Novel Member of the Bcl-2 Family That Promotes Apoptosis. *The EMBO Journal*, **17**, 384-395. http://dx.doi.org/10.1093/emboj/17.2.384

[48] Hui, J., Stangl, K., Lane, W.S. and Bindereif, A. (2003) HnRNP L Stimulates Splicing of the eNOS Gene by Binding to Variable-Length CA Repeats. *Nature Structural & Molecular Biology*, **10**, 33-37. http://dx.doi.org/10.1038/nsb875

[49] Heiner, M., Hui, J., Schreiner, S., Hung, L.H. and Bindereif, A. (2010) HnRNP L-Mediated Regulation of Mammalian Alternative Splicing by Interference with Splice Site Recognition. *RNA Biology*, **7**, 56-64. http://dx.doi.org/10.4161/rna.7.1.10402

[50] Hung, L.H., Heiner, M., Hui, J., Schreiner, S., Benes, V. and Bindereif, A. (2008) Diverse Roles of hnRNP L in Mammalian mRNA Processing: A Combined Microarray and RNAi Analysis. *RNA*, **14**, 284-296. http://dx.doi.org/10.1261/rna.725208

[51] Rossbach, O., Hung, L.H., Khrameeva, E., Schreiner, S., Konig, J., Curk, T., Zupan, B., Ule, J., Gelfand, M.S. and Bindereif, A. (2014) Crosslinking-Immunoprecipitation (iCLIP) Analysis Reveals Global Regulatory Roles of hnRNP L. *RNA Biology*, **11**, 146-155. http://dx.doi.org/10.4161/rna.27991

[52] Cao, W., Razanau, A., Feng, D., Lobo, V.G. and Xie, J. (2012) Control of Alternative Splicing by Forskolin through hnRNP K during Neuronal Differentiation. *Nucleic Acids Research*, **40**, 8059-8071. http://dx.doi.org/10.1093/nar/gks504

[53] Kornblihtt, A.R., Schor, I.E., Allo, M., Dujardin, G., Petrillo, E. and Munoz, M.J. (2013) Alternative splicing: A Pivotal Step between Eukaryotic Transcription and Translation. *Nature Reviews Molecular Cell Biology*, **14**, 153-165. http://dx.doi.org/10.1038/nrm3525

[54] Caceres, J.F. and Kornblihtt, A.R. (2002) Alternative Splicing: Multiple Control Mechanisms and Involvement in Human Disease. *Trends in Genetics*, **18**, 186-193. http://dx.doi.org/10.1016/S0168-9525(01)02626-9

[55] Manabe, T., Ohe, K., Katayama, T., Matsuzaki, S., Yanagita, T., Okuda, H., Bando, Y., Imaizumi, K., Reeves, R., Tohyama, M. and Mayeda, A. (2007) HMGA1a: Sequence-Specific RNA-Binding Factor Causing Sporadic Alzheimer's Disease-Linked Exon Skipping of Presenilin-2 Pre-mRNA. *Genes to Cells*, **12**, 1179-1191. http://dx.doi.org/10.1111/j.1365-2443.2007.01123.x

[56] Manabe, T., Katayama, T., Sato, N., Gomi, F., Hitomi, J., Yanagita, T., Kudo, T., Honda, A., Mori, Y., Matsuzaki, S.,

Imaizumi, K., Mayeda, A. and Tohyama, M. (2003) Induced HMGA1a Expression Causes Aberrant Splicing of Presenilin-2 Pre-mRNA in Sporadic Alzheimer's Disease. *Cell Death and Differentiation*, **10**, 698-708. http://dx.doi.org/10.1038/sj.cdd.4401221

[57] Ji, Z., Lee, J.Y., Pan, Z., Jiang, B. and Tian, B. (2009) Progressive Lengthening of 3' Untranslated Regions of mRNAs by Alternative Polyadenylation during Mouse Embryonic Development. *Proceedings of the National Academy of Sciences of the United States of America*, **106**, 7028-7033. http://dx.doi.org/10.1073/pnas.0900028106

[58] Ji, Z. and Tian, B. (2009) Reprogramming of 3' Untranslated Regions of mRNAs by Alternative Polyadenylation in Generation of Pluripotent Stem Cells from Different Cell Types. *PLoS ONE*, **4**, e8419. http://dx.doi.org/10.1371/journal.pone.0008419

[59] Smibert, P., Miura, P., Westholm, J.O., Shenker, S., May, G., Duff, M.O., Zhang, D., Eads, B.D., Carlson, J., Brown, J.B., Eisman, R.C., Andrews, J., Kaufman, T., Cherbas, P., Celniker, S.E., Graveley, B.R. and Lai, E.C. (2012) Global Patterns of Tissue-Specific Alternative Polyadenylation in Drosophila. *Cell Reports*, **1**, 277-289.

[60] Meyer, B.E. and Malim, M.H. (1994) The HIV-1 Rev Trans-Activator Shuttles between the Nucleus and the Cytoplasm. *Genes & Development*, **8**, 1538-1547. http://dx.doi.org/10.1101/gad.8.13.1538

[61] Richard, N., Iacampo, S. and Cochrane, A. (1994) HIV-1 Rev Is Capable of Shuttling between the Nucleus and Cytoplasm. *Virology*, **204**, 123-131. http://dx.doi.org/10.1006/viro.1994.1516

[62] Heaphy, S., Dingwall, C., Ernberg, I., Gait, M.J., Green, S.M., Karn, J., Lowe, A.D., Singh, M. and Skinner, M.A. (1990) HIV-1 Regulator of Virion Expression (Rev) Protein Binds to an RNA Stem-Loop Structure Located within the Rev Response Element Region. *Cell*, **60**, 685-693. http://dx.doi.org/10.1016/0092-8674(90)90671-Z

[63] Askjaer, P., Jensen, T.H., Nilsson, J., Englmeier, L. and Kjems, J. (1998) The Specificity of the CRM1-Rev Nuclear Export Signal Interaction Is Mediated by RanGTP. *The Journal of Biological Chemistry*, **273**, 33414-33422. http://dx.doi.org/10.1074/jbc.273.50.33414

[64] Taniguchi, I., Mabuchi, N. and Ohno, M. (2014) HIV-1 Rev Protein Specifies the Viral RNA Export Pathway by Suppressing TAP/NXF1 Recruitment. *Nucleic Acids Research*, **42**, 6645-6658. http://dx.doi.org/10.1093/nar/gku304

[65] Li, Y., Bor, Y.C., Misawa, Y., Xue, Y., Rekosh, D. and Hammarskjold, M.L. (2006) An Intron with a Constitutive Transport Element Is Retained in a Tap Messenger RNA. *Nature*, **443**, 234-237.

[66] Aihara, Y., Fujiwara, N., Yamazaki, T., Kambe, T., Nagao, M., Hirose, Y. and Masuda, S. (2011) Enhancing Recombinant Protein Production in Human Cell Lines with a Constitutive Transport Element and mRNA Export Proteins. *Journal of Biotechnology*, **153**, 86-91. http://dx.doi.org/10.1016/j.jbiotec.2011.03.024

[67] Gruter, P., Tabernero, C., von Kobbe, C., Schmitt, C., Saavedra, C., Bachi, A., Wilm, M., Felber, B.K. and Izaurralde, E. (1998) TAP, the Human Homolog of Mex67p, Mediates CTE-Dependent RNA Export from the Nucleus. *Molecular Cell*, **1**, 649-659. http://dx.doi.org/10.1016/S1097-2765(00)80065-9

[68] Liker, E., Fernandez, E., Izaurralde, E. and Conti, E. (2000) The Structure of the mRNA Export Factor TAP Reveals a *cis* Arrangement of a Non-Canonical RNP Domain and an LRR Domain. *The EMBO Journal*, **19**, 5587-5598. http://dx.doi.org/10.1093/emboj/19.21.5587

[69] Tabernero, C., Zolotukhin, A.S., Valentin, A., Pavlakis, G.N. and Felber, B.K. (1996) The Posttranscriptional Control Element of the Simian Retrovirus Type 1 Forms an Extensive RNA Secondary Structure Necessary for Its Function. *Journal of Virology*, **70**, 5998-6011.

[70] Ernst, R.K., Bray, M., Rekosh, D. and Hammarskjold, M.L. (1997) A Structured Retroviral RNA Element That Mediates Nucleocytoplasmic Export of Intron-Containing RNA. *Molecular and Cellular Biology*, **17**, 135-144.

[71] Wickramasinghe, V.O., McMurtrie, P.I., Mills, A.D., Takei, Y., Penrhyn-Lowe, S., Amagase, Y., Main, S., Marr, J., Stewart, M. and Laskey, R.A. (2010) mRNA Export from Mammalian Cell Nuclei Is Dependent on GANP. *Current Biology*, **20**, 25-31. http://dx.doi.org/10.1016/j.cub.2009.10.078

[72] Wickramasinghe, V.O., Andrews, R., Ellis, P., Langford, C., Gurdon, J.B., Stewart, M., Venkitaraman, A.R. and Laskey, R.A. (2014) Selective Nuclear Export of Specific Classes of mRNA from Mammalian Nuclei Is Promoted by GANP. *Nucleic Acids Research*, **42**, 5059-5071. http://dx.doi.org/10.1093/nar/gku095

[73] Xu, N., Chen, C.Y. and Shyu, A.B. (1997) Modulation of the Fate of Cytoplasmic mRNA by AU-Rich Elements: Key Sequence Features Controlling mRNA Deadenylation and Decay. *Molecular and Cellular Biology*, **17**, 4611-4621.

[74] Bakheet, T., Frevel, M., Williams, B.R., Greer, W. and Khabar, K.S. (2001) ARED: Human AU-Rich Element-Containing mRNA Database Reveals an Unexpectedly Diverse Functional Repertoire of Encoded Proteins. *Nucleic Acids Research*, **29**, 246-254. http://dx.doi.org/10.1093/nar/29.1.246

[75] Zhang, W., Wagner, B.J., Ehrenman, K., Schaefer, A.W., DeMaria, C.T., Crater, D., DeHaven, K., Long, L. and Brewer, G. (1993) Purification, Characterization, and cDNA Cloning of an AU-Rich Element RNA-Binding Protein, AUF1. *Molecular and Cellular Biology*, **13**, 7652-7665.

[76] Carballo, E., Lai, W.S. and Blackshear, P.J. (1998) Feedback Inhibition of Macrophage Tumor Necrosis Factor-α Production by Tristetraprolin. *Science*, **281**, 1001-1005. http://dx.doi.org/10.1126/science.281.5379.1001

[77] Antic, D., Lu, N. and Keene, J.D. (1999) ELAV Tumor Antigen, Hel-N1, Increases Translation of Neurofilament M mRNA and Induces Formation of Neurites in Human Teratocarcinoma Cells. *Genes & Development*, **13**, 449-461. http://dx.doi.org/10.1101/gad.13.4.449

[78] Keene, J.D. (1999) Why Is Hu Where? Shuttling of Early-Response-Gene Messenger RNA Subsets. *Proceedings of the National Academy of Sciences of the United States of America*, **96**, 5-7. http://dx.doi.org/10.1073/pnas.96.1.5

[79] Stoecklin, G., Colombi, M., Raineri, I., Leuenberger, S., Mallaun, M., Schmidlin, M., Gross, B., Lu, M., Kitamura, T. and Moroni, C. (2002) Functional Cloning of BRF1, a Regulator of ARE-Dependent mRNA Turnover. *The EMBO Journal*, **21**, 4709-4718. http://dx.doi.org/10.1093/emboj/cdf444

[80] Yokoshi, M., Li, Q., Yamamoto, M., Okada, H., Suzuki, Y. and Kawahara, Y. (2014) Direct Binding of Ataxin-2 to Distinct Elements in 3' UTRs Promotes mRNA Stability and Protein Expression. *Molecular Cell*, **55**, 186-198. http://dx.doi.org/10.1016/j.molcel.2014.05.022

[81] Lai, W.S., Carballo, E., Strum, J.R., Kennington, E.A., Phillips, R.S. and Blackshear, P.J. (1999) Evidence that Tristetraprolin Binds to AU-Rich Elements and Promotes the Deadenylation and Destabilization of Tumor Necrosis Factor *alpha* mRNA. *Journal of Molecular Cell Biology*, **19**, 4311-4323.

[82] Hudson, B.P., Martinez-Yamout, M.A., Dyson, H.J. and Wright, P.E. (2004) Recognition of the mRNA AU-Rich Element by the Zinc Finger Domain of TIS11d. *Nature Structural & Molecular Biology*, **11**, 257-264.

[83] Fabian, M.R., Frank, F., Rouya, C., Siddiqui, N., Lai, W.S., Karetnikov, A., Blackshear, P.J., Nagar, B. and Sonenberg, N. (2013) Structural Basis for the Recruitment of the Human CCR4-NOT Deadenylase Complex by Tristetraprolin. *Nature Structural & Molecular Biology*, **20**, 735-739. http://dx.doi.org/10.1038/nsmb.2572

[84] Sandler, H., Kreth, J., Timmers, H.T. and Stoecklin, G. (2011) Not1 Mediates Recruitment of the Deadenylase Caf1 to mRNAs Targeted for Degradation by Tristetraprolin. *Nucleic Acids Research*, **39**, 4373-4386. http://dx.doi.org/10.1093/nar/gkr011

[85] Liu, J. (2008) Control of Protein Synthesis and mRNA Degradation by microRNAs. *Current Opinion in Cell Biology*, **20**, 214-221. http://dx.doi.org/10.1016/j.ceb.2008.01.006

[86] Cannell, I.G., Kong, Y.W. and Bushell, M. (2008) How do microRNAs Regulate Gene Expression? *Biochemical Society Transactions*, **36**, 1224-1231. http://dx.doi.org/10.1042/BST0361224

[87] Ha, M. and Kim, V.N. (2014) Regulation of microRNA Biogenesis. *Nature Reviews Molecular Cell Biology*, **15**, 509-524. http://dx.doi.org/10.1038/nrm3838

[88] Brummer, A. and Hausser, J. (2014) MicroRNA Binding Sites in the Coding Region of mRNAs: Extending the Repertoire of Post-Transcriptional Gene Regulation. *BioEssays*, **36**, 617-626. http://dx.doi.org/10.1002/bies.201300104

[89] Jing, Q., Huang, S., Guth, S., Zarubin, T., Motoyama, A., Chen, J., Di Padova, F., Lin, S.C., Gram, H. and Han, J. (2005) Involvement of microRNA in AU-Rich Element-Mediated mRNA Instability. *Cell*, **120**, 623-634. http://dx.doi.org/10.1016/j.cell.2004.12.038

3

TNFα and IL1β Stimulate Differential Gene Expression in Endometrial Stromal Cells

Abha J. Chalpe[1], Chad D. Law[1], Jennifer N. Dumdie[1], Keith A. Hansen[2], Kathleen M. Eyster[1,2*]

[1]Division of Basic Biomedical Sciences, Sanford School of Medicine, The University of South Dakota, Vermillion, SD, USA
[2]Department of Obstetrics and Gynecology, Sanford School of Medicine, The University of South Dakota, Sioux Falls, SD, USA
Email: abhachalpe@gmail.com, clawmdit@gmail.com, jdumdie@ucsd.edu, kahansen@usd.edu, *Kathleen.Eyster@usd.edu

Abstract

The purpose of this study was to test the hypothesis that specific macrophage-secreted cytokines cause gene expression changes in endometrial stromal cells that reproduce the effects of macrophages in the development of endometriosis. Telomerase-immortalized human endometrial stromal cells (T-HESC) were treated with tumor necrosis factor α (TNFα, 5 ng/ml) and interleukin 1β (IL1β, 1 ng/ml). Differential expression of 249 genes was identified by DNA microarray. Ontologies such as peptidases, cell adhesion, cell death/cell cycle, growth factors, cytoskeletal organization, defense/immune system, signal transduction, and transcriptional regulation which are related to the development of endometriosis were represented by these genes. The up-regulation of interleukin 8 (IL8), interleukin 6 (IL6), IL1β and matrix metalloproteinase 3 (*MMP*3) in response to TNFα ± ILIβ in T-HESC cells was confirmed by real time RT-PCR. TNFα ± ILIβ did not affect the migration or invasion of T-HESC cells. This study reinforces our previous investigations on communication between cells of the immune system and endometrial stromal cells and their potential role in the development of endometriosis.

Keywords

Cytokines, Endometriosis, Endometrium, Endometrial Stromal Cells, Microarray, Gene Expression

*Corresponding author.

1. Introduction

Endometriosis is an inflammatory disease in which endometrial tissue implants and grows outside the uterus [1]. The inflammatory nature of the disease, and the large number of genes related to the immune system/inflammation that are up-regulated in endometriosis [2] have led to the concept that factors from the immune system exacerbate the development of endometriosis instead of destroying ectopic endometrial tissue [3]-[8].

Factors secreted by macrophages and other immune system cells that are implicated in the development of inflammatory diseases such as endometriosis include the cytokines. It has been proposed that cytokines may promote neovascularization and attachment of endometrial cells to the peritoneum in the process of development of endometriosis [1] [4] [7]-[11].

Previous studies from our laboratory have demonstrated that reciprocal communication occurs between macrophages/monocytes and endometrial stromal cells in cell culture [5] [6]. These studies demonstrated that factors secreted by macrophages/monocytes caused differential gene expression in telomerase-immortalized human endometrial stromal cells (T-HESC) and vice-versa. In the current project we tested the hypothesis that specific cytokines and growth factors secreted by macrophages (tumor necrosis factor α (TNFα), interleukin 1β (IL1β), interleukin 6 (IL6), and interleukin 8 (IL8), and the growth factor, transforming growth factor β (TGFβ)) cause gene expression changes in T-HESC cells that reproduce the effects of macrophage conditioned medium. We also tested whether TNFα and IL1β increased the migratory and invasive properties of T-HESC cells.

This study used human telomerase-immortalized endometrial stromal cells [5] [6] [12] as a model of the early stages of endometriosis. Endometrial stromal cells are considered to be a critical cell type in the establishment of endometriosis lesions [13]. Simplifying our model to include only this single cell type allowed us to more clearly analyze the effects of cytokines and their potential role in endometriosis in a carefully controlled environment.

2. Materials and Methods

2.1. Experimental Design

Concentration-response curves were carried out to identify the concentrations of TNFα, IL1β, TGFβ, IL8, and IL6 that achieved the best response from cultured endometrial stromal cells. DNA microarrays were then used to analyze differential gene expression in endometrial stromal cells in response to TNFα $\pm$ IL1β. The ability of TNFα $\pm$ IL1β to modify migration and invasion of endometrial stromal cells was also assessed using Boyden chambers.

2.2. Cell Culture

The T-HESC cell line [12] was used for all experiments. T-HESC were obtained from American Type Cell Culture (ATCC, Manassas, VA) (CRL-4003). No ethical permissions were required for this study since the study was carried out in a commercially available cell line. The cells were routinely maintained in Dulbecco's modified Eagle's medium (DMEM, Sigma, St. Louis, MO) as described [6]. When the cells reached 80% confluence they were starved for 24 hours in starvation medium (DMEM+ ITS+ puromycin+ penicillin/streptomycin) before treatment with cytokines or growth factors.

2.3. Concentration-Response Curves for Cytokines

Three concentrations were tested for each cytokine and growth factor: TNFα (0.05, 0.5, and 5 ng/ml), IL1β (0.01, 0.1, and 1 ng/ml), IL8 (25, 150, and 500 ng/ml), IL6 (1, 5, and 10 ng/ml) and TGFβ (0.6, 1.2 and 10 ng/ml). All cytokines and TGFβ were obtained from Cell Sciences (Canton, MA). T-HESC cells were treated with individual cytokines and combinations of cytokines for 48 hours or 70 hours in the absence of FBS. After treatment, RNA was isolated from T-HESC and utilized for real time RT-PCR and DNA microarray analysis.

2.4. RNA Isolation and Quantification

For RNA isolation, T-HESC cells were washed twice with 2 ml phosphate buffered saline (PBS), and 1 ml Tri reagent (Molecular Research Center, Cincinnati, OH) was added to each culture flask. RNA was purified using RNeasy mini kit columns (Qiagen, Valencia, CA) and quantified using the RNA 6000 Nano LabChip in an Agilent 2100 Bioanalyzer (Agilent Technologies, Santa Clara, CA) as described [14].

2.5. Microarray Analysis

Codelink Whole Human Genome Bioarrays (Applied Microarrays, Tempe, AZ) were used to perform microarray analysis of gene expression as described [14]. Statistical analysis of the microarray data was carried out using Gene Spring 7.0 software (Agilent, Santa Clara, CA).

2.6. Real Time Reverse Transcription-Polymerase Chain Reaction (Real Time RT-PCR)

Real time RT-PCR was used to measure the expression levels of IL8, *MMP3*, IL1*B*, and IL6 for the analyses of concentration-response curves when T-HESC cells were treated with TNFα ± IL1β. These genes were chosen based on their fold expression in the microarray analysis. Primers and TaqMan probes were obtained as Assays on Demand from Applied Biosystems/Life Technologies (Foster City, CA) (Hs00174097_m1 for IL1*B*, Hs01567913_g1 for IL8, Hs00174131_m1 for IL6, and Hs00968308_m1 for *MMP3*). The expression of target genes was normalized to the housekeeping gene, glyceraldehyde 3-phosphate dehydrogenase (*GAPDH*). The results were analyzed using qBase software by the delta CT relative quantification method [15].

2.7. Invasion and Migration in Response to Cytokines

Boyden chambers (BD Biosciences, Bedford, MA) were used to analyze the migratory and invasive activities of T-HESC in response to TNFα ± IL1β in two sets of experiments. The first experiment consisted of treating T-HESC cells with TNFα (5 ng/ml) ± IL1β (1 ng/ml) for 48 hours before plating on the Boyden chambers. The treated cells were dyed using a fluorescent CellTracker Probe (Molecular Probes, Eugene, OR) for cell counting after migration/invasion. The treated and dyed cells were placed on the control (migration) and invasion inserts of the Boyden chambers in medium containing cytokines. Each chamber was seeded with 2.5×10^4 cells. Treatment of the cells with TNFα (5 ng/ml) ± IL1β (1 ng/ml) continued throughout the incubation of the cells in the Boyden chamber. The cells were incubated in the chambers for 22 hours; they migrated or invaded in response to the chemoattractant, fibronectin (25 μg/ml, BD Biosciences, Bedford, MA) in the bottom chamber. After 22 hours of incubation the membranes were dissected free from the chambers, fixed in 4% paraformaldehyde for 2 minutes at room temperature and mounted on microscopy slides using immersion oil. The migration and invasion of T-HESC cells treated with TNFα ± IL1β was compared to vehicle-treated control cells. This experiment was repeated 5 times with different passages of T-HESC cells. In the second experiment, TNFα ± IL1β were used as chemoattractants; that is, TNFα ± IL1β were placed at the bottom of the Boyden chambers in place of fibronectin. The untreated cells were dyed with CellTracker dye, then seeded in the Boyden chambers at 2.5×10^4 cells/chamber. The chambers were incubated and processed as in the first experiment. In both experiments fluorescent cells that migrated/invaded were counted using ImageJ software (National Institutes of Health, Bethesda, MD).

2.8. Statistical Analyses

Data from the real-time RT-PCR experiments were analyzed using qBase software [15], and Graph Pad Prism 4.0 (San Diego, CA) software was used to perform analysis of variance (ANOVA) on real time RT-PCR data. Newman Keuls was used as the post hoc test. Gene Spring 7.0 software (Agilent) was used to perform ANOVA on the microarray data. ANOVA was also used to analyze on the data from cell invasion and migration experiments using the GraphPad Prism 4.0 software (San Diego, CA).

3. Results

Concentration-response curves identified 5 ng/ml TNFα and 1 ng/ml IL1β as the optimal concentrations of these cytokines for response in T-HESC cells after 48 hours of treatment (data not shown). In contrast, T-HESC cells did not respond to TGFβ, IL8, or IL6, either singly or in combination (data not shown).

DNA microarray identified 249 genes to be differentially expressed in the analysis of T-HESC cells treated with TNFα ± IL1β (**Table 1**). The gene ontologies that featured in this microarray data were peptidases, cell adhesion, cell death/apoptosis, cell cycle, growth factors, cytoskeletal organization, channels/carriers, enzymes/metabolism, defense/immune system, receptors and ligands, signal transduction, transcriptional regulation, cancer related, vesicle trafficking, chaperonins and other. The genes with a two-fold or greater change in expression and p value of 0.05 or less when compared to control were considered significant. As recommended by minimum information about microarray experiment (MIAME) standards [16], the entire data set for these microarrays has been deposited in National Center for Biotechnology Information (NCBI) Gene Expression Omnibus (GEO; www.ncbi.nlm.nih.gov/geo). The data can be accessed through GEO Series accession number GSE40007.

Four of the differentially expressed genes were chosen for confirmation by real time RT-PCR. IL1*B* was significantly up-regulated in response to TNFα + IL1β (**Figure 1(a)**, **Figure 1(b)**). The expression of IL8 showed

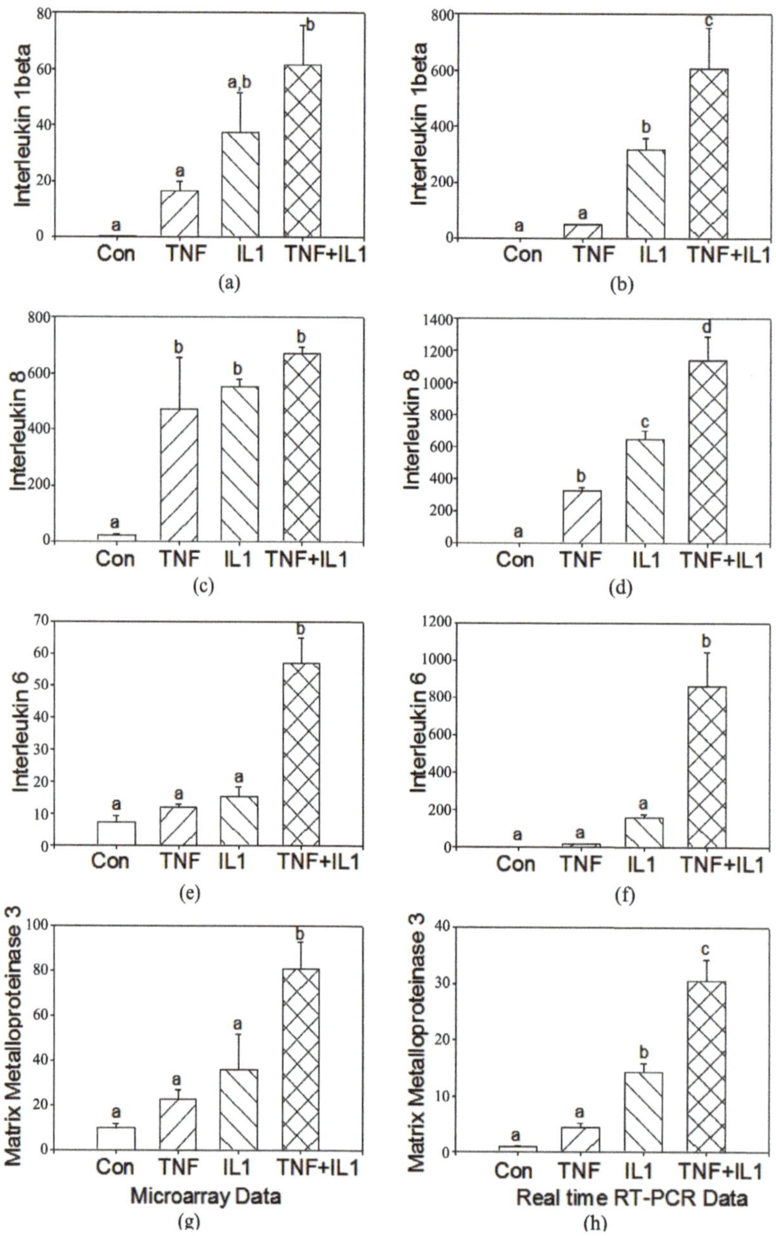

Figure 1. Differential expression of interleukin 1β (a, b), IL8 (c, d), IL6 (e, f), and matrix metalloproteinase 3 (g, h) in human telomerase-immortalized endometrial stromal cells (T-HESC) in response to treatment with TNFα (TNF) ± IL1β (IL1) for 48 hours. Data are shown from DNA microarray analysis (a, c, e, and g), and from real time RT-PCR analysis (b, d, f, and h). Data are the mean ± S.E., experimental n = 3. Bars with different letter superscripts denote that the data for those groups are significantly different from each other (ANOVA, Newman Keuls post hoc test, $p < 0.05$).

Table 1. Differentially expressed genes in human telomerase-immortalized endometrial stromal cells (T-HESC) treated with the cytokines TNFα ± IL1β.[1]

ACCN#[2]	Gene symbol	Con	TNF	IL1β	TNF + IL1β
	Peptidases/related				
NM_003817	ADAM7	1.17	1.58	1.33	2.45
NM_004390	CTSH	3.51	7.22	2.62	5.86
NM_004079	CTSS	2.53	27.38	11.63	27.64
NM_002422	MMP3	10.00	22.59	93.66	81.07
NM_007289	MME	1.80	3.29	2.96	6.78
NM_006587	CORIN	2.12	0.77	0.95	0.65
NM_002575	SERPINB2	1.40	18.72	9.97	17.18
NM_002974	SERPINB4	0.13	0.07	0.49	1.65
NM_005025	SERPINI1	1.76	1.02	1.30	0.89
NM_001899	CST4	0.64	1.34	0.86	0.89
NM_002801	PSMB10	13.72	32.52	21.99	27.56
NM_020903	USP29	0.56	1.02	0.92	1.14
NM_002771	PRSS3	2.78	2.31	2.49	1.16
	Cell adhesion				
NM_181847	AMIGO2	11.28	15.73	15.01	27.73
AB002377	ANKRD28	19.84	11.82	10.64	9.65
NM_001332	CTNND2	4.47	3.30	2.33	2.23
NM_138455	CTHRC1	163.71	302.87	205.68	222.93
NM_080680	COL11A2	0.79	1.62	0.87	1.87
NM_001856	COL16A1	1.99	1.90	3.44	7.51
NM_019035	PCDH18	14.10	6.30	6.41	4.56
NM_000885	ITGA4	12.98	7.09	8.80	2.86
NM_004791	ITGBL1	2.16	1.10	1.68	1.16
NM_000873	ICAM2	0.46	1.63	0.64	1.51
BG032839	PCOTH	6.12	5.05	4.76	3.09
NM_015429	ABI3BP	1.17	3.41	1.02	4.29
NM_012198	GCA	0.35	1.10	0.49	0.92
NM_004148	NINJ1	7.08	17.86	12.91	18.19
NM_004196	CDKL1	2.04	1.28	1.39	1.00
	Cell cycle				
NM_022909	CENPH	11.95	8.06	10.15	3.83
NM_001240	CCNT1	3.89	2.62	2.81	1.79
NM_174942	GAS2L3	8.18	4.94	5.25	3.52
NM_144658	DOCK11	3.92	2.46	2.35	1.61

Continued

NM_015714	G0S2	3.35	56.22	26.31	71.78
NM_007069	HRASLS3	8.35	22.60	11.22	15.94
NM_178428	LCE2A	12.49	5.85	7.09	1.43
NM_030915	LBH	66.32	35.75	35.43	17.75
NM_002358	MAD2L1	10.76	8.01	8.35	3.84
NM_032793	MFSD2	0.34	0.42	0.61	1.14
NM_015946	PELO	17.36	16.53	12.80	7.36
NM_001165	BIRC3	0.36	3.74	1.08	2.46
NM_001225	CASP4	14.64	32.62	21.14	26.54
	Growth factors				
NM_001953	ECGF1	1.37	7.04	2.45	4.63
NM_002006	FGF2	20.82	39.09	28.51	53.87
NM_000618	IGF1	0.66	0.80	0.80	1.58
NM_000618	IGF-IB	0.86	1.14	1.06	1.63
NM_030968	C1QTNF1	2.74	10.70	5.58	17.75
	Cytoskeletal				
NM_005159	ACTC	93.67	25.60	45.09	12.87
NM_001069	TUBB2	21.37	20.63	16.15	9.36
NM_003380	VIM	681.53	478.57	440.76	328.93
NM_002276	KRT19	76.84	59.23	36.94	30.65
NM_002273	KRT8	3.31	1.69	1.67	1.18
NM_031957	KRTAP1-5	22.18	7.82	9.75	3.33
AK128036	KRT18	17.73	15.22	8.71	6.09
NM_001747	CAPG	10.17	5.69	6.60	2.84
NM_002381	MATN3	4.25	1.83	2.70	1.45
NM_005379	MYO1A	1.06	1.98	1.61	4.26
NM_031956	NYD-SP14	1.54	6.17	3.93	6.73
NM_153267	MAMDC2	17.89	20.61	15.54	8.96
	Immune system				
NM_001710	BF	0.50	2.34	1.95	7.73
NM_004335	BST2	0.63	5.49	1.50	10.51
NM_031890	CECR6	2.63	4.58	3.16	7.51
NM_001250	CD40	0.34	1.23	0.64	1.03
NM_001779	CD58	4.91	13.38	5.74	8.17
NM_004355	CD74	0.21	1.40	0.22	1.09
NM_002231	CD82	7.35	21.47	17.49	22.45
NM_004233	CD83	0.38	1.07	0.39	0.77

Continued

NM_002985	CCL5	1.23	98.81	2.54	52.84
NM_006273	CCL7	2.17	14.34	15.83	56.62
NM_016557	CCRL1	2.51	5.98	2.74	4.44
NM_001565	CXCL10	0.31	3.68	2.41	4.22
NM_002089	CXCL2	1.28	4.64	47.90	103.75
NM_002090	CXCL3	5.05	9.62	15.48	31.98
NM_000063	C2	2.55	4.57	3.36	6.88
NM_000064	C3	0.15	0.60	0.84	5.66
NM_001312	CRIP2	2.51	1.68	2.15	1.26
NM_031476	CRISPLD2	0.78	0.69	0.90	4.30
NM_018659	CYTL1	0.40	2.48	1.61	2.83
NM_033255	EPSTI1	3.21	11.00	5.63	18.67
NM_005755	EBI3	1.62	11.10	2.32	9.26
NM_024503	HIVEP3	0.50	0.65	0.88	1.93
NM_207585	IFNAR2	4.20	10.06	5.39	9.26
AA463818	IFITM5	1.44	2.24	2.89	4.13
NM_022168	IFIH1	2.56	7.03	2.88	7.36
NM_004031	IRF7	4.35	10.81	6.03	14.29
NM_002201	ISG20	1.25	3.00	2.90	4.70
NM_005101	G1P2	22.54	46.03	30.04	55.91
NM_022873	G1P3	2.59	4.64	2.54	5.88
NM_005532	IFI27	8.08	67.13	17.37	82.11
NM_006332	IFI30	8.81	36.07	17.09	29.23
NM_005533	IFI35	15.02	35.59	20.08	33.11
NM_006417	IFI44	2.79	6.99	3.57	9.81
NM_006820	IFI44L	0.79	3.96	1.18	8.68
NM_000576	IL1B	0.45	11.85	37.22	61.25
NM_004221	IL32	11.52	193.11	38.77	115.47
NM_000600	IL6	7.43	11.84	15.26	57.08
NM_000880	IL7	0.31	0.83	0.95	1.53
NM_000584	IL8	23.31	472.44	550.94	670.07
NM_002189	IL15RA	4.43	10.04	8.12	15.60
NM_181079	IL21R	0.50	0.91	0.57	1.99
NM_002462	MX1	5.53	17.16	5.53	23.43
NM_002463	MX2	1.80	2.71	1.93	4.20
NM_005746	PBEF1	4.75	7.95	13.02	24.14
NM_014059	RGC32	21.18	32.92	19.81	6.53

Continued

NM_030754	SAA2	2.30	4.05	3.48	5.15
AK023341	NAMPT	1.02	1.90	3.61	6.39
NM_012449	STEAP1	0.97	2.11	2.31	3.77
NM_145006	SUSD3	5.36	9.93	6.54	5.28
NM_006307	SRPX	27.45	14.69	19.06	13.67
NM_006573	TNFSF13B	0.74	3.93	0.99	3.71
NM_002546	TNFRSF11B	4.30	14.52	8.98	18.19
NM_006291	TNFAIP2	1.91	3.05	6.21	13.26
NM_006290	TNFAIP3	1.18	5.37	3.02	10.03
NM_007115	TNFAIP6	47.67	62.33	69.42	157.64
	Receptors/ligands				
NM_000675	ADORA2A	1.07	1.51	2.57	7.14
NM_001146	ANGPT1	10.93	4.20	4.81	8.33
NM_020350	AGTRAP	13.86	28.25	17.23	29.54
NM_001621	AHR	1.48	3.82	4.05	5.43
NM_001878	CRABP2	35.59	6.20	16.08	4.08
NM_000739	CHRM2	2.86	1.57	1.27	1.12
NM_001957	EDNRA	11.15	4.51	8.84	10.56
NM_198569	GPR126	0.70	1.58	0.54	1.83
NM_001505	GPR30	1.86	1.25	1.66	0.80
NM_000815	GABRD	1.94	4.40	2.64	4.79
NM_005264	GFRA1	4.81	15.75	4.25	10.28
NM_152430	OR51E1	1.08	2.30	2.82	2.75
NM_005037	PPARG	7.59	3.96	3.46	3.16
NM_000955	PTGER1	1.88	0.77	0.82	0.47
NM_005854	RAMP2	1.92	1.49	1.11	0.93
NM_206963	RARRES1	1.31	2.01	1.96	4.02
BX648828	ROBO2	1.17	6.74	2.38	7.88
NM_003268	TLR5	0.49	0.85	0.93	1.03
	Signaling				
AK055561	ERN1	1.59	3.15	2.02	3.28
NM_017729	EPS8L1	1.57	5.05	1.95	2.90
BC040073	H19	2.82	1.80	1.57	1.41
NM_006558	KHDRBS3	15.22	13.64	10.60	7.06
NM_005949	MT1F	0.41	0.58	0.64	2.55
NM_005204	MAP3K8	1.15	2.13	4.20	6.50
NM_198270	NHS	7.07	10.27	8.96	18.16

Continued

NM_013956	NRG1	1.70	5.07	5.29	8.27
NM_002502	NFKB2	1.34	2.12	1.90	4.18
NM_003738	PTCH2	1.51	2.49	1.99	5.21
NM_018492	PBK	1.81	0.95	1.22	0.49
NM_002667	PLN	2.70	15.38	6.40	8.46
NM_002668	PLP2	76.69	52.09	54.19	27.74
BX647593	RABL3	18.94	14.52	12.39	9.19
NM_002928	RGS16	0.46	3.83	6.18	1.83
NM_003702	RGS20	13.19	14.61	9.23	5.72
NM_005613	RGS4	4.22	1.95	2.65	1.34
NM_002924	RGS7	3.40	3.24	1.49	0.97
NM_031286	SH3BGRL3	27.48	21.54	19.37	12.19
NM_017720	STAP2	1.78	6.79	2.04	3.24
NM_031244	SIRT5	2.39	1.78	1.32	1.14
NM_005905	SMAD9	1.47	0.43	0.60	0.41
NM_021967	SERF1A	6.09	11.14	8.27	9.34
U81001	SNRPN	7.40	4.43	6.43	3.18
NM_198538	SBSN	1.61	2.55	3.14	6.50
NM_006528	TFPI2	59.73	196.06	114.78	270.54
NM_024873	TNIP3	0.39	2.32	0.97	3.26
NM_052864	TIFA	1.53	3.52	2.01	3.05
NM_006398	UBD	0.22	1.73	0.27	1.85
NM_021116	ADCY1	1.00	1.01	1.62	2.95
	Transcription				
NM_004024	ATF3	6.45	8.28	14.12	16.94
NM_024812	BAALC	3.13	5.66	3.64	3.61
NM_005195	CEBPD	4.79	8.43	14.27	26.51
NM_001822	CHN1	91.85	70.70	81.88	41.69
NM_012242	DKK1	76.96	77.46	39.54	43.95
NM_016323	HERC5	1.49	5.11	2.18	6.65
NM_017912	HERC6	7.24	19.79	9.99	32.31
NM_021063	HIST1H2BD	1.99	3.51	2.07	4.07
NM_003516	HIST2H2AA	7.32	24.50	12.07	30.15
NM_003528	HIST2H2BE	5.77	16.37	8.30	22.28
NM_002165	ID1	1.68	0.34	1.05	0.21
NM_014583	LMCD1	2.62	5.57	2.64	5.61
NM_001290	LDB2	3.71	2.55	2.21	1.84

Continued

NM_002402	MEST	10.88	1.49	4.91	0.95
NM_002449	MSX2	0.71	1.38	1.14	1.50
NM_006167	NKX3-1	2.36	5.11	6.71	12.12
NM_002824	PTMS	65.65	47.54	37.46	30.93
AK095843	ZNF883	1.09	0.70	0.43	0.52
NM_003222	TFAP2C	7.27	7.80	13.62	22.43
NM_025079	ZC3H12A	5.76	9.67	14.77	22.79
NM_002583	PAWR	2.35	2.02	1.46	1.17
	Cancer related				
NM_002350	LYN	1.73	2.50	1.95	3.59
NM_014751	MTSS1	5.12	9.45	5.68	7.91
NM_152858	WTAP	19.85	47.39	38.36	69.56
NM_012258	HEY1	1.29	0.90	1.35	2.74
NM_002129	HMGB2	132.06	85.51	81.03	61.72
NM_198389	PDPN	0.44	1.35	1.52	2.48
NM_203401	STMN1	8.08	8.28	7.18	3.66
T92525	TM4SF1	1.94	3.83	1.77	3.02
	Vesicle traffic				
NM_004209	SYNGR3	0.70	1.60	1.09	1.67
NM_003764	STX11	1.29	3.95	2.20	5.03
NM_001233	CAV2	37.41	24.89	28.18	17.15
NM_177478	FTMT	13.60	37.36	23.54	43.41
N47412	FTH1	96.62	185.95	122.52	180.14
NM_005723	TSPAN5	29.76	24.58	23.11	14.73
	Channels/carriers				
NM_002977	SCN9A	34.40	19.34	17.66	12.17
NM_017585	SLC2A6	0.88	1.64	1.14	2.28
NM_199329	SLC43A3	2.35	4.54	6.64	14.16
NM_182936	SLC8A3	19.70	41.98	23.74	43.99
NM_005630	SLCO2A1	0.18	1.11	0.38	2.48
NM_007332	TRPA1	0.55	5.45	3.54	15.81
NM_017680	ASPN	2.04	0.16	0.76	0.30
NM_005603	ATP8B1	3.99	4.03	2.63	1.91
NM_130463	ATP6V1G2	2.55	1.52	1.83	1.19
NM_018672	ABCA5	2.32	4.71	2.28	5.70
	Enzymes/Metab				
NM_016816	OAS1	1.97	4.33	2.22	6.06

Continued

NM_003733	OASL	0.21	0.91	0.27	1.65
NM_033068	ACPT	1.86	3.02	2.84	3.72
NM_000693	ALDH1A3	1.59	9.74	6.50	25.74
NM_001628	AKR1B1	21.52	103.85	46.04	137.31
NM_001732	BTN1A1	21.64	15.49	13.32	9.76
BC052289	CPA4	1.94	0.59	0.90	0.42
NM_003851	CREG1	7.58	11.64	7.89	14.67
NM_018371	CSGALNACHT1	0.55	0.36	0.91	2.28
NM_016246	DHRS10	6.88	12.81	7.58	8.38
NM_013989	DIO2	2.38	3.51	4.94	8.04
NM_001935	DPP4	1.24	1.62	2.53	4.60
NM_005803	FLOT1	45.35	19.03	23.97	15.37
NM_000167	GK	2.93	3.70	2.93	5.69
NM_031302	GLT8D2	15.55	7.90	10.25	10.28
NM_025193	HSD3B7	1.56	1.18	1.04	0.70
NM_005525	HSD11B1	0.75	39.11	22.01	97.54
NM_016048	ISOC1	2.82	6.06	4.14	5.38
NM_003937	KYNU	0.48	3.73	1.45	4.40
NM_032857	LACTB	31.98	57.66	40.03	40.52
NM_194436	LDHD	2.19	3.47	3.36	4.44
NM_015907	LAP3	9.60	29.99	14.72	33.31
NM_002543	OLR1	3.23	10.12	3.24	5.65
NM_175886	PRPS1L1	9.75	5.67	5.78	3.98
NM_002573	PAFAH1B3	20.92	19.58	16.95	10.21
NM_017554	PARP14	3.41	5.16	3.72	6.56
NM_004878	PTGES	1.66	3.87	4.24	13.26
NM_016147	PME-1	84.44	79.05	57.95	42.37
NM_032957	RTEL1	5.59	23.26	15.09	10.62
NM_014509	SERHL2	1.96	1.56	1.25	0.87
W01427	NR2F2	10.63	7.36	7.78	4.56
NM_003896	ST3GAL5	3.70	2.90	3.38	1.41
NM_006645	STARD10	7.31	13.29	7.21	12.01
NM_021199	SQRDL	7.49	20.23	10.97	16.50
NM_000636	SOD2	4.43	69.46	49.27	118.66
NM_013251	TAC3	0.34	6.33	1.02	4.26
NM_016381	TREX1	12.84	25.77	13.25	19.95

Continued

NM_001071	TYMS	5.72	4.39	4.28	2.49
	Other				
NM_025004	CCDC15	3.02	2.68	2.25	1.16
NM_004764	PIWIL1	1.61	2.57	2.61	3.43
NM_138419	FAM54A	1.32	0.87	0.89	0.47
NM_144617	HSPB6	9.70	3.07	4.10	2.67
W60905	PPIA-like	0.83	0.98	1.15	1.65

[1]Values were determined by DNA microarray and are the mean of an experimental n of 3 passages of cells. GeneSpring 7.0 software was used to perform ANOVA on microarray data. [2]Abbreviations used: Genbank accession number (ACCN#), control (Con), tumor necrosis factor α (TNF), interleukin 1β (IL1β), transcriptional regulation (transcription), enzymes/metabolism (enzymes/metab).

a robust up-regulation in response to TNFα and IL1β (**Figure 1(c)**, **Figure 1(d)**). IL6 was significantly up-regulated in response to TNFα + IL1β compared to TNFα or IL1β alone (**Figure 1(e)**, **Figure 1(f)**). The expression of *MMP*3 was significantly increased in T-HESC cells treated with TNFα + IL1β (**Figure 1(g)**, **Figure 1(h)**) as demonstrated by microarray analysis and real time RT-PCR.

Since T-HESC cells are under the influence of TNFα + IL1β for a total of 70 hours in the Boyden chambers during migration and invasion experiments, gene expression was measured in T-HESC cells at 70 hours. **Figure 2** shows the significant up-regulation of IL8, IL6 and IL1*B* in T-HESC cells in response to IL1β alone and to TNFα + IL1β compared to control and TNFα alone at 70 hours as measured by real time RT-PCR.

The migration and invasion of T-HESC cells as measured with Boyden chambers were not significantly changed in response to treatment with TNFα or IL1β alone or in combination (**Figure 3(a)-(b)**) (n = 5). Similarly, TNFα and IL1β did not act as chemoattractants to stimulate the migration or invasion of T-HESC cells (**Figure 3(c)-(d)**).

4. Discussion

In this project we investigated the effects of five macrophage-secreted factors on gene expression in a human endometrial stromal cell line (T-HESC). TNFα and IL1β stimulated differential gene expression in T-HESC cells as shown by DNA microarray and real time RT-PCR. Thus, the most important finding of this study is that TNFα ± IL1β partially reproduced the effect of factors secreted by macrophages (macrophage conditioned medium) on T-HESC cells. In contrast, T-HESC cells did not respond to TGFβ, IL6, or IL8. This finding suggests that endometrial stromal cells may not be the primary target for these macrophage-secreted factors.

Cellular processes involved in the development of endometriosis include cell migration, invasion, survival, adhesion, proliferation, and angiogenesis [9]. Many of the gene ontologies identified in this study are clearly associated with cellular functions that are necessary for the establishment of endometriosis. For example, peptidases and their regulators are important to cellular invasion through the extracellular matrix, and cytoskeletal organization is associated with the ability of cells to migrate and invade. Genes in the ontology of cell cycle are involved in proliferation, and the ontology of cell adhesion is important to the ability of endometrial cells to adhere to the peritoneal organs or the peritoneal wall. The ontologies of growth factors, receptors and ligands, signal transduction, and transcriptional regulation are all related to the regulation of cellular functions involved in the development of endometriosis. Moreover, the ontology of defense/immune system is relevant to the inflammatory response that has been implicated in endometriosis [3] [8] [17].

The four genes chosen for confirmation by real time RT-PCR following microarray analysis, IL8, IL6, *MMP*3 and IL1*B*, can be designated inflammatory markers as they are all implicated in inflammation and are up-regulated in endometriosis [6], as well as in T-HESC treated with TNFα + IL1β. These genes have been extensively studied in endometriosis [3] [17]-[21]. Increased numbers of macrophages are found in the peritoneal fluid of women with endometriosis compared to that of women without endometriosis [17]. Macrophages secrete IL8 which has been shown to be increased in women with endometriosis [22]. These data associate IL8 with the pathogenesis of endometriosis. IL1β is known to induce the expression of IL6 and IL8 in T-HESC cells [23]-[25]. This study confirms the literature reports. Ponce and coworkers [26] have demonstrated that the expression of

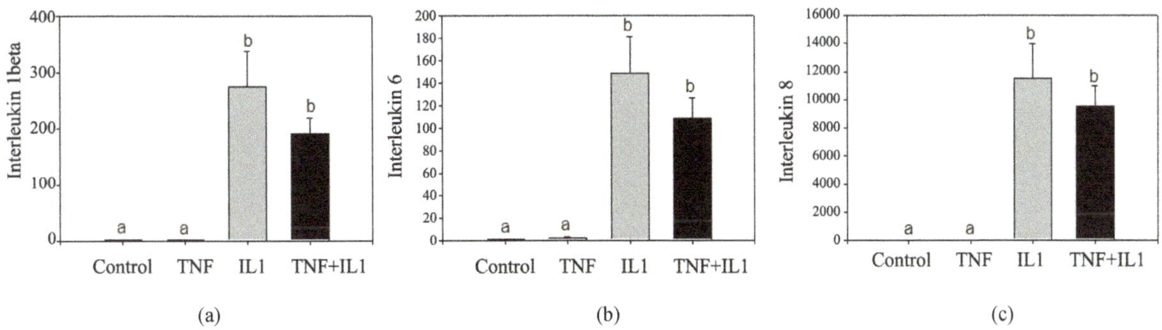

Figure 2. Differential gene expression of IL1β (a); IL6 (b); and IL8 (c) in human telomerase-immortalized endometrial stromal cells in response to treatment with TNFα (TNF) ± IL1β (IL1) for 70 hours as measured by real time RT-PCR. This time period corresponds to the time at which the cell invasion and migration assays were carried out. Data are the mean ± S.E., experimental n = 3. Bars with different letter superscripts denote that the data for those groups are significantly different from each other (ANOVA, Newman Keuls post hoc test, $p < 0.05$).

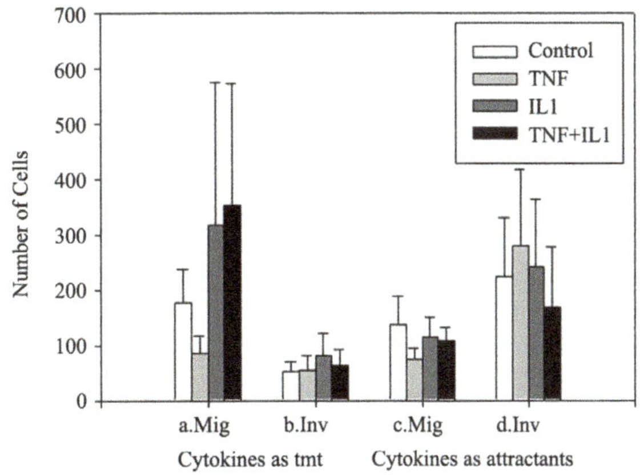

Figure 3. Migration and invasion of human telomerase-immortalized endometrial stromal cells (T-HESC) under control conditions or treated with TNFα, IL1β, or TNFα + IL1β in Boyden chambers. Groups a and b illustrate migration (mig) and invasion (inv) in response to treatment (tmt) with cytokines during the migration or invasion process. Groups c and d show migration and invasion when cytokines were placed in the bottom chamber as chemoattractants. Data are the mean ± S.E., experimental n = 5. No significant differences were identified among the groups (ANOVA, $p < 0.05$).

IL6 mRNA is down-regulated in endometriosis tissue compared to normal endometrium during the late-secretory phase of the menstrual cycle. On the other hand, Fassbender and coworkers [27] identified increased expression of IL6 in macroscopically normal endometrium from patients with endometriosis. The reports in the literature and our study demonstrate the dynamic expression of IL6 in endometrial stromal cells and endometriosis. MMP3 was up-regulated in ectopic endometrial tissue compared to eutopic endometrium in women with endometriosis [5] [28]. The up-regulation of MMP3 in our study indicates that MMP3 expression is regulated by TNFα and IL1β in endometrial stromal cells. These cytokines may also be responsible for the up-regulation of MMP3 in endometriosis.

The treatment of T-HESC cells with the combination of TNFα + IL1β for 48 hours substantially increased the expression of IL1*B*, IL8, IL6 and *MMP*3. The effect of the combined cytokines on the expression of IL1*B*, IL6, and *MMP*3 appeared to be synergistic, whereas the effect on IL8 expression appeared to be additive. In contrast, treatment of the cells with the combination of TNFα + IL1β for 70 hours did not result in a further increase in expression of the genes tested. Rather the combination of cytokines showed a small, albeit statistically insignificant, decline in expression levels compared to IL1β alone. It is unclear whether the change in response between 48 hours and 70 hours was due to loss of activity of TNFα or due to the attenuation of the response of the cells to the treatment.

The expression of the IL8 and IL6 genes was up-regulated in response to TNFα + IL1β in T-HESC cells, but the cells did not respond to treatment with IL8 or IL6. We tested higher concentrations of IL8 (25, 150, 500 ng/ml) compared to the concentrations reported in the literature [29] and still did not observe a response by T-HESC cells. It is possible that cultured T-HESC cells do not have receptors for IL8 or IL6. It is also possible that a different set of genes is affected in T-HESC cells in response to IL8 and IL6 than the ones tested in this study (IL8 and *MMP*3).

We had hypothesized that factors secreted by macrophages would increase the migratory and invasive properties of T-HESC cells [1]. Two separate experiments were designed to examine the effects of TNFα and IL1β on migration and invasion when endometrial stromal cells are bathed in the pool of cytokines (as the treatment of T-HESC cells) or when endometrial stromal cells are attracted towards a source of cytokines (TNFα $\pm$ IL1β as chemoattractants) in endometriosis. However, TNFα and IL1β did not affect the migration or invasion of T-HESC cells in this study in either experimental paradigm. TNFα and IL1β are not specifically known as chemokines so their inability to act as chemoattractants was not surprising. The lack of an effect of these two cytokines on migration and invasion of cultured T-HESC cells does not negate their effect on other cellular functions needed for the establishment of endometriosis.

5. Conclusion

In conclusion, TNFα and IL1β partially reproduced the effect of macrophage conditioned medium on gene expression in T-HESC cells. However, TNFα and IL1β failed to demonstrate an effect on the migration and invasion of T-HESC cells. Thus other cytokines, in combination with TNFα and IL1β, chemokines, and other growth factors, are expected to fully duplicate the effect of factors secreted by macrophages on endometrial stromal cells.

Acknowledgements

This work was supported by the Wesley R. Parke award to Abha J. Chalpe and by internal funds from the Division of Basic Biomedical Sciences, the Sanford School of Medicine Research Funds, and the Department of Obstetrics & Gynecology, Sanford School of Medicine of The University of South Dakota. The Genomics Core was funded by NIH P20GM103443.

References

[1] Kyama, C.M., Mihalyi, A., Simsa, P., Falconer, H., Fulop, V., Mwenda, J.M., Peeraer, K., Tomassetti, C., Meuleman, C. and D'Hooghe, T.M. (2009) Role of Cytokines in the Endometrial-Peritoneal Cross-Talk and Development of Endometriosis. *Front Biosci* (*Elite Ed*), **1**, 444-454.

[2] Eyster, K.M., Klinkova, O., Kennedy, V. and Hansen, K.A. (2007) Whole Genome Deoxyribonucleic Acid Microarray Analysis of Gene Expression in Ectopic versus Eutopic Endometrium. *Fertility and Sterility*, **88**, 1505-1533. http://dx.doi.org/10.1016/j.fertnstert.2007.01.056

[3] Siristatidis, C., Nissotakis, C., Chrelias, C., Iacovidou, H. and Salamalekis, E. (2006) Immunological Factors and Their Role in the Genesis and Development of Endometriosis. *Journal of Obstetrics and Gynaecology Research*, **32**, 162-170. http://dx.doi.org/10.1111/j.1447-0756.2006.00373.x

[4] Banu, S.K., Lee, J., Starzinski-Powitz, A. and Arosh, J.A. (2008) Gene Expression Profiles and Functional Characterization of Human Immortalized Endometriotic Epithelial and Stromal Cells. *Fertility and Sterility*, **90**, 972-987. http://dx.doi.org/10.1016/j.fertnstert.2007.07.1358

[5] Eyster, K.M., Hansen, K.A., Winterton, E., Klinkova, O., Drappeau, D. and Mark-Kappeler, C.J. (2010) Reciprocal Communication between Endometrial Stromal Cells and Macrophages. *Reproductive Sciences*, **17**, 809-822. http://dx.doi.org/10.1177/1933719110371854

[6] Klinkova, O., Hansen, K.A., Winterton, E., Mark, C.J. and Eyster, K.M. (2010) Two-Way Communication between Endometrial Stromal Cells and Monocytes. *Reproductive Sciences*, **17**, 125-136. http://dx.doi.org/10.1177/1933719109348922

[7] Bersinger, N.A., Günthert, A.R., McKinnon, B., Johann, S. and Mueller, M.D. (2011) Dose-Response Effect of Interleukin (IL)-1β, Tumour Necrosis Factor (TNF)-α, and Interferon-γ on the *in Vitro* Production of Epithelial Neutrophil Activating Peptide-78 (ENA-78), IL-8, and IL-6 by Human Endometrial Stromal Cells. *Archives of Gynecology and Obstetrics*, **283**, 1291-1296. http://dx.doi.org/10.1007/s00404-010-1520-3

[8] Maybin, J.A., Critchley, H.O.D. and Jabbour, H.N. (2011) Inflammatory Pathways in Endometrial Disorders. *Molecular and Cellular Endocrinology*, **335**, 42-51. http://dx.doi.org/10.1016/j.mce.2010.08.006

[9] Giudice, L.C. and Kao, L.C. (2004) Endometriosis. *Lancet*, **364**, 1789-1799. http://dx.doi.org/10.1016/S0140-6736(04)17403-5

[10] Jones, C.J.P., Nardo, L.G., Littam P. and Fazleabas, A.T. (2009) Ultrastructure of Ectopic Peritoneal Lesions from Women with Endometriosis, Including Observations on the Contribution of Coelomic Mesothelium. *Reproductive Sciences*, **16**, 43-55. http://dx.doi.org/10.1177/1933719108324891

[11] Jovanović, M. and Vićovac, L. (2009) Interleukin-6 Stimulates Cell Migration, Invasion and Integrin Expression in HTR-8/SVneo Cell Line. *Placenta*, **30**, 320-328. http://dx.doi.org/10.1016/j.placenta.2009.01.013

[12] Krikun, G., Nor, G., Alvero, A., Guller, S., Schatz, F., Sapi, E., Rahman, M., Caze, R., Qumsiyeh, M. and Lockwood, C.J. (2004) A Novel Immortalized Human Endometrial Stromal Cell Line with Normal Progestational Response. *Endocrinology*, **145**, 2291-2296. http://dx.doi.org/10.1210/en.2003-1606

[13] Mai, K.T., Yazdi, H.M., Perkins, D.G. and Parks, W. (1997) Pathogenetic Role of the Stromal Cells in Endometriosis and Adenomyosis. *Histopathology*, **30**, 430-442. http://dx.doi.org/10.1046/j.1365-2559.1997.4910725.x

[14] Eyster, K.M. and Brannian, J.D. (2009) Gene Expression Profiling in the Aging Ovary. *Methods in Molecular Biology*, **590**, 71-89. http://dx.doi.org/10.1007/978-1-60327-378-7_5

[15] Hellemans, J., Mortier, G., De Paepe, A., Speleman, F. and Vandesompele, J. (2007) qBase Relative Quantification Framework and Software for Management and Automated Analysis of Real-Time Quantitative PCR Data. *Genome Biology*, **8**, R19. http://dx.doi.org/10.1186/gb-2007-8-2-r19

[16] Brazma, A., Hingamp, P., Quackenbush, J., Sherlock, G., Spellman, P., Stoeckert, C., Aach, J., Ansorge, W., Ball, C.A., Causton, H.C., *et al.* (2001) Minimum Information about a Microarray Experiment (MIAME)—Toward Standards for Microarray Data. *Nature Genetics*, **29**, 365-371. http://dx.doi.org/10.1038/ng1201-365

[17] Bondza, P.K., Maheux, R. and Akoum, A. (2009) Insights into Endometriosis-Associated Endometrial Dysfunctions: A Review. *Frontiers in Bioscience*, **1**, 415-428.

[18] Osteen, K.G., Keller, N.R., Feltus, F.A. and Melner, M.H. (1999) Paracrine Regulation of Matrix Metalloproteinase Expression in the Normal Human Endometrium. *Gynecologic and Obstetric Investigation*, **48**, 2-13. http://dx.doi.org/10.1159/000052863

[19] Braundmeier, A.G. and Nowak, R.A. (2006) Cytokines Regulate Matrix Metalloproteinases in Human Uterine Endometrial Fibroblast Cells through a Mechanism that Does Not Involve Increases in Extracellular Matrix Metalloproteinase Inducer. *American Journal of Reproductive Immunology*, **56**, 201-214. http://dx.doi.org/10.1111/j.1600-0897.2006.00418.x

[20] Guay, S. and Akoum, A. (2007) Stable Inhibition of Interleukin 1 Receptor Type II in Ishikawa Cells Augments Secretion of Matrix Metalloproteinases: Possible Role in Endometriosis Pathophysiology. *Reproduction*, **134**, 525-534. http://dx.doi.org/10.1530/REP-06-0377

[21] Yang, J.H., Wu, M.Y., Chen, M.J., Chen, S.U., Yang, Y.S. and Ho, H.N. (2009) Increased Matrix Metalloproteinase-2 and Tissue Inhibitor of Metalloproteinase-1 Secretion but Unaffected Invasiveness of Endometrial Stromal Cells in Adenomyosis. *Fertility and Sterility*, **91**, 2193-2198. http://dx.doi.org/10.1016/j.fertnstert.2008.05.090

[22] Milewski, L., Dziunycz, P., Barcz, E., Radomski, D., Roszkowski, P.I., Korczak-Kowalska, G., Kamiński, P. and Malejczyk, J. (2011) Increased Levels of Human Neutrophil Peptides 1, 2, and 3 in Peritoneal Fluid of Patients with Endometriosis: Association with Neutrophils, T Cells and IL-8. *Journal of Reproductive Immunology*, **91**, 64-70. http://dx.doi.org/10.1016/j.jri.2011.05.008

[23] Akoum, A., Lawson, C., McColl, S. and Villeneuve, M. (2001) Ectopic Endometrial Cells Express High Concentrations of Interleukin (IL)-8 *in Vivo* Regardless of the Menstrual Cycle Phase and Respond to Oestradiol by Up-Regulating IL-1-Induced IL-8 Expression *in Vitro*. *Molecular Human Reproduction*, **7**, 859-866. http://dx.doi.org/10.1093/molehr/7.9.859

[24] Lebovic, D.I., Bentzien, F., Chao, V.A., Garrett, E.N., Meng, Y.G. and Taylor, R.N. (2000) Induction of an Angiogenic Phenotype in Endometriotic Stromal Cell Cultures by Interleukin-1*β*. *Molecular Human Reproduction*, **6**, 269-275. http://dx.doi.org/10.1093/molehr/6.3.269

[25] Hou, Z., Zhou, J., Ma, X., Fan, L., Liao, L. and Liu, J. (2008) Role of Interleukin-1 Receptor Type II in the Pathogenesis of Endometriosis. *Fertility and Sterility*, **89**, 42-51. http://dx.doi.org/10.1016/j.fertnstert.2007.01.044

[26] Ponce, C., Torres, M., Galleguillos, C., Sovino, H., Boric, M.A., Fuentes, A. and Johnson, M.C. (2009) Nuclear Factor κB Pathway and Interleukin-6 Are Affected in Eutopic Endometrium of Women with Endometriosis. *Reproduction*, **137**, 727-737. http://dx.doi.org/10.1530/REP-08-0407

[27] Fassbender, A., Overbergh, L., Verdrengh, E., Kyama, C.M., Vodolazakaia, A., Bokor, A., Mueleman, C., Peeraer, K., Tomassetti, C., Waelkens, E., *et al.* (2011) How Can Macroscopically Normal Peritoneum Contribute to the Pathogenesis of Endometriosis? *Fertility and Sterility*, **96**, 697-699. http://dx.doi.org/10.1016/j.fertnstert.2011.06.034

[28] Shan, K., Wang, Y., Zhang, J.H., Guo, W., Wang, N. and Li, Y. (2005) The Function of the SNP in the MMP1 and

MMP3 Promoter in Susceptibility to Endometriosis in China. *Molecular Human Reproduction*, **11**, 423-427. http://dx.doi.org/10.1093/molehr/gah177

[29] Sharpe-Timms, K.L., Nabli, H., Zimmer, R.L., Birt, J.A. and Davis, J.W. (2010) Inflammatory Cytokines Differentially Up-Regulate Human Endometrial Haptoglobin Production in Women with Endometriosis. *Human Reproduction*, **25**, 1241-1250. http://dx.doi.org/10.1093/humrep/deq032

4

A Role for Histone Chaperones in Regulating RNA Polymerase II

JoDi Lynn Osborn, Susanna F. Greer

Department of Biology, Georgia State University, Atlanta, USA
Email: sgreer@gsu.edu

Abstract

Transcription is a highly regulated cellular process in which dysfunction leads to disease. One level of regulation is chromatin structure which protects promoters from transcription factor binding. To circumvent this blockade, histone chaperones aid in displacement of nucleosomes. In particular, the histone chaperone complex HUCA, consisting of Hira, Ubn1, Cabin1, and ASF1a, replaces histone variant H3.1 with H3.3 in front of actively transcribing RNA Polymerase II (RNAPII). The 26S proteasome is a major degrader of proteins within the cell and plays both proteolytic and non-proteolytic roles in transcriptional regulation. One major role is the degradation of irreversibly arrested RNAPII. Several interactions between HUCA, the 26S proteasome, and RNAPII have been characterized individually; we now present observations from our lab and others which directly associate elongating RNAPII with the degradation machinery through observations of involvement with the HUCA complex. Our short report presents these ideas and discusses their importance in transcriptional regulation as well as implications in disease manifestation.

Keywords

Histone Chaperones, RNA Polymerase II, 26S Proteasome, Transcriptional Regulation

1. Introduction

Cellular function is highly dependent on regulated gene expression which is controlled by the coordinated efforts of thousands of proteins including transcription factors, cofactors, and chromatin regulators. The coordinated efforts and interactions of these proteins control cell fate and cellular responses to physiological or environmental changes. One level at which protein interaction, and thus gene expression, is ordered is control of transcription factor accessibility to DNA by chromatin (Reviewed in [1]).

Despite over half a century of research in transcriptional regulation, the intricate details of the system have not been fully elucidated. Our proposed model addresses the regulatory interactions between RNA Polymerase II, the histone chaperone complex HUCA, and the 26S Proteasome. We hypothesize that the histone chaperone HUCA tethers the 26S proteasome to elongating RNAPII to facilitate rapid degradation in the case of irreversibly arrested RNAPII. A schematic of the proposed mechanism is provided in **Figure 1**. Disassembly of RNAPII elongation complexes has previously been linked to the ATPase Cdc48 with proposed roles for ubiquitination in dissociating Rpb1 from the complex [2]. The 26S proteasome binds transcribing genes at locations enriched with RNAPII [3], 19S and 20S subunits of the proteasome bind promoters, ORFs, and termination regions [3]-[6]. In *C. elegans*, RNAPII is found localized in "degradation centers" where blocked transcription initiates proteasome mediated degradation [7]. Thus, components of the proteasome are linked with chromatin in multiple and overlapping contexts. Each of these observations suggests that removal of proteins from chromatin and degradation of at least some of these proteins, is likely to be a common theme in the regulation of gene expression.

2. Chromatin Organization and Histone Modifications

The basic unit of chromatin is the nucleosome which consists of 147 base pairs of DNA wrapped around a histone octamer of two dimers each of canonical histones H2A, H2B, H3, and H4 as well as the linker histone H1 [8]. Nucleosomes are packaged into higher order chromatin structures; modulation of which influences transcriptional outcomes. The N-terminal tails of histones undergo multiple covalent modifications including acetylation, methylation, sumoylation, ADP-ribosylation, and ubiquitination which can positively or negatively affect protein binding to DNA [9]. Different combinations of these histone marks dictate whether the chromatin is relaxed, allowing access by transcription factors, or closed, preventing transcription (Reviewed in [10]).

In addition to the impact of histone modifications on chromatin structure, nucleosome positioning poses an inherent block to elongating RNA Polymerase II (RNAPII) [11]. Nucleosomes are arranged in regularly spaced intervals, with the length of intervening linker regions between nucleosomes varying between species and cell types [12]. In general, nucleosomes are depleted at many enhancers and promoters and occupy preferred positions within genes [13]. Nucleosome occupancy differs in that it is a reflection of the depletion of nucleosomes from functional regions of genes and is critical to transcriptional outcomes. Access of RNAPII and other

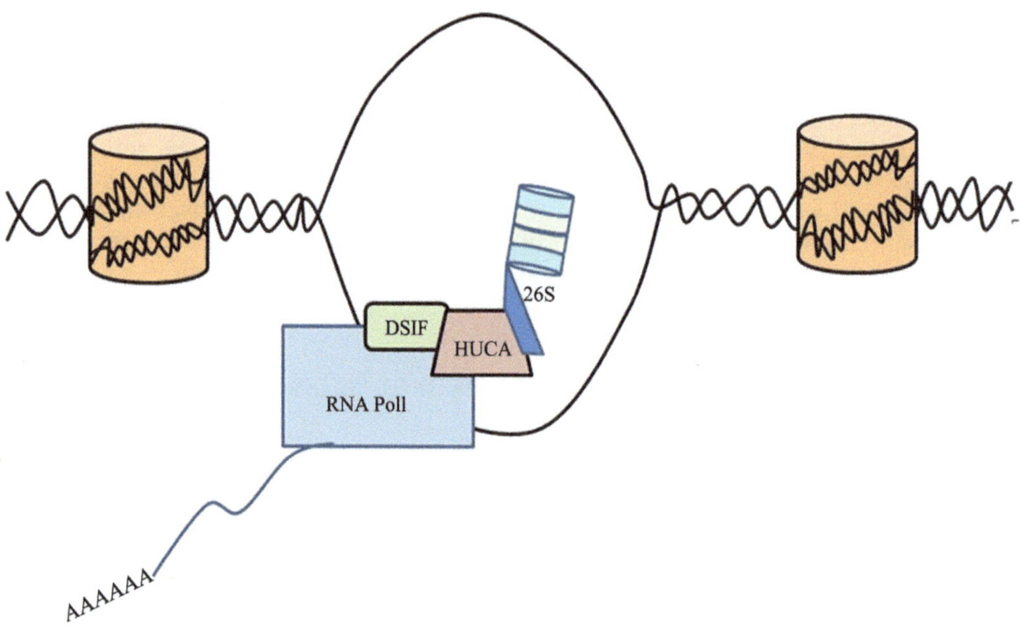

Figure 1. Proposed Mechanism. The histone chaperone HUCA tethers the 26S proteasome to elongating RNA Polymerase II to facilitate rapid degradation in the case of irreversible stalling of the RNA Pol II elongation complex. By keeping the cellular degradation machinery in close proximity, stalled RNA Pol II can be quickly degraded, thus avoiding continual repeat transcription and the accumulation of abnormal proteins within the cell.

DNA-binding proteins to DNA is inhibited by nucleosomes and RNAPII is unable to mediate nucleosome eviction on its own [14] [15]. To assist RNAPII passage, chromatin remodelers and histone chaperones aid in removal of the nucleosome blockade. Chromatin remodelers unwind DNA from nucleosomes in an ATP-dependent manner [16] whereas histone chaperones are enzymes that facilitate exchange of histone variants during transcription in a histone-dependent, ATP-independent manner [17]. In addition to histone modifications and chromatin remodeling, replacement of canonical core histones with specialized variants assists in regulation of the open or closed states of chromatin [18] [19].

3. Histone Chaperones: Multi-Faceted Roles

Because nucleosome assembly and disassembly regulate chromatin dynamics, histone exchange is coordinated by multiple histone chaperones and their many functional partners. Histone chaperones form multi-subunit complexes and perform a multitude of functions depending on binding partners [20]. While their main role is to facilitate nucleosome disassembly and reassembly, their secondary roles lie in a variety of processes such as transcriptional regulation, cellular senescence, histone variant deposition, cell cycle regulation, DNA replication and repair, mRNA processing, and gene silencing [17] [21]. The dynamic role of chromatin itself highlights the need for histone chaperones to serve versatile roles in modulating the many complexes that regulate chromatin dynamics and subsequent pathways. One such example of a multi-faceted histone chaperone is the HUCA complex. HUCA is known to be critical in chromatin reorganization; we now further postulate HUCA to be of equal importance in regulating RNAPII processivity and arrest.

4. HUCA Complex: An Important but Not Fully Understood Histone Chaperone Complex

Histone chaperones work in concert with one another and other proteins to perform a multitude of cellular functions [20]. One such histone chaperone complex is HUCA, which consists of the histone chaperone Histone Regulatory A (HIRA), Ubinuclein 1 (UBN1), Calcineurin binding 1 (CABIN1), and the histone chaperone Anti-silencing factor 1a (ASF1a) [22]-[24]. Human HIRA is an evolutionarily conserved fusion of Hir1 and Hir2 which form a nucleosome remodeling complex in lower eukaryotes [25]. The N-terminal Hir1-like domain consists of seven WD40 repeats [26]. Immediately following the di-peptide WD repeats is a 37 amino acid domain known as the B-domain [22]. The C-terminal Hir2-like domain consists of a leucine zipper (LXXLL motif) [27]. It is via these functional domains that HIRA acts as a scaffold around which other complex subunits form [22] [28]. UBN1 is an ortholog of the yeast protein Histone promoter control 2 (Hpc2) and contains an evolutionarily conserved Hpc2-related domain (HRD) [28]. The N-terminal region of UBN1 binds the WD40 repeats of HIRA as well as histone tails [23]. CABIN1 is a functional ortholog of Hir3p but the only structural similarity is 30 tetratricopeptide repeats (TPR) at the N-terminus [18]. The TPR domain of CABIN1 interacts with the C-terminus of HIRA [24]. ASF1a has an evolutionarily conserved immunoglobulin fold at its N-terminus and a divergent, species specific, C-terminus [29]. ASF1a binds HIRA via interaction of its N-terminal sequence with the B domain of HIRA [22]. For a schematic of the HUCA complex and its components, see **Figure 2**. HIRA binds the promoters of histone genes and prevents their transcription outside of S-phase [30] [31]. The mechanism of histone promoter silencing is through blockade of the SWI/SNF complex which is required for transcription of histone genes during S-phase [32]. This was initially shown for histones H2A and H2B but is not the case for histone H3.3 which is transcribed throughout the entire cell cycle [33]. HIRA simultaneously binds DNA and RNAPII while depositing histone H3.3 during transcription [34]. The ability of HIRA to bind DNA non-specifically has been implicated in prevention of cryptic transcription in *Schizosaccharomyces pombe* [35]. Heterogenous HIRA expression leads to DiGeorge Syndrome in humans and has been implicated in developmental impediments in *Drosophila melanogaster*, *Gallus gallusdomesticus*, and *Mus musculus* [36]-[38]. Complete knockout of HIRA is lethal as early as embryonic development day 10 [38].

In addition to its role in the histone chaperone HUCA complex, ubinuclein 1 has been posited to be a histone chaperone in its own right [39]. UBN1 has also been shown to act as a scaffold protein and to play a role in cell-cell adhesion in canine kidney cells [40] [41]. CABIN1 has recently been classified as a member of the HUCA complex and implicates regulation of this histone chaperone complex by intracellular calcium [24] [42]. ASF1a, which is also a histone chaperone, binds H3.1, H3.2, and H3.3 and plays roles in nucleosome eviction in both DNA-replication and transcription [22] [43]. In addition to its role in nucleosome eviction, ASF1 has been

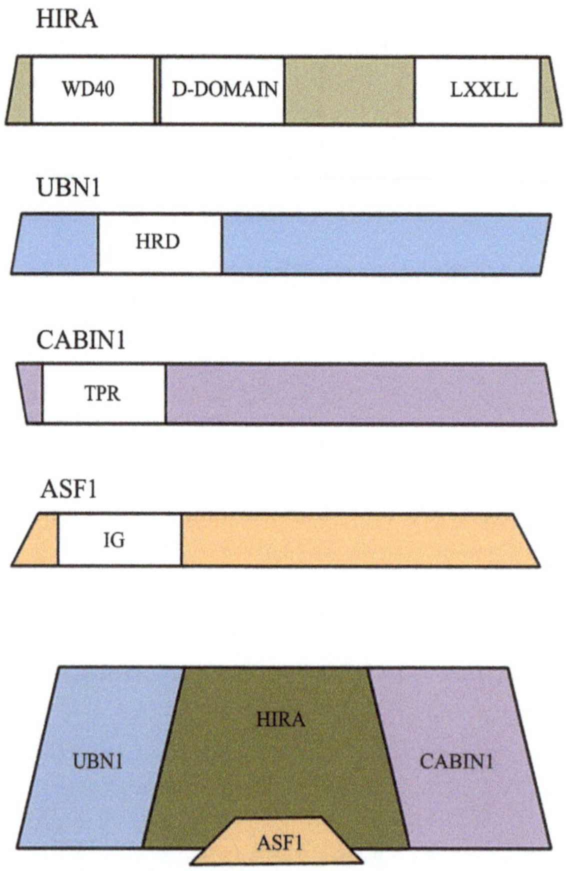

Figure 2. Schematic of the HUCA Complex. HUCA is composed of the histone chaperone Histone Regulatory A (HIRA), Ubinuclein 1 (UBN1), calcineurin binding 1 (CABIN1), and the histone chaperone Anti-silencing Factor 1 (ASF1). HIRA acts as a scaffold around which the other complex subunits form. The N-terminal domain of UBN1 binds the N-terminal WD repeats of HIRA. The TPR domain of CABIN1 interacts with the C-terminus of HIRA and ASF1 binds the B-domain of HIRA.

shown to play roles in regulating acetylation in both histones H3 and H4 in *Saccharomyces cerivisiae* [20]. Despite the elucidation of various roles of HUCA components, specific regulatory roles of the complex as a whole are not understood.

5. Transcriptional Elongation by RNA Polymerase II (RNAPII)

RNAPII transcribes mRNA from a DNA template in all protein-encoding genes. Transcription relies on a coordinated network of histone modifying enzymes and histone chaperones to allow elongating RNAPII to traverse the chromatin landscape. In particular, HUCA deposits the histone variant H3.3 in front of elongating RNPII to facilitate transcription through nucleosome complexes [34]. Transcription proceeds in three distinct stages: initiation, elongation, and termination [44]. The pre-initiation complex (PIC) which forms at the gene promoter, contains RBP1 and the general transcription factors (GTFs) TFIIB, TFIID, TFIIE, and TFIIH [45]. Other factors are also present including DRB Sensitivity Inducing Factor (DSIF) and Negative Elongation Factor (NELF) which positively [46] and negatively [47] regulate RNAPII, respectively. The C-terminal domain (CTD) of RBP1 consists of 52 repeats of the heptapeptide Y-S_2-P-S_5-T-P-S [48]. Phosphorylation status of the serine residues within the CTD correlates with transcriptional activity. Serine 5 phosphorylation (Ser5ph) by the cyclin dependent kinase 7 (cdk7) subunit of TFIIH leads to transcriptional initiation by recruitment of factors that facilitate escape from the promoter [49]. Initial transcription can be non-productive until the formation of an 8 to 9 bp hybrid of the DNA template and the nascent mRNA [50]. Serine 2 is phosphorylated (ser2ph) by the cdk9 subunit of Positive Transcription Elongation Factor b (P-TEFb) which stimulates productive elongation [51], in part by triggering the release of NELF [52].

6. The Three States of RNAPII: Paused, Stalled, and Arrested

Regulation of chromatin states by histone remodelers and histone chaperones is only one level at which transcription elongation can be controlled. In addition to blocking RNAPII access to DNA, RNAPII processivity regulates transcription. Early studies of heat shock proteins showed that RNAPII is present at the promoter prior to induction [53]. Subsequent studies have shown that RNAPII is *paused* at the promoters of other stimulus-responsive genes [54]. Pausing is mediated by DRB-Sensitivity Inducing Factor (DSIF) and Negative Elongation Factor (NELF). NELF preferentially binds hypophosphorylated RNAPII and is required for promoter-proximal pausing [55]. The two major causes of RNAPII *stalling* are lesions in the DNA template and nucleotide misincorporation. Lesions in the template cause RNAPII to stall until transcription-coupled DNA repair can occur [56]. The stalled RNAPII remains bound to the DNA template and once the lesion has been excised and repaired, elongation can resume [57] [58]. Nucleotide misincorporation stalls RNAPII by causing it to backtrack one base pair [59] [60]. TFIIS stimulates RNAPII to hydrolytically cleave the misincorporated nucleotide [61]. After cleavage, the newly formed 3' end is prepared for the addition of the appropriate nucleotide, a process which occurs rapidly and has been termed "kinetic proofreading" [62]. Pausing and stalling of RNAPII are mediated by both histone chaperones and the 26S proteasome. In the event that stalled RNAPII cannot resume elongation, RNAPII becomes irreversibly *arrested* and is poly-ubiquitinated and tagged for degradation by the 26S proteasome [63] [64].

7. Ubiquitination of RNAPII and Degradation by the 26S Proteasome

Arrested RNAPII is tagged for degradation by the addition of four ubiquitin moieties linked by lysine 48 (K48-linkage) [65]. Poly-ubiquitination is the concerted effort of three types of enzymes: ubiquitin-activating (E1), ubiquitin-conjugating (E2), and ubiquitin-ligase (E3) [65]. RNAPII poly-ubiquitination is unique in that it occurs in a sequential manner involving two different E3 ligases. Neural precursor cell Expressed Developmentally Downregulated 4 (NEDD4) mono-ubiquitinates RNAPII [66] and then Elongin/Cullin converts this mono-ubiquitination to poly-ubquitination [67] [68]. Ubiquitination of RNAPII occurs on its largest subunit Rpb1 and only occurs when the polymerase is irreversibly arrested (reviewed in [69]).

8. The 26S Proteasome: Degrader of Proteins

The 26S proteasome is the major non-lysosomal player in degradation of damaged or unnecessary proteins within the cell including RNAPII [65]. The 26S proteasome also plays roles in transcription initiation by controlling the location and quantity of available transcription factors [70]. The 26S proteasome is composed of three distinct sub-complexes: the 19S regulatory particle (RP) lid, the 19S RP base, and the 20S catalytic particle (CP) [71]. The 19S RP lid is composed of 9 unique subunits: Regulatory Particle Non-ATPase 3 (Rpn3), Rpn5-9, Rpn11-12 and Rpn15 [71]. The 19S base consists of a hexameric ring of AAA ATPases (RPT1-6) and three non-ATPase subunits and is connected to the 19S RP lid by the linker protein Rpn10 [71] [72]. The 20S catalytic particle (CP) is a barrel-shaped complex composed of four heptameric rings composed of either α or β subunits stacked in an $\alpha\beta\beta\alpha$ conformation [71].

9. Interactions between HUCA, the 26S Proteasome, and RNAPII

Several interactions between HUCA, the 26S proteasome, and RNAPII have been characterized. Liquid chromatography tandem mass spectrometry revealed that all six ATPase subunits of the 19S RP base co-purify with HIRA in *S. pombe* [73]. The interaction between the 19S ATPases and HIRA is thought to counteract the repressive functions of HIRA on histone genes, but this role has yet to be further studied [73]. We have further verified interactions between the 19S ATPases and HIRA in co-immunoprecipitation assays in mammalian cells (Osborn and Greer, unplublished). ChIP-seq density profiles followed by co-immunoprecipitation show that multiple subunits of the HUCA complex bind to both the initiating and elongating forms of RNAPII [74]. The importance of the interaction between HUCA and RNAPII is underscored by the fact that in yeast cells depleted of HIRA, RNAPII recruitment is impaired [75]. HIRA binds to both subunits of DSIF [76] which travels throughout the coding region along with RNAPII [77]. Additionally, HIRA interacts directly with the elongation factors SPT6 and SPT16 [76] [78]. Finally, the ASF1a subunit of HUCA interacts directly with the bromodomain of CCG1, the largest subunit of TFIID [79].

Recent studies from our lab and others are the first to directly associate elongating RNAPII with the degradation machinery through observations of the involvement of the HUCA chaperone complex. We propose that the HUCA complex tethers the 26S Proteasome to elongating RNAPII to enable rapid recognition and degradation of RNAPII by the proteasome. In addition to non-proteolytic roles in transcription (reviewed in [80]), tethering of the 26S proteasome to elongating RNAPII would facilitate its rapid degradation upon arrest. Recent observations of a novel Ubiquitin-independent proteasome pathway suggest that elongating RNAPII stalls at Topoisomerase IIβ-DNA cleavage complexes where blocked RNAPII serves as a damage signal and associated 19S ATPases are activated as an early event during the encounter for detecting the Topoisomerase IIβ roadblocks [81] [82]. The 19S ATPases have also been proposed to unfold non-covalent protein-DNA nucleosome complexes during elongation; we suggest the ATPases and HUCA serve these and much earlier roles in allowing consistent tethering of RNAPII to the degradation machinery.

10. Concluding Remarks

While the molecular basis for interactions among stalled elongation, proteasomal degradation, and HUCA remains to be elucidated, understanding roles for 19S ATPases in mediating degradation of RNAPII will contribute therapeutically to multiple disease scenarios. In Cockayne syndrome, irreversibly arrested RNAPII leads to transcription anomalies resulting in phenotypes ranging from subtle to severe [83]. Additionally CAG repeat diseases such as Huntington disease and Spinocerebellar Ataxias are caused, in part, by repeated transcription by arrested RNAPII [84]. While diseases such as Cockayne Syndrome and CAG repeat disorders have transcriptional arrest as an underlying cause, the molecular mechanisms of decreased transcript fidelity leading to disease manifestation have not been further investigated. In addition, several cancer therapeutics including etoposide/VP-16 and doxorubicin stabilize topoisomerase IIβ-DNA cleavage complexes and trigger degradation of topoisomerase IIβ, the large subunit of RNAPII, and exposure of DNA damage [85] [86]. These drugs and the DNA damage they cause are associated with severe side effects; better understanding of the mechanisms by which these drugs initiate degradation will contribute to their enhanced clinical application.

Acknowledgements

J.L.O. was supported by a Graduate Fellowship from the Molecular Basis of Disease Area of Focus at Georgia State University. Grant support provided by a Research Scholar Grant from the American Cancer Society to S.F.G.

References

[1] Petesch, S.J. and Lis, J.T. (2012) Overcoming the Nucleosome Barrier during Transcript Elongation. *Trends in Genetics*, **28**, 285-294. http://dx.doi.org/10.1016/j.tig.2012.02.005

[2] Verma, R., Oania, R., Fang, R., Smith, G.T. and Deshaies, R.J. (2011) Cdc48/p97 Mediates UV-Dependent Turnover of RNA Pol II. *Molecular Cell*, **41**, 82-92. http://dx.doi.org/10.1016/j.molcel.2010.12.017

[3] Auld, K.L., Brown, C.R., Casolari, J.M., Komili, S. and Silver, P.A. (2006) Genomic Association of the Proteasome Demonstrates Overlapping Gene Regulatory Activity with Transcription Factor Substrates. *Molecular Cell*, **21**, 861-871. http://dx.doi.org/10.1016/j.molcel.2006.02.020

[4] Collins, G.A. and Tansey, W.P. (2006) The Proteasome: A Utility Tool for Transcription? *Current Opinion in Genetics & Development*, **16**, 197-202. http://dx.doi.org/10.1016/j.gde.2006.02.009

[5] Gillette, T.G., Gonzalez, F., Delahodde, A., Johnston, S.A. and Kodadek, T. (2004) Physical and Functional Association of RNA Polymerase II and the Proteasome. *Proceedings of the National Academy of Sciences of the United States of America*, **101**, 5904-5909. http://dx.doi.org/10.1073/pnas.0305411101

[6] Krogan, N.J., Lam, M.H., Fillingham, J., Keogh, M.C., Gebbia, M., Li, J., *et al.* (2004) Proteasome Involvement in the Repair of DNA Double-Strand Breaks. *Molecular Cell*, **16**, 1027-1034. http://dx.doi.org/10.1016/j.molcel.2004.11.033

[7] Scharf, A., Grozdanov, P.N., Veith, R., Kubitscheck, U., Meier, U.T. and von Mikecz, A. (2011) Distant Positioning of

Proteasomal Proteolysis Relative to Actively Transcribed Genes. *Nucleic Acids Research*, **39**, 4612-4627. http://dx.doi.org/10.1093/nar/gkr069

[8] Luger, K., Mader, A.W., Richmond, R.K., Sargent, D.F. and Richmond, T.J. (1997) Crystal Structure of the Nucleosome Core Particle at 2.8 A Resolution. *Nature*, **389**, 251-260. http://dx.doi.org/10.1038/38444

[9] Bannister, A.J. and Kouzarides, T. (2011) Regulation of Chromatin by Histone Modifications. *Cell Research*, **21**, 381-395. http://dx.doi.org/10.1038/cr.2011.22

[10] Zentner, G.E. and Henikoff, S. (2013) Regulation of Nucleosome Dynamics by Histone Modifications. *Nature Structural & Molecular Biology*, **20**, 259-266. http://dx.doi.org/10.1038/nsmb.2470

[11] Kireeva, M.L., Walter, W., Tchernajenko, V., Bondarenko, V., Kashlev, M. and Studitsky, V.M. (2002) Nucleosome Remodeling Induced by RNA Polymerase II: Loss of the H2A/H2B Dimer during Transcription. *Molecular Cell*, **9**, 541-552. http://dx.doi.org/10.1016/S1097-2765(02)00472-0

[12] Kelly, T.K., Liu, Y., Lay, F.D., Liang, G., Berman, B.P. and Jones, P.A. (2012) Genome-Wide Mapping of Nucleosome Positioning and DNA Methylation within Individual DNA Molecules. *Genome Research*, **22**, 2497-506. http://dx.doi.org/10.1101/gr.143008.112

[13] Fenouil, R., Cauchy, P., Koch, F., Descostes, N., Cabeza, J.Z., Innocenti, C., *et al.* (2012) CpG Islands and GC Content Dictate Nucleosome Depletion in a Transcription-Independent Manner at Mammalian Promoters. *Genome Research*, **22**, 2399-2408. http://dx.doi.org/10.1101/gr.138776.112

[14] Chang, C.H. and Luse, D.S. (1997) The H_3/H_4 Tetramer Blocks Transcript Elongation by RNA Polymerase II *in Vitro*. *The Journal of Biological Chemistry*, **272**, 23427-23434. http://dx.doi.org/10.1074/jbc.272.37.23427

[15] Izban, M.G. and Luse, D.S. (1992) Factor-Stimulated RNA Polymerase II Transcribes at Physiological Elongation Rates on Naked DNA but Very Poorly on Chromatin Templates. *The Journal of Biological Chemistry*, **267**, 13647-13655.

[16] Hota, S.K. and Bartholomew, B. (2011) Diversity of Operation in ATP-Dependent Chromatin Remodelers. *Biochimica et Biophysica Acta (BBA)—Gene Regulatory Mechanisms*, **1809**, 476-487. http://dx.doi.org/10.1016/j.bbagrm.2011.05.007

[17] Eitoku, M., Sato, L., Senda, T. and Horikoshi, M. (2008) Histone Chaperones: 30 Years from Isolation to Elucidation of the Mechanisms of Nucleosome Assembly and Disassembly. *Cellular and Molecular Life Sciences*, **65**, 414-444. http://dx.doi.org/10.1007/s00018-007-7305-6

[18] Tagami, H., Ray-Gallet, D., Almouzni, G. and Nakatani, Y. (2004) Histone H3.1 and H3.3 Complexes Mediate Nucleosome Assembly Pathways Dependent or Independent of DNA Synthesis. *Cell*, **116**, 51-61. http://dx.doi.org/10.1016/S0092-8674(03)01064-X

[19] Szenker, E., Ray-Gallet, D. and Almouzni, G. (2011) The Double Face of the Histone Variant H3.3. *Cell Research*, **21**, 421-434. http://dx.doi.org/10.1038/cr.2011.14

[20] Avvakumov, N., Nourani, A. and Cote, J. (2011) Histone Chaperones: Modulators of Chromatin Marks. *Molecular Cell*, **41**, 502-514. http://dx.doi.org/10.1016/j.molcel.2011.02.013

[21] Burgess, R.J. and Zhang, Z. (2010) Histones, Histone Chaperones and Nucleosome Assembly. *Protein & Cell*, **1**, 607-612. http://dx.doi.org/10.1007/s13238-010-0086-y

[22] Tang, Y., Poustovoitov, M.V., Zhao, K., Garfinkel, M., Canutescu, A., Dunbrack, R., *et al.* (2006) Structure of a Human ASF1a-HIRA Complex and Insights into Specificity of Histone Chaperone Complex Assembly. *Nature Structural & Molecular Biology*, **13**, 921-929. http://dx.doi.org/10.1038/nsmb1147

[23] Tang, Y., Puri, A., Ricketts, M.D., Rai, T.S., Hoffmann, J., Hoi, E., *et al.* (2012) Identification of an Ubinuclein 1 Region Required for Stability and Function of the Human HIRA/UBN1/CABIN1/ASF1a Histone H3.3 Chaperone Complex. *Biochemistry*, **51**, 2366-2377. http://dx.doi.org/10.1021/bi300050b

[24] Rai, T.S., Puri, A., McBryan, T., Hoffman, J., Tang, Y., Pchelintsev, N.A., *et al.* (2011) Human CABIN1 Is a Functional Member of the Human HIRA/UBN1/ASF1a Histone H3.3 Chaperone Complex. *Molecular and Cellular Biology*, **31**, 4107-4118. http://dx.doi.org/10.1128/MCB.05546-11

[25] Lamour, V., Lecluse, Y., Desmaze, C., Spector, M., Bodescot, M., Aurias, A., *et al.* (1995) A Human Homolog of the *S. cerevisiae* HIR_1 and HIR_2 Transcriptional Repressors Cloned from the DiGeorge Syndrome Critical Region. *Human Molecular Genetics*, **4**, 791-799. http://dx.doi.org/10.1093/hmg/4.5.791

[26] Lorain, S., Demczuk, S., Lamour, V., Toth, S., Aurias, A., Roe, B.A., *et al.* (1996) Structural Organization of the WD Repeat Protein-Encoding Gene HIRA in the DiGeorge Syndrome Critical Region of Human Chromosome 22. *Genome Research*, **6**, 43-50. http://dx.doi.org/10.1101/gr.6.1.43

[27] Magnaghi, P., Roberts, C., Lorain, S., Lipinski, M. and Scambler, P.J. (1998) HIRA, a Mammalian Homologue of *Saccharomyces cerevisiae* Transcriptional Co-Repressors, Interacts with Pax3. *Nature Genetics*, **20**, 74-77.

http://dx.doi.org/10.1038/1739

[28] Banumathy, G., Somaiah, N., Zhang, R., Tang, Y., Hoffmann, J., Andrake, M., *et al.* (2009) Human UBN1 Is an Ortholog of Yeast Hpc2p and Has an Essential Role in the HIRA/ASF1a Chromatin-Remodeling Pathway in Senescent Cells. *Molecular and Cellular Biology*, **29**, 758-770. http://dx.doi.org/10.1128/MCB.01047-08

[29] Daganzo, S.M., Erzberger, J.P., Lam, W.M., Skordalakes, E., Zhang, R., Franco, A.A., *et al.* (2003) Structure and Function of the Conserved core of Histone Deposition Protein Asf1. *Current Biology*, **13**, 2148-2158. http://dx.doi.org/10.1016/j.cub.2003.11.027

[30] Prochasson, P., Florens, L., Swanson, S.K., Washburn, M.P. and Workman, J.L. (2005) The HIR Corepressor Complex Binds to Nucleosomes Generating a Distinct Protein/DNA Complex Resistant to Remodeling by SWI/SNF. *Genes & Development*, **19**, 2534-2539. http://dx.doi.org/10.1101/gad.1341105

[31] Osley, M.A. and Lycan, D. (1987) Trans-Acting Regulatory Mutations That Alter Transcription of *Saccharomyces cerevisiae* Histone Genes. *Molecular and Cellular Biology*, **7**, 4204-4210.

[32] Dimova, D., Nackerdien, Z., Furgeson, S., Eguchi, S. and Osley, M.A. (1999) A Role for Transcriptional Repressors in Targeting the Yeast Swi/Snf Complex. *Molecular Cell*, **4**, 75-83. http://dx.doi.org/10.1016/S1097-2765(00)80189-6

[33] Wunsch, A.M. and Lough, J. (1987) Modulation of Histone H3 Variant Synthesis during the Myoblast-Myotube Transition of Chicken Myogenesis. *Developmental Biology*, **119**, 94-99. http://dx.doi.org/10.1016/0012-1606(87)90210-7

[34] Ray-Gallet, D., Quivy, J.P., Scamps, C., Martini, E.M., Lipinski, M. and Almouzni, G. (2002) HIRA Is Critical for a Nucleosome Assembly Pathway Independent of DNA Synthesis. *Molecular Cell*, **9**, 1091-1100. http://dx.doi.org/10.1016/S1097-2765(02)00526-9

[35] Anderson, H.E., Wardle, J., Korkut, S.V., Murton, H.E., Lopez-Maury, L., Bahler, J., *et al.* (2009) The Fission Yeast HIRA Histone Chaperone Is Required for Promoter Silencing and the Suppression of Cryptic Antisense Transcripts. *Molecular and Cellular Biology*, **29**, 5158-5167. http://dx.doi.org/10.1128/MCB.00698-09

[36] Llevadot, R., Scambler, P., Estivill, X. and Pritchard, M. (1996) Genomic Organization of TUPLE1/HIRA: A Gene Implicated in DiGeorge Syndrome. *Mammalian Genome*: *Official Journal of the International Mammalian Genome Society*, **7**, 911-914. http://dx.doi.org/10.1007/s003359900268

[37] Loppin, B., Bonnefoy, E., Anselme, C., Laurencon, A., Karr, T.L. and Couble, P. (2005) The Histone H3.3 Chaperone HIRA Is Essential for Chromatin Assembly in the Male Pronucleus. *Nature*, **437**, 1386-1390. http://dx.doi.org/10.1038/nature04059

[38] Roberts, C., Sutherland, H.F., Farmer, H., Kimber, W., Halford, S., Carey, A., *et al.* (2002) Targeted Mutagenesis of the *Hira* Gene Results in Gastrulation Defects and Patterning Abnormalities of Mesoendodermal Derivatives Prior to Early Embryonic Lethality. *Molecular and Cellular Biology*, **22**, 2318-2328. http://dx.doi.org/10.1128/MCB.22.7.2318-2328.2002

[39] Balaji, S., Iyer, L.M. and Aravind, L. (2009) HPC2 and Ubinuclein Define a Novel Family of Histone Chaperones Conserved Throughout Eukaryotes. *Molecular BioSystems*, **5**, 269-275. http://dx.doi.org/10.1039/b816424j

[40] Aho, S., Lupo, J., Coly, P.A., Sabine, A., Castellazzi, M., Morand, P., *et al.* (2009) Characterization of the Ubinuclein Protein as a New Member of the Nuclear and Adhesion Complex Components (NACos). *Biology of the Cell*, **101**, 319-334. http://dx.doi.org/10.1042/BC20080072

[41] Gruffat, H., Lupo, J., Morand, P., Boyer, V. and Manet, E. (2011) The Nuclear and Adherent Junction Complex Component Protein Ubinuclein Negatively Regulates the Productive Cycle of Epstein-Barr Virus in Epithelial Cells. *Journal of Virology*, **85**, 784-794. http://dx.doi.org/10.1128/JVI.01397-10

[42] Elsaesser, S.J. and Allis, C.D. (2010) HIRA and Daxx Constitute Two Independent Histone H3.3-Containing Predeposition Complexes. *Cold Spring Harbor Symposia on Quantitative Biology*, **75**, 27-34. http://dx.doi.org/10.1101/sqb.2010.75.008

[43] Schwabish, M.A. and Struhl, K. (2006) Asf1 Mediates Histone Eviction and Deposition during Elongation by RNA Polymerase II. *Molecular Cell*, **22**, 415-422. http://dx.doi.org/10.1016/j.molcel.2006.03.014

[44] Shandilya, J. and Roberts, S.G. (2012) The Transcription Cycle in Eukaryotes: From Productive Initiation to RNA Polymerase II Recycling. *Biochimica et Biophysica Acta* (*BBA*)—*Gene Regulatory Mechanisms*, **1819**, 391-400. http://dx.doi.org/10.1016/j.bbagrm.2012.01.010

[45] Orphanides, G., Lagrange, T. and Reinberg, D. (1996) The General Transcription Factors of RNA Polymerase II. *Genes & Development*, **10**, 2657-2683. http://dx.doi.org/10.1101/gad.10.21.2657

[46] Wada, T., Takagi, T., Yamaguchi, Y., Ferdous, A., Imai, T., Hirose, S., *et al.* (1998) DSIF, a Novel Transcription Elongation Factor That Regulates RNA Polymerase II Processivity, Is Composed of Human Spt_4 and Spt_5 Homologs. *Genes & Development*, **12**, 343-356. http://dx.doi.org/10.1101/gad.12.3.343

[47] Yamaguchi, Y., Takagi, T., Wada, T., Yano, K., Furuya, A., Sugimoto, S., *et al.* (1999) NELF, a Multisubunit Com-

plex Containing RD, Cooperates with DSIF to Repress RNA Polymerase II Elongation. *Cell*, **97**, 41-51. http://dx.doi.org/10.1016/S0092-8674(00)80713-8

[48] Corden, J.L. (1990) Tails of RNA Polymerase II. *Trends in Biochemical Sciences*, **15**, 383-387. http://dx.doi.org/10.1016/0968-0004(90)90236-5

[49] Ohkuma, Y. and Roeder, R.G. (1994) Regulation of TFIIH ATPase and Kinase Activities by TFIIE during Active Initiation Complex Formation. *Nature*, **368**, 160-163. http://dx.doi.org/10.1038/368160a0

[50] Liu, X., Bushnell, D.A., Silva, D.A., Huang, X. and Kornberg, R.D. (2011) Initiation Complex Structure and Promoter Proofreading. *Science*, **333**, 633-637. http://dx.doi.org/10.1126/science.1206629

[51] Peterlin, B.M. and Price, D.H. (2006) Controlling the Elongation Phase of Transcription with P-TEFb. *Molecular Cell*, **23**, 297-305. http://dx.doi.org/10.1016/j.molcel.2006.06.014

[52] Ivanov, D., Kwak, Y.T., Guo, J. and Gaynor, R.B. (2000) Domains in the SPT5 Protein That Modulate Its Transcriptional Regulatory Properties. *Molecular and Cellular Biology*, **20**, 2970-2983. http://dx.doi.org/10.1128/MCB.20.9.2970-2983.2000

[53] Lis, J. (1998) Promoter-Associated Pausing in Promoter Architecture and Postinitiation Transcriptional Regulation. *Cold Spring Harbor Symposia on Quantitative Biology*, **63**, 347-356. http://dx.doi.org/10.1101/sqb.1998.63.347

[54] Gilchrist, D.A., Fromm, G., dos Santos, G., Pham, L.N., McDaniel, I.E., Burkholder, A., *et al.* (2012) Regulating the Regulators: The Pervasive Effects of Pol II Pausing on Stimulus-Responsive Gene Networks. *Genes & Development*, **26**, 933-944. http://dx.doi.org/10.1101/gad.187781.112

[55] Muse, G.W., Gilchrist, D.A., Nechaev, S., Shah, R., Parker, J.S., Grissom, S.F., *et al.* (2007) RNA Polymerase Is Poised for Activation across the Genome. *Nature Genetics*, **39**, 1507-1511. http://dx.doi.org/10.1038/ng.2007.21

[56] Donahue, B.A., Yin, S., Taylor, J.S., Reines, D. and Hanawalt, P.C. (1994) Transcript Cleavage by RNA Polymerase II Arrested by a Cyclobutane Pyrimidine Dimer in the DNA Template. *Proceedings of the National Academy of Sciences of the United States of America*, **91**, 8502-8506. http://dx.doi.org/10.1073/pnas.91.18.8502

[57] Tornaletti, S. and Hanawalt, P.C. (1999) Effect of DNA Lesions on Transcription Elongation. *Biochimie*, **81**, 139-146. http://dx.doi.org/10.1016/S0300-9084(99)80046-7

[58] Toulme, F., Guerin, M., Robichon, N., Leng, M. and Rahmouni, A.R. (1999) *In Vivo* Evidence for Back and Forth Oscillations of the Transcription Elongation Complex. *The EMBO Journal*, **18**, 5052-5060. http://dx.doi.org/10.1093/emboj/18.18.5052

[59] Zenkin, N., Yuzenkova, Y. and Severinov, K. (2006) Transcript-Assisted Transcriptional Proofreading. *Science*, **313**, 518-520. http://dx.doi.org/10.1126/science.1127422

[60] Nudler, E. (2012) RNA Polymerase Backtracking in Gene Regulation and Genome Instability. *Cell*, **149**, 1438-1445. http://dx.doi.org/10.1016/j.cell.2012.06.003

[61] Jeon, C. and Agarwal, K. (1996) Fidelity of RNA Polymerase II Transcription Controlled by Elongation Factor TFIIS. *Proceedings of the National Academy of Sciences of the United States of America*, **93**, 13677-13682. http://dx.doi.org/10.1073/pnas.93.24.13677

[62] Larson, M.H., Zhou, J., Kaplan, C.D., Palangat, M., Kornberg, R.D., Landick, R., *et al.* (2012) Trigger Loop Dynamics Mediate the Balance between the Transcriptional Fidelity and Speed of RNA Polymerase II. *Proceedings of the National Academy of Sciences of the United States of America*, **109**, 6555-6560. http://dx.doi.org/10.1073/pnas.1200939109

[63] Ratner, J.N., Balasubramanian, B., Corden, J., Warren, S.L. and Bregman, D.B. (1998) Ultraviolet Radiation-Induced Ubiquitination and Proteasomal Degradation of the Large Subunit of RNA Polymerase II. Implications for Transcription-Coupled DNA Repair. *The Journal of Biological Chemistry*, **273**, 5184-5189. http://dx.doi.org/10.1074/jbc.273.9.5184

[64] Somesh, B.P., Reid, J., Liu, W.F., Sogaard, T.M., Erdjument-Bromage, H., Tempst, P., *et al.* (2005) Multiple Mechanisms Confining RNA Polymerase II Ubiquitylation to Polymerases Undergoing Transcriptional Arrest. *Cell*, **121**, 913-923. http://dx.doi.org/10.1016/j.cell.2005.04.010

[65] Hershko, A. and Ciechanover, A. (1998) The Ubiquitin System. *Annual Review of Biochemistry*, **67**, 425-479. http://dx.doi.org/10.1146/annurev.biochem.67.1.425

[66] Anindya, R., Aygun, O. and Svejstrup, J.Q. (2007) Damage-Induced Ubiquitylation of Human RNA Polymerase II by the Ubiquitin Ligase Nedd4, but Not Cockayne Syndrome Proteins or BRCA1. *Molecular Cell*, **28**, 386-397. http://dx.doi.org/10.1016/j.molcel.2007.10.008

[67] Yasukawa, T., Kamura, T., Kitajima, S., Conaway, R.C., Conaway, J.W. and Aso, T. (2008) Mammalian Elongin A Complex Mediates DNA-Damage-Induced Ubiquitylation and Degradation of Rpb1. *The EMBO Journal*, **27**, 3256-3266. http://dx.doi.org/10.1038/emboj.2008.249

[68] Harreman, M., Taschner, M., Sigurdsson, S., Anindya, R., Reid, J., Somesh, B., *et al.* (2009) Distinct Ubiquitin Ligases Act Sequentially for RNA Polymerase II Polyubiquitylation. *Proceedings of the National Academy of Sciences of the United States of America*, **106**, 20705-20710. http://dx.doi.org/10.1073/pnas.0907052106

[69] Wilson, M.D., Harreman, M. and Svejstrup, J.Q. (2013) Ubiquitylation and Degradation of Elongating RNA Polymerase II: The Last Resort. *Biochimica et Biophysica Acta (BBA)—Gene Regulatory Mechanisms*, **1829**, 151-157. http://dx.doi.org/10.1016/j.bbagrm.2012.08.002

[70] Kwak, J., Workman, J.L. and Lee, D. (2011) The Proteasome and Its Regulatory Roles in Gene Expression. *Biochimica et Biophysica Acta (BBA)—Gene Regulatory Mechanisms*, **1809**, 88-96. http://dx.doi.org/10.1016/j.bbagrm.2010.08.001

[71] Bedford, L., Paine, S., Sheppard, P.W., Mayer, R.J. and Roelofs, J. (2010) Assembly, Structure, and Function of the 26S Proteasome. *Trends in Cell Biology*, **20**, 391-401.

[72] Glickman, M.H., Rubin, D.M., Coux, O., Wefes, I., Pfeifer, G., Cjeka, Z., *et al.* (1998) A Subcomplex of the Proteasome Regulatory Particle Required for Ubiquitin-Conjugate Degradation and Related to the COP9-Signalosome and eIF3. *Cell*, **94**, 615-623. http://dx.doi.org/10.1016/S0092-8674(00)81603-7

[73] Anderson, H.E., Kagansky, A., Wardle, J., Rappsilber, J., Allshire, R.C. and Whitehall, S.K. (2010) Silencing Mediated by the Schizosaccharomyces Pombe HIRA Complex Is Dependent upon the Hpc2-Like Protein, Hip4. *PlOS ONE*, **5**, e13488. http://dx.doi.org/10.1371/journal.pone.0013488

[74] Ray-Gallet, D., Woolfe, A., Vassias, I., Pellentz, C., Lacoste, N., Puri, A., *et al.* (2011) Dynamics of Histone H3 Deposition *in Vivo* Reveal a Nucleosome Gap-Filling Mechanism for H3.3 to Maintain Chromatin Integrity. *Molecular Cell*, **44**, 928-941. http://dx.doi.org/10.1016/j.molcel.2011.12.006

[75] Chujo, M., Tarumoto, Y., Miyatake, K., Nishida, E. and Ishikawa, F. (2012) HIRA, a Conserved Histone Chaperone, Plays an Essential Role in Low-Dose Stress Response via Transcriptional Stimulation in Fission Yeast. *The Journal of Biological Chemistry*, **287**, 23440-23450. http://dx.doi.org/10.1074/jbc.M112.349944

[76] DeSilva, H., Lee, K. and Osley, M.A. (1998) Functional Dissection of Yeast Hir1p, a WD Repeat-Containing Transcriptional Corepressor. *Genetics*, **148**, 657-667.

[77] Andrulis, E.D., Guzman, E., Doring, P., Werner, J. and Lis, J.T. (2000) High-Resolution Localization of Drosophila Spt5 and Spt6 at Heat Shock Genes *in Vivo*: Roles in Promoter Proximal Pausing and Transcription Elongation. *Genes & Development*, **14**, 2635-2649. http://dx.doi.org/10.1101/gad.844200

[78] Formosa, T., Ruone, S., Adams, M.D., Olsen, A.E., Eriksson, P., Yu, Y., *et al.* (2002) Defects in SPT16 or POB3 (yFACT) in *Saccharomyces cerevisiae* Cause Dependence on the Hir/Hpc Pathway: Polymerase Passage May Degrade Chromatin Structure. *Genetics*, **162**, 1557-1571.

[79] Chimura, T., Kuzuhara, T. and Horikoshi, M. (2002) Identification and Characterization of CIA/ASF1 as an Interactor of Bromodomains Associated with TFIID. *Proceedings of the National Academy of Sciences of the United States of America*, **99**, 9334-9339. http://dx.doi.org/10.1073/pnas.142627899

[80] Bhat, K.P. and Greer, S.F. (2011) Proteolytic and Non-Proteolytic Roles of Ubiquitin and the Ubiquitin Proteasome System in Transcriptional Regulation. *Biochimica et Biophysica Acta (BBA)—Gene Regulatory Mechanisms*, **1809**, 150-155. http://dx.doi.org/10.1016/j.bbagrm.2010.11.006

[81] Ishizuka, T., Satoh, T., Monden, T., Shibusawa, N., Hashida, T., Yamada, M., *et al.* (2001) Human Immunodeficiency Virus Type 1 Tat Binding Protein-1 Is a Transcriptional Coactivator Specific for TR. *Molecular Endocrinology*, **15**, 1329-1343. http://dx.doi.org/10.1210/mend.15.8.0680

[82] Satoh, T., Ishizuka, T., Tomaru, T., Yoshino, S., Nakajima, Y., Hashimoto, K., *et al.* (2009) Tat-Binding Protein-1 (TBP-1), an ATPase of 19S Regulatory Particles of the 26S Proteasome, Enhances Androgen Receptor Function in Cooperation with TBP-1-Interacting Protein/Hop2. *Endocrinology*, **150**, 3283-3290. http://dx.doi.org/10.1210/en.2008-1122

[83] Dianov, G.L., Houle, J.F., Iyer, N., Bohr, V.A. and Friedberg, E.C. (1997) Reduced RNA Polymerase II Transcription in Extracts of Cockayne Syndrome and Xeroderma Pigmentosum/Cockayne Syndrome Cells. *Nucleic Acids Research*, **25**, 3636-3642. http://dx.doi.org/10.1093/nar/25.18.3636

[84] Salinas-Rios, V., Belotserkovskii, B.P. and Hanawalt, P.C. (2011) DNA Slip-Outs Cause RNA Polymerase II Arrest *in Vitro*: Potential Implications for Genetic Instability. *Nucleic Acids Research*, **39**, 7444-7454. http://dx.doi.org/10.1093/nar/gkr429

[85] Lyu, Y.L., Kerrigan, J.E., Lin, C.P., Azarova, A.M., Tsai, Y.C., Ban, Y., *et al.* (2007) Topoisomerase IIβ Mediated DNA Double-Strand Breaks: Implications in Doxorubicin Cardiotoxicity and Prevention by Dexrazoxane. *Cancer Research*, **67**, 8839-8846. http://dx.doi.org/10.1158/0008-5472.CAN-07-1649

[86] Zhang, S., Liu, X., Bawa-Khalfe, T., Lu, L.S., Lyu, Y.L., Liu, L.F., *et al.* (2012) Identification of the Molecular Basis of Doxorubicin-Induced Cardiotoxicity. *Nature Medicine*, **18**, 1639-1642. http://dx.doi.org/10.1038/nm.2919

5

Toxicological Effects of Chlorpyrifos and Lead on the Aquatic Snail *Helisoma duryi*

Norah Basopo*, **Thamsanqa Ngabaza**

Department of Applied Biology and Biochemistry, National University of Science and Technology, Bulawayo, Zimbabwe

Email: *norah.basopo@nust.ac.zw

Abstract

Aquatic reservoirs remain the ultimate sink of chemical pollutants emanating from anthropogenic activities such as agriculture, mining and industry. Freshwater biota undoubtedly is at risk from the adverse effects of these water pollutants and there is therefore, a need to monitor effects of these chemical pollutants in order to safeguard the health of aquatic biota. We investigated the oxidative stress effects of chlorpyrifos and lead on the freshwater snail *Helisoma duryi* to assess the potential of using this enzyme system as a biondicator of exposure to environmental pollutants. Groups of snails were exposed to 5 ppb lead acetate and 25 ppb chlorpyrifos for 7 days after which half of the snails were sacrificed and the other half were allowed to recover in clean water and sacrificed after another 7 days. Post mitochondrial fractions were used to measure the activities of the following antioxidant enzymes: superoxide dismutase, catalase, glutathione peroxidase, glutathione S-transferase and diphosphotriphosphodiaphorase. Both pollutants enhanced the activities of all the antioxidant enzymes suggesting a defensive mechanism by the snail to combat the oxidative stress due to the organophosphate chlopryrifos and metal pollutant lead. There was a significant recovery of the antioxidant defense system of the snails allowed to recover in clean water shown by the reduced alteration of the antioxidant enzyme activities of the snails allowed to recover for 7 days. This suggests the need to minimize exposure of aquatic biota to chemical pollutants and remediate the polluted water reservoirs in order to safeguard the health of aquatic life.

Keywords

Pollutants, Snails, Oxidative-Stress, Organophosphates, Metals

*Corresponding author.

1. Introduction

Anthropogenic activities undertaken by man in his quest to improve his livelihood have resulted in the production and release of chemicals in quantities that the environment cannot cope with. Increase in urbanization and industrial activities as well as uncontrolled exploitation of arable land has undoubtedly resulted in significant discharge of chemical pollutants and their byproducts which find their way into aquatic reservoirs which are the ultimate sink for environmental pollutants. Thousands of organic and inorganic materials such as pesticides, fertilizers, detergents and metals contaminate natural aquatic ecosystems daily and affect aquatic biota such as snails and fish. Many developing countries, including Zimbabwe, depend on agriculture to boost their economies; thus pesticides are likely to represent an important source of xenobiotics in contaminated rivers. Pesticides such as organophosphates, carbamates and pyrethroids are extensively used in agriculture and public health. In agriculture pesticides such as chlorpyrifos are used in the control of pests of cotton, corn and fruit trees such as apples and oranges [1]. Chlorpyrifos and other organophosphates are chemicals designed to target the nervous system of pest organisms in particular they inhibit the activities of cholinesterases, the enzymes required to ensure the smooth transmission of nervous impulses. Cholinesterases are thus termed useful biomarkers of exposure to environmental pollutants such as organophosphates and carbamates since these two classes of compounds share a common mechanism of toxicity [2]. Literature reports have however, shown that chlorpyrifos and other organophosphates are non-target organ/tissue specific. Wang *et al.* [3] reported the inhibition of the glutathione based antioxidant enzyme glutathione S-transferase (GST) of the goldfish (*Carassius auratus*) after exposure to chlorpyrifos and a carbamate isoprocarb supporting the fact that organophosphates and carbamates do not only target AChEs. Ferrari *et al.* [4] also observed alterations of antioxidant enzyme activities after exposure to organophosphates. The authors observed enhancement of superoxide dismutase (SOD) and GST in amphibian larvae exposed to the organophosphate, azinphos-methyl.

Mining and industrial activities also release byproducts such as metals and organic compounds which have stressful effects on the environment. In aquatic ecosystems these pollutants affect the wellbeing of aquatic biota. Netpae *et al.* [5] reported an increase in the activities of SOD, GST and glutathione reductase of the freshwater clam (*Corbicula fluminea*) after exposure to the heavy metal copper. Several studies have demonstrated that heavy metals can promote the formation of reactive oxygen species (ROS) in aquatic organisms [6]-[8]. Low degradability and bioaccumulation at different trophic levels through aquatic food webs make them major stress factors to aquatic organisms [2]. There is a need therefore to come up with early warning signals that can be used as preventive measures that protect the health of aquatic biota.

The objective of this study was to evaluate in laboratory conditions the oxidative stress effects of an organophosphorus insecticide, chlorpyrifos and the metal lead, as single entities on the aquatic snail species *Helisoma duryi.*

2. Materials and Methods

2.1. Chemicals

All enzymes, substrates, pesticides and chemicals were bought from either from Sigma Chemical Company or Aldrich Chemical Company, Germany. All other laboratory reagents, used in this study, were of analytical (ANALAR) grade.

2.2. Snail Breeding and Exposure

Snails were bred in outdoor cement aquaria containing tap water and fed on fresh garden lettuce according to the method of [9]. Juvenile snails (12 per group) were exposed to 25 ppb of chlopyrifos or 5 ppb of lead acetate for 7 days after which half of the snails were sacrificed and the other half placed in clean water for another 7 days. The exposures were performed in duplicates. After the exposure period the snails were sacrificed and post mitochondrial fractions prepared.

2.3. Preparation of Homogenates

Six whole snails from each experimental group were pooled and homogenized in (5X weight of soft tissue) ml of ice-cold homogenization buffer (0.1 M potassium phosphate pH 7.4). The homogenates were centrifuged at

10,000 × g for 10 minutes and the resultant supernatant (S-10) fraction stored at −80°C until analyzed. Protein content in post mitochondrial fractions of the snails was determined using the method of [10] using bovine serum albumin as the calibration standard and absorbance at 750 nm.

All enzymes assays described below were performed on S-10 fractions of the whole snail homogenates.

2.4. Biochemical Assays

2.4.1. Superoxide Dismutase

Superoxide dismutase (E.C.1.15.1.1) activity in the S-10 fractions was measured following the method of [11]. Xanthine and xanthine oxidase were used to generate superoxide anion radicals, which react with 2-(4-indophenyl)-3(4-nitro-phenyl)-5-phenyl tetrazolium chloride (NBT) to form a red formazan dye. One enzyme unit of superoxide dismutase is defined as the amount, which inhibits the NBT reaction by 50%. Specific activity was defined as units/mg protein.

2.4.2. Catalase

Catalase (E.C. 1.11.1.6) activity was measured in S-10 fractions according to the method of [12] and expressed as nanomoles of hydrogen peroxide decomposed/min/mg protein.

2.4.3. Glutathione Peroxidase

Glutathione peroxidase (EC.1.11.1.9) activity was determined in S-10 fractions according to the method of [13] and was expressed as micromoles GSH oxidized/min/mg protein.

2.4.4. Glutathione S-Transferase

Glutathione S-transferase (EC.2.5.1.18) activity was measured in S-10 fractions according to the method of [14], and expressed as millimoles of 1-chloro-2-4-dintrobenzene-glutathione (CDNB-GSH) conjugate formed/min/mg protein.

2.4.5. NAD(P)H Quinone Oxidoreductase

NAD(P)H quinone oxidoreductase (EC.1.6.99.2) activity was measured in S-10 fractions following the method of [15] and was expressed as millimoles decomposed DCPIP/min/mg protein.

2.5. Statistical Analysis

Statistical differences between control and experimental groups were analyzed using the Dunnet test. Differences were considered significant at $^{*}p < 0.05$.

3. Results

3.1. Superoxide Dismutase Activity

Superoxide dismutase activity was significantly increased ($p < 0.05$) in both lead and chlorpyrifos exposed snails when compared to the controls, unexposed snails (**Figure 1**). Chlorpyrifos increased SOD activity by 93% while lead caused an increase of 165% in the enzyme activity. There was a significant difference in SOD activity between the exposed snails and the snails allowed to recover.

3.2. Catalase Activity

Catalase activity was significantly increased ($p < 0.05$) in all pollutant exposed snails when compared to the controls (**Figure 2**). Increases of 155% and 227% in enzyme activities were observed in snails exposed to chlorpyrifos and lead respectively. In the recovery period there were significant differences in enzyme activity between the exposed snails and the snails allowed to recover.

3.3. Glutathione Peroxidase

Glutathione peroxidase activity was significantly increased ($p < 0.05$) in lead and chloryrifos exposed snails

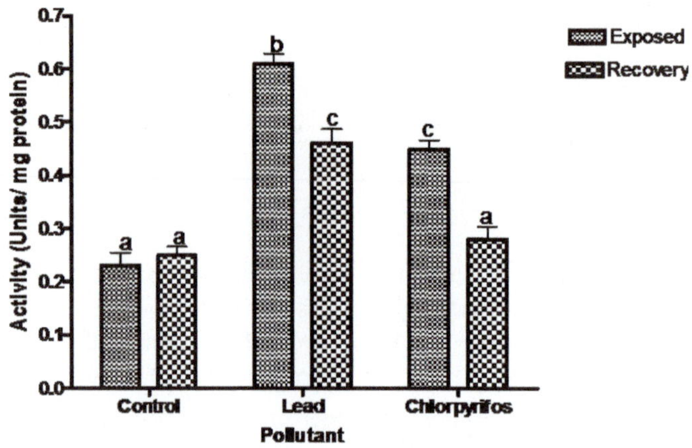

Figure 1. Effect chlorpyrifos and lead on superoxide dismutase activity of the freshwater snail *H. duryi.* = exposed and = recovery. Significance of results was ascertained at $p < 0.05$ and different letters express significant differences and bars with the same letters indicate that there is no significant difference.

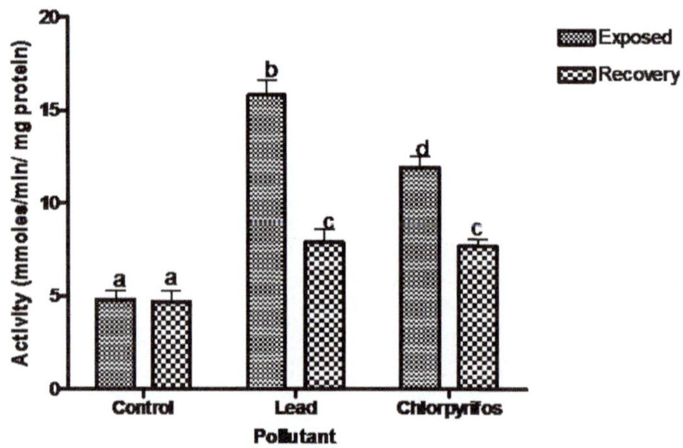

Figure 2. Effect chlorpyrifos and lead on catalase activity of the freshwater snail *H. duryi.* = exposed and = recovery. Significance of results was ascertained at $p < 0.05$ and different letters express significant differences and bars with the same letters indicate that there is no significant difference.

when compared to the controls (**Figure 3**). Activation of enzyme activities of 22% and 30% were observed in *H. duryi* snails exposed to chlorpyrifos and lead respectively. In the recovery period there was significant difference between the exposed snails and the snails allowed to recover.

3.4. Glutathione S-Transferase

Glutathione S-transferase (GST) activity was significantly increased ($p < 0.05$) in lead and chloryrifos exposed snails when compared to the controls (**Figure 4**). Chlorpyrifos activated GST activity by 66% while lead caused an increase in enzyme activity of 51%. In the recovery period there was significant difference between the exposed snails and the snails allowed to recover.

3.5. NAD(P)H Quinone Oxidoreductase

The activity of NAD(P)H quinone oxidoreductase was significantly increased ($p < 0.05$) in pollutant exposed snails when compared to the controls (**Figure 5**). Increases in NAD(P)H quinone oxidoreductase of 40% and 34% were observed in chlorpyrifos and lead treated snails respectively. In the recovery period there were no significant differences between the exposed snails and the snails allowed to recover.

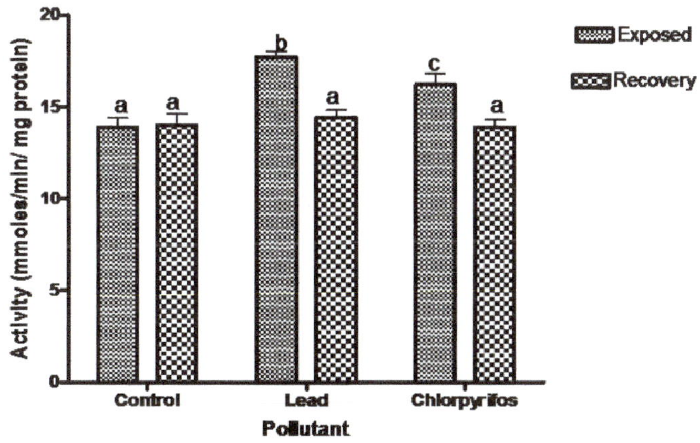

Figure 3. Effect chlorpyrifos and lead on glutathione peroxidase activity of the freshwater snail *H. duryi.* = exposed and = recovery. Significance of results was ascertained at $p < 0.05$ and different letters express significant differences and bars with the same letters indicate that there are no significant differences.

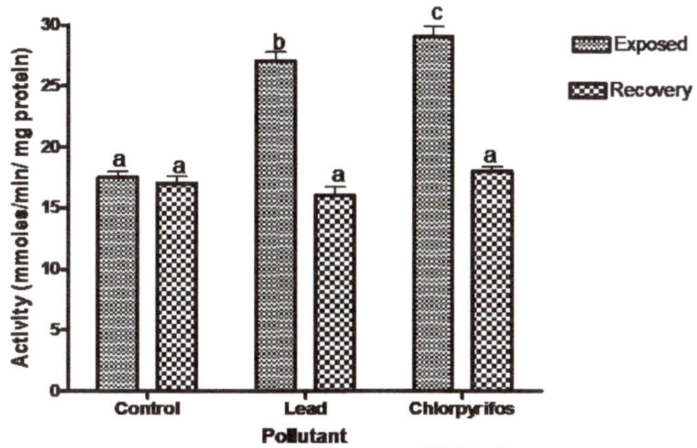

Figure 4. Effect chlorpyrifos and lead on glutathione S-transferase activity of the freshwater snail *H. duryi.* = exposed and = recovery. Significance of results was ascertained at $p < 0.05$ and different letters express significant differences and bars with the same letters indicate that there are no significant differences.

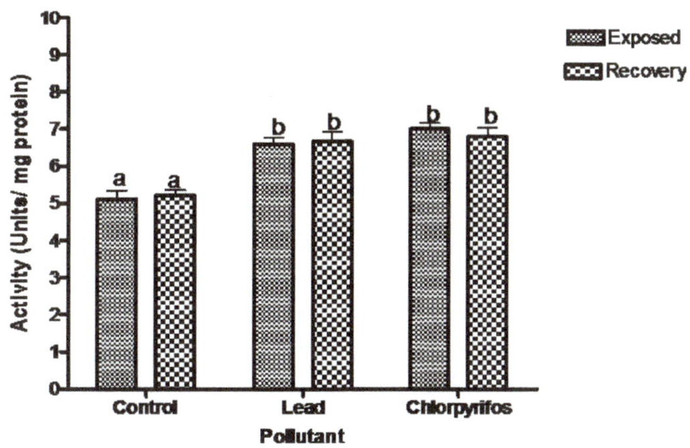

Figure 5. Effect chlorpyrifos and lead on NAD(P)H quinone oxidoreductase activity of the freshwater snail *H. duryi.* = exposed and = recovery. Significance of results was ascertained at $p < 0.05$ and different letters express significant differences and bars with the same letters indicate that there are no significant differences.

4. Discussion

Pollutants such as metals occur naturally in the environment. However, due to man's daily activities to improve his livelihood, their levels in ecosystems increase to levels that cause adverse effects to other organisms. Metals like lead are introduced in aquatic ecosystems as effluent from mining and industrial activities. Pesticides which are used by man to improve the quality and quantity of crops find their way in aquatic environments via aerial drifts during spraying, leaching and as runoffs after heavy rains. In fact literature shows that of the pesticides applied by farmers only 50% gets to the target organisms [16] and the rest end up on nontarget organisms where they usually cause deleterious effects on these organisms. Literature reports show that metals and pesticides induce oxidative stress in cells of living organisms and this result from an imbalance between the generation and elimination of reactive oxygen species [17].

Metals like lead probably cause production of reactive oxygen species (ROS) which if not eliminated effectively attack macromolecules such as lipids, proteins and carbohydrates in living organisms. The induction of oxidative stress in biological systems by lead is reported by several researchers however, mechanism of ROS induction is unknown [18]-[20]. The actual mechanism of elimination of ROS like superoxide anion radical is generally by the dismutation activity of SOD producing hydrogen peroxide. Our results showed that lead activated the activities of antioxidant enzymes in the range 30% to 227% (**Figures 1-5**) depending on the enzyme. The activation of the antioxidant enzymes observed in the present study suggests that lead has oxidative stress inducing effects which the snail's defense system counteracted by increasing the activities of the enzymes that protect living organisms from oxidative stress. Superoxide dismutases are a group of metalloenzymes that catalyse the dismutation of superoxide anion radicals ($O_2^{\cdot-}$) to form hydrogen peroxide (H_2O_2) and oxygen. Activation of SOD of 165% in lead exposed snails was observed in the present study. Altered SOD activity has been shown to be indicative of exposure to pollutants that induce oxidative stress [21]. The observed activation of SOD in pollutant exposed snails suggest that the detoxification of the pollutants by the snail's defense system probably produces superoxide anion radicals as byproducts.

Our results also showed activations of CAT and GPX of 227% and 30% respectively in snails exposed to lead (**Figure 2** and **Figure 3**). Induction of CAT and/or GPx is a protective response by the snail's system to combat the oxidative stress effects of hydrogen peroxide a product of the dismutation of superoxide anion radical produced during normal metabolism of pesticides [22]. Our results are consistent with the findings of [23] who also observed elevated CAT and GPX activities in snails exposed to metals such as lead, cadmium, iron and copper. Radwan *et al.* [18] also reported activation of CAT, GPX and GST in *Theba pisana* snails exposed to the heavy metals; lead, zinc, copper and cadmium. Induced catalase activity has been regarded as an adaptive response to reactive anion radicals generated during metabolism of various pollutants such as metal residues [5]. Catalase and GPX protect cells from H_2O_2 generated within them and it plays an important role in the acquisition of tolerance to oxidative stress [24]. The higher activations of CAT activity when compared to GPX activity observed imply that CAT is probably, the major enzyme detoxifying the reactive hydrogen peroxide a product of the dismutation of SOD. In fact the linked reactions involving SOD and catalase (CAT) provide the first line of defense against oxidative stress [25].

Activated GST activity shows the important role of this enzyme in protecting cells against oxidative stress [26]. Increased levels of GST activity of 51%, observed in this study, in snails exposed to the heavy metal lead reflect and suggest that the snail defense system managed to positively respond by attempting to protect itself against xenobiotic induced toxicity.

Activation of NAD(P)H quinone oxidoreductase, a detoxification enzyme, that reduces reactive quinones and quinone-imines to less reactive and less toxic hydroquinones in living organisms has been shown to be a defense strategy by living organisms to guard against the oxidative effects of reactive quinones which can be formed during detoxification reactions. In the present study we observed significant ($p < 0.05$) activation of NAD(P)H quinone oxidoreductase in all snails exposed to chlorpyrifos or the heavy metal, lead (**Figure 5**). The activation of NAD(P)H quinone oxidoreductase suggest the production of quinone like structures as intermediates or by-products in the detoxification mechanisms of studied environmental pollutants. Another probable reason for the activation of NAD(P)H quinone oxidoreductase activity in pollutant exposed snails is that the enzyme maybe involved in the detoxification of other ROS in the metabolism of xenobiotics. Activation of NAD(P)H quinone oxidoreductase as a defense mechanism has been reported in other polluted exposed aquatic organisms. Porte *et al.* [27] observed an increase in microsomal NADPH dependent quinone oxidoreductase in organic pollutant exposed *Mytilus galloprovincialis*.

The organophosphate chlorpyrifos is an insecticide, acaricide and miticide used to control foliage and soil-borne insect pests on a variety of food and feed crops. It controls pests such as cutworms, aphids, root borers, beetles, armyworms, spider mites and pink bollworms in crops that include cotton, soya beans, wheat, corn, citrus and tree nuts [28]. Organophosphorus pesticides like chlorpyrifos are normally associated with neurotoxicity and in particular this group of insecticides inhibits the activity of acetylcholinesterase thereby causing accumulation of acetylcholine molecules and affecting the normal functioning of nerves. Literature reports have however, showed that most pesticides, organophosphates included, affect organs/tissues other than their target organs/tissues [29]. Literature also shows that there is limited information on effects of pesticides on antioxidant enzymes in invertebrates such as snails. Salama *et al.* [30] reported an increase in levels of GSH and increased oxidative stress in the land snail *Helix aspersa* exposed to chlorpyrifos. Leomanni *et al.* [31] reported induction of the antioxidant enzymes SOD, CAT, GPX and glutathione reductase in the aquatic snail species *Cantareus apertus* exposed to the carbamate carbaryl.

In the present study chlorpyrifos activated the activities of SOD, CAT, GPX, GST and DTD in all pesticide treated snails. Activation ranges of 22% - 155% were observed depending on the antioxidant enzyme analyzed. Chlorpyrifos is a degradable pesticide, and a number of environmental forces may be active in its breakdown. In soil, water, plants and animals, the major pathway of degradation involves cleavage of the phosphorus ester bond to yield 3,5,6-trichloro-2-pyridinol (TCP). This conversion to TCP is a detoxification step and the TCP can be further degraded via microbial activity and photolysis to carbon dioxide and organic matter. During normal metabolism of chlorpyrifos ROS may be produced and biological systems use antioxidant enzyme systems to protect themselves from effects of ROS. The results observed in the present study imply that probably in the metabolism of chlorpryifos in the freshwater snails *H. duryi*, ROS were indeed produced and probably that caused the defense mechanism of the snails to be activated (indicated by activation of CAT, GPX and GST) probably as an adaptive mechanism to overcome the oxidative stress induced effects of chlorpyrifos. Our results are supported by Khalil [32] who also reported induction of CAT, GPX and GST in the freshwater snail *Lanistes carinatus* exposed to the organophosphorus insecticide chlorpyrifos.

All exposed snails significantly recovered from the effects of the two pollutants chlorpyrifos and lead when allowed to recover in clean water for 7 days. The antioxidant enzymes of snails exposed to chlopryrifos recovered by 18% - 90% depending on the antioxidant enzyme. While, the snails exposed to lead recovered by 49% - 88% depending on the antioxidant enzymes in the recovery period. There are few studies that have shown how contaminated aquatic organisms can recover their biochemical status when the pollutants are removed. Ma *et al.* [33] reported activation of antioxidant enzymes SOD, CAT and GST of *Physa acuta* exposed to the pesticide abamectin and the enzyme activities returned to control levels during the recovery period. Recovery of altered biochemical endpoint parameters has also, been reported in other aquatic organisms exposed to water pollutants. A significant recovery in all enzymatic and non-enzymatic endpoints was noticed after 7 days of recuperation period in fish (*Oreochromis mossambicus*) exposed to the organophosphorus insecticide profenofos [34].

In the present study there was no significant recovery observed in the NAD(P)H quinone oxidoreductase activities for all pollutant exposed snails. The results suggest that probably NAD(P)H quinone oxidoreductase is an active enzyme in the scavenging of ROS during the detoxification of environmental pollutants. Our results are supported by Siegel *et al.* [35] who have shown that apart from catalyzing the reduction of quinones, NAD(P)H quinone oxidoreductases also have $O_2^{\cdot-}$ scavenging properties.

5. Conclusion

Our results have shown that the antioxidant enzyme system of *H. duryi* snail's is sensitive to the studied pollutants indicated by the consistent activations of the studied antioxidant enzymes and as such has a potential of being used as a non-specific biological indicator of exposure to environmental pollutants.

Acknowledgements

This research was supported by the International Programme in Chemical Sciences (IPICS), Uppsala University, Sweden and the Research Board, NUST, Bulawayo, Zimbabwe.

References

[1] Testai, E., Buratti, F.M. and Consiglio, E.D. (2010) Chlorpyrifos. In: Krieger, R.I., Doull, J., van Hemmen, J.J., Hodg-

son, E., Maibach, H.I., Ritter, L., Ross, J. and Slikker, W., Eds., *Handbook of Pesticide Toxicology*, Vol. 2, Elsevier, Burlington, 1505-1526. http://dx.doi.org/10.1016/b978-0-12-374367-1.00070-7

[2] Nunes, B. (2011) The Use of Cholinesterases in Ecotoxicology. *Reviews of Environmental Contamination and Toxicology*, **212**, 29-59. http://dx.doi.org/10.1007/978-1-4419-8453-1_2

[3] Wang, C., Lu, G. and Cui, J. (2012) Responses of AChE and GST Activities to Insecticide Coexposure in *Carassius auratus*. *Environmental Toxicology*, **27**, 50-57. http://dx.doi.org/10.1002/tox.20612

[4] Ferrari, A., Lascano, C., Pechen de D'Angelo, A.M. and Venturino, A. (2011) Effects of Azinphos Methyl and Carbaryl on *Rhinella arenarum* Larvae Esterases and Antioxidant Enzymes. *Comparative Biochemistry Physiology C and Toxicological Pharmacology*, **153**, 34-39. http://dx.doi.org/10.1016/j.cbpc.2010.08.003

[5] Netpae, T., Phalaraksh, C. and Wongkham, W. (2012) Antioxidant Enzyme Activities and DNA Damage as Biomarker of Copper Effect on *Corbicula fluminea*. *Electronic Journal of Biology*, **8**, 19-23.

[6] Company, R., Serafim, A., Bebianno, M.J., Cosson, R., Shillito, B. and Fiala-Médioni, A. (2004) Effect of Cadmium, Copper and Mercury on Antioxidant Enzyme Activities and Lipid Peroxidation in the Gills of the Hydrothermal Vent Mussel *Bathymodiolus azoricus*. *Marine Environmental Research*, **58**, 377-381. http://dx.doi.org/10.1016/j.marenvres.2004.03.083

[7] Guidi, P., Frenzilli, G., Benedetti, M., Bernardeshi, M., Falleni, A., Fattorini, D., Reooli, F., Scarcelli, V. and Nigro, M. (2010) Antioxidant, Genotoxic and Lysosomal Biomarkers in the Freshwater Bivalve (*Uni pictorum*) Transplanted in a Metal Polluted River Basin. *Aquatic Toxicology*, **100**, 75-83. http://dx.doi.org/10.1016/j.aquatox.2010.07.009

[8] Bianco, K., Yusseppone, M.S., Otero, S., Luquet, C., Rios de Molina, M.C. and Kristoff, G. (2013) Cholinesterase and Neurotoxicity as Highly Sensitive Biomarkers for an Arganophosphate Insecticide in a Freshwater Gastropod (*Chilina gibbosa*) with Low Sensitivity Carboxylesterases. *Aquatic Toxicology*, **26-35**, 144-145. http://dx.doi.org/10.1016/j.aquatox.2013.09.025

[9] Naik, Y.S. and Hasler, J.A. (2002) An Economical Method for the Maintenance of Schistosome Life-Cycle Using Outdoor Facilities. *Discovery Innovation*, **14**, 221-224.

[10] Lowry, O.H., Rosebrough, N.J., Farr, A.L. and Randall, R.J. (1951) Protein Measurement with Folin Phenol Reagent. *Journal of Biological Chemistry*, **193**, 265-275.

[11] Sun, Y., Larry, W.O. and Ying, L. (1988) A Simple Method for Clinical Assay of Superoxide Dismutase. *Clinical Chemistry*, **34**, 497-500.

[12] Claiborne, A. (1989) Catalase Activity. In: Greeenwald, A.R., Ed., *Handbook of Methods for Methods for Oxygen Radical Research*, CRC Press Inc., Boca Raton, 283-284.

[13] Flohe, L. and Gunzler, W.A. (1984) Assays of Glutathione Peroxidase. *Methods in Enzymology*, **105**, 114-121. http://dx.doi.org/10.1016/S0076-6879(84)05015-1

[14] Habig, W.H., Pabst, M.J. and Jakoby, W.B. (1974) Glutathione S-Transferases. *Journal of Biological Chemistry*, **349**, 7130-7139.

[15] Lind, C., Cadenas, E., Hochstein, P. and Ernster, L. (1990) DT Diaphorase: Purification, Properties and Function. *Methods in Enzymology*, **186**, 287-301. http://dx.doi.org/10.1016/0076-6879(90)86122-C

[16] Pimentel, D. and Burgess, M. (2012) Small Amounts of Pesticides Reaching Target Insects. *Environment Development and Sustainability Journal*, **14**, 1-2. http://dx.doi.org/10.1007/s10668-011-9325-5

[17] Patra, R.C., Rautray, A.K. and Swarup, D. (2011) Oxidative Stress in Lead and Cadmium Toxicity and Its Amelioration. *Veterinary Medicine International Journal*, **2011**, Article ID: 457327. http://dx.doi.org/10.4061/2011/457327

[18] Radwan, M.A., El-Gendy, K.S. and Gad, A.F. (2010) Oxidative Stress Biomarkers in the Digestive Gland of *Theba pisana* Exposed to Heavy Metals. *Archives of Environmental Contamination Toxicology*, **58**, 828-835. http://dx.doi.org/10.1007/s00244-009-9380-1

[19] Rajkumar, J.S.I. and Milton, M.C.J. (2011) Biochemical Changes Induced by Cadmium, Copper, Lead and Zinc Exposure to Perna Viridis under Long Term Toxicity Test. *International Journal of Pharmacology and Biosciences*, **2**.

[20] Nowakowska, A., Laciak, M. and Capula, M. (2011) Organ Profile of the Antioxidant Defense System and Accumulation of Metals in *Helix aspersa* Snails. *Polish Journal of Environmental Studies*, **21**, 1369-1375.

[21] Alhifi, M.A., Khan, M.Z. and Gholem V.S. (2005) Effect of Dimethoate 30% EC on Antioxidant Enzymes in Chick Embryo Brain, Liver and Heart. *International Medical Journal*, **4**.

[22] El-Gendy, K.S., Radwan, M.A. and Gad, A.F. (2009) *In Vivo* Evaluation of Oxidative Stress Biomarkers in the Land Snail, *Theba pisana* Exposed to Copper-Based Pesticides. *Chemosphere*, **77**, 339-344. http://dx.doi.org/10.1016/j.chemosphere.2009.07.015

[23] El-Shenawy, N. and Amaal, M. (2012) Using Enzymatic and Nonenzyamatic Antioxidant Defense Systems of the Land Snail *Eobonia vermiculata* as Biomarkers of Terrestrial Heavy Metal Pollution. *Ecotoxicology and Environmen-*

tal Safety, **84**, 347-354. http://dx.doi.org/10.1016/j.ecoenv.2012.08.014

[24] Verlecar, X.N., Jena, K.B. and Chainy, G.B.N. (2008) Modulation of Antioxidant Defences in Digestive Gland of *Perna viridis* on Mercury Exposures. *Chemosphere*, **71**, 1977-1985.

[25] Leveelahti, L., Rytkönen, K.T., Renshaw, G.M.C. and Nikinmaa, M. (2013) Revisiting Redox-Active Antioxidant Defenses in Response to Hypoxic Challenge in Both Hypoxia-Tolerant and Hypoxia-Sensitive. *Fish Species Fish Physiology and Biochemistry*, **40**, 183-191. http://dx.doi.org/10.1007/s10695-013-9835-1

[26] Vontas, J.G., Small, G.J. and Hemingway, J. (2001) Glutathione S-Transferases as Antioxidant Defense Agents Confer Pyrethroid Resistance in *Nilaparvata lugens*. *Biochemical Journal*, **357**, 65-72. http://dx.doi.org/10.1042/bj3570065

[27] Porte, C.S., Albaiges, M.J. and Livingstone, D.R. (1991) Responses of Mixed Function Oxygenase and Antioxidase Enzyme System of Mytilus sp. to Organic Pollution. *Comparative Biochemistry and Physiology*, **100C**, 183-186.

[28] U.S. EPA (U.S. Environmental Protection Agency) (1999) Chlorpyrifos: HED Preliminary Risk Assessment for the Reregistration Eligibility Decision (RED) Document. Chemical No. 059101. Barcode: D260163, Case: 818975, Submission: S568580.

[29] Valavanidis, A., Vlahogianni, T., Dassenakis, M. and Scoullos, M. (2006) Molecular Biomarkers of Oxidative Stress in Aquatic Organisms in Relation to Toxic Environmental Pollutants. *Ecotoxicology and Environmental Safety*, **64**, 178-189. http://dx.doi.org/10.1016/j.ecoenv.2005.03.013

[30] Salama, A.K., Khaled, A.O., Nabia, A., Saber, Salim S.A. (2005) Oxidative Stress Induced by Different Pesticides in Land Snails *Helix aspersa*. *Pakistan Journal of Biological Sciences*, **8**, 92-96. http://dx.doi.org/10.3923/pjbs.2005.92.96

[31] Leomanni, A., Schettino, T., Calisi, A., Gorbi, S., Mezzelani, M., Regoli, F. and Lionetto, M.G. (2015) Antioxidant and Oxidative Stress Related Responses in the Mediterranean Land Snail *Cantareus apertus* Exposed to the Carbamate Pesticide Carbaryl. *Comparative Biochemistry and Physiology C*, **168**, 20-27. http://dx.doi.org/10.1016/j.cbpc.2014.11.003

[32] Khalil, A.M. (2015) Toxicological Effects and Oxidative Stress Responses in Freshwater Snail *Lanistes carinatus* Following Exposure to Chlorpyrifos. *Ecotoxicology and Environmental Safety Volume*, **116**, 137-142. http://dx.doi.org/10.1016/j.ecoenv.2015.03.010

[33] Ma, J., Zhou, C., Li, Y. and Li, X. (2014) Biochemical Responses to the Toxicity of the Biocide Abamectin on the Freshwater Snail *Physa acuta*. *Ecotoxicology and Environmental Safety*, **101**, 31-33. http://dx.doi.org/10.1016/j.ecoenv.2013.12.009

[34] Kavitha, P. and Venkateswara Rao, J. (2009) Sub-Lethal Effects of Profenofos on Tissue-Specific Antioxidative Responses in a Euryhyaline Fish, *Oreochromis mossambicus*. *Ecotoxicology and Environmental Safety*, **72**, 1727-1733. http://dx.doi.org/10.1016/j.ecoenv.2009.05.010

[35] Siegel, D., Gustafson, D.L., Dehn, D.L., Han, J.Y., Boonchoong, P., Berliner, L.J. and Ross, D. (2004) NAD(P)H: Quinone Oxidoreductase 1: Role as a Superoxide Scavenger. *Molecular Pharmacology*, **65**, 1238-1247. http://dx.doi.org/10.1124/mol.65.5.1238

6

Glycation Induced Physicochemical Changes in Low-Density Lipoprotein and Its Role in Promoting Cholesterol Accumulation in Macrophages along with Antiglycation Effect of Aminoguanidine

D. S. Jairajpuri[1*], S. Fatima[2], Z. S. Jairajpuri[3]

[1]Department of Medical Biochemistry, College of Medicine and Medical Sciences, Arabian Gulf University, Manama, Kingdom of Bahrain
[2]Department of Biochemistry, Faculty of Life Science, Aligarh Muslim University, Aligarh, India
[3]Department of Pathology, Hamdard Institute of Medical Sciences and Research, Hamdard University, New Delhi, India
Email: *jairajpurids@gmail.com, shamila_3@rediffmail.com, zeebasj@rediffmail.com

Abstract

The present study aimed at investigating physicochemical changes in modified LDL by sugars specifically fructose due to recent reports on its involvement in cardiovascular diseases and also glucose and their role in subsequent *in vitro* accumulation of cholesterol in macrophages. Antiglycation action of aminoguanidine was also investigated. LDL isolated from human blood was incubated with fructose or glucose and aminoguanidine where indicated. The physicochemical changes in modified LDL were detected by electrophoretic, spectroscopic and chemical analysis. Accumulation of cholesterol and its inhibiton in human monocyte-derived macrophages incubated with modified LDL was determined by HPLC. Results showed increased relative electrophoretic mobility, hyperchromicity at 280 nm, development of AGE fluorescence, decrease in free amino groups and increased carbonyl content in glycated LDL as compared to native LDL. Also total cholesterol accumulated in macrophages was more for glycated LDL as compared to native LDL. The magnitude of changes was more prominent in case of fructose as compared to glucose. Aminoguanidine showed remarkable restriction of glycation-induced alterations in LDL and also in accumulation of cholesterol in macrophages. The study thus proclaims that LDL-AGEs formed by fructose may contribute to accelerated initiation of diabetes induced atherosclerosis via foam cells generation and

*Corresponding author.

aminoguanidine may have therapeutic potential against it.

Keywords

Glycation, LDL-AGEs, D-Fructose, Aminoguanidine, Diabetes-Induced Atherosclerosis

1. Introduction

Non-enzymatic glycation involves a complex series of sequential reactions between reducing sugars and nucleophilic groups of various biomolecules like proteins, lipids and nucleic acids, giving rise to fluorescent and/or colored adducts known as AGEs [1]. AGEs, which are also termed as glycotoxins, are well-known triggers of excess reactive oxygen species and abnormally high oxidative stress. They have been shown to accumulate in the circulation and in various tissues with normal process of ageing [2] and also under number of pathological conditions like atherosclerosis [3], osteoarthritis, retinopathy [4] and cancer.

Majorly glycation reactions focus on those mediated by glucose (glucation), since glucose is the most abundant monosaccharide in blood and tissues and its levels are elevated remarkably in diabetes [5]. However, recent years researchers have witnessed a remarkable interest in the glycation reactions mediated by fructose (fructation) as it is far more reactive than glucose [6]. Local increases in fructose concentrations have been demonstrated in diabetic-like conditions in peripheral nerves, blood vessels and erythrocytes [7].

Lipoproteins like other proteins are susceptible to glycation-induced modifications resulting in formation of highly heterogeneous and complex lipid-AGEs [8] which affect the structural and functional attributes of lipoproteins [9]. Glycation reactions have been linked with the development of diabetes-associated cardiovascular diseases [10]. LDL is referred to as a prime target for oxidative modifications contributing to different processes that can be considered proatherogenic. This has led to the hypothesis of possible role of LDL-AGEs in increasing artherosclerotic risk of patients with diabetes and hypercholesterolemia [11]. Metabolic abnormalities associated with glycation of LDL include diminished recognition of LDL by the classic LDL receptor; increased covalent binding of LDL in vessel walls; enhanced uptake of LDL by macrophages, thus stimulating foam cell formation; increased platelet aggregation; formation of LDL-immune complexes; and generation of oxygen free radicals, resulting in oxidative damage to both the lipid and protein components of LDL [12].

With growing implications of glycation in number of pathological conditions, currently several strategies are being employed to control glycation. For the present study, AG has been chosen as the inhibitor, which is a highly reactive nucleophilic reagent that reacts with many biological molecules (pyridoxal phosphate, pyruvate, glucose etc.). AG is known to efficiently trap dicarbonyl intermediates to form substituted triazines preventing AGE formation and thus inhibiting many glycation induced chances in different biomolecules [13]. Also some studies on animals have shown positive effect of AG in retarding the process of glycation [14]. Panagiotopoulos *et al.* explored the inhibitory action of AG on accumulation of AGEs and on the development of artherosclerosis. It was observed in the study on male New Zealand white cross rabbits fed on a high cholesterol diet, that increased doses of AG (25, 50 and 100 mg/kg AG per body weight) prevented AGE accumulation and plaque formation in the aortic arch, thoracic and abdominal aorta in a concentration dependent manner [14].

Recent evidences on LDL-AGEs as a prime link for diabetes-associated cardiovascular diseases have shifted the focus of glycobiology on studies involving glycation of LDL with different compounds like glucose [15], ribose [16], glycolaldehyde [17], etc. Studies have also shown LDL-AGEs induced accelerated formation of cholesterol-laden foam cells which play a critical role in disease development [18]. Also the antagonistic role of various compounds like carnosine against LDL glycation and in turn preventing foam cell formation has been investigated [17] [19]. However, to best of our knowledge, none have been performed by using fructose and aminoguanidine.

Thus on the basis of above explanations, the present study focuses on examining the physicochemical changes due to *in vitro* treatment of LDL with glucose or fructose. The possible inhibitory effect of AG on physicochemical changes in glycated LDL and also on its *in vitro* accumulation of cholesterol in HMDM will also be investigated.

2. Materials and Methods

2.1. Materials

All the chemicals used during the study like D-glucose, D-fructose, AG, DETAPAC, TNBS, DNPH as well as salts for PBS were procured from Sigma Aldrich (St Louis MO, USA). Precast 1% agarose gels and PD10 columns were purchased from Helena Laboratories (Mt. Waverly, Vic). All chemicals were of analytical grade and all solvents of HPLC grade. Solutions were prepared using Nanopure water (Milli Q system, Millipore-Waters, Lane Cove, Australia).

2.2. *In Vitro* Glycation of LDL with Glucose and Fructose

LDL was isolated from healthy volunteers with normal basal metabolic rate and between age group of 25 - 32 years old working in HIMSR, Jamia Hamdard with normoglycaemic and normolipidaemic profiles [18]. Blood was collected after taking approval from the Jamia Hamdard Ethics and Research Committee as well as signed informed consent from the volunteers. Purified LDL (1 mg/ml) in 100 mM PBS, pH 7.2 was incubated with fructose or glucose at a final concentration of 25 mM. Where indicated 25 μM AG was included. The mixtures were sterilized by filtration through a 0.2 μm pore-size nitrocellulose filter and then incubated at 37°C for 21 days under sterile aerobic conditions. LDL incubated with 25 μM DETAPAC under the same conditions served as control. The added reagents (DETAPAC, fructose, glucose, AG) were removed post incubation by PD10 chromatography.

2.3. Isolation and Culturing of HMDM

Monocytes were isolated from white cell concentrates by counter-current elutriation and then incubated (1×10^5/ml) in 12-well plates (Costar, Corning, USA) in serum-free RPMI-1640 for 2 hours. The plates were then washed and adhering cells were incubated (5% CO_2 and 37°C) in RPMI for 10 days to give HMDM [20].

2.4. Electrophoretic Analysis

Changes in overall LDL particle charge in control as well as treated samples were quantified by electrophoretic mobility on agarose gels [18].

2.5. Absorbance Measurements

The UV absorption measurements of control LDL and treated samples were obtained by measuring the ultraviolet absorption profile between the wavelength ranges of 200 - 400 nm on a Shimadzu spectrophotometer using a cuvette of 1.0 cm pathlength. Three hundred micrograms of samples in a total volume of 1.0 ml was taken for the analysis.

Percent increase in UV absorbance was calculated to be

$$= \left[\frac{\text{Absorbance}(\text{glycated} - \text{LDL}) - \text{Absorbance}(\text{control} - \text{LDL})}{\text{Absorbance}(\text{glycated} - \text{LDL})}\right] \times 100 \tag{1}$$

2.6. Fluorescence Measurements

Similarly control sample, sugar incubated samples with or without AG were studied by measuring fluorescence at 25°C ± 0.2°C on a Hitachi F2000 spectrofluorometer (Tokyo, Japan). The samples were excited at 370 nm excitation and an emission was recorded in the range of 350 - 500 nm. The development of AGE fluorescence of treated samples was recorded with emission at 450 nm.

Percent AGE fluorescence was calculated to be

$$= \left[\frac{\text{AGE Flourescence}(\text{glycated} - \text{LDL}) - \text{AGE Flourescence}(\text{control} - \text{LDL})}{\text{AGE Flourescence}(\text{glycated} - \text{LDL})}\right] \times 100 \tag{2}$$

2.7. Amino Group Estimation

The amino groups of control and treated samples were determined by the method of Habeeb and Hiramoto [21] with slight modification. Suitable aliquots of sample were dissolved in 1.0 ml of 0.1 M sodium tetraborate buffer, pH 9.3, 25 µl of 0.3 M TNBS added and the tubes agitated instantly to ensure complete mixing and allowed to stand for 30 minutes at room temperature. Absorbance of the yellow colour developed was recorded at 420 nm against a reagent blank. Glycine was used as a standard amino acid. The results were expressed as total number of amino groups per molecule of LDL.

2.8. Determination of Protein Bound Carbonyl Groups

Protein-bound carbonyl groups were estimated according to the protocol of Levine *et al.* by using DNPH. The samples were read at 379 nm and the results were expressed as the number of nanomoles of carbonyl per mg of sample LDL using a $\varepsilon_{379\ nm} = 22{,}000\ M^{-1}\cdot cm^{-1}$ [22].

2.9. Cellular Cholesterol-Loading Studies

HMDM were exposed to control as well as treated LDL samples (150 µg/ml) in media containing 10% lipoprotein-deficient serum. After 72 hours, cells were washed with 20 mM PBS and lysed in water. Total cholesterol concentration was quantified by HPLC [23].

2.10. Data Analysis

Data analysis was performed using one-way analysis of variance using the Newman-Keuls multiple comparison test. $P \leq 0.05$ was taken as significant.

3. Results and Discussion

In order to compare the physicochemical changes between native LDL and modified LDL (incubated in the absence as well as presence of AG) electrophoretic, spectroscopic and chemical analysis was carried out.

3.1. Electrophoretic Analysis

Incubation of LDL with 25 mM glucose or fructose resulted in a significant increase in REM and thus indicative of an overall decrease in positive charge (**Table 1**). This decrease in charge is more prominent in case of fructose treated sample as compared to one treated with glucose. On the other hand, presence of 25 µM AG decreased particle modification as evident by low REM for samples with sugar + AG in comparison to those with sugar only (**Table 1**).

The greater shift in REM observed with sugar treated LDL compared to samples with AG represents the formation of an adduct that contributes to charge modification which is more in case of fructose than in glucose. This is expected due to high reactivity of fructose [24]. On the other hand, charge modification for samples with AG was fairly less than samples with only sugars, thus highlighting the ability of AG to prevent formation of

Table 1. Change in LDL particle charge. Overall LDL particle charge was determined by REM for sugar- modified LDL both in the absence or presence of AG and that incubated with DETAPAC (control).

SAMPLES	REM
Control (LDL + DETAPAC)	1.3 + 0.2
LDL + Glucose	3.6 + 0.4[#]
LDL + Fructose	4.2 + 0.4[#]
LDL + Glucose + AG	2.1 + 0.2[#]
LDL + Fructose + AG	2.7 + 0.4[#]

Results are means + SEM from 3 - 4 separate experiments each with 3 samples per treatment. [#]P < 0.001 compared with control.

AGEs significantly [25]. AG has been reported to react with sugars and aldehydes and detoxify aldehyde-/sugar-modified proteins [26].

3.2. Absorbance and Fluorescence Studies

The possible sugar-induced structural changes in the LDL were examined with respect to changes in the chromophoric and fluorophoric properties. In both the parameters, *i.e.* UV absorbance and fluorescence, an increase was observed in the sugar-treated samples as compared to the control sample. However, the changes in the spectroscopic properties of the samples incubated with sugars along with AG were significantly less as compared to the sample treated with sugars alone.

3.2.1. UV Absorbance Measurements

The magnitude of hyperchromicity was more in case of fructose (75% + 3) as compared to glucose (65% + 3). The presence of AG in LDL samples showed small increase in the absorbance, as a mere 15% + 2 hyperchromicity in case glucose + AG and 22% + 3 hyperchromicity in case of fructose + AG samples (**Figure 1**). Percent hyperchromicity seen in the UV region can be due to sugar induced unfolding of the LDL molecules resulting in exposure of the chromophoric aromatic residues [27]. The large variations observed in the hyperchromicity among fructose treated and glucose treated LDL are suggestive of major differences in conformational alterations resulting from sugar-mediated reactions. AG was highly effective in restricting sugar-induced alterations as evident by marked decrease in the sugar-mediated hyperchromicity in samples were it was included. AG is

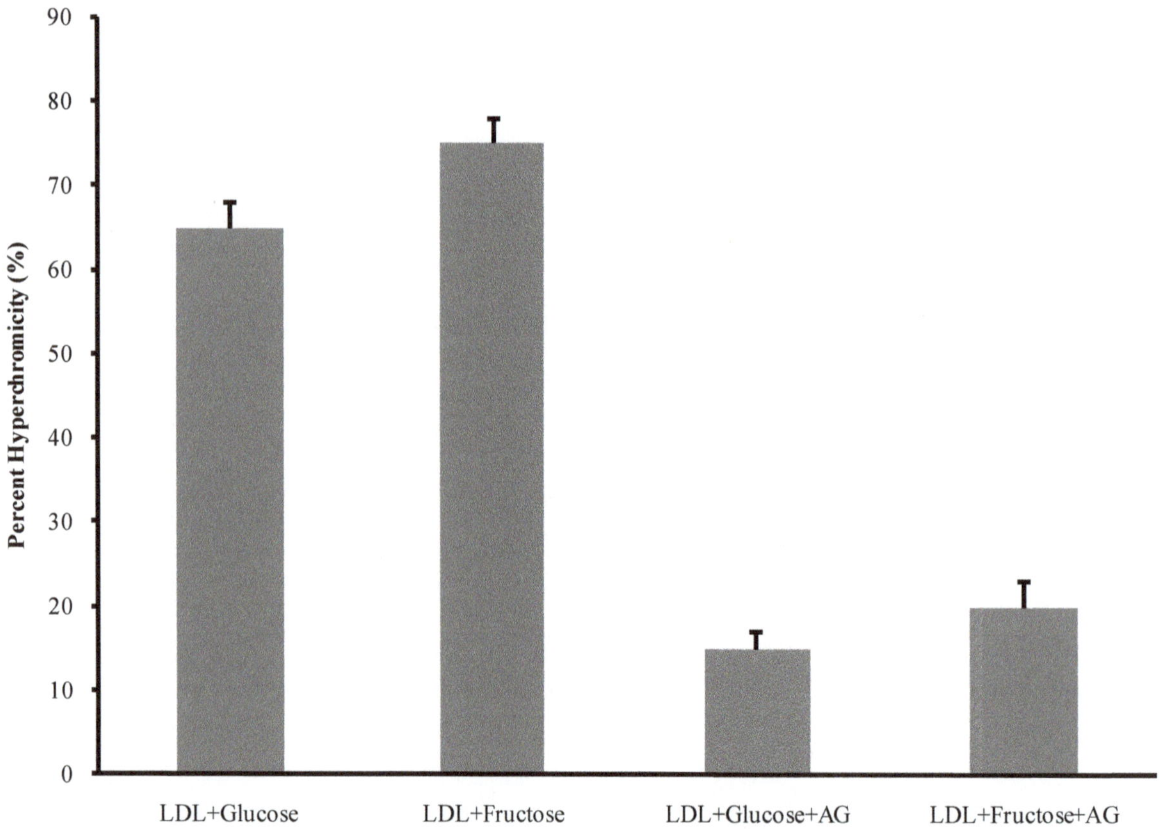

Figure 1. Percent Hyperchromicity. Absorbance in UV region was measured for sugar-modified LDL both in absence or presence of AG and that incubated with DETAPAC (control). LDL (1 mg/ml) was incubated with glucose/fructose (25 mM) either in the absence or in the presence of aminoguanidine (25 μM) and that incubated DETAPAC was taken as control. Incubation was for 21 days at 37°C. Percent hyperchromicity was calculated according to the formula stated in the materials and methods section (Equation (1)). Results are expressed as the means + SEM from 3 - 4 separate experiments each with 3 samples per treatment. $P < 0.05$ for glucose + AG and fructose + AG are compared with glucose treated and fructose treated samples respectively.

considered as a putative antiglycation agent resulting in prevention of formation of AGEs by reacting rapidly with α, β-dicarbonyl compounds [28] [29].

3.2.2. Fluorescence Measurements

Development of AGE fluorescence (370 nm excitation/450 nm emission) is a well established measure of protein glycation/AGE formation [16]. Similar to UV absorbance, the magnitude of new fluorescence developed was more in case of fructose (83% + 3) as compared to glucose (70% + 2). Also as expected, AG was highly effective in restricting sugar-induced alterations as samples showed negligible development of new fluorescence. Percent new fluorescence was 5% + 1 only for both glucose and fructose treated LDL samples in the presence of AG (**Figure 2**). Development of AGE fluorescence is attributed to formation of heterogeneous cyclic structures during the process of glycation which make adducts fluoresce [30]. The fluorophores relative contribution to AGE fluorescence can vary widely depending on the site and extent of glycation [27]. This is why the percent development of AGE fluorescence is seen exponentially high in case of fructose as compared to glucose [31]. Modified LDL with AG, however, showed only negligible development of AGE fluorescence thus confirming antiglycation effect of AG [25].

3.3. Chemical Modifications

The extent of glycation in the treated samples was also determined chemically by estimation of free amino groups and carbonyl group content. In general the results obtained were complementary to the observations in

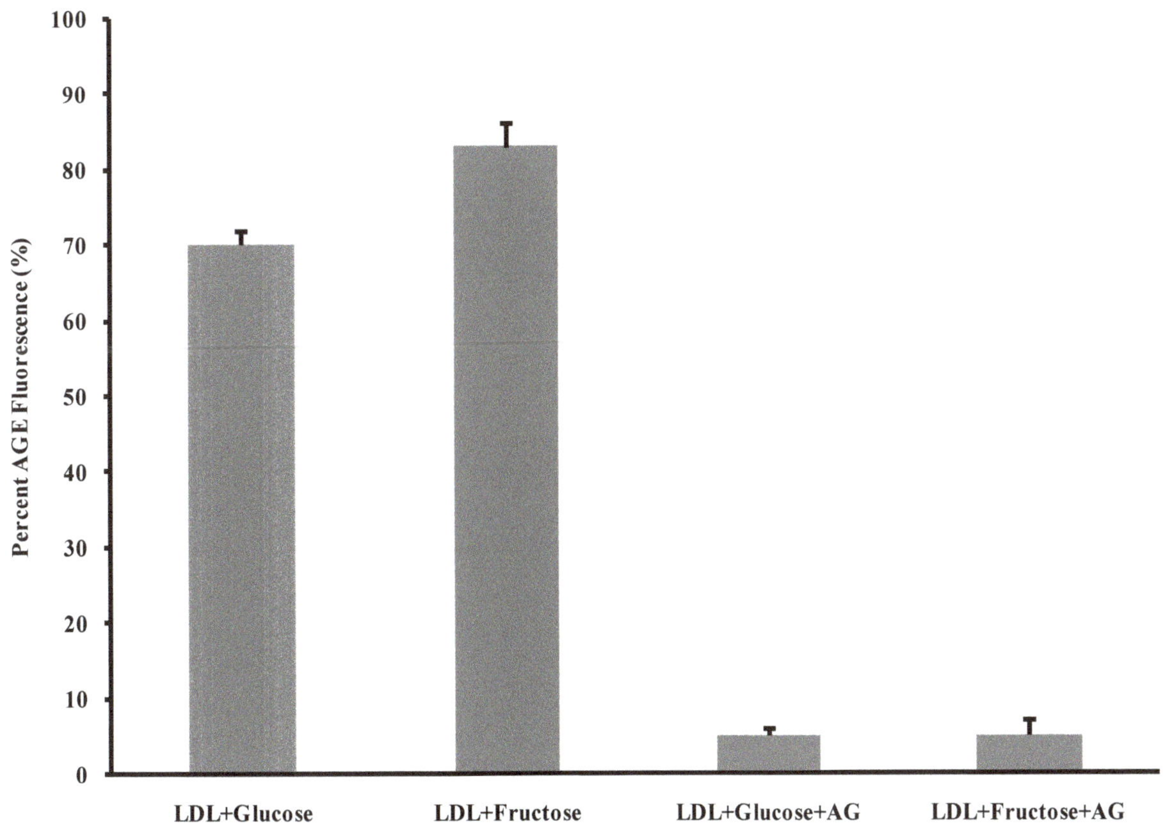

Figure 2. Percent AGE Fluorescence. Development of fluorescence was measured for sugar-modified LDL both in absence or presence of AG and that incubated with DETAPAC (control). LDL (1 mg/ml) was incubated with glucose/fructose (25 mM) either in the absence or in the presence of aminoguanidine (25 μM) and that incubated DETAPAC was taken as control. Incubation was for 21 days at 37˚C. Percent AGE fluorescence was calculated according to the formula stated in the materials and methods section (Equation (2)). Results are expressed as the means + SEM from 3 - 4 separate experiments each with 3 samples per treatment. $P < 0.05$ for glucose + AG and fructose + AG are compared with glucose treated and fructose treated samples respectively.

spectroscopic studies and electrophoretic analysis confirming sugar induced LDL glycation in the treated samples. As compared to the control sample of LDL, a decrease in the free amino groups and an increase in the carbonyl content were seen in sugar treated samples. However, marked restriction of chemical modifications of LDL was seen in samples which had AG along with sugars in them.

3.3.1. Determination of Free Amino Groups

Figure 3 shows the amino groups of native LDL and those in treated samples accessible to reaction with TNBS under the conditions used. Control LDL revealed about 40 + 2 free amino groups per molecule of LDL whereas a steep decrease in the free amino groups was seen in LDL treated with glucose (26 + 1 per molecule of LDL) and fructose (18 + 2 per molecule of LDL). Samples with AG showed less reduction in free amino groups as 35 + 2 and 32 + 2 amino groups per molecule of LDL in glucose + AG and fructose + AG samples respectively. Decrease in free amino groups in modified LDL molecules observed is not unexpected since glycation begins with the reaction of sugars mainly with the free amino groups present of various biomolecules and supports the earlier observation of Lapolla *et al.* [32], that exposed amino groups may play an important role in determining the susceptibility to glycation by the sugars.

3.3.2. Determination of Carbonyl Content

Sugar-treated LDL samples were further analyzed for their carbonyl content by reaction with DNPH. Results are exhibited in **Figure 4**. As evident, fructose caused exponential increase in the formation of carbonyl groups in the LDL molecule as compared to glucose. The carbonyl content was 30.33 + 1.01 nmol·mg^{-1} LDL and 22.14 + 1.22 nmol·mg^{-1} LDL for fructose and glucose respectively. Carbonyl group content is a consequence of glycation,

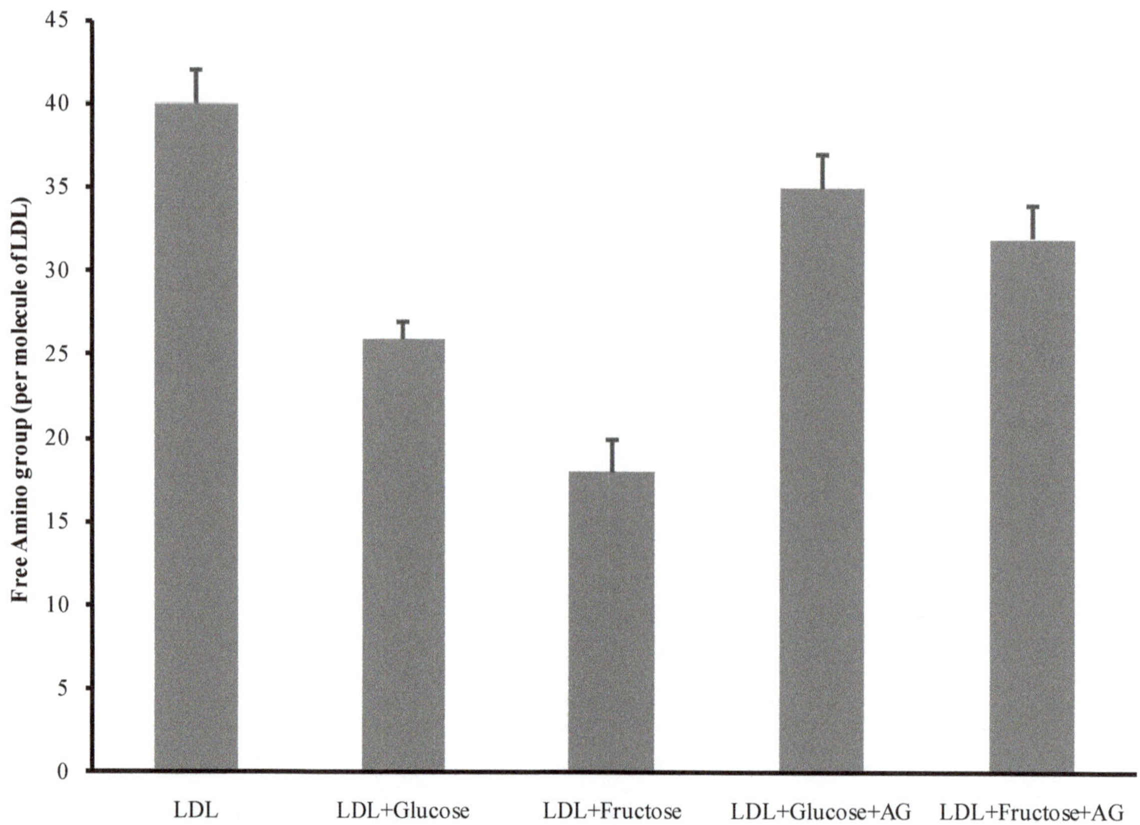

Figure 3. Free Amino Group Content. Free amino groups were measured for sugar-modified LDL both in absence or presence of AG and that incubated with DETAPAC (control). LDL (1 mg/ml) was incubated with glucose/fructose (25 mM) either in the absence or in the presence of aminoguanidine (25 μM) and that incubated DETAPAC was taken as control. Incubation was for 21 days at 37°C. Results are expressed as the means + SEM from 3 - 4 separate experiments each with 3 samples per treatment. $P < 0.05$ for glucose + AG and fructose + AG are compared with glucose treated and fructose treated samples respectively.

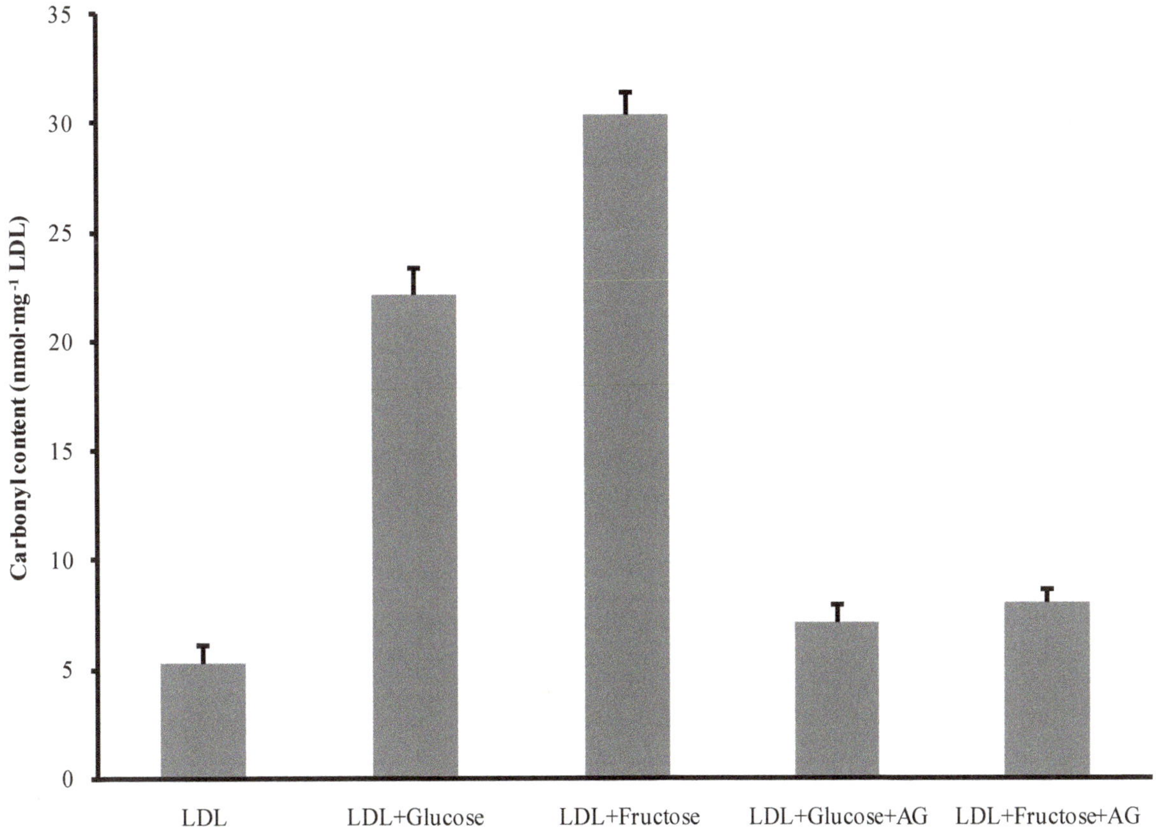

Figure 4. Total Carbonyl Content. Total carbonyl content was measured for sugar-modified LDL both in absence or presence of AG and that incubated with DETAPAC (control). LDL (1 mg/ml) was incubated with glucose/fructose (25 mM) either in the absence or in the presence of aminoguanidine (25 µM) and that incubated DETAPAC was taken as control. Incubation was for 21 days at 37˚C. Results are expressed as the means + SEM from 3 - 4 separate experiments each with 3 samples per treatment. $P < 0.05$ for glucose + AG and fructose + AG are compared with glucose treated and fructose treated samples respectively.

thus LDL incubated with DETAPAC alone show negligible carbonyl group content of 5.24 + 0.84 $nmol·mg^{-1}$ LDL. LDL incubated with respective sugars in the presence of AG showed generation of very small amount of carbonyl content as 7.12 + 0.77 $nmol·mg^{-1}$ LDL in case of glucose and 8.01 + 0.54 $nmol·mg^{-1}$ LDL in case of fructose. DNPH is a standard method for detecting carbonyl groups and can be used to follow changes on glycation of lipoproteins [16]. The carbonyls detected under the conditions are clearly the result of sugar-induced glycation [2], since there is little DNPH-reactivity in the LDL incubated with DETAPAC (control). It is well recognized that carbonyl groups are incorporated in proteins as a consequence of glycation and protein carbonyl content is considered as a reliable measure of glycation [33]. Also low levels of carbonyl group content in AG treated samples are again in agreement with previous studies which have identified it as forming discrete adducts with carbonyl compounds during glycation, confirming its role as true scavenger of carbonyl compounds [13] [26].

3.4. Total Cholesterol Content in HMDM

Incubation of HMDM with sugar treated LDL (150 µg) for 48 hours resulted in a significant accumulation of cellular cholesterol as evident in **Figure 5**. In accordance to earlier observations fructose treated LDL resulted in far more cholesterol accumulation (220 + 4 nmoles/mg cell protein) as compared to glucose treated LDL (170 + 3 nmoles/mg cell protein). However, the samples with AG during the LDL modification phase showed significantly low levels of cholesterol accumulation by HMDM as the amount of total cholesterol obtained in case of glucose + AG (110 + 3 nmoles/mg cell protein) and fructose + AG (116 + 4 nmoles/mg cell protein) was not significantly different from that of control (95 + 3 nmoles/mg cell protein). The total cholesterol content obtained post incubation of HMDM with sugar-modified LDL strongly embarked the implications of LDL glycation

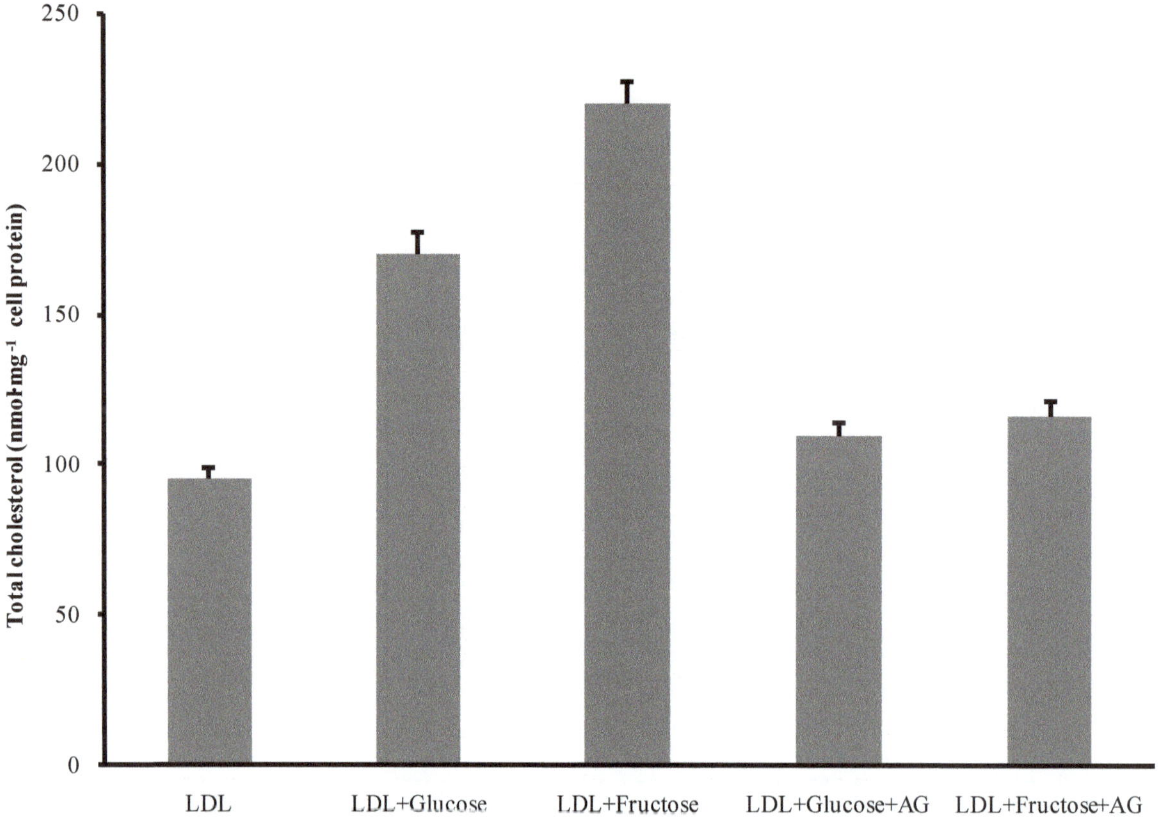

Figure 5. Total Cholesterol Concentration Present in HMDM. Total cholesterol content in HMDM was measured after incubation with sugar-modified LDL both in absence or presence of AG and that incubated with DETAPAC (control). LDL (1 mg/ml) was incubated with glucose/fructose (25 mM) either in the absence or in the presence of aminoguanidine (25 μM) and that incubated DETAPAC was taken as control. Incubation was for 21 days at 37°C. HMDM were exposed to control as well as treated LDL samples (150 μg/ml) for 72 hours and then total cholesterol content was quantified by HPLC. Results are expressed as the means + SEM from 3 - 4 separate experiments each with 3 samples per treatment. $P < 0.05$ for glucose + AG and fructose + AG are compared with glucose treated and fructose treated samples respectively.

on macrophagial accumulation of cholesterol, in turn resulting in the formation of foam cells; a prominent hallmark of atherosclerosis [34].

The accumulation observed more in case of fructose is due to modification of LDL to an exaggerated extent owing to high reactivity of fructose [35]. However, LDL-modified by both the sugars in the presence of AG were able to achieve almost complete inhibition of lipid loading even when complete prevention of LDL modification was not achieved, as evident in physicochemical analysis. This indicated that all that was required to prevent lipid loading was reduction of damage below a key threshold related to modification of key residues on apolipoprotein B-100 [17]. This has potential therapeutic significance.

It is now recognized, that *in vitro* glycation of LDL can induce intracellular cholesterol accumulation in macrophages resulting in foam cell formation which is a characteristic of atherosclerosis [23]. Fructose has been taken as the prime sugar for the study due to numerous reports on its increased consumption in the form of high fructose corn syrup in processed foods [36] and also it has been recently implicated as a risk factor for cardiovascular diseases [37]. Moreover during diabetic conditions excessive blood glucose levels automatically result in high levels of fructose in tissues/organs were polyol pathway is active [38]. Fructose is known to be almost 300 times more reactive as compared to glucose due to presence of greater number of reactive open chain structures [39], thus causing far more AGE formation and oxidative stress than glucose [6].

AG is a well established therapeutic agent for the prevention of glycation process which is a characteristic of normal aging as well as different pathological conditions like diabetes, atherosclerosis [28]. It reacts rapidly with different sugars and dicarbonyl compounds to prevent the formation of AGEs. Adducts formed are substituted 3-amino-1, 2, 4-triazine derivatives [13]. Inhibition of disease mechanisms, especiallly vascular complications in experimental diabetes, by AG has provided evidence that accumulation of AGEs is a risk factor for disease progression and AG is effective in preventing it [28] [29] and is currently undergoing clinical trials to be used as a therapeutic drug for AGE associated pathologies.

4. Conclusion

The study gives strong evidence that fructose-induced alterations in LDL are far more prominent than that by glucose. This is of high significance owing to numerous reports about the role of fructose in lipid metabolism and also in cardiovascular disorders [37] [39]. However, the key finding from the data obtained is that aminoguanidine is effective at very low concentrations to combat glycation even with a sugar as potentially reactive as fructose. Thus, aminoguanidine may serve as effective scavengers of glycation agents hence of protein glycation and as potential therapeutic agents to inhibit diabetes-induced atherosclerosis.

Conflict of Interest

There is no conflict of interest that can be perceived as prejudicing the impartiality of the present research study.

Acknowledgements

The authors are grateful to the Aligarh Muslim University (Aligarh, India) and Hamdard Institute of Medical Sciences and Research, Hamdard University (New Delhi, India) for the research facilities. Acknowledgment is also extended to College of Medicine and Medical Sciences, Arabian Gulf University (Manama, Bahrain) for other facilities.

References

[1] Cerami, A. and Ulrich, P. (2001) Pharmaceutical Intervention of Advanced Glycation Endproducts. *Novartis Foundation Symposia*, **235**, 202-212.

[2] Anguizola, J., Matsuda, R., Barnaby, O.S., Hoy, K.S., Wa, C.L., DeBolt, E., Koke, M. and Hage, D.S. (2013) Review: Glycation of Human Serum Albumin. *Clinica Chimica Acta*, **425**, 64-76. http://dx.doi.org/10.1016/j.cca.2013.07.013

[3] Yokoyama, H., Araki, S., Haneda, M., Matsushima, M., Kawai, K., Hirao, K., Oishi, M., Sugimoto, K., Sone, H., Maegawa, H. and Kashiwagi, A. (2012) Chronic Kidney Disease Categories and Renal-Cardiovascular Outcomes in Type 2 Diabetes without Prevalent Cardiovascular Disease: A Prospective Cohort Study (JDDM25). *Diabetologia*, **55**, 1911-1918. http://dx.doi.org/10.1007/s00125-012-2536-y

[4] Yamagishi, S., Nakamura, K. and Matsui, T. (2006) Advanced Glycation End Products (AGEs) and Their Receptor (RAGE) System in Diabetic Retinopathy. *Current Drug Discovery Technologies*, **3**, 83-88. http://dx.doi.org/10.2174/157016306776637555

[5] Ghosh-Moulick, R., Bhattacharya, J., Roy, S., Basak, S. and Dasgupta, A.K. (2007) Compensatory Secondary Structure Alterations in Protein Glycation. *Biochimica et Biophysica Acta (BBA)—Proteins and Proteomics*, **1774**, 233-242. http://dx.doi.org/10.1016/j.bbapap.2006.11.018

[6] Sakai, M., Oumomi, M. and Kasuga, M. (2002) Experimental Studies on the Role of Fructose in the Development of Diabetic Complications. *Kobe Journal of Medical Sciences*, **48**, 125-136.

[7] Poulsom, R., Boot-Handford, R.P. and Heath, H. (1983) The Effects of Long-Term Treatment of Streptozotocin-Diabetic Rats with an Aldose Reductase Inhibitor. *Experimental Eye Research*, **37**, 507-515. http://dx.doi.org/10.1016/0014-4835(83)90027-1

[8] Gasser, A. and Forbes, J.M. (2008) Advanced Glycation: Implications in Tissue Damage and Disease. *Protein and Peptide Letters*, **15**, 385-391. http://dx.doi.org/10.2174/092986608784246515

[9] Stitt, A.W. (2003) The Role of Advanced Glycation in the Pathogenesis of Diabetic Retinopathy. *Experimental and Molecular Pathology*, **75**, 95-108. http://dx.doi.org/10.1016/S0014-4800(03)00035-2

[10] Rashid, I., Brown, B.E., van Reyk, D.M. and Davies, M.J. (2006) The Roles of Glycation, Glycoxidation and Advanced Glycation End-Product Formation in Diabetes-Induced Atherosclerosis. In: Cheema, S.K., Ed., *Biochemistry of Artherosclerosis*, Springer Verlag, New York, 247-283. http://dx.doi.org/10.1007/0-387-36279-3_12

[11] Zimmermann, R., Panzenböck, U., Wintersperger, A., Levak-Frank, S., Graier, W., Glatter, O., Fritz, G., Kostner, G.M. and Zechner, R. (2001) Lipoprotein Lipase Mediates the Uptake of Glycated LDL in Fibroblasts, Endothelial Cells, and Macrophages. *Diabetes*, **50**, 1643-1653. http://dx.doi.org/10.2337/diabetes.50.7.1643

[12] Lyons, T.J. (1993) Glycation and Oxidation: A Role in the Pathogenesis of Atherosclerosis. *American Journal of Car-*

diology, **71**, 26-31. http://dx.doi.org/10.1016/0002-9149(93)90142-Y

[13] Majd, A.A., Goodarzi, M.T., Hassanzadeh, T., Tavilani, H. and Karimi, J. (2014) Aminoguanidine Partially Prevents the Reduction in Liver Pyruvate Kinase Activity in Diabetic Rats. *Advances in Biomedical Research*, **3**, 260-265. http://dx.doi.org/10.4103/2277-9175.148233

[14] Panagiotopoulos, S., Brien, R.C.O., Bucala, R., Cooper, M.E. and Jerums, G. (1998) Aminoguanidine Has an Anti-Atherogenic Effect in the Cholesterol-Fed Rabbit. *Atherosclerosis*, **136**, 125-131. http://dx.doi.org/10.1016/S0021-9150(97)00192-5

[15] Rabbani, N., Chittari, M.V., Bodmer, C.W., Zehnder, D., Ceriello, A. and Thornalley, P.J. (2010) Increased Glycation and Oxidative Damage to Apolipoprotein B100 of LDL Cholesterol in Patients with Type 2 Diabetes and Effect of Metformin. *Diabetes*, **59**, 1038-1045. http://dx.doi.org/10.2337/db09-1455

[16] Ahmad, S., Akhter, F., Moinuddin, Shahab, U. and Khan, M.S. (2013) Studies on Glycation of Human Low Density Lipoprotein: A Functional Insight into Physic-Chemical Analysis. *International Journal of Biological Biomolecules*, **62**, 167-171.

[17] Rashid, I., van Reyk, D.M. and Davies, M.J. (2007) Carnosine and Its Constituents Inhibit Glycation of Low-Density Lipoproteins That Promotes Foam Cell Formation *in Vitro*. *FEBS Letters*, **581**, 1067-1070. http://dx.doi.org/10.1016/j.febslet.2007.01.082

[18] Brown, B.E., Dean, R.T. and Davies, M.J. (2005) Glycation of Low-Density Lipoproteins by Methylglyoxal and Glycoaldehyde Gives Rise to the *in Vitro* Formation of Lipid-Laden Cells. *Diabetologia*, **48**, 361-369. http://dx.doi.org/10.1007/s00125-004-1648-4

[19] Younis, N.N., Soran, H., Pemberton, P., Charlton-Menys, V., Elseweidy, M.M. and Durrington, P.N. (2013) Small Dense LDL Is More Susceptible to Glycation than More Buoyant LDL in Type 2 Diabetes. *Clinical Science* (*London*), **124**, 343-349. http://dx.doi.org/10.1042/CS20120304

[20] Garner, B., Dean, R.T. and Jessup, W. (1994) Human Macrophage-Mediated Oxidation of Low-Density Lipoprotein Is Delayed and Independent of Superoxide Production. *Journal of Biochemistry*, **301**, 421-428.

[21] Habeeb, A.F.S.A. and Hiramoto, R. (1968) Reaction of Proteins with Glutaraldehyde. *Archives of Biochemistry and Biophysics*, **126**, 16-26. http://dx.doi.org/10.1016/0003-9861(68)90554-7

[22] Levine, R.L., Williams, J., Stadtman, E.R. and Shacter, E. (1994) Carbonyl Assays for Determination of Oxidatively Modified Proteins. *Methods in Enzymology*, **233**, 346-357. http://dx.doi.org/10.1016/S0076-6879(94)33040-9

[23] Brown, B.E., Mahroof, F.M., Cook, N.L., van Reyk, D.M. and Davies, M.J. (2006) Hydrazine Compounds Inhibit Glycation of Low-Density Lipoproteins and Prevent the *in Vitro* Formation of Model Foam Cells from Glycolaldehyde-Modified Low-Density Lipoproteins. *Diabetologia*, **49**, 775-783. http://dx.doi.org/10.1007/s00125-006-0137-3

[24] Jairajpuri, D.S., Fatima, S. and Saleemuddin, M. (2008) Complexing of Glucose Oxidase with Anti-Glucose Oxidase Antibodies or the F(ab)'$_2$/F(ab)' Fragments Derived There from Protects Both the Enzyme and Antibody/Antibody Fragments. *Biochemistry*, **73**, 1235-1241. http://dx.doi.org/10.1134/S0006297908110102

[25] Labieniec-Watala, M., Siewiera, K. and Jozwiak, Z. (2011) Resorcylidene Aminoguanidine (RAG) Improves Cardiac Mitochondrial Bioenergetics Impaired by Hyperglycaemia in a Model of Experimental Diabetes. *International Journal of Molecular Sciences*, **12**, 8013-8026. http://dx.doi.org/10.3390/ijms12118013

[26] Price, D.L., Rhett, P.M., Thorpe, S.R. and Baynes, J.W. (2001) Chelating Activity of Advanced Glycation End-Product Inhibitors. *Journal of Biological Chemistry*, **276**, 48967-48972. http://dx.doi.org/10.1074/jbc.M108196200

[27] Bachetti, T., Masciangelo, S., Armeni, T., Bicchiega, V. and Ferretti, G. (2014) Glycation of High Density Lipoprotein by Methylglyoxal: Effect on HDL-Paraoxonase Activity. *Metabolism*, **63**, 307-311. http://dx.doi.org/10.1016/j.metabol.2013.10.013

[28] Koh, G., Lee, D. and Woo, J. (2010) 2-Deoxy-D-Ribose Induces Cellular Damage by Increasing Oxidative Stress and Protein Glycation in a Pancreatic Beta-Cell Line. *Metabolism—Clinical and Experimental*, **59**, 325-332. http://dx.doi.org/10.1016/j.metabol.2009.07.028

[29] Thornalley, P.J. (2003) Use of Aminoguanidine (Pimagedine) to Prevent the Formation of Advanced Glycation Endproducts. *Archives of Biochemistry and Biophysics*, **419**, 31-40. http://dx.doi.org/10.1016/j.abb.2003.08.013

[30] Ansari, N.A., Moinuddin, A.K. and Ali, A. (2009) Preferential Recognition of Amadori-Rich Lysine Residues by Serum Antibodies in Diabetes Mellitus: Role of Protein Glycation in the Disease Process. *Human Immunology*, **70**, 417-424. http://dx.doi.org/10.1016/j.humimm.2009.03.015

[31] Jakus, V. and Rietbrock, N. (2004) Advanced Glycation End-Products and the Progress of Diabetic Vascular Complications. *Physiological Research*, **53**, 131-142.

[32] Lapolla, A., Molin, L. and Traldi, P. (2013) Protein Glycation in Diabetes as Determined by Mass Spectrometry. *International Journal of Endocrinology*, **2013**, 1-11. http://dx.doi.org/10.1155/2013/412103

[33] Beal, M.F. (2002) Oxidatively Modified Proteins in Aging and Disease. *Free Radical Biology and Medicine*, **32**, 797-803. http://dx.doi.org/10.1016/S0891-5849(02)00780-3

[34] Li, A.C. and Glass, C.K. (2002) The Macrophage Foam Cell as a Target for Therapeutic Intervention. *Nature Medicine*, **8**, 1235-1242. http://dx.doi.org/10.1038/nm1102-1235

[35] Zhang, Y.H., An, T., Zhang, R.C., Zhou, Q., Huang, Y. and Zhang, J. (2013) Very High Fructose Intake Increases Serum LDL-Cholesterol and Total Cholesterol: A Meta-Analysis of Controlled Feeding Trials. *Journal of Nutrition*, **143**, 1391-1398. http://dx.doi.org/10.3945/jn.113.175323

[36] Bray, G.A., Nielsen, S.J. and Popkin, B.M. (2004) Consumption of High-Fructose Corn Syrup in Beverages May Play a Role in the Epidemic of Obesity. *American Journal of Clinical Nutrition*, **79**, 537-543.

[37] Aeberli, I., Zimmermann, M.B., Molinari, L., Lehmann, R., L'Allemand, D., Spinas, G.A. and Berneis, K. (2007) Fructose Intake Is a Predictor of LDL Particle Size in Overweight Schoolchildren. *American Journal of Clinical Nutrition*, **86**, 1174-1178.

[38] Miyazawa, N., Kawasaki, Y. and Fuji, J. (1998) Immunological Detection of Fructated Proteins *in Vivo* and *in Vitro*. *Biochemistry Journal*, **336**, 101-107.

[39] Bantle, J.P., Raatz, S.K., Thomas, W. and Georgopoulos, A. (2000) Effects of Dietary Fructose on Plasma Lipids in Healthy Subjects. *American Journal of Clinical Nutrition*, **72**, 1128-1134.

Abbreviations

AGEs: Advanced glycated end-products;
AG: Aminoguanidine;
DETAPAC: Diethyleneaminepenta acetic acid;
DNPH: 2, 4-dinitrophenylhydrazine;
HMDM: Human monocyte-derived macrophages;
LDL: Low density lipoprotein;
LDL-AGEs: low density lipoprotein advanced glycation end products;
PBS: Phosphate buffer saline;
REM: relative electrophoretic mobility;
RPMI: Roswell Park Memorial Institute;
TNBS: Trinitrobenzene sulphonic acid.

7

Interaction of Hemoglobin with Binuclearcationic Tetranitrosyl Iron Complex with Penicillamine. Cations Binding Sites

Lidia Syrtsova, Natalia Sanina, Boris Psikha, Ildar Tukhvatullin, Natal'ja Shkondina, Olesia Pokidova, Alexander Kotelnikov

Institute of Problems of Chemical Physics of the Russian Academy of Sciences, Moscow Region, Russian Federation
Email: syrtsova@icp.ac.ru

Abstract

In this paper, the kinetics of the interaction of the nitrosyl iron complex with the ligands penicillamine $[Fe_2(SC_5H_{11}NO_2)_2(NO)_4]SO_4 \cdot 5H_2O$ (*I*) with deoxyhemoglobin (Hb) was studied. The kinetic modeling method defined the number of binding (*I*) molecules and equilibrium constant of the coupling reaction of (*I*) with Hb (K_s). At equimolar concentrations of (*I*) and Hb (2×10^{-5} M), the Hb molecule binds only one (*I*) with K_s equal to 4.3×10^7 M^{-1}. When increasing the (*I*) concentration, the number of binding sites of Hb increases and K_s decreases. These results are analyzed in accordance with the data on the existence of cations binding sites in Hb.

Keywords

Biomedicine, Iron-Sulfur Cluster, Ligand Binding, Heme, Nitric Oxide

1. Introduction

In the last ten years, it has been well established that nitric oxide, NO, having a wide spectrum of biological activities and the ability to affect various body systems [1] [2], is actively involved in the process of carcinogenesis [3]. Depending on the chemical characteristics and the local concentration *in vivo*, NO may act on different biotargets to stimulate the tumor generation process or *vice versa*, exercising its inhibition [4]-[6]. In this regard, the development of experimental approaches to treatment of neoplastic diseases that are based on NO-therapy requires fundamental study correlations, investigating the "Structure-Activity" of exogenous compounds generating NO in physiological solutions, in order to establish the molecular and genetic mechanisms of their action on target mammalian cells and for the synthesis of compounds with improved properties (low toxicity, greater bioavailability, etc.). Biomimetics of nitrosyl cell intermediates, in particular, synthetic models of the active centers of nitrosyl iron-sulfur proteins, are promising compounds for the treatment of neoplastic diseases [7]. We have previously shown, for the first time, that the anionic and neutral nitrosyl iron complexes (NICs), in the absence of additional activation, decompose with release of NO in protic media containing Hb [8] [9], and that the reaction rate constants depend on the molecular structure of the complexes. It was found that the NICs nitrosylated Hb, interacting with the heme at the 6th free coordination site. The resulting complex, HbNO, is a depot of NO. It not only provides a storage form of NO (since the lifetime of free NO in the cell is seconds), but also determines the prolonged action of the NICs as donors of NO. It was also found that NIC nitrosylated not only Hb but also ferri- and ferrocytochrome [10]. It is known [11] that there is a center in Hb for binding of anions formed by positively charged functional amino acid residues that can bind, for example, diphosphoglycerate (DPG). As a result of allosteric binding, DPG affects Hb affinity for O_2. Recently, we have identified the binding sites of cations in Hb [12]. It was found that a cationic NIC (**Figure 1**) with cysteamine ligands ***(II)*** $[Fe_2(S(CH_2)_2NH_3)_2(NO)_4]SO_4{\cdot}2.5H_2O$ (CCDC 663194), was a promising anticancer NO-donor agent, apoptosis inducer in human tumor cells [13], bound with two negatively charged cavities on the Hb surface. These issues are important for the metabolism of cationic NICs and also for investigating the first discovered sites of cation binding in such a physiologically relevant protein as Hb. So, we determined the goal of this work to be the investigation of the binding of Hb with other cationic complex, ***(I)*** (**Figure 1**) with larger size than ***(II)***, in order to reliably verify the existence of cations binding sites in Hb. According to the X-ray diffraction data, the complex ***(I)*** contains two protonated NH_3-groups in the penicillamine ligands: $[Fe_2(SC_5H_{11}NO_2)_2(NO)_4]SO_4{\cdot}5H_2O$ (CCDC 680286).

2. Experimental

2.1. Materials

We used bovine Hb (Serva, Germany), $Na_2HPO_4{\cdot}6H_2O$ and $NaH_2PO_4{\cdot}H_2O$ (MP Biomedicals, Germany). The water was purified by distillation in a Bi/Duplex distiller (Germany). ***(II)*** was synthesized using the known method [14]. Synthesis of ***(I)*** is described in [15]. It has been obtained by reaction of dissolved in water ferrous

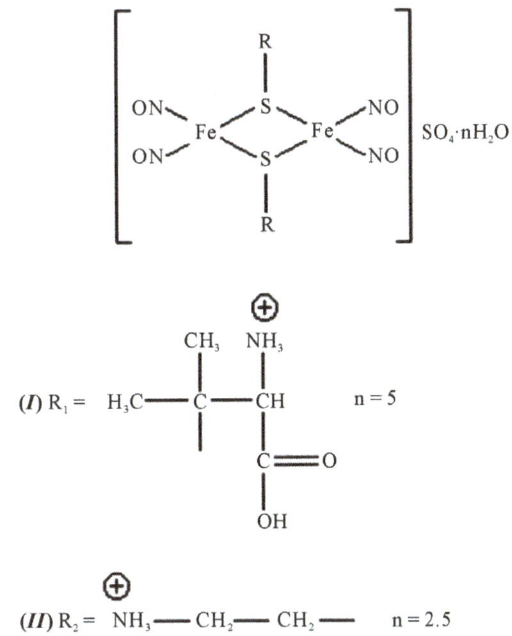

Figure 1. Chemical structures of the tetranitrosyl iron complexes ***(I)*** and ***(II)*** [14] [15].

sulphate (*II*) with an aqueous solution of D-penicillamine in the molar ratio 1:3. The reaction was performed using standard vacuum line and Schlenk technology under argon. Previously, oxygen has been removed from the water by triple freezing and vacuum pumping. To the dry mixture, containing 0.42 g (1.5 mmol) of $FeSO_4 \cdot 7H_2O$ and 0.68 g (4.5 mmol) D-penicillamine poured 10 ml water, prepared as described above, and nitric oxide have been passed through the resulting deep purple solution at room temperature. The fine red needles have been appeared on the walls of the reaction vessel after 10 - 12 min, gradually filling the entire volume of the solution. The mixture was kept 3 days at 6°C - 8°C, filtered and dried in vacuum under argon. The filtered product is stable in the solid phase in the air for a long time (months). Yield is 198 mg (20%). Elemental analysis of the polycrystalline powders obtained was performed at the Analytical Center of the Institute of Problems of Chemical Physics of the Russian Academy of Sciences. For (*I*): $Fe_2S_3N_6C_{10}H_{32}O_{17}$. Found %: Fe, 15.62; S, 13.34; N, 11.89; C, 16.80, H, 4.62. Calculated, %: Fe, 15.64; S, 13.42; N, 11.72; C, 16.76, O, 37.99; H, 4.47; IR: ν/cm^{-1} = 1771 (s, NO); 1723 (s, NO).

2.2. Operation Technique in Inert Gas Atmosphere

Operation technique in inert gas atmosphere has been described earlier [8].

2.3. Preparation of Hb Solution

A homogeneous solution of bovine Hb was prepared from commercial Hb (a mixture of oxygenated hemoglobin (HbO_2) and methemoglobin according to a known procedure [9].

2.4. Kinetics of Hb Reaction with the Complex (*I*)

To a weighed sample of the (*I*) in a vessel filled with nitrogen, an anaerobic 0.05 M phosphate buffer, pH 7.0, was added so as to prepare a solution of the complex with a concentration of 6×10^{-4} M. Then the solution was stirred for 15 min under a nitrogen stream until the complex was completely dissolved. An aliquot (0.1: 0.5; 0.75; 1 ml) of the solution was drawn under a nitrogen stream and transferred to the anaerobic sample cell and the reference cell (the volume 4 ml, the optical path length 1 cm) containing an anaerobic phosphate buffer, pH 7.0 in an amount sufficient for obtaining the final (*I*) concentration 2×10^{-5}; 10^{-4}; 1.5×10^{-4} and 2×10^{-4} M, respectively. Then the phosphate buffer, pH 7.0, was added to the reference and sample cells in amounts required to make the volume of the reaction solution after the insertion of Hb into the sample cell equal to 3.0 ml. The reaction was initiated by adding an Hb solution with the initial concentration of 4.5×10^{-4} M into the sample cell. The concentration of Hb in the sample cell was $(1.94 - 2.1) \times 10^{-5}$ M. Then the difference absorption spectra were recorded at certain time intervals. The absorption spectra were measured until Hb was completely transformed into HbNO, *i.e.*, until the spectra stopped to change.

2.5. Absorption Spectra

Absorption spectra were recorded at 25°C using a Specord M-40 spectrophotometer equipped with an interface for computer-aided registration of spectra and thermostatic cuvette holder. Amount of Hb and HbNO was evaluated spectrophotometrically. For this purpose absorption spectra were factored by components using program Mathcad 11 Enterprise Edition as described in the paper [8].

2.6. For Kinetic Modeling

We considered the assumed reaction scheme describing the interaction of (*I*) with Hb. The rate constants were determined by the least squares method based on the numerical solution of the corresponding system of differential equations. The concentrations of NO or the HbNO were determined after the resolution of the absorption spectra into components (the spectra of Hb and HbNO), and used as experimental data.

2.7. The Analysis of Hb Surface

The analysis of Hb surface was performed using the program PyMOL [16]. To view and analyze the surface of Hb, we used X-ray data of bovine Hb [17], retrieved from the database PDB, access code 1HDA (see Ref. [18]). This method allowed us to determine the total electrostatic surface charge. The density of the color depends on its magnitude. Dimensions can be determined with an accuracy of ~10%.

3. Results and Discussion

3.1. Reaction of (*I*) with Hb

In the present work we studied the interaction of (***I***) with Hb. All reactions with complex (***I***) were carried out under nitrogen because NO rapidly reacts with O_2 to give nitrogen oxides (the rate constant is $2 \times 10^6\ M^{-2} \cdot s^{-1}$) [19]. Hb is a trap for NO: the binding rate is close to the diffusion rate [20], the equilibrium constant is $3 \times 10^{10}\ M^{-1}$ [21]. Hb gives a characteristic absorption spectrum, which changes in the course of NO binding. Hence, as was reported in [8] [9], the NO release can be followed from the formation of HbNO. Since all sulfur-nitrosyl iron complexes show absorption in the visible region, we recorded the difference absorption spectra of the control and test systems with Hb containing the complex (***I***) at the same concentrations (see the Experimental section). The changes in the difference absorption spectra with time in the course of the reaction of Hb with the complex (***I***) are displayed in **Figure 2**. We stopped recording the spectra after the latter ceased to change. This was accompanied by a decrease in the absorbance at the maximum at 556 nm in the absorption spectrum of Hb and an increase in the absorbance at 545 and 575 nm suggesting the formation of the HbNO complex. These

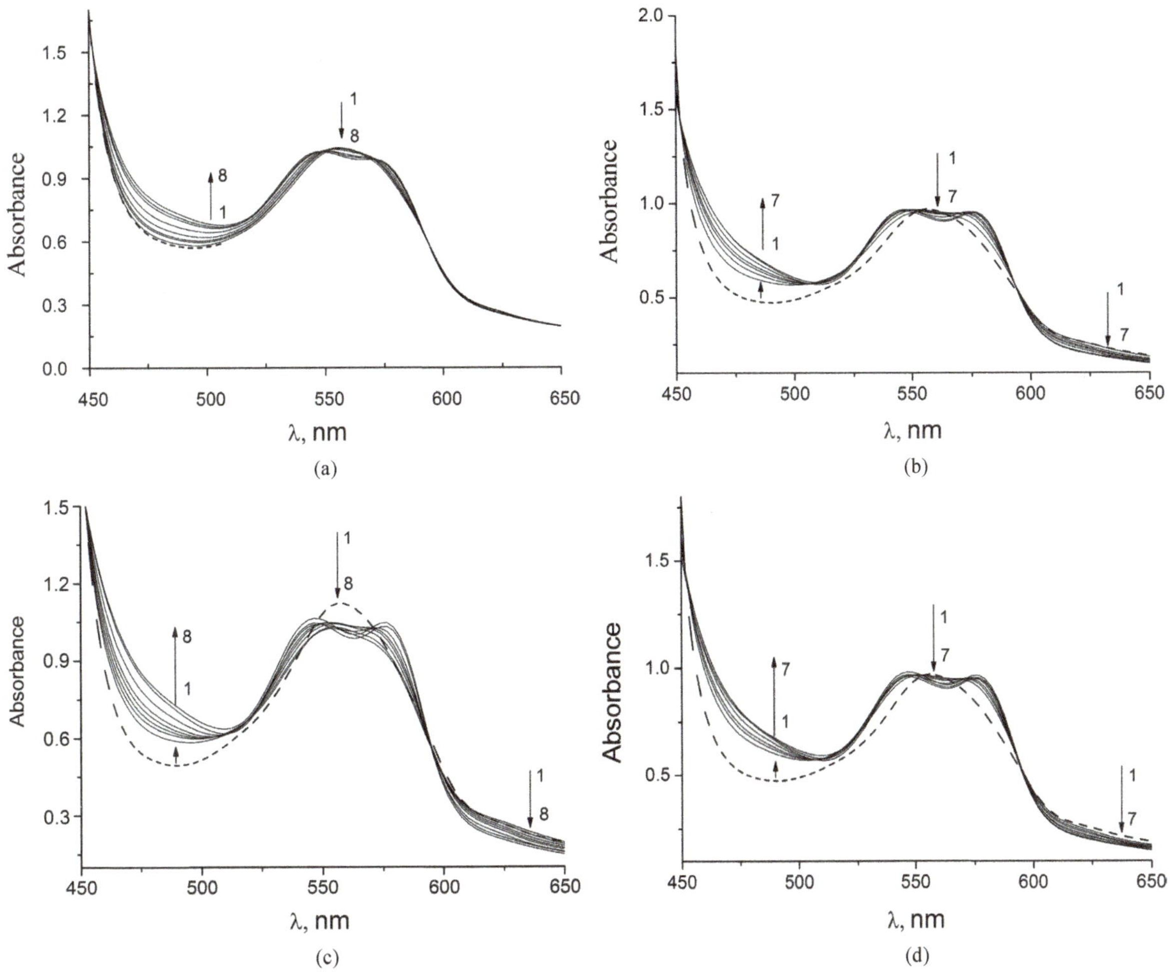

Figure 2. Kinetics of change of difference spectra at the interaction of (***I***): 2×10^{-5} (a), 10^{-4} (b), 1.5×10^{-4} (c) 2×10^{-4} (d) M with Hb 2×10^{-5} (a), 1.96×10^{-5} (b) 2.1×10^{-5} (c), 1.94×10^{-5} (d) M. λ = 450 - 650 nm. Solvent is 0.05 M phosphate buffer, pH 7.0, temperature is 25˚C. Dotted lines are the spectrum of Hb. Spectra 1-8 were registered at 0.16, 0.33, 0.5, 0.66, 0.91, 1.16, 2.16, 3.16 (a), 0.03, 0.07, 0.1, 0.15, 0,33, 1, 2.25, 3.16 (c), spectra 1-7 were registered at 0.03, 0.07, 0.12, 0.27, 0.42, 0.67, 3.83 (b), and 0.05, 0.1, 0.15, 0.25, 1.35, 1.83, 3.92 (d) h after start of reaction. Conditions of reaction: 25˚C, solvent is 0.05 M phosphate buffer, pH 7.0.

spectra have three isosbestic points at 551, 570, and 595 nm. This is evidence that only Hb and HbNO contribute to the absorption spectra, as in the case of the reactions of this class of NICs with Hb investigated in our earlier research [8]. We measured the kinetics of the formation of HbNO by recording the accumulation of HbNO and the deconvolution of the absorption spectra with the use of the MathCad program (**Figure 3**). The rate constants k (**Table 1**) is somewhat less than the constant rate of NO release from ***(I)***, defined by the sensor electrode in 0.05 M phosphate buffer pH 7.0 and a temperature 25°C in a nitrogen atmosphere: $k_1 = (4.6 \pm 0.1) \times 10^{-3}\ s^{-1}$ [22]. In [8] the authors explain why the HbNO formation reaction should "track" the NO released from NIC. We firmly established that Hb usually stabilizes the NIC and the reaction of NO release in the presence of Hb is slower than without Hb, as was found in the determination of the reaction rate of NO release from NIC using the sensor electrode [12]. We have determined that sulfur ligands, which are part of the NICs with excess electron density, contribute to the stabilization of complexes in solution [9]. The greater the electron density on the NIC ligand, the more stable it is in the presence of Hb [23]. This stabilization is due to a known anion-binding center in Hb, located in the cavity between Hb subunits [11]. Recently, we described the presence of cations binding sites on Hb. In the case of the cationic NIC ***(II)***, having thiol ligands cysteamine instead of penicillamines there was significant stabilization of the complex in the presence of Hb. This was attributed to the presence of the cationic sites, negative charge centers, reversible binding ***(II)***, and that these centers were visualized [12]. The size of ***(I)*** was greater than the size of ***(II)***. Thus, using the program PyMOL [16], we found that the size of the ***(I)*** dication is ~375 Å^3, while the size of the ***(II)*** dication is equal to ~265 Å^3 [12].

3.2. Kinetic Modeling of the Reactions of *(I)* with Hb

Consider the reaction of ***(I)*** with Hb. If Hb is present in the medium, it seems, due to the large value of the equilibrium constant of $Hb + NO \rightleftharpoons HbNO$, equal to $3 \times 10^{10}\ M^{-1}$ [21], it shifts the equilibrium decomposition reaction of ***(I)*** towards the formation of HbNO and, being a depot for NO, carries it to the respective targets.

Table 1. The values of the kinetic parameters K_s and n_s, describing the interaction of the complex ***(I)*** with Hb.

No	$[(I)]_0 \cdot 10^5$, M	$[Hb]_0 \cdot 10^5$, M	n_s	$K_s \cdot 10^{-7}$, M^{-1}
1	2	2.0	1.0	4.3
2	10	1.96	5.0	0.80
3	15	2.10	7.0	0.71
4	20	1.94	10.2	0.64

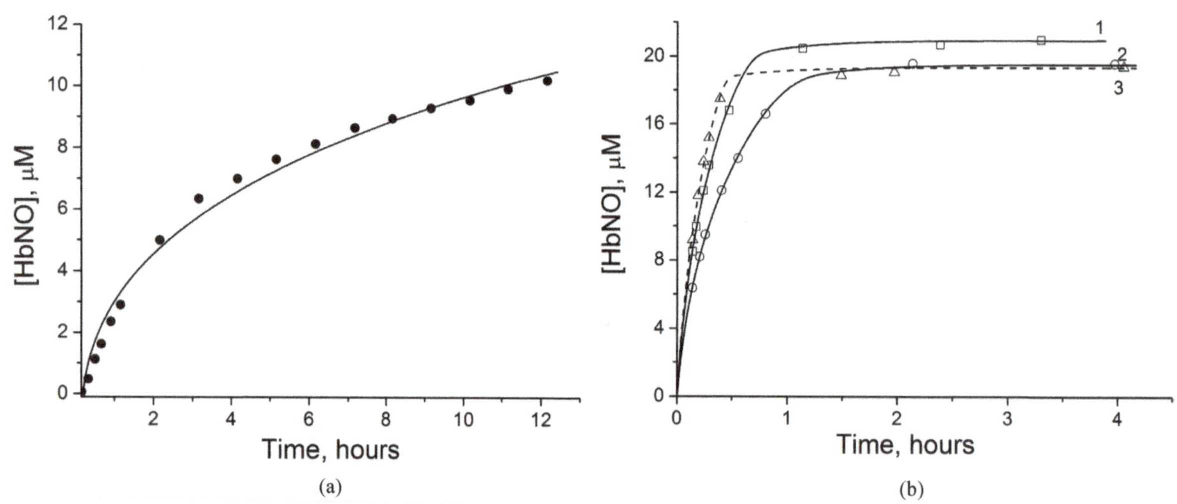

Figure 3. Kinetics of HbNO formation upon interaction of ***(I)*** (a) 2×10^{-5} and (b): 10^{-4} (2), 1.5×10^{-4} (1) and 2×10^{-4} (3) M with Hb: $[Hb]_0$ (a) 2.0×10^5, and (b): 1.96×10^5 (1), 2.1×10^5 (2), 1.94×10^5 (3) M. Points are the experimental data, the lines are calculation on the basis of the reactions of **Figure 2**. Kinetic curve 2 in (b) is based on data in **Figure 2(b)**; the curve 1, **Figure 2**; curve 3, **Figure 2(d)**.

Without the Hb equilibrium decomposition reaction of **(*I*)**, evidently biased toward **(*I*)** decomposition. It is consumed in the reactions of NO, known as the universal controllers necessary for functions of cellular metabolism [1] [2].

In [12], the authors observed a decrease in the processing rate for NO release for cysteamine cationic NIC **(*II*)** in the presence of Hb, which was explained by adsorption of the complex by the Hb molecule, resulting in a substantially reduced capacity of **(*II*)** to evolve NO. For **(*I*)**, a structural analog of **(*II*)**, the reaction scheme corresponds to a model of the given process:

$$(I) \xrightleftharpoons{(k_1,k_{-1})} P_1 + NO \quad (1)$$

$$P_1 \xrightleftharpoons{(k_2,k_{-2})} P_2 + L \quad (2)$$

$$(I) \xrightleftharpoons{(k_3)} P_3 + L \quad (3)$$

$$(I) + S_{Hb} \xrightleftharpoons{(Ks)} (I) - S_{Hb} \quad (4)$$

$$(I) - S_{Hb} \xrightarrow{(k_4)} NO \quad (5)$$

$$Hb + NO \xrightleftharpoons{(k_5,k_{-5})} HbNO \quad (6)$$

Here S_{Hb} are the active binding site of **(*I*)** on Hb surface.

In accordance with this scheme, the free molecules of complex **(*II*)** evolve NO in solution, with reversible reaction rate constants k_1 and k_{-1}. Decomposition of **(*I*)** in water is followed by separation of the penicillamine ligand [22] from molecule **(*I*)** and the product P_1, which is obtained after the separation of NO from **(*I*)**, as seen in reactions (2) and (3). Simultaneously, the molecule is **(*I*)** adsorbed on an Hb macromolecule. It is assumed that the binding sites of **(*I*)** exist on Hb surface (in a certain amount, n_s per Hb molecule), with which the molecule of complex **(*I*)** can reversibly interact with an equilibrium constant, K_s. In the bound state of the complex, molecules evolve NO with a rate constant of k_4. The separated NO molecules reversibly interact with the heme iron of Hb to form the experimentally measured HbNO product.

He values of rate constants $k_1 = 4.6 \times 10^{-3}$ s^{-1}, $k_{-1} = 9.7 \times 10^3$ M^{-1}·s^{-1}, $k_2 = 5.3 \times 10^{-4}$ s^{-1}, $k_{-2} = 0.1$ M^{-1}·s^{-1}, $k_3 = 6.6 \times 10^{-6}$ s^{-1} we determined in paper [22]. The literature contains information on the value of the constant $k_5 = 10^8$ s^{-1} [19] and the equilibrium constant K_5, k_5/k_{-5}, equal to 3×10^{10} M^{-1} [21]. The unknown parameters are K_s, n_s and k_4. Analysis of the inverse problem has shown that by available experimental data, these parameters cannot be uniquely determined. Therefore, in a first approximation, we considered the limiting case when NO release by the bound complex **(*I*)**—S_{Hb} is negligible. This regime corresponds to the value of $k_4 \sim 10^{-8}$ s^{-1} [12], as in the case of **(*II*)**. The problem reduces to the determination with the experimental data [HbNO](t), the equilibrium constant K_s, and the number of active sites on the surface of the hemoglobin, n_s.

The corresponding system of equations is as follows:

$$\frac{d[(I)]}{dt} = \left(-k_1[(I)] + k_{-1}[P_1][NO] - k_3[(I)]\right) - \beta \cdot k_4[(I)S_{Hb}]$$

$$\frac{d[S_{Hb}]}{dt} = -(1-\alpha)\left(-k_1[(I)] + k_{-1}[P_1][NO] - k_3[(I)]\right) - \beta \cdot k_4[(I)S_{Hb}]$$

$$\frac{d[P_1]}{dt} = k_1[(I)] - k_{-1}[P_1][NO] - k_2[P_1] + k_{-2}[P_2][L]$$

$$\frac{d[NO]}{dt} = k_1[(I)] - k_{-1}[P_1][NO] + k_4[(I)S_{Hb}] - k_5[Hb][NO]$$

$$\frac{d[P_2]}{dt} = k_2[P_1] - k_{-2}[P_1][L]$$

$$\frac{d[L]}{dt} = k_2[P_1] - k_{-2}[P_2][L] + k_3[(I)]$$

$$\frac{d[P_3]}{dt} = k_3[(\boldsymbol{I})]$$

$$\frac{d[Hb]}{dt} = -k_5[Hb][NO]$$

$$\frac{d[HbNO]}{dt} = k_5[Hb][NO]$$

$$[(\boldsymbol{I})S_{Hb}](t) = K_s[(\boldsymbol{I})](t)\cdot[S_{Hb}](t)$$

Here $\alpha = \dfrac{1+K_S[(\boldsymbol{I})]}{1+K_S[(\boldsymbol{I})]+K_S[S_{Hb}]}$; $\beta = \dfrac{1}{1+K_S[(\boldsymbol{I})]+K_S[S_{Hb}]}$.

The initial conditions, taking into account the detailed balance in the reaction (4):

$$[(\boldsymbol{I})](0) = \left(-b+\left(b^2+4K_s[(\boldsymbol{I})]_0\right)^{0.5}\right)\Big/(2K_s),$$

where

$$b = 1-K_s\left([(\boldsymbol{I})]_0-[S_{Hb}]_0\right);$$

$$[S_{Hb}](0) = [(\boldsymbol{I})](0)-[(\boldsymbol{I})]_0+[S_{Hb}]_0;$$

$$[Hb](0) = [Hb]_0;$$

$$[P_1](0) = [NO](0) = [P_2](0) = [L](0) = [P_3](0) = [HbNO](0) = 0;$$

Here $[(\boldsymbol{I})]_0$, $[Hb]_0$, $[S_{Hb}]_0 = n_s\cdot[Hb]_0$ are the initial concentrations of the complex (***I***), Hb, and the active sites on the surface S_{Hb}, respectively.

Unknown kinetic parameters K_s and n_s were determined by numerical solution of the inverse task, using the kinetics of accumulation of HbNO in four experiments of the (***I***) interaction with Hb (**Table 1**).

Calculations were carried out at the value of the equilibrium constant $K_5 = 0.5 \times 10^9$ M^{-1}, which provides a better match of theory and experiment than $K_5 = 3 \times 10^{10}$ M^{-1}. This is justified, considering the fact that, in the presence of salts, K_5 decreases, wherein the salt effect increases in the following order: NaCl, KCl, sodium citrate, sodium phosphate [24]. In our case, the experiments were performed in 0.05 M phosphate buffer, pH 7.0. Point values (**Table 1**) satisfactorily describe the experimental data (**Figure 3**). As seen from **Table 1**, increasing the concentration of the complex causes value n_s to increase and K_s to decrease. Apparently, there are various regions of different binding strength on the surface of Hb for binding of the cation complex. At low concentrations of complex (***I***) (number 1 in **Table 1**), its molecules primarily bind with regions, providing a stronger link (large K_s), and while increasing the concentration of $[(\boldsymbol{I})]_0$ during adsorption, they begin to participate in other regions on the Hb surface with lower bond strength. In processing of the experimental data, this leads to a change in the effective values of the parameters: increase of n_s and decrease of K_s.

3.3. Possible Location of (*I*) on the Surface of Hb. Cation Binding Sites

It is known that Hb is composed of four subunits: two α (A, C) and two β (B, D). The α subunit consists of 141 amino acid residues, and the β subunit contains 145 residues. The α subunits are in contact with each other and form a narrow entrance of the through-channel. The β subunits are in contact with each α subunit, but are not in contact with each other and form a wide entrance of this channel. On the whole, Hb is a symmetrical macromolecule with a twofold symmetry axis and a channel running through the structure. It is known that the anion binding site, where DPG is bound, is located in this through-channel [11]. In the through-channel, 8 cationic and 3 anionic pairs of amino acids were located [12]; in general it is positively charged and directly enters the channel, which is also positively charged. Therefore, the complexes (***I***) and (***II***) can not penetrate to this channel. It has been shown in [12], using the program PyMOL, that the surface area of Hb has a local concentration of both negatively and positively charged amino acids , as well as more or less than the neutral portion of the surface.

Among various roughnesses on the Hb surface, there are four large cavities (see **Figure 4**). Two of them (cavities 1) are located in the region of the negatively charged surface from the side of the wide entrance at equal distances (~13 Å) from the hemes of the A and D subunits. The volume of the cavity 1 is ~17 Å^3. Two other large cavities (cavities 2) with volumes of ~850 Å^3 are also negatively charged and are located symmetrically with respect to the cavity 1 at equal distances from the hemes of the A and D subunits (see **Figure 4**). The negative potential of the cavities 1 and the adjacent surfaces is formed by the negatively charged carboxyl groups of the following amino acids: cavity 1 (subunits A, B, D), by A-Glu23, A-Glu27, A-Glu30, A-Asp47, B-Glu5, B-Glu6, B-Glu120, B-Asp128, D-Glu89, D-Asp93, and the terminal aminoacid D-His145; cavity 1 (subunits C, D, B), by C-Glu23, C-Glu27, C-Glu30, C-Asp47, D-Glu5, D-Glu6, D-Glu120, D-Asp128, B-Glu89, B-Asp93 and the terminal amino acid B-His145. The negative potential of the cavities 2 and the adjacent surfaces is formed by the negatively charged carboxyl groups of the following amino acids: cavity 2 (subunits A, C, D), by A-Asp85 and the terminal amino acid A-Arg141, C-Asp6, D-Glu42, D-Asp46, D-Asp51; cavity 2 (subunits C, A, B), by C-Asp85 and the terminal amino acid C-Arg141, A-Asp6, B-Glu42, B-Asp46, B-Asp51. Most likely, the complex ***(II)*** can be located in these cavities and can be bound to the negatively charged functional groups of Hb. Correspondingly, it can be suggested that at least one ***(II)*** molecule can be bound in the cavity 1, and at least three ***(II)*** molecules can be bound in the cavity 2. In total, each Hb molecule can bind at least eight ***(II)*** molecules. Taking into account the results of the kinetic modeling [12], which gave approximately 12 ***(II)***-binding sites per Hb molecule, it can be concluded that there are from 8 to 12 such sites. **Figure 4** shows that, in addition to the cavities 1 and 2, there are adjacent surfaces bearing a negative charge. Apparently, they are also involved in the binding of ***(II)***. The surfaces in the direct vicinity of the hemes are positively charged, the charge gradually decreasing with increasing distance from the heme. In this study, we had a different situation. The volume of ***(I)*** is ~375 Å^3, and only one molecule of ***(I)*** can fit into the large, negatively charged cavity, *i.e.*, to cavity 2. Moreover, in this case, the binding is stronger: $K_s = 4.3 \times 10^7$ M^{-1}. Therefore, the ***(I)*** molecules bind more weakly. The PyMOL program identifies shallow and indistinct regions with mixed charge from a small positive to a negative charge on the surface of the β-subunit at a distance of approximately 10 Å from the heme. On such areas, the binding of ***(I)*** probably occurs in this case, starting from a negative area. So K_s decreases with increasing concentrations of ***(I)***.

4. Conclusion

The results of this work on the interaction of the complex ***(I)*** with Hb fully confirm and complement the findings of [12] of cations binding sites on the Hb surface. The function of these sites in metabolism in general has not yet been established. However, it is clear in the case of NICs. That negative cavities on the surface of Hb are the pools of NICs. Associated with Hb molecules NICs emit NO slightly ($k \sim 10^{-8}$ s^{-1}). Hb may carry adsorbed

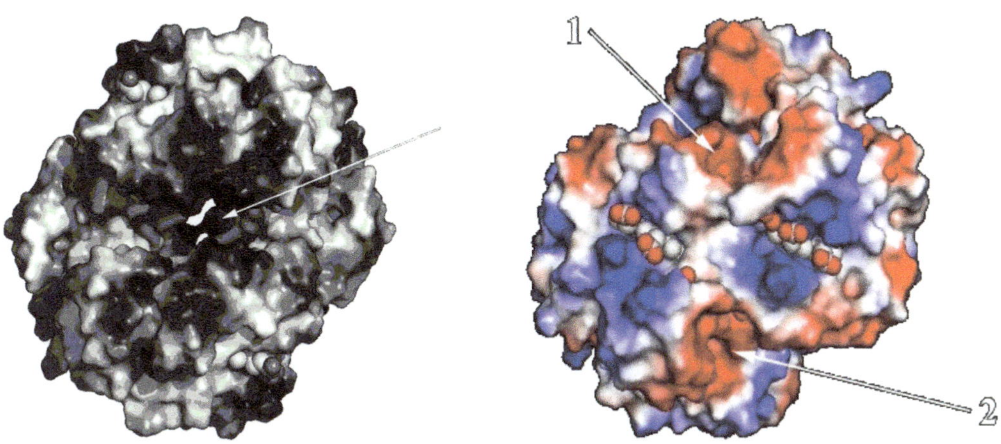

(a) (b)

Figure 4. Analysis of the Hb surface. The regions with a negative surface charge are shown in red (b) or gtey (a), and the regions with a positive surface charge are shown in blue (b) or black (a). The X-ray diffraction data for the bovine Hb molecule [25] are used. (a) The view on to the through-channel (indicated by an arrow); (b) The view onto the cavities 1 and 2. **Figure 4** was taken from paper [12].

therein NICs to other targets. Since NICs associated with Hb reversibly, under decreasing NICs concentration lose contact with it. We can assume that the binding sites of cations can also be carriers of other metabolites with positively charged functional groups.

Acknowledgements

This study was financially supported by the RFBR (Grant No. 13-03-00549).

We wish to thank Dr. A.V. Chudinov for the preparation of program of computer processing of absorption spectra by the least square method using program MathCad.

References

[1] Lancaster, J.R. (2010) Metal-Catalyzed Nitric Oxide Nitrozo Interconversions and Biological Signaling. *Nitric Oxide: Biology and Chemistry*, **22**, 13.

[2] Lewandowska, H., Kalinowska, M., Brzoska, K., Wojciuk, K., Wojciuk, G. and Kruszewski, M. (2011) Nitrosyl Iron Complexes—Synthesis, Structure and Biology. *Dalton Transactions*, **40**, 8273-8289. http://dx.doi.org/10.1039/c0dt01244k

[3] Mocellin, S. (2009) Nitric Oxide: Cancer Target or Anticancer Agent? *Current Cancer Drug Targets*, **9**, 214-236. http://dx.doi.org/10.2174/156800909787581015

[4] Wink, D.A. and Mitchell, J.B. (1998) Chemical Biology of Nitric Oxide: Insights into Regulatory, Cytotoxic, and Cytoprotective Mechanisms of Nitric Oxide. *Free Radical Biology Medicine*, **25**, 434-456. http://dx.doi.org/10.1016/S0891-5849(98)00092-6

[5] Morbidelli, L., Donnini, S. and Ziche, M. (2004) Role of Nitric Oxide in Tumor Angiogenesis. *Cancer Treatment and Research*, **117**, 155-167. http://dx.doi.org/10.1007/978-1-4419-8871-3_11

[6] Prostovit, L.M., Adams, M.A., Lash, G.E., Heaton, J.P. and Graham, C.H. (2002) Oxygen-Mediated Regulation of Tumor Cell Invasiveness. Involvement of a Nitric Oxide Signaling Pathway. *Journal of Biological Chemistry*, **277**, 35730-35737. http://dx.doi.org/10.1074/jbc.M204529200

[7] Sanina, N.A. and Aldoshin, S.M. (2011) Structure and Properties of Iron Nitrosyl Complexes with Functionalized Sulfur-Containing Ligands. *Russian Chemical Bulletin, International Edition*, **60**, 1223-1251. http://dx.doi.org/10.1007/s11172-011-0192-x

[8] Sanina, N.A., Syrtsova, L.A., Shkondina, N.I., Rudneva, T.N., Malkova, E.S., Bazanov, T.A., Kotel'nikov, A.I. and Aldoshin, S.M. (2007) Reaction of Sulfur-Nitrosyl Iron Complexes of "g = 2.03" Family with Hemoglobin (Hb): Kinetics of Hb-NO Formation in Aqueous Solutions. *Nitric Oxide: Biology and Chemistry*, **16**, 181-188. http://dx.doi.org/10.1016/j.niox.2006.10.005

[9] Sanina, N.A., Syrtsova, L.A., Shkondina, N.I., Malkova, E.S., Kotelnikov, A.I. and Aldoshin, S.M. (2007) Hemoglobin-Stabilized Tetranitrosyl Binuclear Iron Complex with Pyridine-2-yl in Aqueus Solutions. *Russian Chemical Bulletin, International Edition*, **56**, 761-766. http://dx.doi.org/10.1007/s11172-007-0114-0

[10] Roudneva, T.N., Syrtsova, L.A., Sanina, N.A., Shkondina, N.I., Kotelnikov, A.I. and Aldoshin, S.M. (2010) Formation of Iron Nitrosyl Complexes of Ferri- and Ferrocytochrome C at Its Interaction with Nitrosyl Iron Complex Cysaconite. *Nitric Oxide: Biology and Chemistry*, **22**, 546.

[11] Arnone, A. (1972) X-Ray Diffraction Study of Binding of 2,3-Diphosphoglycerate to Human Deoxyhaemoglobin. *Nature*, **237**, 146-149. http://dx.doi.org/10.1038/237146a0

[12] Sanina, N.A., Syrtsova, L.A., Psikha, B.L., Tukhvatullin, I.A., Shkondina, N.I., Rudneva, T.N., Kotel'nikov, A.I. and Aldoshin, S.M. (2012) Revealing of the Cation-Binding Sites on the Surface of Hemoglobin in Its Reaction with the NO Donor, the Nitrosyl Iron Complex $\{Fe_2[S(CH_2)_2NH_3]_2(NO)_4\}SO_4 \cdot 2.5H_2O$. *Russian Chemical Bulletin*, **61**, 2350-2355. http://dx.doi.org/10.1007/s11172-012-0330-0

[13] Sanina, N.A., Lysenko, K.A., Zhukova, O.S., Roudneva, T.N., Emel'yanova, N.S. and Aldoshin, S.M. (2011) Water-Soluble Binuclear Nitrosyl Iron Complexes with Natural Aliphatic Thiolyls Possessing Cytotoxic, Apoptotic and NO

Donor Activity. Patent No. US 8, 067, 628 B2.

[14] Roudneva, T.N., Sanina, N.A., Lyssenko, K.A., Aldoshin, S.M., Antipin, M.Y. and Ovanesyan, N.S. (2009) Synthesis and Structure of a Water-Soluble Nitrosyl Iron Complex with Cysteamine Ligand. *Mendeleev Communications*, 19, 253-255. http://dx.doi.org/10.1016/j.mencom.2009.09.006

[15] Sanina, N.A., Serebryakova, L.I., Shul'zhenko, N.S., Pisarenko, O.I., Roudeva, T.N. and Aldosin, S.M. (2012) Application of Iron Binuclear Complex Sulphur-Nitrosyl Complex of the Canionic Type as a Vasodilative Drug. Patent No. RU No 2460531 C1.

[16] De Lano, W.L. (2002) The PyMOL Molecular Graphics System. Version 0.99rc6 Schrödinger, LLC, De Lano Scientific, San Carlos.

[17] Perutz, M.F., Fermi, G., Poyart, C., Pagnier, J. and Kister, J. (1993) A Novel Allosteric Mechanism in Haemoglobin. Structure of Bovine Deoxyhaemoglobin, Absence of Specific Chloride-Binding Sites and Origin of the Chloride-Linked Bohr Effect in Bovine and Human Haemoglobin. *Journal of Molecular Biology*, **233**, 536-545. http://dx.doi.org/10.1006/jmbi.1993.1530

[18] Bernstein, F.C., Koetzle, T.F., Williams, G.J.B., Meyer Jr., E.F., Brice, M.D., Rodgers, J.R., Kennard, O., Shimanouchi, T. and Tasumi, M. (1977) The Protein Data Bank: A Computer-Based Archival File for Macromolecular Structures. *Journal of Molecular Biology*, **112**, 535-542. http://dx.doi.org/10.1016/S0022-2836(77)80200-3

[19] Miles, A.M., Wink, D.A., Cook, J.C. and Grisham, M.B. (1996) Determination of Nitric Oxide Using Fluorescence Spectroscopy. In: Packer, L., Ed., *Methods in Enzymology V*, Part A, Academic Press, New York, 105-121.

[20] Cassoly, R. and Gibson, Q.H. (1975) Conformation, Co-Operativity and Ligand Binding in Human Hemoglobin. *Journal of Molecular Biology*, **91**, 301-313. http://dx.doi.org/10.1016/0022-2836(75)90382-4

[21] Antonini, E. and Brunori, M. (1971) Hemoglobin and Myoglobin in the Reactions with Ligands. In: Neuberger, A. and Tatum, E.L., Eds., *North-Holland Research Monographs. Frontiers of Biology*, Vol. 21, North-Holland Publishing Company, Amsterdam and London, 276.

[22] Syrtsova, L.A., Sanina, N.A., Lyssenko, K.A., Kabachkov, E.N., Psikha, B.L., Shkondina, N.I., Pokidova, O.V., Kotelnikov, A.I. and Aldoshin, S.M. (2014) Reversible Dissociation and Ligand-Glutathione Exchange Reaction in Binuclear Cationic Tetranitrosyl Iron Complex with Penicillamine. *Bioinorganic Chemistry and Applications*, **2014**, Article ID: 641407. http://dx.doi.org/10.1155/2014/641407

[23] Sanina, N.A. Syrtsova, L.A., Chudinova, E.S., Shkondina, N.I., Rudneva, T.N., Kotelrnikov, A.I. and Aldoshin, S.M. (2009) Regularities in the Stabilization by Hemoglobin of Binuclear Iron Complexes $[Fe_2(\mu\text{-N-C-SR})_2(NO)_4]$ Containing Benzimidazolyl and Benzothiazolylthiol Ligands. *Russian Chemical Bulletin*, **58**, 566-571.

[24] Haurowitz, F. and Hardin, R. (1954) Respiratory Proteins. In: Neurath, H. and Bailey, K., Eds., *The Proteins*, Academic Press, New York, 279-344.

[25] Perutz, M.F., Shih, D.T. and Williamson, D. (1994) The Chloride Effect in Human Hemoglobin: A New Kind of Allosteric Mechanism. *Journal of Molecular Biology*, **239**, 555-560. http://dx.doi.org/10.1006/jmbi.1994.1394

Abbreviations

(I)	complex $[Fe_2(SC_5H_{11}NO_2)_2(NO)_4]SO_4{\cdot}5H_2O$
(II)	complex $[Fe_2(S(CH_2)_2NH_3)_2(NO)_4]SO_4{\cdot}2.5H_2O$
NIC	nitrosyl iron complex
Hb	deoxyhemoglobin

8

Structural Implications of Universal Complementarities in Translation—High Accuracy at the Decoding Site

Kozo Nagano

Nagano Research Institute of Molecular Biology, Yokohama, Japan
Email: fwkz4754@nifty.com

Abstract

X-ray structures of transfer RNAs (tRNAs) bound to the whole ribosome do not fully explain the mechanism of translation. The cause of the failure seems to come mainly from a high Mg^{2+} ion concentration compared to that in the living cells. There exists a wide range of nucleotide sequence conservation in tRNA and ribosomal RNAs (rRNAs) of small and large subunits as well as sequence complementarities that seems to explain how high accuracy in translation can be achieved at the decoding site. Conformational transition between U33-folded and U33-extended forms of anticodon loops of tRNAs and G-C pair formation and disruption between C1399 and G1504 of 16S rRNA, etc. play the central role in explaining why E-site tRNA can automatically be expelled when an aminoacyl-tRNA at the A site turns out to be cognate.

Keywords

Ribosomal RNA, Transfer RNA, Spectinomycin, Universal Rule, U33-Folded Form, U33-Extended Form, GTP Hydrolysis, Cognate, Near-Cognate

1. Introduction

Protein synthesis is carried out on the ribosome that is composed of small and large subunits. Amino acid is brought into the interface of the two subunits as a form of aminoacyl-tRNA (aa-tRNA) in cooperation with elongation factor Tu (EF-Tu), complexed with guanosine triphosphate (GTP). Through the intersubunit space tRNAs move along with message-carrying RNA (mRNA) from the entrance toward the exit. There exist three sites for the tRNA-mRNA complex; the first site is for an aa-tRNA (A site), the second for a peptidyl-tRNA (P site), and the third for the exit of a deacylated tRNA (E site). The peptidyl group of P-site tRNA is transferred to the aminoacyl group of aa-tRNA at the A site on the peptidyl transferase center (PTC). The peptidyl-tRNA on the A site and the deacylated tRNA on the P site are forced to move in the next step toward the P and E sites, respectively, by a mechanism called translocation. During the course of translocation, it is known that the classical state of two tRNAs, A/A and P/P state, passes through an intermediate state, A/P and P/E, that is called hybrid state, and finally reaches the classical P/P and E/E state. X-ray crystallographic analyses have clarified the three dimensional (3D) structure of *Thermus thermophilus* 70S ribosome at 5.5 Å resolution that contained bound mRNA of only A- and P-site codons and three deacylated tRNAs bound to A, P, and E sites [1]. Successively, the structure of 70S ribosome containing a model of mRNA and two tRNAs bound to P and E sites was solved at 3.7 Å resolution [2]. It was shown that the adenine base at the CCA end of the deacylated tRNA bound to the E site is recognized by C2394 and G2421 of helix 88 of 23S rRNA (H88) [3]. Nucleotide numberings of 16S and 23S rRNAs follow the convention of [4], while helix numberings of 16S and 23S rRNAs that of [5]. On the other hand, the T-arm, D-arm, and the 3'-acceptor arm of A-site tRNA in the X-ray structure of the 70S ribosome were invisible, while an unintentional E-site tRNA was observed along with P-site tRNA complexed with mRNA [6]. Base pairing of the E-site codon to the anticodon of E-site tRNA was observed in the ribosome structure of a post-initiation complex, although the base-pairing conformation was not a typical Watson-Crick type [7]. At the hybrid state the small subunit is rotated with respect to the large subunit that is called ratcheting or ratchet-like intersubunit rotational movement [8]. It was for the first time visualized by cryo-electron microscopy (cryo-EM) and particle classification procedures that the hybrid state of both A/P and P/E tRNAs in an authentic factor-free ribosome complex during translocation [9]. Such an intersubunit rotation of pre-translocational ribosomes was observed to be spontaneous in the absence of elongation factor G (EF-G) [10]-[12]. It is also important to note that the ribosome binding the peptidyl-tRNA at the P site is locked, or in other words, does not rotate [13]. Accordingly, in order to return by elevating energetically from the classical P/P and E/E state to the classical A/A and P/P state, the step of codon recognition in participation of aa-tRNA·EF-Tu·GTP ternary complex (TC) as well as a universal mechanism of the long helix formation between small and large rRNAs, that was proposed on the basis of universally conserved nucleotide complementarity [14] would be required. Although the 3D arrangements of the three tRNAs in both classical and hybrid states have been clarified, why and how E-site tRNA is automatically expelled from the E site when A-site tRNA turns out to be cognate can be explained neither by X-ray structures nor by cryo-EM structures of the whole ribosome. Moreover, how high accuracy in translation can be achieved has never been explained in the light of the 3D structures of the whole ribosomes. It has been known that the first step of codon-anticodon recognition only expels the TC of noncognate aa-tRNAs with accuracy of about 200/10,000 [15]. The mechanism of kinetic proofreading at the step of GTP hydrolysis was proposed to expel the TC of near-cognate aa-tRNA to enhance the accuracy up to 5/10,000 [16]. It was found, however, long time ago that the accuracy is reduced by the use of high Mg^{2+} concentration [17]. Fig. 1 of [18] has shown that incorporation of near-cognate aa-tRNA into A site sharply increases from 6 mMol of a Mg^{2+} concentration and reaches a plateau at 12 mMol Mg^{2+}. How such a phenomenon could be realized on the 3D structure of the ribosome?

2. Universal Rule of Complementarities of rRNAs and tRNAs

It was found that the region of nucleotides 1064 - 1077 of 16S rRNA in *Esherichia coli* sequence is complementary with the region of nucleotides 2653 - 2666 of 23S rRNA [14], that is called sarcine-ricin loop (SRL) (**Figure 1**).

The 12 nucleotide sequence of 23S rRNA, AGUACGAGAGGA, is the longest universally conserved among all known rRNAs [20]. The nucleotides at the positions 1068-1073 of 16S rRNA constitute the 5'-side of secondary structure, helix 35 of 16S rRNA (h35), while those at the positions 2653 - 2658 and 2663 - 2666 are the head region of H95. The other side of h35 contains the region of universally conserved sequence 1103-1107 CGAGC, that is capable of making base pairs to a part of T-stem loop of almost all tRNAs, GTΨCG or GTΨCA at the positions 53 - 57 in the case of yeast $tRNA^{Phe}$. This relationship holds for all kind of living cells including eubacteria, archaea, chloroplasts, eukarya, and mitochondria [14]. The left-most structure in **Figure 1** plays an important role in GTP hydrolysis, while another structure on the second from the right-most side seems to be that of transition state. Although such a long helix has been observed by neither X-ray analysis nor cryo-EM, a model building technique can show how it looks like (**Figure 2**). In this structure, the possible base pairs, G2655-U1075 and G2659-C1071, are unpaired, but it is consistent with many cases in eukarya, because the

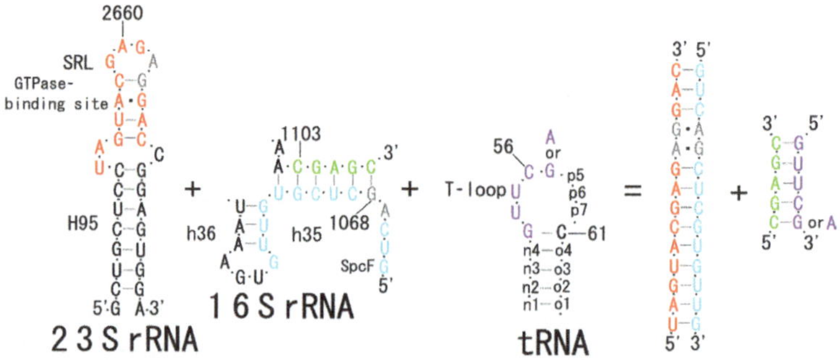

Figure 1. Universal rule of complementarities between 23S and 16S rRNAs and between 16S rRNA and tRNA [14]. SpcF denotes a spectinomycin footprint site [19].

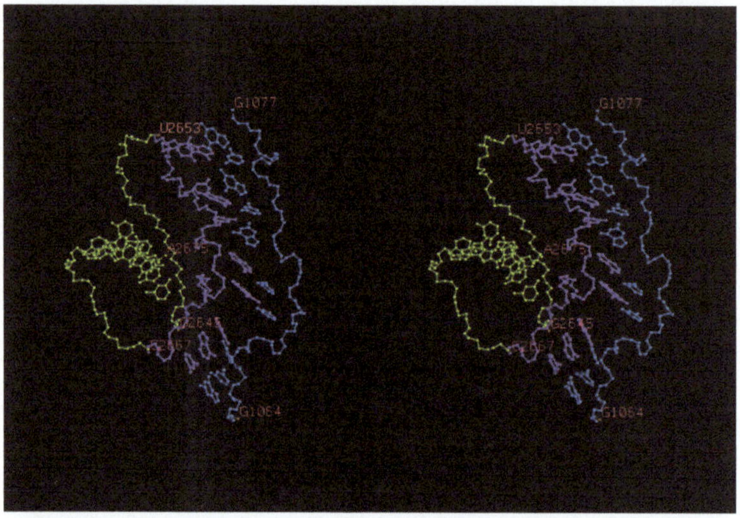

Figure 2. Stereo view of main-chain representation for double-stranded helix formed between SRL-region, U2653-C2666, of 23S rRNA, which is drawn in magenta, and the region G1064-G1077 of 16S rRNA in skyblue. The other region in yellow is G2645-C2652 and C2667-A2675 of H95. Some base pairs in H95, such as 2650-2652/2668-2670, are unfolded. Only main chains and base-pairing bases are drawn for simplicity. Red letters denote the locations of nucleotides, G1064, G1077, G2645, A2675, U2653, and C2667, respectively.

corresponding bases are guanines instead of U1075 and C1071, respectively [14]. Three base pairs in H95, C2652-G2668, C2651-G2669, and U2650-A2670 are unfolded. It is also understandable, because the first one of the three is A-C pair in some cases. Before achieving the structure shown in **Figure 2**, the regions of nucleotides in the 5'-side of h35 must be unfolded. This state can be obtained by the interaction of aa-tRNA in the TC of EF-Tu·GTP with the 3'-side of h35. The region of 1103-1107 of 16S rRNA, CGAGC, is able to interact with the GTΨCG or GTΨCA at the T-loop of aa-tRNA in the TC with EF-Tu·GTP. Such a model-building work started on the basis of atomic coordinates presented by the Protein Data Bank data bases, 2HGR and 2HGU [2]. It is also noteworthy that the region of 1064 - 1068, GUCAG, as well as the nucleotide C1192 of 16S rRNA constitute the binding site of an antibiotic spectinomycin, the conformation of which was clarified by X-ray structure (see Fig. 4 of [21]). Since it has been known that spectinomycin inhibits EF-G dependent translocation [22] [23], the conformational change of the above region of 16S rRNA must have something to do with the mechanism of translocation. Although the binding site of both EF-G· GTP complex and aa-tRNA·EF-Tu·GTP TC on the 30S subunit ribosome could be similar, a recent result of α-sarcin-treated ribosomes has clarified that cleavage of SRL inhibited EF-G binding, GTP hydrolysis, and mRNA-tRNA translocation, while it slightly affected binding of the TC to the ribosome [24]. A big difference in binding of the GTP complexes between EF-G and EF-Tu·tRNA from the viewpoint of the present work is the absence of helix formation of the region 1103 - 1107 of 16S rRNA with the region of EF-G corresponding to the T-loop of tRNAs. Accordingly, both spectinomycin binding to the nucleotides, G1064, C1066 and C1192 of 16S rRNA [21] and splitting of the nucleotide bond between A2660 and G2661 in SRL [24] obstructs the transition from the left-most conformations to the second from the right-most conformation in **Figure 1**. This transition seems to produce movement of the 30S subunit relative to the 50S subunit [8]-[25]. The long helix formation represented by the second from the right-most side in **Figure 1** could be the source of driving force from the pretranslocation (PRE) state to the posttranslocation (POST) one.

3. U33-Extended and U33-Folded Forms of the Anticodon Loop of tRNA

Uridine 33 at the 5'-side corner of the anticodon loop of tRNA is universally conserved except for cytidine 33 in some initiator formylmethionyl-tRNAs. The uridine base of U33 is folded inside the anticodon loop of tRNAs observed so far by X-ray crystallography, and hydrogen-bonded to a phosphate oxygen atom of 36th nucleotide of the anticodon [26]. This structure is referred to as U33-folded form in this paper. This conformation of the codon-anticodon helix of U33-folded form does not seem stable enough. The U33 base can be expected to stick out or to be exposed if the conformation of the codon-anticodon helix assumes that of an ideal A-form RNA. Such a structure could be a little bit unstable because of the exposed U33 base compared to the hydrogen-bonded U33 base of the U33-folded form. And yet, the stability of the U33-extended form heavily depends on the stability of the codon-anticodon helix, and seems to play an important role in enhancing the accuracy of translation. The relative stability between the U33-folded and extended forms must also depend on the concentration of Mg^{2+} of the buffer solution used in the experiments. If the Mg^{2+} concentration is pretty high, short-range interactions would be predominant. It is said that the Mg^{2+} concentration inside muscle cells is about 1 mMol that is about the same as that outside the cells. On the other hand, a Mg^{2+} concentration used for crystallization of whole ribosomes is 10 mMol or much higher in some cases. Crystal making people say that they use the concentration as high as required for obtaining a good crystal. The conformation of the anticodon loop of tRNAs in the living cell must be more dynamic. The difference in the positively charged ions between the living ribosomes and the ribosome crystals are mainly compensated by K^+, if the environment of ribosomes are about the same as the muscle cells. A recent result of X-ray structure of the decoding site has shown that the G-U base pairs of near-cognate codon-anticodon interactions at the first position (between mRNA/Phe and $tRNA_2^{Leu}$) and at the second position (between mRNA/Cys and $tRNA^{Tyr}$) cannot be discriminated from the cognate G-C and A-U base pairs, respectively [27]. The sharp increase of the near-cognate GTPase reaction between Poly (U)-programmed ribosomes and Leu_2 TCs from 6 mMol to 12 mMol (See Fig. 1 of [18]) strongly suggests that deprotonation of N3 and protonation of O4 of uridine base of G-U pair cannot be distinguished from G-C pair [27], and that the conformational change of the anticodon loops from the U33-extended to U33-folded form of the ribosomes do not discriminate the near-cognate Leu_2-tRNA from the cognate Phe-tRNA. **Figure 3(a)** shows a different 3D structure for the near-cognate G-U pair at the first position of the codon-anticodon helix on the basis of the model building technique of the present work, in which the U33 base, which is drawn in magenta, sticks out from the anticodon loop, although a Mg^{2+} hexahydrate still binds between OP2 atom of YG37 and O2 atom of O2' methylated OMC32. The U33-extended form of the anticodon loop would destabilize the loop structure. Besides, deprotonation from the collision of N1-H of guanine base and N3-H of uridine base resulting in the movement of the proton to the O4 atom of the uridine base might be preferable under the Mg^{2+} excess conditions. Fig. 1 of [27] also shows that the P/A-kink of the mRNA is widely open as the kinks obtained by the other X-ray structures of the ribosomes [1]-[6] and supported by binding two clusters of Mg^{2+} hexahydrate. On the other hand, the present study can demonstrate a different type of P/A- and E/P-kinks in **Figure 3(b)**, each of which is supported by one cluster of Mg^{2+} hexahydrate. The difference of the kink structures of the present work from those in the X-ray structures can be rationalized on the basis of the low Mg^{2+} concentration described above. In contrast to **Figure 3(a)**, **Figure 3(c)** and **Figure 3(d)** show the structural difference in the A-site codon-anticodon helix between the cognate case in C and the near-cognate case in D. In these two structures, the bases of both U33 and OMC32 as well as YG37 are exposed in order to represent a delicate difference between the cognate and near-cognate codon-anticodon helices. As a result of the difference, a Mg^{2+} hexahydrate between OP2 atom of YG37 and O2 atom of OMC32 is expelled, giving rise to a slight difference in the interactions between the bases of U33 and A1396. The interactions of A1492 and A1493 to the minor groove of the codon-anticodon helix can be achieved almost the same as those in the X-ray structures [27], but the interaction of G530 is rather difficult without a conformational change in the 530 loop due to the big conformational change in the ASL structure as well as the big shift in the location of h45, which will be described later. The near-cognate G-U base pair at the first bases of the A-site codon in **Figure 3(d)** is quite the same as that of **Figure 3(a)**, but resulting in a delicate but recognizable difference from the cognate case in **Figure 3(c)**, while the structure shown in **Figure 3(a)** cannot be achieved because of its big structural difference from the corresponding one in [27].

The exposed base of U33 in the U33-extended form is supposed to be supported by base-pairing with a certain universally conserved adenine base at the decoding site of 16S rRNA, the secondary structure of which is shown in **Figure 4**. Such base pairs are long-range interactions in contrast to the hydrogen bonded U33 base inside the anticodon loop so far observed in X-ray structures. Accordingly, a transition between U33-folded and U33-extended forms would be more frequent in the physiological condition of the Mg^{2+} concentration. **Figure 4** shows a schematic drawing of secondary structure of 16S rRNA at the decoding site. Although the G-C pair between C1399 and G1504 has been predicted [4] [5] because of its high conservation, it was not confirmed in

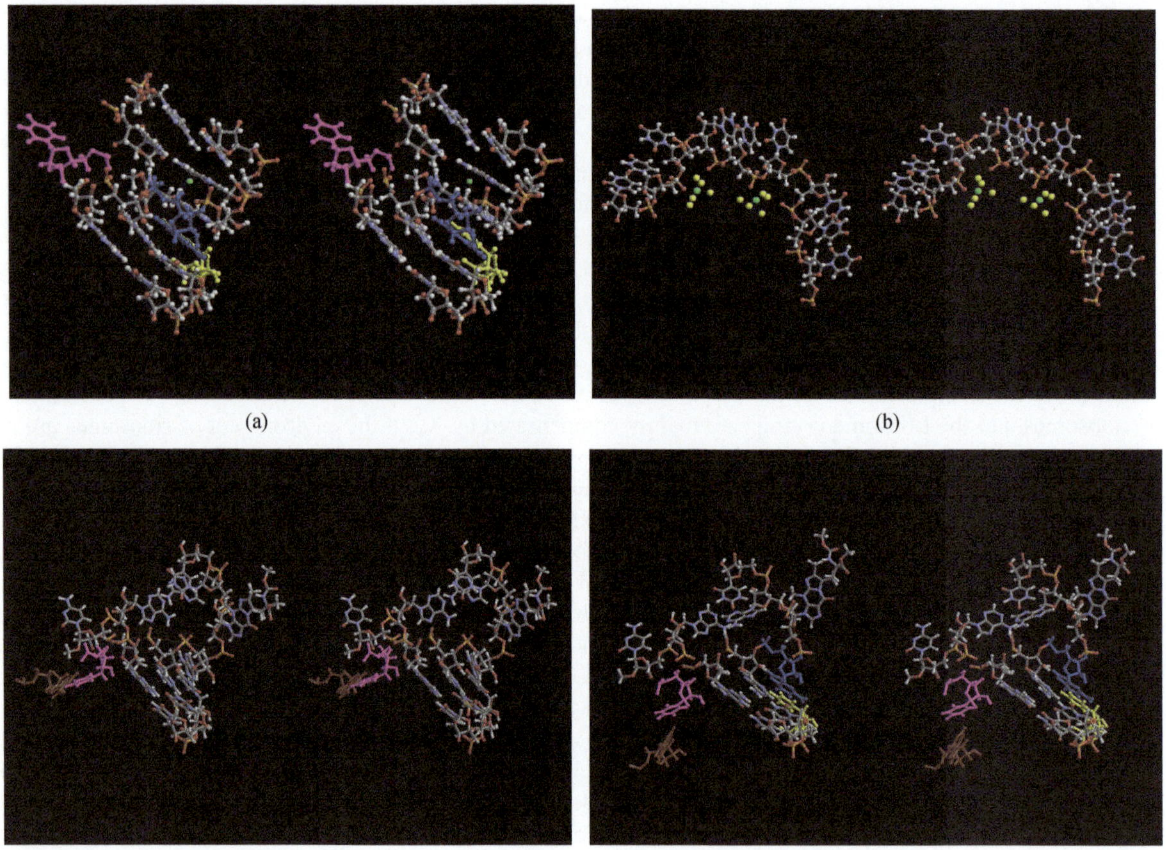

(a) (b) (c) (d)

Figure 3. (a) Virtual picture of stereo view for the near-cognate G-U wobble base pair, which is drawn by blue guanine and yellow uridine, at the first position of the A-site codon-anticodon helix. Mg^{2+} ion in green is kept between O2P atom of YG37 and O2 atom of 2'-O-methylated OMC32. In order to keep the wobble base pair, the base of U33 in magenta must be extended outward and destabilized because of the loss of hydrogen-bonding energy; (b) Stereo view of mRNA kinks used in the present work. Three codons at the A, P, and E sites, that are composed of all uridines, and two clusters Mg^{2+} hexahydrate binding at the P/A- and E/P-kinks, by representing green big ball for Mg^{2+} ions and yellow big balls for water molecules. The sharp kinks are formed by binding one Mg^{2+} hexahydrate at one kink instead of two, which are observed in X-ray structures; (c) Stereo view of cognate A-site codon-anticodon helix and the ASL of the present work, in which U33 of the ASL and A1396 at the decoding site of 16S rRNA are drawn in magenta and brown, respectively. Significance of A1396 will be described in the following Figure. This picture is shown so as to look similar to that in (a) as far as possible; (d) Stereo view of near-cognate A-site codon-anticodon helix and the ASL of the present work, in which G-U wobble base pair at the first base pairs of the A-site codon-anticodon helix is drawn in the same colors as those in (a) and both U33 and A1396 are as those in (c).

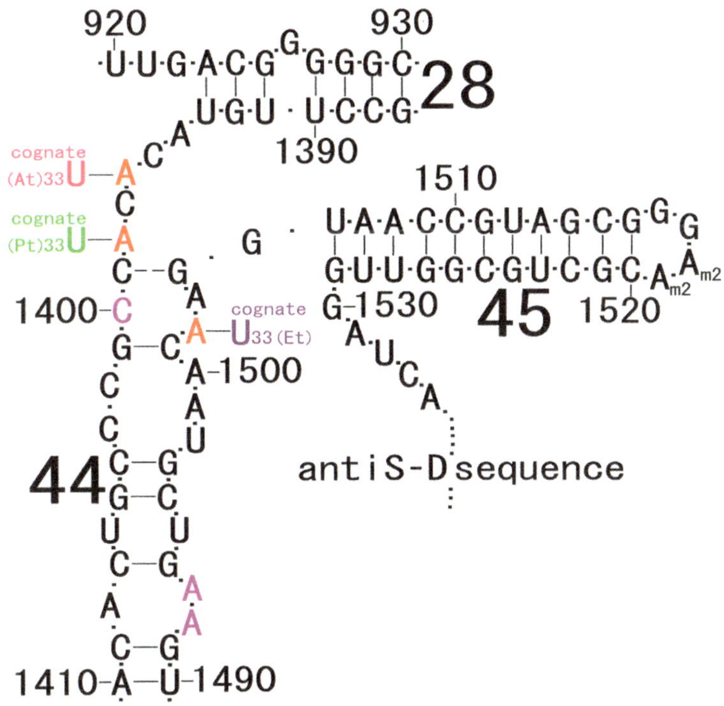

Figure 4. Assignment of three nucleotides, A1396, A1398, and A1502, on the secondary structure of the decoding site of 16S rRNA. Small numerals are residue numbers of 16S rRNA [4], while large numerals denote helix numbers [5]. Base pairs are represented by single bars for both A-U and G-C pairs, while two short bars between C1399 and G1504 represents that the G-C pair is formed to reject either U33 at the A site or at the E site and not formed in the transition state in which all three U33 bases at the A, P, and E sites are base-paired with A1396, A1398, and A1502, respectively. Two A_{m2} at the positions of 1518 and 1519 are universally conserved dimethylated adenines. See text for colored nucleotides at positions, 1400, 1492 and 1493. S-D denotes Shine-Dalgarno.

the crystal structure of 2HGR [2]. On the other hand, G1401-C1501 pair as well as the base stacking of C1400 to the first base of the anticodon of P-site tRNA [28] was confirmed. It has been known that both adenines and cytidines in the decoding site are universally conserved. Accordingly, it was quite natural to think that U33 base of the U33-extended form of P-site tRNA could be base-paired with A1398. It necessarily leads to the assignment of A1396 to the base-pairing partner of the U33-extended form of aa-tRNA at the recognition mode of A site (T site). It seemed possible for either A1502 or A1503 with U33 of the U33-extended form of deacylated tRNA at the E site. The result of the present study has shown that A1502 is most favorable, while A1503 seems a little too far from the U33 of E-site tRNA.

4. A-P and P-E tRNA Docking Pairs in the Pre- and Posttranslocation States, Respectively

The universal rule of nucleotide sequence complementarities (**Figure 1**) was first noticed by a finding that oligonucleotide fragments such as TΨCG or UUCG inhibited the binding of aa-tRNA at the A site [29]. It was found that such a highly conserved sequence exists on the 3'-side of h35 of 16S rRNA, and that the other side of h35 is complementary to the conserved loophead sequence of SRL in 23S rRNA [14]. In addition to the above finding, it was also found that various tRNA fragments containing D-loop inhibited aa-tRNA binding at the A site [29]. This strongly suggested that the T-loop of aa-tRNA making base-pairs with the 3'-side of h35 and the D-loop of aa-tRNA making base-pairs with the T-loop of P-site tRNA are an essential necessary requirement for the aa-tRNA entering the A site. This idea was combined with the α-ε model of [30], resulting in the tRNA docking pair model [31]. In contrast to the X-ray structure of the elbow region of the isolated L-form tRNA, in which G19 is paired with C56 as a Watson-Crick type, but G18 is hydrogen-bonded to Ψ55 in an irregular fashion [26], the universally conserved guanines, G18 and G19, of aa-tRNA at the T site are base-paired with the other universally conserved cytidines, C56 and C61, of P-site tRNA, respectively, and also G18 and G19 of

P-site tRNA with C56 and C61 of E-site tRNA, respectively. Such a big difference in the shape of tRNA between the isolated L-form of X-ray structure [1] [2] [26] and the docking model of the present work also comes from a big difference in the Mg^{2+} concentration, as mentioned above.

The results of X-ray structures of the whole ribosomes have shown that two conserved regions of 16S rRNA, the ridge containing G1338 and A1339 near h29 and the head of 790 loop, constitute barriers that block a translocational movement of P-site bound tRNA toward E site, if we assume that such a movement is just a simple-minded translation from the crystallographic P site to the crystallographic E site. Transition from the X-ray structure to the predicted one of the present study would start from the U33-extended form of aa-tRNA by interaction between U33 (aa-tRNA) and A1396, which is exposed on the X-ray structure, as well as the base-pairing of GTΨCG (or GTΨCA) of aa-tRNA with CGAGC1107. The effector region in the body of elongation factors, both EF-Tu and EF-G, would collide with the long helix composed of h30, h32, and h34 [4] [5], resulting in accomplishment of the long-range interactions between the SRL-region of 23S rRNA and the 5'-side of h35 of 16S rRNA. Under the condition of fixing the location of A76 of both P- and E-site tRNAs, the U33-extended forms make the distance between G18 (P-tRNA) and C56 (E-tRNA) shorter and C56 (P-tRNA) closer to G18 (aa-tRNA), and also, the twisting makes splitting of D-loop from T-loop for both P- and E-site tRNAs at their respective elbow regions. Such a conformational change for both P- and E-site tRNAs allows base-pairings between G18 (aa-tRNA) and C56 (P-tRNA) and between G18 (P-tRNA) and C56 (E-tRNA), and successively, G-C pairs between G19 (aa-tRNA) and C61 (P-tRNA) and between G19 (P-tRNA) and C61 (E-tRNA), respectively. Thus, the A-P-E tRNA docking triplet form can be completed, and finally accomplishes the codon-anticodon helix at the E site. This is a series of conformational changes after unfolding of h35 by the collision with the body of elongation factors, EF-Tu and EF-G, ending up with formation of the codon-anticodon helix at the E site.

Figure 5(a) shows that U33 at the P site (green) is base-paired with A1398 (red), U33 at the E site (purple) is base-paired with A1502 (pink), but U33 at the A site (magenta) is not base-paired with A1396 (brown). C1399-G1504 pair is formed and shown in gold. **Figure 5(b)** shows that U33 at the P site (green) is base-paired with A1398 (red), U33 at the A site (magenta) base-paired with A1396 (brown), but U33 at the E site (purple) is not base-paired with A1502 (pink). C1399-G1504 pair (gold) is formed. On the other hand, **Figure 5(c)** shows that all three U33 at the A, P, and E sites, (magenta, green, and purple, respectively), are base-paired with A1396 (brown), A1398 (red), and A1502 (pink), respectively, at the expense of separation of the bases of C1399 and G1504 (gold). Since both tRNAs at the P and E sites are cognate, in contrast to the case of X-ray structure, aa-tRNA at the T site would take a transition-state conformation, as shown in **Figure 5(c)**, if it is also cognate. During fluctuation of the whole structure, both residues, C1399 and G1504, would become close enough to base-pair, as they are not so far separated. Then, A-U base pair of either A site or E site is broken. Which A-U pair is broken depends on the stability of the codon-anticodon helices at the A and E sites, since the distance between the U33 base and C1400 base, for example, would become larger in the case of less stable anticodon loop. From this reason, both noncognate (before GTP hydrolysis) and near-cognate aa-tRNA (at the step of GTP hydrolysis) are rejected, while only cognate one (after GTP hydrolysis) can enter the A site to expel the deacylated tRNA at the E site from the ribosome. This would be a reasonable explanation of the mechanism of translation concerning how high accuracy could be achieved at the decoding site.

5. How E-Site tRNA Is Expelled Automatically When A-Site tRNA Turns out to Be Cognate

Figure 6(a) shows a stereo view of P-site tRNA (green) surrounded by various regions of 16S and 23S rRNAs. It indicates that the SRL region of 23S rRNA (yellow) is bound together with the region of G1064-G1077 of 16S rRNA (skyblue), taking the conformation (**Figure 2**). It is also important to note that the region of 16S rRNA is extended from the main part of 30S subunit. In order to achieve the above long-range base-pairing interactions, it is essential that the long helix composed of h30, h32 and h34 of 16S rRNA should have bends of a considerable degree around the regions between h32 and h34 as well as between h30 and h32. It necessarily causes a big deviation of the barrier, G1338 and A1339, toward the joint region of h32 and h34 from its original position, and the location of h45 is changed very much. As a result, the universally conserved dimethylated adenines, $A_{m2}1519$ and $A_{m2}1520$, at the loop of h45 makes a contact with the downstream of mRNA. This result can explain why those residues are dimethylated, while the X-ray structure, 2HGR [2], cannot. A1396 approaches the U33 base of aa-tRNA, while h44 of 16S rRNA makes a contact at the inner groove of A-site codon, with un-

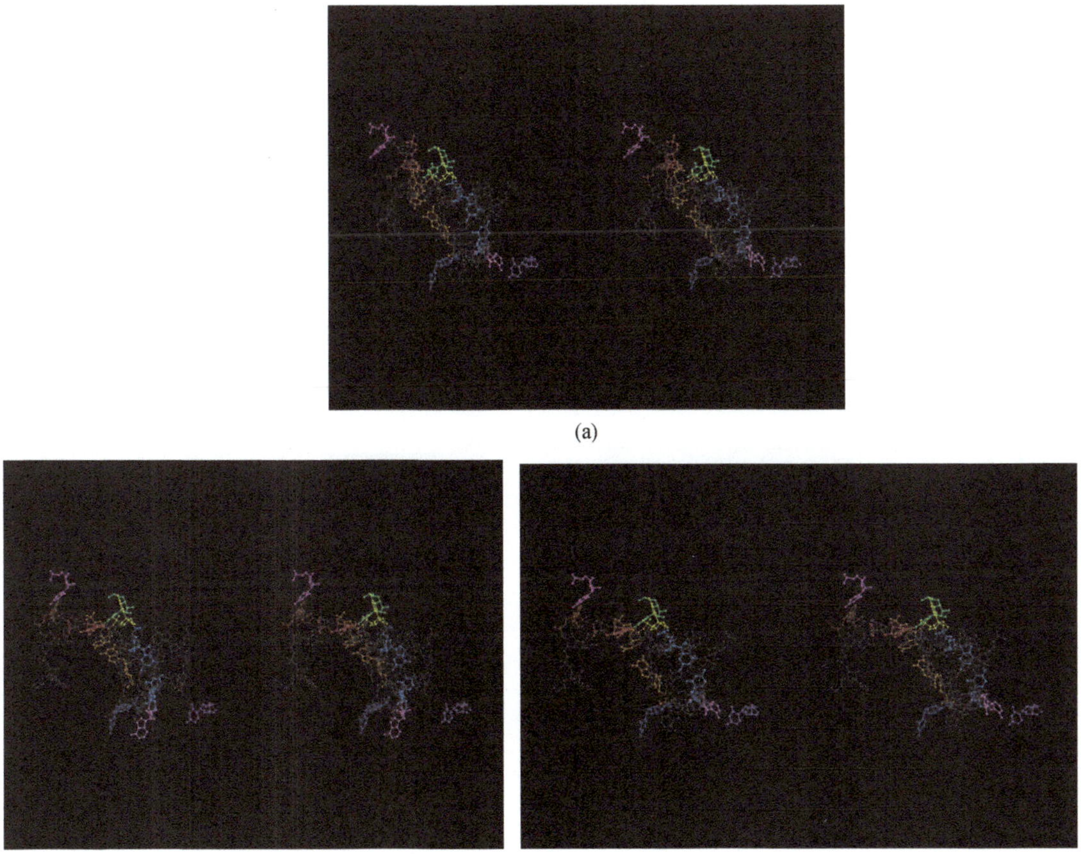

(a)

(b) (c)

Figure 5. Stereo views of full-atom representations for the residues at the decoding site of 16S rRNA. (a) U33 of P-site tRNA (green) is base-paired with A1398 (red), that of E-site tRNA (purple) base-paired with A1502 (pink), but U33 (magenta) at the recognition mode of A site (T site) is not base-paired. A1396 (brown) is far from the U33 base (magenta). G1401-C1501 (skyblue) is always formed. C1399-G1504 (gold) is formed in this case. This structure leads to rejection of either noncognate or near-cognate aa-tRNA from the A site. C1400 is drawn in yellow, while U1506, which starts the h45, is in blue; (b) This picture is very similar to (a), but different from (a) in the following two points: U33 at the A site (magenta) is base-paired with A1396 (brown), and U33 at the E site (purple) is not base-paired with A1502 (pink); (c) This picture is also very similar to (a) and (b), except that all three U33 at the A, P, and E sites are base-paired with A1396, A1398, and A1502, respectively, but C1399-G1504 (gold) is broken.

iversally conserved A1492 and A1493 [32].

Figure 6(b) shows the transition-state conformation of three tRNAs, in which E-site tRNA as well as both A- and P-site tRNAs are usually all cognate. This structure corresponds to the local structure of the decoding site (**Figure 5(c)**). When U33 of the A-site tRNA is base-paired with A1396 and U33 of the E-site tRNA is not base-paired with A1502 (**Figure 5(b)**), the deacylated tRNA at the E site (purple) in the three tRNA binding structure changes its conformation from the U33-extended and docking with P-site tRNA state to the U33-folded and isolated form (**Figure 6(c)**). Here, the deacylated tRNA is just leaving the ribosome in the direction of L1 stalk. **Figure 6(d)** shows what would happen to the T site when the conformations of nucleotides at the decoding site are such as those in **Figure 5(a)**. When the aa-tRNA (magenta) is noncognate, the base-pair of the codon-anticodon helix would not be formed without need of U33-A1396 base-pair formation. In the case of either near-cognate or cognate, the codon-anticodon helix could become an ideal A-form, and the help of U33-A1396 would be very important at the step of GTP hydrolysis. After fluctuation of the conformations at the decoding site between those as in **Figure 5(c)** and **Figure 5(a)**, the helix at the T site would be broken, if the one at the E site is a little stronger. Then, the conformation of the aa-tRNA (magenta) would become a U33-folded form with its isolated L-form elbow region, and leaves the ribosome in the direction of L7/L12 stalk. At the instant, the conformation of the A-site codon would loosen, and become ready for accepting the next anticodon of aa-tRNA

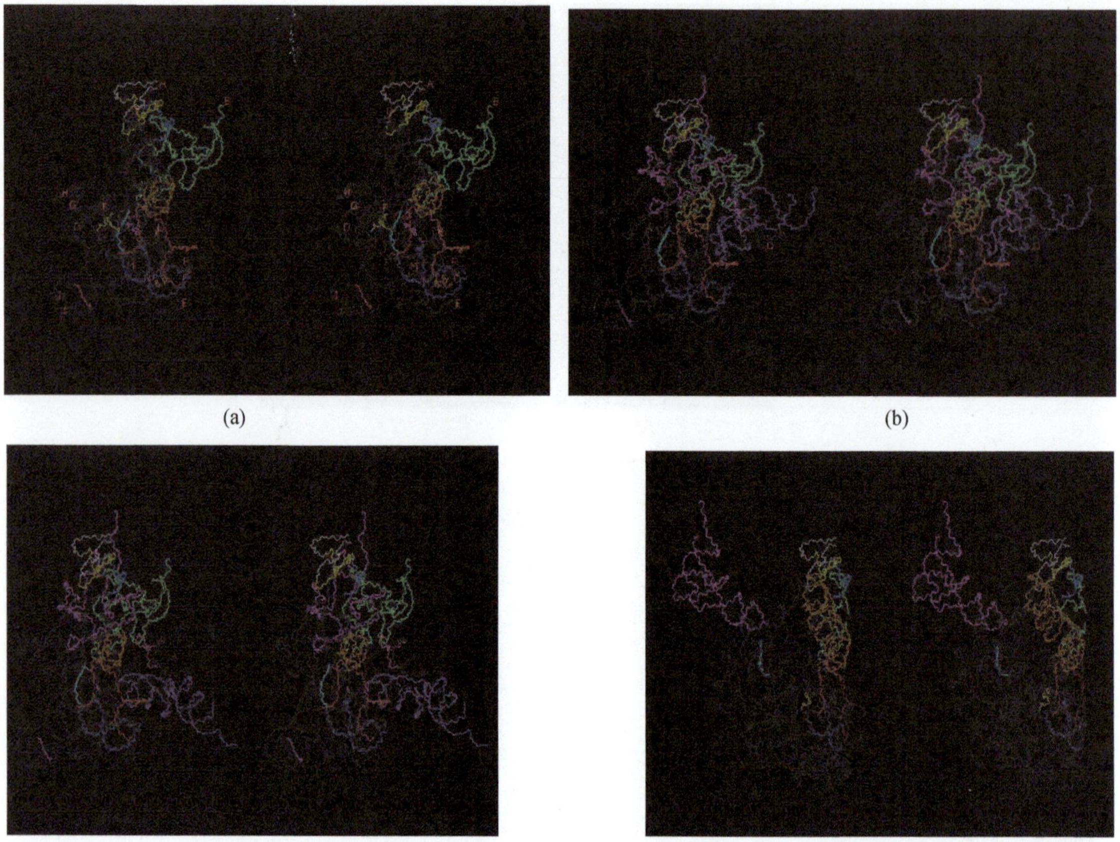

(a) (b)

(c) (d)

Figure 6. Stereo views of main-chain representations for the conformations of tRNAs and mRNA. (a) P-site tRNA (green) in the center. Nucleotides are G2645-C2652 (white) and G2668-A2675 (white), with a label A indicating the location of G2645. The region of U2653-C2667 (yellow) is the SRL of 23S rRNA. Label B denotes the location of A76 of tRNA at the P site. mRNA (red) with label C for its first residue. The following regions of 16S rRNA are shown in different colors: G1064-G1077 (skyblue), C1103-C1107 (cyan), G527-A533 (yellow tint), U1189-A1191 (pink), G1338-A1339 with bases (gold), A1408-A1420 (orange), G1480-A1493 (orange), U1506-G1529 (blue), C1399-G1504 pair with bases (violet), and A1396 (brown). Labels from D to P denote the locations of C545, G922, U956, U991, A1041, G1117, A1180, A1236, G1334, A1350, C1384, C1452, and U1542, respectively; (b) Transition-state three tRNAs, A-site tRNA (magenta), P-site tRNA (green), and E-site tRNA (purple). This structure corresponds to the local picture of the decoding site represented in **Figure 5(c)**. Red label Q denotes the location of C805 of 16S rRNA; (c) Three tRNA binding model, in which E-site tRNA (purple) takes an isolated L-form with its U33-folded anticodon loop. This structure corresponds to the local picture of the decoding site represented in **Figure 5(b)**; (d) Corresponding to the local picture of the decoding site in **Figure 5(a)**, aa-tRNA (magenta) takes an isolated L-form with its U33-folded anticodon loop.

as a TC, that would have probably been waiting for opening the A-site codon by using the 3'-side of h35 and A1394. A1394 is also highly conserved, although it is less conserved than A1396 and A1398.

6. Translocation as a Rotational Movement of A-P tRNA Docking Pair to P-E Pair

The effect of spectinomycin on the EF-G-dependent translocation published in the old papers [22] [23] are mentioned before. Recently, Fredrick and Noller did not observe such an effect during the course of study on sparsomycin-dependent translocation [33]. On the other hand, Peske *et al.* [34] have concluded that EF-G produces conformational rearrangements on the ribosome in the vicinity of the decoding region on the 30S subunit and that spectinomycin has a direct inhibitory effect on tRNA-mRNA movement, that is quite consistent with the basic assumptions of the present work. Comparison of **Figure 6(b)** with **Figure 6(c)** and **Figure 6(d)** suggests that translocation is not a translational movement of tRNAs from the crystallographically obtained P site to the E site but a rotational one around an axis that is close to both A- and E-site tRNAs. If the aminoacyl group remains at the A site and the P-site tRNA is still peptidylated, translocation would not occur because of locking at the P site [13]. Peptidyl transferase reaction results in peptidyl-tRNA at the A site and deacylated tRNA at the P site, the state of which automatically brings about P and E sites on the 50S subunit. This is due to a rotational freedom around the hinge between H14 and H22 (nucleotides AA423) and a rigid binding of two helices, H22 and H88 at a pseudoknot CCAU416/AUGG2410 of 23S rRNA. C2394 and G2421 at the base of H88, which is empty as a result of the process shown in **Figure 6(c)**, could turn to catch the adenine base of the CCA end of deacylated tRNA at the P site [3]. According to chemical modification experiments, 2'-hydroxyl groups of the nucleotides 71 and 76 of the deacylated P-site tRNA can be recognized by nucleotide C1892 in the minor groove of H68 and A2433 and A2434 at the base of H74 of 23S rRNA, respectively [35]. Those residues also constitute part of the E site on the 50S subunit. It would expel the P-loop at the loop of H80 as well as A2450 and its associated nucleotides of the central loop of 50S subunit from the CCA end of P-site tRNA, that is the P site on the 50S subunit, and transfer them to the CCA end of the neighboring A-site tRNA. These are what we call P/E and A/P hybrid states. Such a transition would not occur if either the 3'-end of P-site tRNA is peptidylated or some of the above key nucleotides are chemically modified. This could be an explanation as to why peptidyl-tRNA bound to the P site locks ribosomal functions such as translocation, termination, and recycling [13]. **Figure 7(a)** shows that the foot of H88 (cyan), where C2395 and C2421 are situated, reaches the adenine base of P-sitet RNA (green), while the P-loop of H80 (red) catches the bases of C74 and C75 of A-site tRNA (magenta). This is the 3Dstructural representation of the 3'-end regions of tRNAs at the A/P and P/E hybrid states. When EF-G·GTP complex comes to occupy the space of A-site tRNA, the A-P tRNA docking pair rotates around the axis passing closer to the A-, P- and E-site tRNAs to reach the site of P-E tRNA docking pair at the classical P/P and E/E states. The 3D structural representation of them can be shown as in **Figure 7(b)**.

7. How Elongation Factors Work on the Present Models to Produce GTP Hydrolysis?

The recent result of the crystal structure of the ribosome bound to EF-Tu and aa-tRNA [36] does not seem to fully explain the mechanism of accommodation of aa-tRNA into A site. Starting from the tRNA structure in magenta in **Figure 6(d)**, which is either leaving the ribosome or entering the ribosome as a TC with EF-

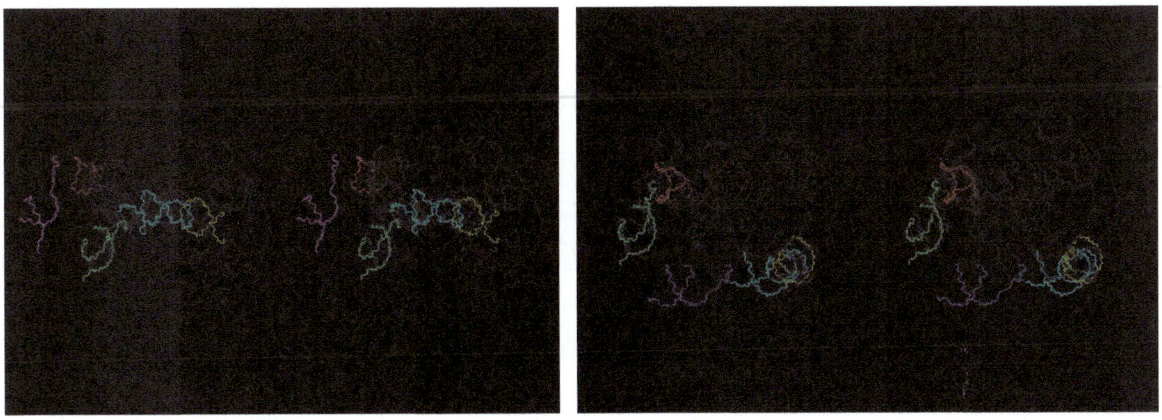

(a) (b)

Figure 7. Stereo views of main-chain representations for the conformations of the 3'-end regions of tRNAs surrounded by the nucleotides at the H80 (P-loop) and H88. It is known that C2422 can bind to the 3'-end adenine base (A76) of deacylated tRNA at either P or E site. The loop of H88, A2407-G2410, is tightly bound with C413-U416 at the loop of H22, which is drawn in yellow, and works as widely movable long rod, by fixing its one end at the position of G424 on the 50S subunit, between the 3'-end region of P-site tRNA (green) in the PRE state, and that of E-site tRNA (purple) in the POST state. (a) The structure of 3'-end regions of two tRNAs on the 50S subunit in the PRE state. The regions of G1-U8 and A66-A76 of A-site tRNA (magenta), G1-U8 and A66-A76 of P-site tRNA (green), G2246-C2258 (H80) (red), C2395-G2421 (H88) (cyan), G406-C421 (H22) (yellow) of 23S rRNA and the other associated regions (grayish white) are drawn. This corresponds to the intermediate structure in the A/P and P/E hybrid states; (b) The structure of 3'-end regions of two tRNAs on the 50S subunit in the POST state. The regions of G1-U8 and A66-A76 of P-site tRNA (green), G1-U8 and A66-A76 of E-site tRNA (purple), G2246-C2258 (H80) (red), C2395-G2421 (H88) (cyan), G406-C421 (H22) (yellow) of 23S rRNA, and the other regions (grayish white) are drawn. This corresponds to the structure in the classical P/P and E/E states.

Tu·GTP, its 3' end region was modified by transplanting 3' end region of Trp-tRNA^{TRP} along with EF-Tu and GDP bound to a cognate codon on the 70S ribosome 2Y18 [37]. GDP structure was changed to GTP by adding Pγ phosphate. The conformation of EF-Tu was changed so that the Pγ atom comes closer to N6 atom of A2660 and Pβ atom closer to N4 atom of C2658. Pγ and Pβ atoms are covered by the region of amino acid residues 20 - 26 (VDHGKTT) in EF-Tu. The G base of GTP moiety is hydrogen bonded to U2656, while the O3' and O2' atoms on its ribose ring are hydrogen bonded to N6 atom of A2657 and O6 atom of G1074 of 16S rRNA, respectively. Such a local structure of GTP binding to the present model at the interface of RSL of 23S rRNA and the 5'-half of h35 of 16S rRNA on the ribosome structure is shown in **Figure 8(a)**. Inspection of this local structure suggests that GTP hydrolysis could occur when the region of EF-Tu along with GTP molecule moves

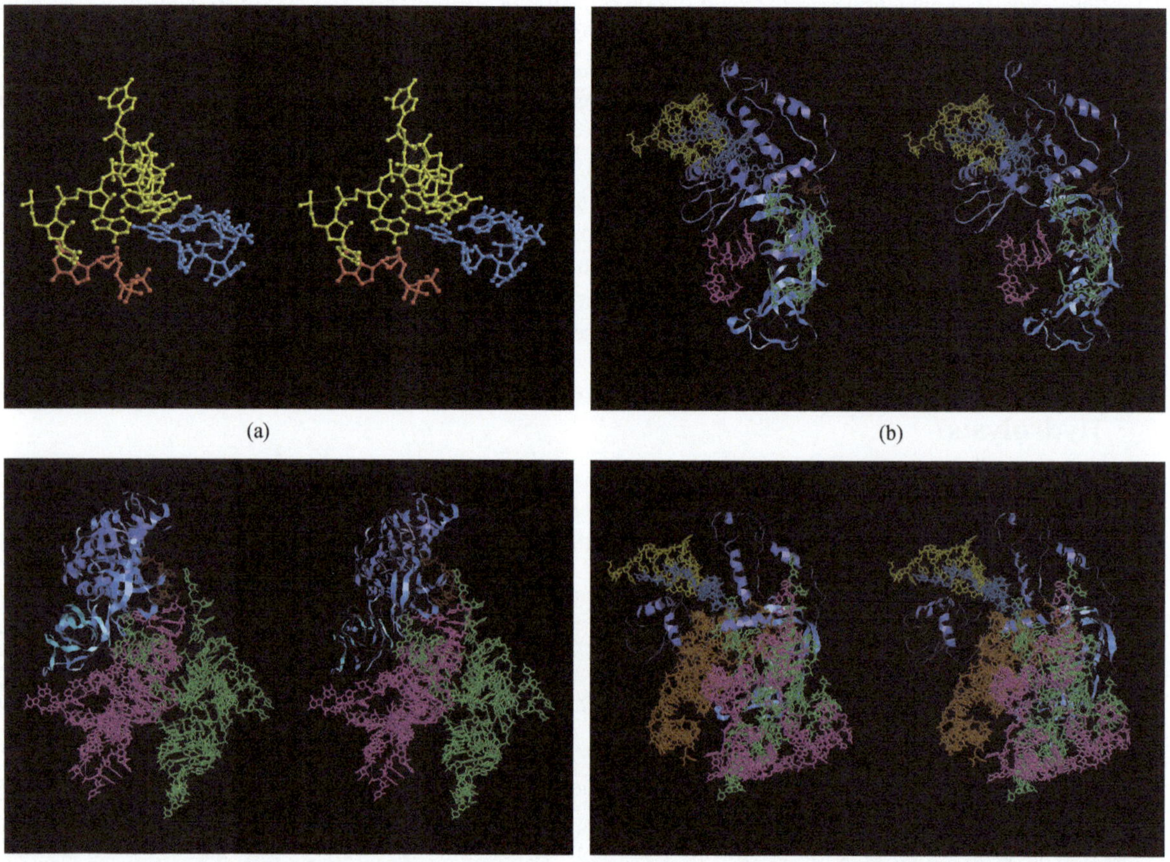

(a) (b) (c) (d)

Figure 8. Stereo views of elongation factors during the course of A-site accommodation and translocation with GTP hydrolysis. (a) Stereo views of main-chain representations of GTP molecule (red) at the border of RSL from U2656 to G2661 (yellow) and 5' half of h35 of 16S rRNA from G1069 to G1072 (skyblue); (b) The same model in (a) is shown from the other direction that is from the 3' ends of P-site tRNA (green) toward the mRNA. When the codon-anticodon interaction is that of cognate, the conformational changes at the ASL and the elbow region of T-site tRNA (magenta) induces the GTP hydrolysis, ending at the conformation of A-site tRNA as shown in **Figure 6(b)-(c)**. Blue ribbon, and the other nucleotide chains represent the peptide chain of EF-Tu from A1 to E405 with VDHGKTT (20-26) missing, RSL region (C2652-C2667 in yellow), 5' side of 16S rRNA (G1064-G1077 in skyblue), T-site tRNA (A5-U8, A14-U15, 5MC49, U68, U72 and A76 in magenta), and P-site tRNA (G45-C56 and C61-G65 in green), respectively. The regions of T- and P-site tRNAs are only shown for those in close contact with EF-Tu. The region of U72-A76 in Trp-tRNA^{TRP} (brown) is taken to connect the main body of T-site tRNA and the EF-Tu polypeptide regions (ribbon) stretching toward GTP binding site by A1-R190 and P-site tRNA by G191-E405; (c) The location and conformation of EF-Tu is shown after the accommodation of A-site tRNA is finished. EF-Tu leaves the ribosome immediately. The rest of the model is the same as that of **Figure 6(c)**; (d) The ancestral form of EF-G bound to the PRE state of ribosome is shown. The body of tRNA in gold corresponds to the domain IV of EF-G. The rest of the ancestral EF-G is the same as that of EF-Tu, which appeared in (b). Since A-site is occupied by a big body of tRNA, the location of GTP binding to the border of RSL of 23S rRNA and 5' half of h35 is shifted by as far as 27 Å from the P-site tRNA.

in the direction from N4 atom of C2658 to N6 atom of A2660 in contrast to the movement vertical to the above. When the distance between the N4 and the N6 atoms becomes larger accompanying the Pβ and Pγ atoms, respectively, NH2 group of the N2 position of G2659 attacks the Pβ-Pγ bond. The total structure of EF-Tu with the related regions of T- and P-site tRNAs is shown in **Figure 8(b)**. After GTP hydrolysis in the structure shown in **Figure 8(b)**, the conformation of EF-Tu returns to the original one observed in the crystal structure of the ribosome bound to EF-Tu and Trp-tRNA^{TRP} 2Y18 [37]. Such a structure is shown in **Figure 8(c)**. The EF-Tu molecule would leave the ribosome soon under the normal condition of low Mg^{2+} concentration. The conformation of the A-site tRNA is the same as those shown in **Figure 6(b)-(c)**. Since the binding of A-site tRNA very close to P-site tRNA, the location of which is almost the same as the other structures shown in **Figure 6**, the cavity between the RSL region and P-site tRNA becomes a little wider, where EF-G is supposed to bind. The structure of tRNA in gold, as shown in **Figure 8(d)**, is a putative ancestral form of EF-G, in which the tRNA-like body in gold represents the domain IV of EF-G [38], while the other part represents the rest of the whole structure of EF-Tu. If the conformational change in the part of EF-Tu occurs after GTP hydrolysis, the structure very much like the one shown in **Figure 8(b)** would be obtained. Because of the exposed form of the domain IV of EF-G, or the tRNA moiety in gold, would be very unstable, the structure would become immediately the same as that in **Figure 8(c)** after translocation. This is exactly what happened in the crystal structure of the ribosome shortly after binding of EF-G [39]. The reason why EF-G dependent translocation is inhibited by binding of spectinomycin ([22] [23] [33]) can be explained very clearly as the result of missing the contribution from the 5' half of h35 in the mechanism of GTP hydrolysis shown in **Figure 8(a)**. In contrast to the EF-G dependent translocation, the A site tRNA accommodation catalyzed by EF-Tu does not seem to be affected by spectinomycin. The reason for this would be that the region of CGAGC1107 on the 3' half of h35 is unfolded by interaction with the GTΨCG region of T-loop of aa-tRNA, or could be due to a slight difference in the distance between Pγ atom and 3' end of P-site tRNA (27 Å).

8. How the First Base of Stop Signals, U, Can Be Recognized?

After translocation, the new A-site codon is open for the next aa-tRNA. Before going to further recognition of aa-tRNA, how is it judged to be either a sense codon or a stop signal such as UAA, UAG or UGA? Three kinds of release factors are known to recognize stop signals as follows: RF1 in prokaryote has the conserved tripeptide motif Pro-Ala-Thr for UAG and UAA, RF2 in prokaryote has the motif Ser-Pro-Phe for UGA and UAA, and a single eukaryotic one eRF1 has the tetrapeptide motif Asn-Ile-Lys-Ser for UAG, UGA, and UAA [40]. It was insisted that a tripeptide Pro-Ala-Thr in prokaryotic RF1 can recognize the stop signals specifically [41]. Furthermore, the above three motifs of class-1 RFs have been believed to recognize stop signals directly as if they are anticodons, and are confirmed for RF2 by cryo-EM [42] [43] and for eRF1 by analogy from its tRNA-like shape of the X-ray structure [44]. However, a big question still remains unsolved as to how the first base of stop signals, U, is recognized [40]. Here, a hypothetical mechanism is presented of base-pairing of stop signals with a conserved nucleotide sequence UUA1092 at the loop of h37. If the codon is UGG, the UGG/UUA base pairs would not be stable enough because of two consecutive wobble base pairs. It would be more favorable for such a transient helix UUA/stop signal to allow one of the tri- or tetrapeptide motifs to come close enough, and to withdraw immediately after one of the release factors achieves a direct interaction with the stop signal. It seems to be a common mechanism to all living kingdoms except for mitochondria of some protista (*Trypanosoma brucei* and *Plasmodium falciparum*) and animalia [14] because of missing h37 in their nucleotide sequence of small subunit rRNAs. UGA is used as the codon for Trp-tRNA^{Trp} of animalia mitochondria (See Table 12 of [45]). The evolution of tRNA^{Trp} seems to have been developing in relation with that of stop signals and also release factors. **Figure 9** shows that the UUA sequence at the loophead of h37 (pink) is situated at a good location of finding one of the stop signals in the A-site codon (red) before any release factor comes to form a more stable complex.

9. Negative Cooperativity between A and E Sites

Nierhaus and his coworkers have insisted that a deacylated tRNA bound to E site and cognate to the E-site codon allosterically and negatively regulates affinity for a cognate tRNA entering to A site [46] [47]. Since any proofreading mechanism in translation other than kinetic one [16] has never been proposed, the present study has started to provide a theoretical background for the attractive proposal [46]. The present mechanism of transi-

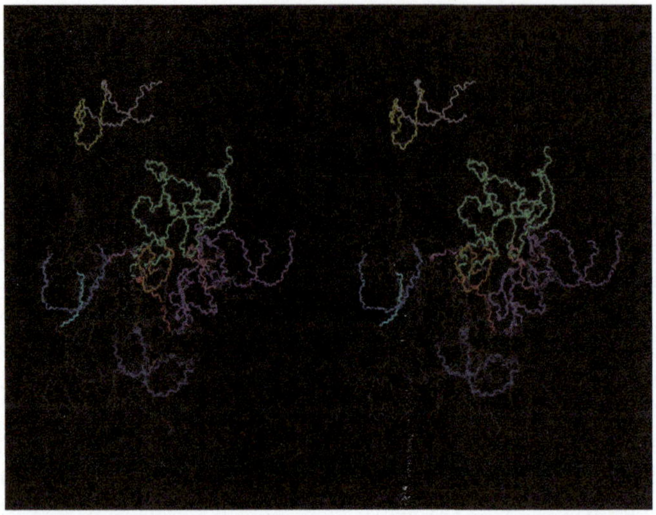

Figure 9. Stereo views of main-chain representations for the structures of the regions changed from those drawn in **Figure 6(c)** shortly after translocation. The base-pairing interaction between the SRL region and the 5'-end region of h35 disappears. The hypothetical interaction between the region of UUA1902 (pink) of h37 and a stop signal UAA (red) is shown. Only three codons for A-, P-, and E-site anticodons are drawn in red. One reason for the hypothesis is that h37 is missing for animalia mitochondria accompanied with a change in UGA codon usage for $tRNA^{TRP}$. The other regions are as in **Figure 6(a)** and **Figure 6(b)**.

tion-state three-tRNA structure represents another aspect of kinetic proofreading mechanism [16], because it is accompanied by a step of GTP hydrolysis. In the present model-building works, the three-tRNA binding model at the transition state (**Figure 6(b)**) strongly supports the negative cooperativity between A and E sites, because A- and E-site tRNAs are very close to each other, although the two tRNAs do not have direct contacts with each other. However, the experimental findings of Nierhaus and his coworkers [46]-[50] described above have recently been refused to accept by many research groups [51]-[54]. Chen *et al.* [53] analysed how E-site tRNA is dissociated early in protein synthesis. They denoted the two options as the 2-3-2 tRNA and 2-1-2 tRNA pathways, where the integers indicate the number of tRNAs occupying the ribosome during the conversion of the POST complex to the PRE complex. On the 2-1-2 pathway E-site tRNA leaves the ribosome before aa-tRNA is accommodated with the A site. In this study, the model of the 2-3-2 pathway was presented, while the conformation of the tRNA anticodon loop of the 2-1-2 pathway would be the U33-folded form, and both the codon-anticodon base-pairings at the E site and the docking G-C pairings between P- and E-site tRNAs would be lost. Since the P-site tRNA slippage in the analysis of +1 programmed ribosomal frameshifting (PRF) [55] would mean a considerable degree of dynamic behavior of the anticodon loops at the decoding site, the competition of U33-extended and U33-folded forms would also be a realistic problem depending on various factors, such as Mg^{2+} concentration, conformation of the nascent polypeptide, specific nucleotide sequence of mRNA, etc. The transition from the U33-extended form to the U33-folded form and vice versa at the P-site tRNA under the condition that the codon-anticodon helix is kept at the E site and that the A-site codon is open would cause a P-site tRNA slippage even if the +1 frame is near-cognate [56]. In the absence of the codon-anticodon interactions at the E site, −1 PRF could also occur in a comparable probability with that of +1 PRF. If the transition gives rise to a noncognate +1 PRF, it would soon come back to the original zero frame. Even if such a transition is rather frequent, it does not show any detectable change at the decoding site, if the frameshifting always returns to the original frame. Accordingly, the findings so far obtained about the PRFs are quite consistent with the 3D structural models presented in this study. A Mg^{2+} hexahydrate ($Mg(H_2O)_6{}^{2+}$) was found to bind to a phosphate oxygen of 37th nucleotide of the anticodon loop of yeast phenylalanine tRNA (6tna) [26]. It is located between O2 atom of O-methyl cytosine at the 32nd position (OMC32) and O3' atom of 36th nucleotide of the anticodon loop. This would be a result of hydrogen bonding between O4 atom of U33 and OP2 atom of the 36th nucleotide. However, if the Mg^{2+} concentration is getting higher and higher, U33-folded form would become prevailing. The distance between two phosphoric oxygen atoms of a kink of mRNA between the U33-folded forms of two adjacent codon-anticodon helices would be fixed around the range near 8 Å by intervention of a Mg^{2+} hydrate. This would make the anticodon of the E-site tRNA unable to form a helix with the E-site codon.

Nierhaus and his coworkers have been using 4.5 mMol Mg^{2+} and some amount of polyamines as a buffer of physiological condition [57]. Petropoulos and Green [54] used the same buffer, and yet refused the negative cooperativity between A- and E sites [46]. Chen *et al.* [53] observed that the 2-3-2 pathway only occurred shortly after initiation and followed by consecutive 2-1-2 pathways and that the translation pathway is mainly determined by the PRE conformational states (either classical or hybrid). They used TAM[15] buffer, which contains 15 mMol $MgAc_2$ [53]. It seems to me that they are using too high Mg^{2+} concentration, even if they lowered it from 15 to 7 mMol [53]. Every group should discuss with each other in terms of accuracy in translation. I insist that the 2-3-2 pathway would prevail if the PRE states are those of U33-extended and A-P tRNA docking pair with a much less Mg^{2+} concentration. G18-G19 as well as C56 and C61 in tRNA sequence are highly conserved for all living kindoms [14]. Proofreading mechanism in translation should be explored under a condition of 1 mMol Mg^{2+} concentration.

10. Conclusion

The mechanism of translation was described in detail in the preceding sections. Several important points, however, still remain untouched, and should be concluded here. The point 1: The locations of tRNAs for A, P, and E sites on the X-ray structures of the whole ribosomes were distantly separated from each other, and barriers seemed to bother smooth translocation from P- to E-sites, if the translocation could occur as a simple-minded translational movement [1] [2] [6] [7]. In translocation, A-site tRNA must jump into P site, cooperatively with the previous P-site tRNA jumping into E site. It is hard to explain mechanistically, but, more difficult is to imagine how E-site tRNA is automatically expelled when a new tRNA at the A site turns out cognate. The present paper described such a mechanism in detail. Here, I can only explain very easily how X-ray structures of the whole ribosome can be obtained from my models of A-, P- and E-site tRNAs. When the Mg^{2+} concentration exceeds 6 mMol, the local conformations of the above three tRNAs at the anticodon loops, and their elbow regions change drastically and break the G-C pairs made between two neighboring tRNAs. In order to understand rationally the process of translocation in the X-ray structure, we must shift to our models of translocation, carry out the rotational movement of translocation, and again return to the posttranslocation X-ray structure. The formation of the long helix made by the SRL region of 23S rRNA and 5' side of h35 of 16S rRNA, which is referred to as GTPase-binding site, could only be achieved under the low Mg^{2+} concentration, and would disappear under the condition of crystallization of X-ray structure probably because of less flexibility of rRNAs. The point 2: Another point to be subjugated was deprotonation from N3 and protonation of O4 in the uridine base of wobble base pair under the condition of Mg^{2+} concentration higher than 6 mMol making it impossible to discriminate between G-C and G-U pairs, in other words, between cognate and near cognate codon-anticodon interactions [27]. It would be essential to adopt a low Mg^{2+} concentration to avoid the problem, and to have a possible interaction of U33-extended form of the anticodon loop of tRNAs with the very highly conserved nucleotide sequences at the decoding site. The point 3: The mechanical movement of translocation was interpreted as a rotational movement, while hybrid states such as A/P and P/E were able to be achieved without need of surmounting a high energy barrier as a rotational movement of a long rod of helical rRNAs made by a rigid binding of H22 and H88 under a condition that P-site tRNA is deacylated. The point 4: GTP hydrolysis results in a big conformational change in EF-Tu with a right hand (amino acid sequence from Ala1 to Arg190) pushing the GTPase-binding site and with a left hand (from Gly191 to Glu405) pulling the whole body of P-site tRNA. The final shape of A-site tRNA is just like a sitting-up of the T-site tRNA. On the other hand, GTP hydrolysis in translocation by means of EF-G produces a little different result from that of EF-Tu. This could be due to a slight difference in amino acid sequence between EF-Tu and EF-G (one amino acid insertion) and result in a different rotational movement with EF-G's left hand pushing the whole body of A-site tRNA toward P-site tRNA and with its right hand pushing the GTP-ase binding site to break the long helix made by the SRL region of 23S rRNA and 5'-side of h35 of 16S rRNA. The big distortion introduced by formation of the GTPase-binding site is cancelled and returns to the original state. This movement of the whole ribosome represents exactly where the driving force of the rachet-like movement of the ribosome comes from.

11. Methods

Programs and Atomic Coordinates

The programs for manipulating 3D models in the present work were written in FORTRAN95 by the present author. The atomic coordinates of the original forms of tRNAs and rRNAs were obtained from the Protein Data Bank. The predicted model for the virtual case of near-cognate codon-anticodon helix with high Mg^{2+} concentration (**Figure 3(a)**) was built by use of atomic coordinates 6tna [26]. In order to draw colored stereo pictures, a program called RasMol [58] was used.

References

[1] Yusupov, M.M., *et al.* (2001) Crystal Structure of the Ribosome at 5.5 Å Resolution. *Science*, **292**, 883-896. http://dx.doi.org/10.1126/science.1060089

[2] Korosterev, A., Trakhanov, S., Laurberg, M. and Noller, H.F. (2006) Crystal Structure of a 70S Ribosome-tRNA Complex Reveals Functional Interactions and Rearrangements. *Cell*, **126**, 1065-1077. http://dx.doi.org/10.1016/j.cell.2006.08.032

[3] Sergiev, P.V., *et al.* (2005) Function of the Ribosomal E-Site: A Mutagenesis Study. *Nucleic Acids Research*, **33**, 6048-6056. http://dx.doi.org/10.1093/nar/gki910

[4] Gutell, R.R. (1993) The Simplicity behind the Elucidation of Complex Structure in Ribosomal RNA. In: Nierhaus, K.H., Franceschi, F., Subramanian, A.R., Erdmann, V.A. and Wittmann-Liebold, B., Eds., *The Translational Apparatus*, *Structure*, *Function*, *Regulation*, *Evolution*, Plenum Press, New York, 477-488.

[5] Brimacombe, R. (1995) The Structure of Ribosomal RNA: A Three-Dimensional Jigsaw Puzzle. *European Journal of Biochemistry*, **230**, 365-383. http://dx.doi.org/10.1111/j.1432-1033.1995.0365h.x

[6] Selmer, M., *et al.* (2006) Structure of the 70S Ribosome Complexed with mRNA and tRNA. *Science*, **313**, 1935-1942. http://dx.doi.org/10.1126/science.1131127

[7] Yusupova, G., Jenner, L., Rees, B., Moras, D. and Yusupov, M. (2006) Structural Basis for Messenger RNA Movement on the Ribosome. *Nature*, **444**, 391-394. http://dx.doi.org/10.1038/nature05281

[8] Frank, J. and Agrawal, R.K. (2000) A Ratchet-Like Intersubunit Reorganization of the Ribosome during Translocation. *Nature*, **406**, 318-322. http://dx.doi.org/10.1038/35018597

[9] Agirrezabala, X., Lei, J., Brunelle, J.L., Ortiz-Meoz, R.F., Green, R. and Frank, J. (2008) Visualization of the Hybrid State of tRNA Binding Promoted by Spontaneous Ratcheting of the Ribosome. *Molecular Cell*, **32**, 190-197. http://dx.doi.org/10.1016/j.molcel.2008.10.001

[10] Kim, H.D., Puglisi, J. and Chu, S. (2007) Fluctuations of tRNAs between Classical and Hybrid States. *Biophysical Journal*, **104**, 13661-13665.

[11] Munro, J.B., Altman, R.B., O'Connor, N. and Blanchard, S.C. (2007) Identification of Two Distinct Hybrid State Intermediates on the Ribosome. *Molecular Cell*, **25**, 505-517. http://dx.doi.org/10.1016/j.molcel.2007.01.022

[12] Cornish, P.V., Ermolenko, D.N., Noller, H.F. and Ha, T. (2008) Spontaneous Intersubunit Rotation in Single Ribosomes. *Molecular Cell*, **30**, 578-588. http://dx.doi.org/10.1016/j.molcel.2008.05.004

[13] Valle, M., Zavialov, A., Sengupta, J., Rawat, U., Ehrenberg, M. and Frank, J. (2003) Locking and Unlocking of Ribosomal Motions. *Cell*, **114**, 123-134. http://dx.doi.org/10.1016/S0092-8674(03)00476-8

[14] Nagano, K. and Nagano, N. (2007) Mechanism of Translation Based on Intersubunit Complementarities of Ribosomal RNAs and tRNAs. *Journal of Theoretical Biology*, **245**, 644-668. http://dx.doi.org/10.1016/j.jtbi.2006.11.004

[15] Lotfield, R.B. and Vanderjagt, D. (1972) The Frequency of Errors in Protein Biosynthesis. *Biochemical Journal*, **128**, 1353-1356.

[16] Hopfield, J.J. (1974) Kinetic Proofreading: A New Mechanism for Reducing Errors in Biosynthetic Processes Requiring High Specificity. *Proceedings of the National Academy of Sciences of the United States of America*, **71**, 4135-4139. http://dx.doi.org/10.1073/pnas.71.10.4135

[17] Szer, W. and Ochoa, S. (1964) Complexing Ability and Coding Properties of Synthetic Polynucleotides. *Journal of Molecular Biology*, **8**, 823-834. http://dx.doi.org/10.1016/S0022-2836(64)80163-7

[18] Thompson, R.C., Dix, D.B., Gerson, R.B. and Karim, A.M. (1981) Effect of Mg^{2+} Concentration, Polyamines, Streptomycin, and Mutations in Ribosomal Proteins on the Accuracy of the Two-Step Selection of Aminoacyl-tRNAs in Protein Biosynthesis. *Journal of Biological Chemistry*, **256**, 6676-6681.

[19] Moazed, D. and Noller, H.F. (1974) Interaction of Antibiotics with Functional Sites in 16S Ribosomal RNA. *Nature*, **327**, 389-394. http://dx.doi.org/10.1038/327389a0

[20] Egebjerg, J., Larsen, N. and Garrett, R.A. (1990) Structural Map of 23S rRNA. In: Hill, E.E., Dahlberg, A., Garrett, R.A., Moore, P.B., Schlesinger, D. and Warner, J.R., Eds., *The Ribosome Structure*, *Function and Evolution*, American Society for Microbiology, Washington DC, 168-179.

[21] Carter, A.P., Clemons, W.M., Brodersen, D.E., Morgan-Warren, R.J., Wimberly, B.T. and Ramakrishnan, V. (2000)

Functional Insights from the Structure of the 30S Ribosomal Subunit and Its Interactions with Antibiotics. *Nature*, **407**, 340-348. http://dx.doi.org/10.1038/35030019

[22] Bollen, B., Davies, J., Ozaki, M. and Mizushima, S. (1969) Ribosomal Protein Conferring Sensitivity to the Antibiotic Spectinomycin in *Escherichia coli*. *Science*, **165**, 85-86. http://dx.doi.org/10.1126/science.165.3888.85

[23] Sigmund, C.D., Ettayebi, M. and Morgan, E.A. (1984) Antibiotic Resistance Mutations in 16S and 23S Ribosomal RNA Genes of *Escherichia coli*. *Nucleic Acids Research*, **12**, 4653-4663. http://dx.doi.org/10.1093/nar/12.11.4653

[24] Garcia-Ortega, L., Alvarez-Garcia, E., Gavilanes, J.G., Martinez-del-Pozo, A. and Joseph, S. (2010) Cleavage of the Sarcin-Ricin Loop of 23S rRNA Differentially Affects EF-G and EF-Tu Binding. *Nucleic Acids Research*, **38**, 4108-4119. http://dx.doi.org/10.1093/nar/gkq151

[25] Horan, L.H. and Noller, H.F. (2007) Intersubunit Movement Is Required for Ribosomal Translocation. *Proceedings of the National Academy of Sciences of the United States of America*, **104**, 4881-4885. http://dx.doi.org/10.1073/pnas.0700762104

[26] Holbrook, S.R., Sussman, J.L., Warrant, R.W. and Kim, S.H. (1978) Crystal Structure of Yeast Phenylalanine Transfer RNA II. Structural Features and Functional Implications. *Journal of Molecular Biology*, **123**, 631-660.

[27] Demeshkina, N., Jenner, L., Westhof, E., Yusupov, M. and Yusupova, G. (2012) A New Understanding of the Decoding Principle on the Ribosome. *Nature*, **484**, 256-259. http://dx.doi.org/10.1038/nature10913

[28] Prince, J.B., Taylor, B.H., Thurlow, D.L., Ofengand, J. and Zimmermann, R.A. (1982) Covalent Crosslinking of $tRNA_1^{Val}$ to 16S RNA at the Ribosomal P Site: Identification of Crosslinked Residues. *Proceedings of the National Academy of Sciences of the United States of America*, **79**, 5450-5454. http://dx.doi.org/10.1073/pnas.79.18.5450

[29] Sprinzl, M., Wagner, T., Lorenz, S. and Erdmann, V.A. (1976) Regions of tRNA Important for Binding to the Ribosomal A and P Sites. *Biochemistry*, **15**, 3031-3039. http://dx.doi.org/10.1021/bi00659a015

[30] Spahn, C.M. and Nierhaus, K.H. (1998) Models of the Elongation Cycle: An Evaluation. *Biological Chemistry*, **379**, 753-772.

[31] Nagano, K. and Nagano, N. (1997) Transfer RNA Docking Pair Model in the Ribosomal Pre- and Post-Translocational States. *Nucleic Acids Research*, **25**, 1254-1264. http://dx.doi.org/10.1093/nar/25.6.1254

[32] Ogle, J.M., Brodersen, D.E., Clemons Jr., W.M., Tarry, M.J., Carter, A.P. and Ramakrishnan, V. (2001) Recognition of Cognate Transfer RNA by the 30S Ribosomal Subunit. *Science*, **292**, 897-902. http://dx.doi.org/10.1126/science.1060612

[33] Fredrick, K. and Noller, H.F. (2003) Catalysis of Ribosomal Translocation by Sparsomycin. *Science*, **300**, 1159-1162. http://dx.doi.org/10.1126/science.1084571

[34] Peske, F., Savelsbergh, A., Katunin, V.I., Rodnina, M.V. and Wintermeyer, W. (2004) Conformational Changes of the Small Ribosomal Subunit during Elongation Factor G-Dependent tRNA-mRNA Translocation. *Journal of Molecular Biology*, **343**, 1183-1194. http://dx.doi.org/10.1016/j.jmb.2004.08.097

[35] Feinberg, J.S. and Joseph, S. (2001) Identification of Molecular Interactions between P-Site tRNA and the Ribosome Essential for Translocation. *Proceedings of the National Academy of Sciences of the United States of America*, **98**, 11120-11125. http://dx.doi.org/10.1073/pnas.211184098

[36] Schmeing, T.M., Voorhees, R.M., Kelley, A.C., Gao, Y.G., Murphy IV, F.V., Weir, J.R. and Ramakrishnan, V. (2009) The Crystal Structure of the Ribosome Bound to EF-Tu and Aminoacyl-tRNA. *Science*, **326**, 688-694. http://dx.doi.org/10.1126/science.1179700

[37] Schmeing, T.M., Voorhees, R.M., Kelley, A.C. and Ramakrishnan, V. (2011) How Mutations in tRNA Distant from the Anticodon Affect the Fidelity of Decoding. *Nature Structural & Molecular Biology*, **18**, 432-436. http://dx.doi.org/10.1038/nsmb.2003

[38] AEvarsson, A., Brazhnikov, E., Garber, M., Zheltonosova, J., Chirgadze, Y., al-Karadaghi, S., *et al.* (1994) Three Dimensional Structure of the Ribosomal Translocase: Elongation Factor G from *Thermus thermophilus*. *EMBO Journal*, **13**, 3669-3677.

[39] Gao, Y.G., Selmer, M., Dunham, C.M., Weixlbaumer, A., Kelley, A.C. and Ramakrishnan, V. (2009) The Structure of the Ribosome with Elongation Factor G Trapped in the Posttranslocational State. *Science*, **326**, 694-699. http://dx.doi.org/10.1126/science.1179709

[40] Kisselev, L.L. and Buckingham, R.H. (2000) Translational Termination Comes of Age. *Trends in Biochemical Sciences*, **25**, 561-566. http://dx.doi.org/10.1016/S0968-0004(00)01669-8

[41] Ito, K., Uno, M. and Nakamura, Y. (2000) A Tripeptide "Anticodon" Deciphers Stop in Messenger RNA. *Nature*, **403**, 680-684.

[42] Rawat, U.B., Zavialov, A.V., Sengupta, J., Valle, M., Grassucci, R.A., Linde, J., *et al.* (2003) A Cryo-Electron Microscopic Study of Ribosome-Bound Termination Factor RF2. *Nature*, **421**, 87-90. http://dx.doi.org/10.1038/nature01224

[43] Klaholz, B.P., Pape, T., Zavialov, A.V., Myasnikov, A.G., Orlova, E.V., Vestergaard, B., *et al.* (2003) Structure of the *Escherichia coli* Ribosomal Termination Complex with Release Factor 2. *Nature*, **421**, 90-94. http://dx.doi.org/10.1038/nature01225

[44] Song, H., Mugnier, P., Das, A.K., Webb, H.M., Evans, D.R., Tuite, M.F., *et al.* (2000) The Crystal Structure of Human Eukaryotic Release Factor eRF1-Mechanism of Stop Codon Recognition and Peptidyl-tRNA Hydrolysis. *Cell*, **100**, 311-321. http://dx.doi.org/10.1016/S0092-8674(00)80667-4

[45] Osawa, S., Jukes, T.H., Watanabe, K. and Muto, A. (1992) Recent Evidence for Evolution of the Genetic Code. *Microbiological Reviews*, **56**, 229-264.

[46] Nierhaus, K.H. (1990) The Allosteric Three-Site Model for the Ribosomal Elongation Cycle: Features and Future. *Biochemistry*, **29**, 4997-5008. http://dx.doi.org/10.1021/bi00473a001

[47] Di Giacco, V., Marquez, Y., Qin, Y., Pech, M., Triana-Alonso, F.J., Wilson, D.N. and Nierhaus, K.H. (2011) Shine-Dalgarno Interaction Prevents Incorporation of Noncognate Amino Acids at Codon Following the AUG. *Proceedings of the National Academy of Sciences of the United States of America*, **108**, 16980-16985.

[48] Geigenmüller, U. and Nierhaus, K.H. (1990) Significance of the Third tRNA Binding Site, the E-Site, on *E. coli* Ribosomes for the Accuracy of Translation: An Occupied E-Site Prevents the Binding of Non-Cognate Aminoacyl-tRNA to the A-Site. *EMBO Journal*, **9**, 4527-4533.

[49] Schilling-Bartetzko, S., Bartetzko, A. and Nierhaus, K.H. (1992) Kinetic and Thermodynamic Parameters for tRNA Binding to the Ribosome and for the Translocation Reaction. *Journal of Biological Chemistry*, **267**, 4703-4712.

[50] Nierhaus, K.H. (2006) Decoding Errors and the Involvement of the E-Site. *Biochimie*, **88**, 1013-1019. http://dx.doi.org/10.1016/j.biochi.2006.02.009

[51] Semenkov, Y.P., Rodnina, M.V. and Wintermeyer, W. (1996) The "Allosteric Three-Site Model" of Elongation Cannot Be Confirmed in a Well-Defined Ribosome System from *Escherichia coli. Proceedings of the National Academy of Sciences of the United States of America*, **93**, 12183-12188. http://dx.doi.org/10.1073/pnas.93.22.12183

[52] Uemura, S., Aitken, C.E., Korlach, J., Flusberg, B.A., Turner, S.W. and Puglisi, J.D. (2010) Real-Time tRNA Transit on Single Translating Ribosomes at Codon Resolution. *Nature*, **464**, 1012-1017. http://dx.doi.org/10.1038/nature08925

[53] Chen, C., Stevens, B., Kaur, J., Smilansky, Z., Cooperman, B.S. and Goldman, Y.E. (2011) Allosteric vs. Spontaneous Exit-Site (E-Site) tRNA Dissociation Early in Protein Synthesis. *Proceedings of the National Academy of Sciences of the United States of America*, **108**, 16980-16985. http://dx.doi.org/10.1073/pnas.1106999108

[54] Petropoulos, A.D. and Green, R. (2012) Further *in Vitro* Exploration Fails to Support the Allosteric Three-Site Model. *Journal of Biological Chemistry*, **287**, 11642-11648. http://dx.doi.org/10.1074/jbc.C111.330068

[55] Liao, P.Y., Gupta, P., Petrov, A.N., Dinman, J.D. and Lee, K.H. (2008) A New Kinetic Model Reveals the Synergistic Effect of E-, P- and A-Sites on +1 Ribosomal Frameshifting. *Nucleic Acids Research*, **36**, 2619-2629. http://dx.doi.org/10.1093/nar/gkn100

[56] Sundararajan, A., Michaud, W.A., Qian, Q., Stahl, G. and Farabaugh, P.J. (1999) Near-Cognate Peptidyl-tRNAs Promote +1 Programmed Translational Frameshifting in Yeast. *Molecular Cell*, **4**, 1005-1015. http://dx.doi.org/10.1016/S1097-2765(00)80229-4

[57] Bartetzko, A. and Nierhaus, K.H. (1988) Mg^{2+}/NH^{4+}/Polyamine System for Polyuridine-Dependent Polyphenylalanine Synthesis with Near *in Vivo* Characteristics. *Methods in Enzymology*, **164**, 650-658. http://dx.doi.org/10.1016/S0076-6879(88)64075-4

[58] Bernstein, H.J. (2000) Recent Changes of RasMol, Recombining the Variants. *Trends in Biochemical Sciences*, **25**, 453-455. http://dx.doi.org/10.1016/S0968-0004(00)01606-6

9

Glycosidically Bound Volatile Compounds of *Satureja montana* L., *S. cuneifolia* Ten., *S. subspicata* Vis. and Endemic *S. visianii* Šilić

Valerija Dunkić*, Ivana Radovanović, Nada Bezić, Elma Vuko

Department of Biology, Faculty of Science, University of Split, Split, Croatia
Email: *dunkic@pmfst.hr

Abstract

In this paper, the glycoconjugated volatile compounds of four Croatian *Satureja* species (*Satureja montana* L., *S. cuneifolia* Ten., *S. subspicata* Vis. and endemic *S. visianii* Šilić) were investigated. Content and composition of these compounds were examined depending on the stage of plant development. GC and GC–MS analysis of volatile aglycones revealed twenty-one compounds. Thymoquinone, geraniol and carvacrol were detected in all vegetative phases of the investigated plants. Other quantitatively important aglycones were eugenol and thymol of *S. montana*, phenyl ethyl alcohol, benzene acetaldehyde, borneol, α-terpineol, thymol and eugenol of *S. cuneifolia*, phenyl ethyl alcohol, benzene acetaldehyde, terpinen-4-ol, α-terpineol and β-ionone of *S. subspicata* and camphor, thymol and 8α-acetoxylemolol of *S. visianii*. Moderate similarity in the chemical composition of essential oils and volatile aglycones of investigated plant species indicate that many biologically active compounds are glycosylated and accumulate as non-volatile glycosides.

Keywords

***Satureja montana*, *S. cuneifolia*, *S. subspicata*, *S. visianii*, Glicosydically Bound Volatiles, Free Aglycones**

1. Introduction

The plants of the genus *Satureja* (Lamiaceae) are often aromatic herbs and shrubs widely distributed in the Mediterranean area, Asia and boreal America. In the Mediterranean region of Croatia four *Satureja* species are reported: *Satureja montana* L., *S. cuneifolia* Ten., *S. subspicata* Vis. and an endemic *S. visianii* Šilić [1]. These

*Corresponding author.

annual or perennial semi-bushy plants inhabit arid, sunny, stony and rocky habitats along the Adriatic coast. Among these plants, winter savory (*S. montana*) and wild savory (*S. cuneifolia*) are the most common in this part of Croatia [2].

In numerous Lamiaceae species, many secondary metabolites, including important biologically active compounds, are glycosylated and accumulate as non-volatile glycosides. These compounds are soluble in water, nonvolatile and odourless. Connected with the fact that non-volatile glycosides can release volatile flavour compounds (aglycones) by acid or enzymatic hydrolysis, these plant metabolites are a possible source of aroma. Free and glycosidically bound volatiles are secondary metabolites important for the plant survival, reproduction and chemotaxonomy since they occur in adaptations as strong phylogenetic and ecological components. Chemical composition, antimicrobial activity and other possible applications in phytotherapy of *Satureja* essential oils were previously widely investigated [2]-[8]. Despite that, we observed a relative paucity of data on the composition of glycosidically bound volatile compounds in this group. Analyses of glycosidically bound volatiles and antioxidant activity of free volatile compounds were previously reported for *S. montana* L. [9] [10]. Monoterpenes are usually present as main constituents of the aglycones as well as the main components in essential oil composition of aerial parts of *Satureja* plants [5] [11] [12]. In the literature data, we found no other published results dealing with the glycosidically bound volatiles of other *Satureja* species. Considering all mentioned above, our goal was to study the composition of glycosidically bound volatile compounds of four Croatian *Satureja* species (*S. montana, S. cuneifolia, S. subspicata* and endemic species *S. visianii*) and compare composition of the obtained free and glycosidically bound volatiles with the composition of the essential oil.

2. Material and Methods

2.1. Plant Material

Three *Satureja* species (*S. montana, S. cuneifolia* and *S. subspicata*) were collected during 2014 at the Kozjak Mountain (near the city of Split, Croatia), prior to flowering (in July, leaves and stalks) in the course of flowering (in September, flowering tops, leaves and stalks) and after flowering (in November, leaves and stalks). *Satureja visianii* was collected in the same period on the peninsula of Pelješac, Croatia. Voucher specimens are deposited in the herbarium at the Department of Biology, Faculty of Science, University of Split, Croatia [No. FNSST 2014: 11 (A, B, C), 12 (A, B, C), 13 (A, B, C), 14 (A, B, C)].

2.2. Isolation of Glycosidically Bound Volatile Compounds

The glycosides of volatile compounds were isolated from 100 g of fresh plant material by extraction at room temperature (24 h). Extraction was effected by percolation with 500 ml ethyl acetate. As internal standard, 500 µg of octyl-β-D-glucoside was added to ethyl acetate. 20 mg β-glucosidase from almonds (Fluka) was added to the glycosidic solution, along with 3 ml pentane for trapping liberated aglycones. Hydrolysis was carried out at 37°C for 72 hours.

2.3. Gas Chromatography-Mass Spectrometry

Gas chromatography (GC) analyses were performed on a gas chromatograph (model 3900; Varian Inc., Lake Forest, CA, USA) equipped with flame ionization detector (FID), mass spectrometer (MS) (model 2100T; Varian Inc.), capillary column VF-5ms (30 m × 0.25 mm i.d., coating thickness 0.25 µm). The individual peaks were fixed by comparison of their retention indices, and/or authentic samples, as well as by comparing their mass spectra with literature data [13].

3. Results and Discussion

The content of aglycones obtained from fresh plant material varied from 1.8 to 7.6 $mg \cdot kg^{-1}$ according to the stage of plant development. The content of these compounds was the highest during the flowering period of plant, and decreased in the period after flowering. GC-MS analysis of the aglycones revealed twenty-one compound, representing 82.6% - 97.2% of the total aglycone fraction (**Table 1**).

Based on the incidence throughout all the vegetative stages of plant development, the main aglycones in investigated *Satureja* species are thymoquinone (2.4% - 56.9%), geraniol (2.2% - 16.9%) and carvacrol (3.4% -

Table 1. Phytochemical composition (%) of the glycosidically bound volatile compounds of *Satureja montana* L., *S. cuneifolia* Ten., *S. subspicata* Vis. and *S. visianii* Šilić.

Component	RI	*S. montana*			*S.cuneifolia*			*S. subspicata*			*S. visianii*		
		Month (mg·kg^{-1})			Month (mg·kg^{-1})			Month (mg·kg^{-1})			Month (mg·kg^{-1})		
		7.	9.	11.	7.	9.	11.	7.	9.	11.	7.	9.	11.
		(4.0)	(7.6)	(3.2)	(4.0)	(7.6)	(3.2)	(2.2)	(2.7)	(1.8)	(2.2)	(2.7)	(1.8)
1-Octen-3-ol	974	-	-	-	0.7	-	4.2	-	-	1.3	0.2	0.8	1.2
3*E*-Hexenioc acid	983	-	-	-	0.9	-	0	1.8	-	-	-	-	-
2*E*-Hexenioc acid	1005	-	-	-	1.1	-	-	-	-	-	-	0.4	-
Linalool	1099	-	-	-	8.4	-	6.6	-	-	0.6	0.3	1.1	0.9
2,2-dimethyl-3,4-octadienal	1103	-	-	0.9	3.5	-	2.3	-	-	-	-	-	-
Phenyl ethyl alcohol	1106	-	2.3	0.8	1.3	12.7	4.4	5.5	7.3	19.7	2.2	1.6	1.0
Benzene acetaldehyde	1036	-	-	-	8.5	5.0	2.5	5.9	20.3	4.6	-	-	-
Camphor	1143	-	-	-	-	-	-	-	-	-	29.5	35.3	31.2
Borneol	1165	1.4	-	0.9	11.0	19.3	4.4	-	-	-	-	-	-
Terpinen-4-ol	1184	-	-	2.2	3.6	-	-	4.2	8.5	9.2	4.2	4.7	4.6
α-Terpineol	1186	-	-	4.1	7.3	9.8	10.6	31.7	2.7	2.5	3.4	-	-
Myrtenol	1197	-	-	0.3	-	2.1	6.2	-	-	-	0.9	8.6	-
Thymoquinone	1248	45.3	56.9	16.8	19.4	14.4	12.8	6.3	18.6	4.7	8.8	2.4	3.5
Geraniol	1249	2.2	6.8	16.9	4.5	7.7	9.3	8.9	6.3	4.9	3.3	3.8	2.9
3,7-dimethyl-1,5-octadien-3,7-diol	1270	-	-	-	1.0	-	1.2	-	t	-	-	-	-
Thymol	1290	1.4	0.6	11.3	1.3	8.8	4.5	-	-	1.2	15.7	5.2	13.8
Carvacrol	1298	42.6	14.1	12.6	13.9	7.0	3.4	9.2	7.1	26.9	6.4	6.3	3.5
Eugenol	1356	0.7	8.9	17.8	1.2	5.7	6.5	-	13.5	9.2	-	-	1.9
β-Ionone	1487	-	2.3	2.5	-	-	6.0	23.7	8.1	3.9	-	-	-
8*a*-Acetoxylemolol	1792	-	-	1.9	-	-	-	-	-	-	7.7	17.2	22.3
Z-Nuciferol acetate	1829	-	-	1.1	1.3	-	2.9	-	-	0.8	-	-	-
Total		93.6	91.9	90.1	88.9	92.5	87.8	97.2	92.4	89.5	82.6	87.4	86.8

42.6%) (**Table 1**). Thymoquinine was also reported as a dominant aglycone (20.7%) of *S. montana* in the study by Radonić and Miloš [10]. Other quantitatively important aglycones of *S. montana* were eugenol (17.8%) and thymol (11.3%) as a dominant compounds in the period after flowering (**Table 1**). Other quantitatively important aglycones in *S. cuneifolia* were phenyl ethyl alcohol, benzene acetaldehyde, borneol, α-terpineol, thymol and eugenol (**Table 1**). These compounds were detected in all vegetative phases. Aside from thymoquinone, geraniol and carvacrol, quantitatively important aglycones detected in all vegetative phases of *S. subspicata* were phenyl ethyl alcohol, benzene acetaldehyde, terpinen-4-ol, α-terpineol and β-ionone. In addition to above listed dominant aglycones, other quantitatively important components in all vegetative phases of the *S. visianii* were camphor, 8a-acetoxylemolol and thymol (**Table 1**).

Aliphatic alcohols, 2-phenylethanol, benzyl alcohol, eugenol, linalool, geraniol, nerol and α-terpineol can, more or less, be considered common in aglycone fraction of Lamiaceae family [14] and the eugenol was found to be the main aglycone in most plants of this family [10] [14] [15]. Eugenol was the main aglycone of *S. montana* in the period after flowering (**Table 1**). *S. subspicata* and *S. cuneifolia* contained significant concentration of eugenol during and after flowering, but among aglycones of endemic *S. visianii* eugenol was detected in a low percentage in the period after flowering (**Table 1**). Composition of essential oils of above listed *Satureja* species was reported in **Table 1** in our previous paper [2]. Comparison of the chemical composition of volatile aglycones (**Table 1**) with those of essential oils [2] reveals ten common compounds (1-octene-3-ol, linalool, camphor, borneol, terpinen-4-ol, α-terpineol, myrtenol, geraniol, thymol and carvacrol) (**Table 1**). Among two major constituent of the aglycone fraction of *Satureja* species, thymoquinone and carvacrol, only the latter has been detected among the free compounds in the essential oil [2]. These results are consistent with the fact that the free and glycosidically bound plant secondary metabolites have different biosynthetic pathways which establish differences in their chemical composition [14]. The results of our study show moderate similarity in the chemical composition of essential oil and free volatile aglycones of investigated plant species. Regarding the substantial interest for aromatic plants and their possible applications, this paper reveals the presence of compounds whose effect may been overseen due to the fact that they are present in glycosidic form.

Acknowledgements

We acknowledge "The Ministry of Science and Technology of the Republic of Croatia".

References

[1] Šilić, Č. (1979) Monografija rodova *Satureja* L., *Calamintha* Miller, *Micromeria* Bentham, *Acinos* Miller i *Clinopodium* L. u flori Jugoslavije. Zemaljski muzej BiH, Sarajevo.

[2] Bezić, N., Šamanić, I., Dunkić, V., Besendorfer, V. and Puizina, J. (2009) Essential Oil Composition and Internal Transcribed Spacer (ITS) Sequence Variability of Four South-Croatian *Satureja* Species (Lamiaceae). *Molecules*, **14**, 925-938. http://dx.doi.org/10.3390/molecules14030925

[3] Dunkić, V., Kremer, D., Dragojević Müler, I., Stabentheinerd, E., Kuzmić, S., Jurišić Grubešić, R., Vujić, L., Kosalec, I., Randić, M., Srečec, S. and Bezić, N. (2012) Chemotaxonomic and Micromorphological Traits of *Satureja montana* L. and *S. subspicata* Vis. (Lamiaceae). *Chemistry and Biodiversity*, **9**, 2825-2842. http://dx.doi.org/10.1002/cbdv.201100376

[4] Čavar, S., Maksimović, M., Šolić, M.E., Jerković-Mujkić, A. and Bešta, R. (2008) Chemical Composition and Antioxidant and Antimicrobial Activity of Two *Satureja* Essential Oils. *Food Chemisty*, **111**, 648-653. http://dx.doi.org/10.1016/j.foodchem.2008.04.033

[5] Miloš, M., Radonić, A., Bezić, N. and Dunkić, V. (2001) Localities and Seasonal Variations in the Chemical Composition of Essential Oils of *Satureja montana* L. and *S. cuneifolia* Ten. *Flavour and Fragrance Journal*, **16**, 157-160. http://dx.doi.org/10.1002/ffj.965

[6] Angelini, L.G., Carpanese, G., Cioni, P.L., Morelli, I., Macchia, M. and Flamini, G. (2003) Essential Oils from Mediterranean Lamiaceae as Weed Germination Inhibitors. *Journal of Agricultural and Food Chemistry*, **51**, 6158-6164. http://dx.doi.org/10.1021/jf0210728

[7] Skočibušić, M., Bezić, N. and Dunkić, V. (2006) Phytochemical Composition and Antimicrobial Activities of the Essential Oils from *Satureja subspicata* Vis. Growing in Croatia. *Food Chemistry*, **96**, 20-28. http://dx.doi.org/10.1016/j.foodchem.2005.01.051

[8] Ciani, M., Menghini, L., Mariani, F., Pagiotti, R., Menghini, A. and Fatichenti, F. (2000) Antimicrobial Properties of Essential Oil of *Satureja montana* L. on Patho-Genic and Spoilage Yeasts. *Biotechnology Letters*, **22**, 1007-1010. http://dx.doi.org/10.1023/A:1005649506369

[9] Mastelić, J. and Jerković, I. (2003) Gas Chromatography-Mass Spectrometry Analysis of Free and Glycoconjugated Aroma Compounds of Seasonally Collected *Satureja montana* L. *Food Chemistry*, **80**, 135-140. http://dx.doi.org/10.1016/S0308-8146(02)00346-1

[10] Radonić, A. and Miloš, M. (2003) Chemical Composition and Antioxidant Test of Free and Glycosidically Bound Volatile Compounds of Savory (*Satureja montana* L. Subsp. Montana) from Croatia. *Nahrung/Food*, **47**, 236-237. http://dx.doi.org/10.1002/food.200390055

[11] Mahmoud, S.S. and Croteau, R.B. (2002) Strategies for Transgenic Manipulation of Monoterpene Biosynthesis in Plants. *Trends in Plant Science*, **7**, 366-373. http://dx.doi.org/10.1016/S1360-1385(02)02303-8

[12] Wink, M. (2003) Evolution of Secondary Metabolites from an Ecological and Molecular Phylogenetic Perspective. *Phytochemistry*, **64**, 3-19. http://dx.doi.org/10.1016/S0031-9422(03)00300-5

[13] Adams, R.P. (2007) Identification of Essential Oil Components by Gas Chromatography/Mass Spectroscopy. 4th Edition, Allured Publishing, Carol Stream.

[14] Stahl-Biskup, E., Intert, F., Holthuijzen, J., Stengele, M. and Schulz, G. (1993) Glycosidically Bound Volatiles. A Review. 1986-1991. *Flavour and Fragrance Journal*, **8**, 61-80. http://dx.doi.org/10.1002/ffj.2730080202

[15] Miloš, M., Mastelić, J. and Jerković, I. (2000) Chemical Composition and Glycosidically Bound Volatile Compounds from Oregano (*Origanum vulgare* L. spp. *hirtum*). *Food Chemistry*, **71**, 79-83. http://dx.doi.org/10.1016/S0308-8146(00)00144-8

10

Polyvinylalcohol Stabilizes Anthocyanins of Red Wine in the Solid Phase but Not in the Aqueous Phase

Wataru Sakamoto[1*], Takashi Kanehira[2], Hirohisa Hongou[2], Kozo Asano[3]

[1]Serotec Laboratory, Ebetsu, Japan
[2]Division of Preventive Dentistry, Graduate School of Dental Medicine, Hokkaido University, Sapporo, Japan
[3]Division of Applied Bioscience, Research Group of Molecular Bioscience, Graduate School of Agriculture, Hokkaido University, Sapporo, Japan
Email: [*]w-sakamoto@serotec.co.jp, [*]sakamoto@den.hokudai.ac.jp

Abstract

Monomeric anthocyanins exhibit color variations at different pH, but they are highly unstable at physiological pH. To improve the stability, the effects of polyvinylalcohol on the anthocyanins of young red wine prepared from Cambell Early grapes were examined in aqueous solution and in the solid phase on a cotton swab® and inert materials. Cyanidin-3-glucoside and malvidin-3-glucoside represented approximately 43.6% of total anthocyanins in the red wine. The anthocyanins decreased to 11.3% of the initial amount in aqueous solution at physiological pH after storage for 60 days at room temperature. On the other hand, from 66.9% to 87.2% remained in the solid phase on a cotton swab® and polypropylene membrane in the presence of polyvinylalcohol. The stability of the anthocyanins increased with the concentration of polyvinylalcohol in the solid phase but not in the aqueous solution. The anthocyanins were found to be condensed and immobilized in the complexes of anthocyanins and polyvinylalcohol on the polypropylene membrane by phase-contrast microscopy analysis. The anthocyanins in the cotton swab® displayed color variations when dipped into different pH buffer solutions. These results suggest that anthocyanins from red wine stabilized in the solid phase by polyvinylalcohol can be used as a visual indicator of pH.

Keywords

Red Wine, Anthocyanins, Stability, Aqueous Phase, Solid Phase, Polyvinylalcohol

[*]Corresponding author.

1. Introduction

Anthocyanins are natural, water-soluble, nontoxic pigments displaying a variety of colors from orange to blue according to the pH in aqueous solution [1]. In general, the main monomeric anthocyanins in red wines are the 3-O-monoglucosides of the six free anthocyanins, pelargonidin-3-glucoside, cyanidin-3-glucoside, delphinidin-3-glucoside, peonidin-3-glucoside, petunidin-3-glucoside and malvidin-3-glucoside [2]. Their ionic nature enables changes of the molecular structure according to the prevailing pH, resulting in different colors and hues at different pH values [1]-[3]. Thus, the changes of color of anthocyanins are thought to be useful as a visual indicator of pH, although there is no report on the visualization of clinical pH in body fluids such as saliva and nasal secretion using anthocyanins. This is because monomeric anthocyanins in red wine are unstable depending on various factors such as their structure, concentration, the composition of the solution, pH value, oxygen, storage temperature and time [4] [5]. Recently, Bimpilas *et al.* reported that during 1 year of storage monomeric anthocyanins declined by almost tenfold, probably due to polymerization reactions and copigmentation [4]. Therefore, we tried to extract anthocyanins from young red wine made using Cambell Early grapes in order to obtain an abundance of monomeric anthocyanins as reported by Mateus *et al* [5]. On the other hand, it is well known that polyvinylalcohol, which contains many hydroxyl groups, protects their active ingredients from moisture, oxygen and other environmental components [6]. In addition, it is easy to prepare, has good mechanical properties, excellent chemical resistance, and film-forming ability [7]-[9]. Conte *et al.* recently reported a new technique for the immobilization of lysozyme, active against gram-positive bacteria, without comprising its activity [10]. Interestingly, Singh and Mishra recently demonstrated that the optimized methods of extraction of two primary emitting pigments from their corresponding plant sources were simple, cheap and fairy green using anthocyanin, curcumin and polyvinylalcohol [11]. However, it is not yet clarified by which mechanism polyvinylalcohol stabilizes anthocyanins in the solid phase, although we previously reported that anthocyanins from red wine stabilized by polyvinylalcohol [12]. Therefore, the combination of polyvinylalcohol and young red wine was expected to stabilize anthocyanins, probably due to formation of hydrogen bonds from the hydroxyl group of polyvinylalcohol and protonated flavylium cations of the anthocyanin [13]. We first compared the stability of anthocyanins in the aqueous phase and the solid phase and attempted to stabilize anthocyanins in the solid phase using a combination of young red wine and polyvinylalcohol. Considering these points, we finally attempted to prepare stable anthocyanins as a visual of indicator pH in a cotton swab® based on the changes of color of the anthocyanins from young red wine.

2. Materials and Methods

2.1. Chemicals and Reagents

All chemicals and reagents used in this study were of analytical or HPLC grade. Malvidin-3-glucoside chloride and cyanidin-3-glucoside chloride were purchased as standards from Tokiwa Phytochemical Co. (Chiba, Japan).

2.2. Extraction of Anthocyanins from Grapes

Extraction of anthocyanins was carried out by winemaking using Cambell Early grapes. After washing with tap water, a total mass of 10 kg of grapes was crushed and the resultant 8 L of grape must was transferred to a plastic fermentation jar. Alcohol fermentation was continued for two weeks at room temperature (RT) with twice daily pumping. After removing skins and seeds by gauze filtration, fermentation was further performed for two weeks at RT. After centrifugation at 2000 rpm for 10 min, the red wine (pH 3.5) was stored at −20˚C in a freezer until used as anthocyanin solution. The content of alcohol and pH were measured by enzymatic determination using a Roche Kit (R-Biopharm AG, Darmstadt, Germany) and a Horiba pH meter (F-52; Horiba Co., Kyoto, Japan), respectively.

2.3. Measurement of Total Anthocyanin

The concentration of monomeric anthocyanin was measured using the pH differential method [14], except as otherwise described. The anthocyanin in 96-well microplates was quantified at 540 nm, using cyanidin-3-glucoside chloride as the standard, because the wavelengths of the maximum absorbances for anthocyanin species had been reported to be 498 nm to 540 nm at pH 2.0 [14]. To do this, an aliquot of a 100 μL sample was added to a microtube (1.5 mL; Ina-Optica Co., Osaka, Japan) containing 400 μL of methanol in 1% hydrochloric acid. After vortexing for 10 sec, it was centrifuged at 8000 rpm (Kubota 1120, Kubota Ltd., Osaka, Japan) for 3 min at RT. Then 300 μL of the supernatant was added to a 96-well microplate (NunclonTM 167008; Nunc A/S, Denmark) and the total anthocyanin content was quantified at 540 nm using a microplate reader (Labsystems-Multiskan MS, Dainippon Pharm. Co., Japan).

2.4. HPLC Analysis

Analysis of anthocyanins was performed using a Shimadzu HPLC system (Shimadzu Corp., Kyoto, Japan) equipped with an LC-10ADvp HPLC pump, SIL-10ADvp automated sample injector, CTO-10Avp thermostatted column compartment, and SPD-10Avp detector. The column (ODS-120T; column size 25 × 0.46 cm, Toyo Soda Co., Japan) set at 40˚C was eluted isocratically with 15% acetonitrile in 1% formic acid at a flow rate of 1.0 ml/min and anthocyanins were detected at 520 nm.

2.5. Stability of Total Anthocyanin in Aqueous Solution and the Solid Phase

The anthocyanin concentrations at native pH (pH3.5) and pH 7.0 adjusted with 4.0 M NaOH were 383 μg/ml and 330 μg/ml, respectively. Samples were stored in microtubes and kept at RT without daylight. They were obtained at regular time intervals within a period of 60 days and the anthocyanins were measured by the microplate method. The stability of anthocyanin was expressed as the percentage of residual anthocyanin compared to the total anthocyanin content of each starting material. Regarding the stability of the anthocyanins in the solid phase, we used the anthocyanin of a 70 μL (23 μg anthocyanin) sample that penetrated a cotton swab® (length: 13.0 ± 1.0 mm; diameter: 4.2 ± 0.1 mm; cellulose content: 40 mg; capacity of volume retention: 120 ± 10 μL; Sanritsu Co., Japan) and Bell oasis® (inert polymer; diameter: 6.2 ± 0.1 mm; thickness 3.0 mm; capacity of volume retention: 100 ± 5 μL; Teijin Company, Osaka, Japan). We also used a polypropyrene membrane (square: 2 × 2 cm; thickness: 0.2 mm; Lihit Lab Co., Osaka, Japan) that did not have the capacity for volume retention, unlike the materials described above. Therefore, the anthocyanin solution was directly put onto the surface of the polypropyrene membrane (diameter: 8.4 ± 0.2 mm) and dried at room temperature. Unless otherwise stated, they were kept at room temperature for 60 days. Prior to evaluation of the stability of the anthocyanins, the anthocyanin in the solid phase was extracted using 3 ml of methanol in 1% hydrochloric acid three times every other day at room temperature. The stability was expressed as described above.

2.6. Effect of Polyvinylalcohol on the Stability of Anthocyanin

Polyvinylalcohol (polyvinylalcohol-1500, Wako Pure Chemical Co., Japan) was dissolved in distilled water at 60˚C for one hour to obtain homogeneous 37.5 mg/ml solutions. After dilution with 10% ethanol, 400 μl of polyvinylalcohol (1.5 mg/ml - 37.5 mg/ml) was added to 200 μl of red wine solution (330 μg/ml anthocyanin at pH 7.0) in the experiment of aqueous solution and stored at room temperature. In the solid phase, 100 μl of polyvinylalcohol solution (12.5 mg/ml - 37.5 mg/ml) was penetrated into them and dried at room temperature for 3 days, and then 70 μl of red wine (23 μg anthocyanin at pH 7.0) was penetrated into the solid phases. Regarding the polypropylene membrane, the each solution of polyvinylalcohol and anthocyanin was directly put onto the surface (diameter: 8.4 ± 0.2 mm), respectively.

2.7. Analysis of Surface Morphology of Polyvinylalcohol-Red Wine Complex on the Polypropylene Membrane

First, 100 μl of polyvinylalcohol solution (12.5 mg/ml) was applied to a square (2 × 2 cm) polypropylene membrane and dried at room temperature for 3 days, and then 70 μl of red wine (14 μg of anthocyanin) was coated on the surface of the dried polyvinylalcohol membrane. The polyvinylalcohol-red wine complex was observed at ×400 magnification by phase contrast microscopy using an FC-300Z/OL (Olympus, Tokyo, Japan) and photographed using Fuji film (Neopan Across 100; Fuji Film Co., Tokyo, Japan).

2.8. Preparation of Cotton Swab® as a pH Indicator

A cotton swab® was impregnated with 100 μl of polyvinylalcohol solution (12.5 mg/ml) and dried at room temperature for 3 days. It was further impregnated with 70 μL (20.6 μg anthocyanin) of red wine and again allowed to dry for 1 hour at RT. The cotton swab® was lightly wrapped with aluminum foil until to use.

2.9. Color Changes of Anthocyanins by pH

Color changes of anthocyanins in the cotton swab® by pH were examined as follows. The cotton swab® was dipped into 100 μL of 0.15 M phosphate buffer (pH 5.5 - 8.0) and/or 0.15 M carbonate/biscarbonate buffer (pH 8.5 - 9.5). After 2 minutes, visual color was photographed by PENTAX Digital Camera K-X (Hoya Corporation, Japan).

2.10. Statistical Analysis

Spearman's test and Kruskal-Wallis one-way analysis of variance were used for statistical analysis. Data were expressed as mean ± standard deviation (SD). Differences were considered significant at $p < 0.05$.

3. Results

3.1. HPLC Analysis of Anthocyanins in Red Wine

After fermentation for four weeks, the red wine from the Cambell Early Grapes contained 404 ± 45 μg/ml anthocyanin and 8.7% ± 0.5% alcohol, with a pH of 3.5 ± 0.2. To determine the components of anthocyanin in the red wine, HPLC analysis was carried out using the authentic standards cyaniding-3-glucoside and malvidin-3-glucoside. As shown in **Figure 1**, the red wine showed eight peaks, of which two were identified as cyanidin-

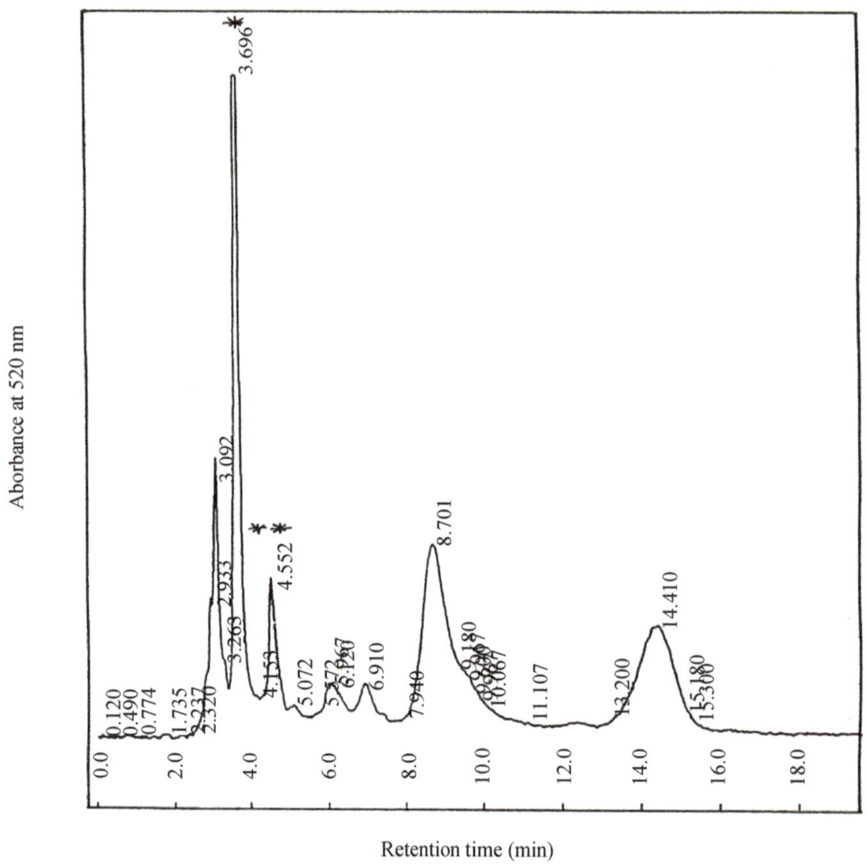

Figure 1. HPLC analysis of anthocyanins in red wine. Red wine (total anthocyanin: 5220 ng) was applied to an ODS-120T column. The anthocyanins were detected at 520 nm. * and ** were identified as cyanidin-3-glucoside and malvidin-3-glucoside from the retention times of their standards. Other experimental conditions are described in the text.

3-glucoside and malvidin-3-glucoside by comparison with the authentic standards (retention times of 3.6 min and 4.5 min, respectively). Cyanidin-3-glucoside accounted for 36.6% of the total anthocyanin and malvidin-3-glucoside 7.0%. Two anthocyanins could not be identified, one with a retention time of 8.7 min (22.8% of the total) and another with one of 14.4 min (17.9%).

3.2. Stability of Total Anthocyanin of Red Wine in Aqueous Solution

To investigate the stability of the total anthocyanin content in the aqueous solution at the native pH (pH 3.5) and pH 7.0, its changes during storage at RT up to 60 days were investigated. Anthocyanin decreased slowly but continuously during storage at low pH (pH 3.5). As shown in **Figure 2**, increasing the pH of the solution to 7.0 hastened the degradation of anthocyanin during the 60 days. Regarding the stability on the solid phase, it was investigated using the anthocyanin solution at pH 7, because the anthocyanins were more unstable at pH 7.0 than at the native pH (3.5) in the aqueous solution as described above. **Table 1** showed that the residual anthocyanin in the aqueous phase was 11.3% ± 3.0% of total anthocyanin after storage for 60 days at RT, but the solid phase ranged from 40.2% to 53.8% of total anthocyanin (**Table 2**). These results showed that the solid phase was more stable than the aqueous one ($p < 0.05$), irrespective of the type of solid phase material.

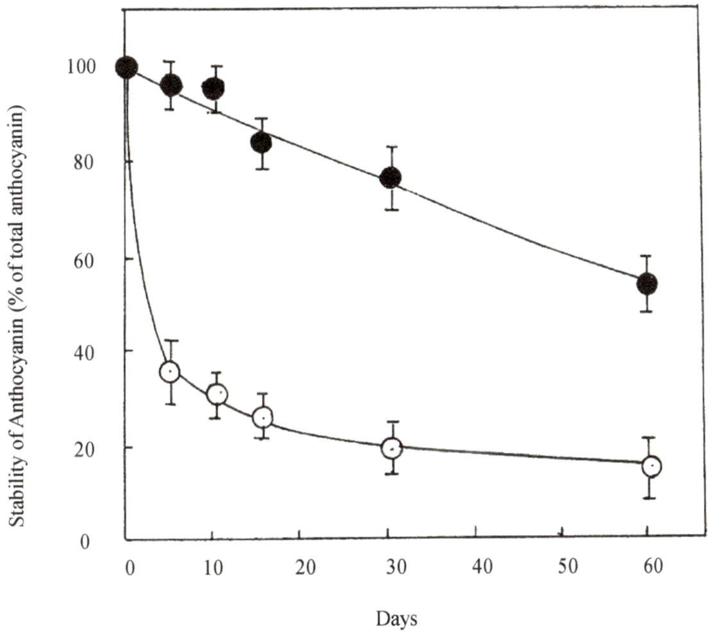

Figure 2. Stability of anthocyanin of red wine in aqueous solution. The stability was investigated during storage at RT. ●: native red wine (pH 3.5, total anthocyanin content 383 μg/ml); ○: red wine adjusted to pH 7.0 (total anthocyanin content 330 μg/ml). Other experimental conditions are described in the text. The data are expressed as the percentage of residual anthocyanin from total anthocyanin obtained in three experiments.

Table 1. Effect of polyvinylalcohol on the stability of anthocyanin in the aqueous phase.

Polyvinylalcohol	Residual anthocyanin (% of total anthocyanin)	
	30 days	60 days
0 mg/ml	18.4 ± 4.0	11.3 ± 3.0
1.0	21.9 ± 0.5	12.6 ± 1.0
8.3	25.8 ± 3.0	14.4 ± 1.7
25	31.1 ± 6.7	17.9 ± 3.3

Experimental conditions were described in the text. The data (mean ± SD) are expressed as the percentage of residual anthocyanin from total anthocyanin after storage of 30 and 60 days in three experiments.

Table 2. Effect of polyvinylalcohol on the stability of anthocyanins in solid phase on various materials.

	Polyvinylalcohol (mg/solid phase)			
Solid phase	0	0.25	1.25	3.75
	Residual anthocyanin (% of total anthocyanin)			
Cotton swab®	40.2 ± 2.2	45.6 ± 1.4	62.7 ± 3.8*	66.9 ± 3.5*
Inert polymer				
Polypropyrene membrane	44.5 ± 2.8	67.2 ± 6.8	80.0 ± 4.4*	87.2 ± 9.0*
Bell oasis®	53.8 ± 7.8	87.3 ± 9.5	97.9 ± 6.8*	101.7 ± 14.3*

100 μl of polyvinylalcohol solution (1.25 mg/ml - 37.5 mg/ml) was penetrated into the solid phases (cooton swab® and Bell oasis®) or applied on the surface of polypropyrene membrane (diameter: 8.4 ± 0.2 mm). After dry at room temperature for 3 days, 70 μl of red wine (23 μg of anthocyanin at pH 7.0)) was penetrated the solid phases or applied on the surface of polypropyrene membrane. After storage for 60 days at room temeperature, residual anthocyanin was measured. The data (mean ± SD) are expressed as the percentage of residual anthocyanin from total anthocyanin in three experiments. *$P < 0.05$ vs .each non-addition of polyvinylalcohol.

3.3. Effect of Polyvinylalcohol on the Stability of Anthocyanin in Aqueous Solution and in the Solid Phase

To investigate the effect of polyvinylalcohol on the stability of anthocyanin in aqueous solution, it was added to the anthocyanin solution, which was then kept at room temperature for 60 days. As shown in **Table 1**, polyvinylalcohol did not significantly affect the stability of anthocyanin in the aqueous solution, when it increased up to the concentrtion of 25 mg/ml. In contrast, the stability of the anthocyanin in the solid phase was increased with the concentration of polyvinylalcohol, irrespective of the type of solid phase material (**Table 2**). In the presence of 1.25 mg polyvinylalcohol, the stability of anthocyanins remained 62.7% ± 3.8% in the cotton swab®, 80.0% ± 4.4% in the polypropyrene membrane, and 97.9% ± 6.8% in the Bell oasis® after storage of 60 days. Interestingly, the stability was well maintained after storage for 180 days, showing that it remained 61.5% ± 6.5% in the cotton swab®, 78.0% ± 5.4% in the polypropyrene membrane, and 95.0% ± 11.1% in the Bell oasis®, whereas it remained 40.8% ± 5.0%, 44.0% ± 8.7%, and 52.6% ± 8.7% in the absence of the polyvinylalcohol, respectively (data not shown).

3.4. Analysis of Surface Morphology of the Polyvinylalcohol-Red Wine Complex on the Polypropylene Membrane

To determine the mechanism by which the polyvinylalcohol increased the stability of anthocyanin in the solid phase, the surface morphology of the polyvinylalcohol-red wine complexes was analyzed by phase contrast microscopy. As shown in **Figure 3**, a heterogeneous porous structure was produced by the polyvinylalcohol (1.25 mg) and red wine (14 μg of anthocyanin) on the polypropylene membrane. Anthocyanin (dark area) was condensed and immobilized in the polymer structure when it was applied to the polyvinylalcohol-coated film and dried at room temperature for 5 days. On the other hand, the structure of polyvinylalcohol alone distributed all around interconnected pores on the film, and the red wine alone was uniformly covered with small particles.

3.5. Relationship between Color Changes of Anthocyanins and pH in the Cotton Swab®

Having shown that polyvinylalcohol stabilizes anthocyanins in the solid phase, we next attempted to supply stable anthocyanins as a visual sensor of pH indicator in the cotton swab®. As shown in **Figure 4**, the complex of anthocyanin and polyvinylalcohol on the cotton swab® displayed a variety of color variations according to the prevailing pH (5.5 - 8.0), when it was dipped into buffer solutions with pHs from 5.5 to 9.0. However, no further changes occurred at pH higher than 8.0. The color of anthocyanin at pH 5.5 - 8.0 was stable at room temperature without light during 2 months but was not when the pH was 8.5 or higher (data not shown).

4. Discussion

We previously reported that the combination of polyvinylalcohol and red wine successfully stabilized anthocyanins of red wine in the solid phase materials such as cotton swab® and inert polymers [12]. This is the first report to

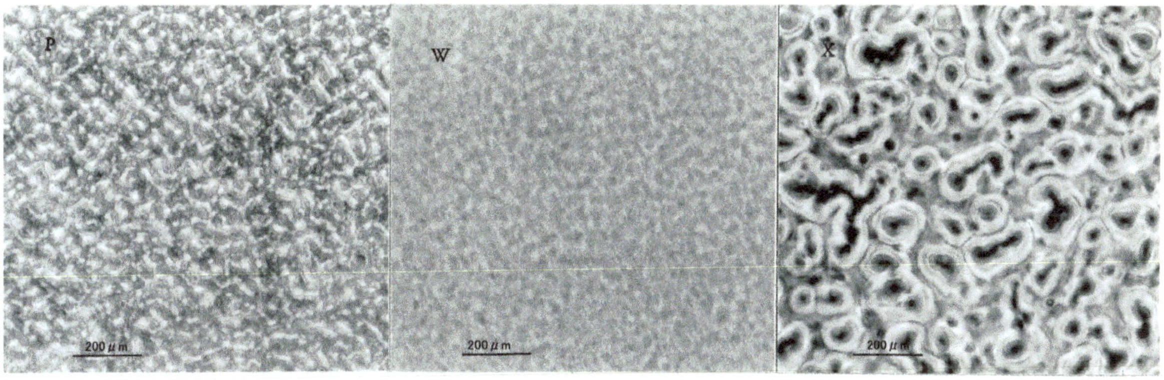

Figure 3. Phase-cntrast microscopy of polyvinylalcohol-red wine complex. 100 μl of polyvinylalcohol (PVA) solution (12.5 mg/ml) was applied on a square (2 × 2 cm) polypropylene membrane and dried at room temperature for 3 days, after which 70 μl of red wine (14 μg of anthocyanin) was applied to the surface (diameter: 8.4 ± 0.2 mm). (P) PVA only; (W) red wine only; (X) PVA + red wine. Bar: 200 μm.

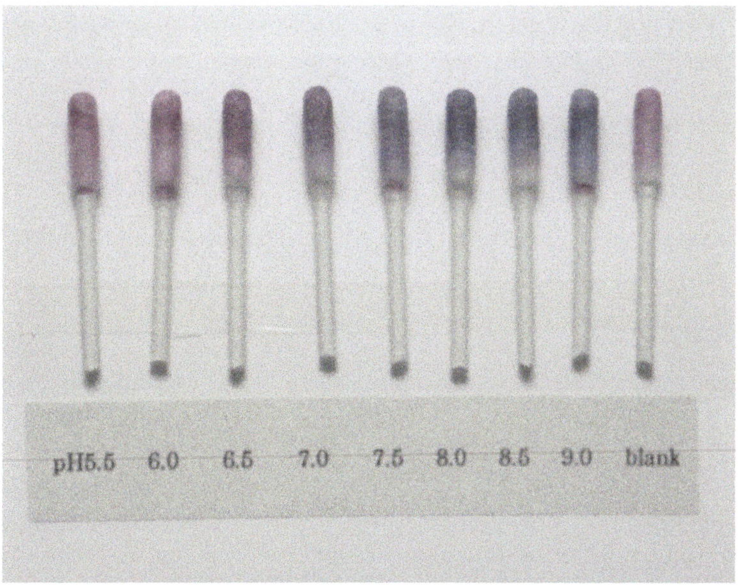

Figure 4. Relationship between color changes of anthocyanin and pH solution in the cotton swab®. The cotton swab® was dipped into 100 μl of 0.15 M phosphate (pH 5.5 - 8.0) and/or 0.15 M carbonate/bicarbonate buffer (pH 8.5 - 9.5), and then a photograph was taken. Blank: native cotton swab® as a pH indicator. Other experimental conditions are described in the text.

be analyzed by phase contrast microscopy of the surface morphology in the polyvinylalcohol-red wine complexes by which mechanism polyvinylalcohol stabilizes anthocyanins of young red wine. Anthocyanins are highly unstable and easily susceptible to degradation, because the three species of water (H^+, OH^- and H_2O) are highly reactive towards them. To overcome this problem, several investigators attempted to stabilize anthocyanins by polyacylation, metal-binding, and their interaction with polysaccharides [15]-[19]. However, the stabilization was not sufficient as it depended on the storage period, pH and temperature. Bimpilas *et al.* reported that, during 1 year of storage, monomeric anthocyanins declined by almost tenfold, probably due to polymerization reactions and copigmentation [4]. Therefore, we prepared young red wine from Cambell Early grapes to obtain an abundance of monomeric anthocyanins, finding that monomeric anthocyanins such as cyanidin-3-glucoside and malvidin-3-glucoside represented approximately 43.6% of total anthocyanins, unlike in commercial wine for storage [4]. We next attempted to stabilize the anthocyanins in the solid phase using the young red wine and polyvinyl alcohol on materials such as a cotton swab® and inert materials. Unlike in aqueous solution, anthocyanin in the solid phase was stable in storage for 180 days at room temperature. Interestingly, the stability of the anthocyanins increased with the concentration of polyvinylalcohol in the solid phase but not in the aqueous solution. To determine the mechanism by which this occurred, we analyzed the surface morphology of polyvinylalcohol-anthocyanin complexes by phase-contrast microscopy. Interestingly, the anthocyanins in the complexes were found to be condensed and immobilized in polyvinylalcohol polymer structures. This result seemed to reflect the finding of He *et al.* that the stability of anthocyanins was affected by their concentrations and surroundings [13]. Therefore, monomeric anthocyanins in red young wine seemed to be condensed by polyvinylalcohol into the solid phase, probably due to the formation of hydrogen bonds from the hydroxyl group of polyvinylalcohol and protonated flavylium cations of the anthocyanin. However, further studies will be necessary to clarify the exact mechanism for the stabilization of anthocyanins at different pHs. Our project goal is to supply stable anthocyanins as a visual indicator for clinical measurement of pH, because anthocyanins display a variety of colors with variations in pH [1]-[4] and are safe to use [19] [20]. Recently, using a pH meter or pH tape, several investigators have shown that salivary pH plays important roles in oral diseases [21] and systemic diseases [22], and that nasal pH becomes more alkaline in allergic rhinitis [23]. However, these measurements remain difficult to perform at frequent intervals with a pH meter, and saliva must be collected prior to the measurement of salivary pH, because the chemical components of the pH indicator tape may be carcinogenic [24]. As expected, the cotton swab® revealed a variety of colors at different pHs due to changes of the molecular structure caused by the ionic nature of anthocyanins [1] [3]. These results suggest that anthocyanins from red wine stabilized in the solid phase by polyvinylalcohol can be used as a visual indicator of pH. Fortunately, the stability of anthocyanins in the presence of 1.25 mg of polyvinylalcohol on a cotton swab® remained 62.7% ± 3.8% and 61.5% ± 6.5% after storage for 60 days and 180 days, respectively. Therefore, further studies expect to be clarified relationship between clinical measurement of pH and an assay system using anthocyanins stablized in the solid phase by the complex of young red wine and polyvinylalcohol.

Acknowledgements

This work was supported by JSPS KAKENHI funds from a Grant-in-Aid for Scientific Research (No. 26463153) to coauthor T.K.

Conflict of Interest

The authors declare that there is no conflict of interest regarding the publication of this paper.

References

[1] Eiro, M.J. and Heinonen, M. (2002) Anthocyanin Color Behavior and Stability during Storage: Effect of Intermolecular Copigmentation. *Journal of Agricultural and Food Chemistry*, **50**, 7461-7466. http://dx.doi.org/10.1021/jf0258306

[2] Mazza, G. and Francis, F.J. (1995) Anthocyanins in Grapes and Grape Products. *Critical Reviews in Food Science and Nutrition*, **35**, 341-371. http://dx.doi.org/10.1080/10408399509527704

[3] Lopes, P., Richard, T., Saucier, C., Teissedre, P.L., Monti, J.P. and Glories, Y. (2007) Anthocyanone A: A Quinine Methide Derivative Resulting from Malvidin 3-*O*-Glucoside Degradation. *Journal of Agricultural and Food Chemistry*, **55**, 2698-2704. http://dx.doi.org/10.1021/jf062875o

[4] Bimpilas, A., Tsimogiannis, D., Brouma, K.B., Lymperopoulou, T. and Oreopoulou, V. (2015) Evolution of Phenolic Compounds and Metal Content of Wine during Alcoholic Fermentation and Storage. *Food Chemistry*, **178**, 164-171. http://dx.doi.org/10.1016/j.foodchem.2015.01.090

[5] Mateus, N., Sivia, A.M.S., Vercauteren, J. and de Freitas, V.J. (2001) Occurrences of Anthocyanin-Derived Pigments in Red Wines. *Journal of Agricultural and Food Chemistry*, **49**, 4836-4840. http://dx.doi.org/10.1021/jf001505b

[6] Saxena, S.K. (2003) Polyvinyl Alcohol (PVA) Chemical and Technical Assessment. *The Joint FAO/WHO Expert Committee on Food Additives*, 61*st Meeting*, Rome, 10-19 June 2003, 1-3.

[7] Kim, J.O., Park, J.K., Kim, J.H., Jin, S.G., Yong, C.S., Li, D.X., *et al.* (2008) Development of Polyvinylalcohol-Sodium Alginate Gel-Matrix-Based Wound Dressing System Containing Nitrofurazone. *International Journal of Pharmaceutics*, **359**, 79-86. http://dx.doi.org/10.1016/j.ijpharm.2008.03.021

[8] Srinivasa, P.C., Ramesh, M.N., Kumar, K.R. and Tharanathan, RN. (2003) Properties and Sorption Studies of Chitosan-Polyvinylalcohol Blend Films. Carbohydrate Polymers *Carbohyd Polym*, **53**, 431-438. http://dx.doi.org/10.1016/S0144-8617(03)00105-X

[9] Danwanichakuk, P. and Sirikhajornnam, P. (2013) An Investigation of Chitosan-grafted-poly(vinyl alcohol) as an Electrolyte Membrane. *Journal of Chemistry*, **2013**, Article ID: 642871.

[10] Conte, A., Buonocore, G.G., Bevilacqua, A., Sinigaglla, M. and Nobile, M.A.D. (2006) Immobilization of Lysozyme on Polyvinylalcohol Films for Active Packaging Applications. *Journal of Food Protection*, **69**, 866-870.

[11] Singh, V. and Mishra, A.K. (2015) White Light Emission from Vegetable Extracts. *Scientific Reports*, **5**, Article ID: 11118. http://dx.doi.org/10.1038/srep11118

[12] Sakamoto, W., Kanehira, T., Hongou, H., Matsuda, A., Mune, M., *et al.* (2015) Application of Color Changes of Anthocyanins in Red Wine and Red Currant Extract as a Clinical pH Indicator. *Journal of Diagnostic Techniques and Biomedical Analysis*, **2**, 2.

[13] He, F., Liang, N.N., Mu, L., Pan, Q.H., Wang, J., Reeves, M.J. and Duan, C.-Q. (2012) Anthocyanins and Their Variation in Red Wines. I. Monomeric Anthocyanins and Their Color Expression. *Molecules*, **17**, 1571-1601. http://dx.doi.org/10.3390/molecules17021571

[14] Lee, J., Durst, R.W. and Wrolstad, R.E. (2005) Determination of Total Monomeric Anthocyanin Pigment Content of Fruit Juices, Beverages, Natural Colorants, and Wines by the pH Differential Method: Collaborative Study. *Journal of AOAC International*, **88**, 1269-1278.

[15] Lewis, C.E., Walker, J.R.L. and Lancaster, J.E. (1995) Effect of Polysaccharides on the Colour of Anthocyanins. *Food Chemistry*, **54**, 315-319. http://dx.doi.org/10.1016/0308-8146(95)00026-F

[16] Tachibana, N., Kimura, Y. and Ohno, T. (2014) Examination of Molecular Mechanism for the Enhanced Thermal Stability of Anthocyanins by Metal Cations and Polysaccharides. *Food Chemistry*, **143**, 452-458. http://dx.doi.org/10.1016/j.foodchem.2013.08.017

[17] Buchweitz, M., Speth, M., Kammerer, D.R. and Carle, R. (2013) Stabilisation of Strawberry (*Fragaria x ananassa* Duch.) Anthocyanins by Different Pectins. *Food Chemistry*, **141**, 2998-3006. http://dx.doi.org/10.1016/j.foodchem.2013.04.117

[18] Nour, V., Trandarfir, I. and Ionica, M.E. (2011) Ascorbic Acid, Anthocyanins, Organic Acids and Mineral Content of Some Black and Red Currant Cultivars. *Fruits*, **66**, 353-362. http://dx.doi.org/10.1051/fruits/2011049

[19] Kruger, M.J., Davies, N., Myburgh, K.H. and Lecour, S. (2014) Proanthocyanins, Anthocyanins and Cardiovascular Diseases. *Food Research International*, **59**, 41-52. http://dx.doi.org/10.1016/j.foodres.2014.01.046

[20] Pojer, E., Mattivi, F., Johnson, D. and Stockley, C.S. (2013) The Case for Anthocyanin Consumption to Promote Human Health: A Review. *Comprehensive Reviews in Food Science and Food Safety*, **12**, 483-508. http://dx.doi.org/10.1111/1541-4337.12024

[21] Baliga, S., Muglikar, S. and Kate, R. (2013) Salivary pH: A Diagnostic Biomarker. *Journal of Indian Society of Periodontology*, **17**, 461-465. http://dx.doi.org/10.4103/0972-124X.118317

[22] Mahesh, D.R., Komli, G., Jayanthi, K., Dinesh, D., Saikavitha, T.V., *et al.* (2014) Evaluation of Salivary Flow Rate, pH and Buffer in Pre, Post and Post Menopausal Women on HRT. *Journal of Clinical and Diagnostic Research*, **8**, 233-236.

[23] England, R.J., Homer, J.J., Knight, L.C. and Ell, S.R. (1999) Nasal pH Measurement: A Reliable and Repeatable Parameter. *Clinical Otolaryngology and Allied Science*, **24**, 67-68. http://dx.doi.org/10.1046/j.1365-2273.1999.00223.x

[24] Cooper, G.S., Longnecker, M.P. and Peters, R.K. (2004) Ovarian Cancer Risk and Use of Phenolphthalein-Containing Laxatives. *Pharmacoepidemiology and Drug Safety*, **13**, 35-39. http://dx.doi.org/10.1002/pds.824

11

The Unexpected Existence of Coding and Non-Coding Fragments along the Eukaryotic Gene

Pietro Volpe[1,2]

[1]Department of Biology, University of Rome "Tor Vergata", Rome, Italy
[2]National Institute of Biostructures and Biosystems, Rome, Italy
Email: volpe@uniroma2.it

Abstract

The pathways leading to synthesis and post-synthetic modification of DNA employed methionine as donor of atoms: the carbon that came from its –CH_3 served for DNA replication and repair either in bacteria or humans; its entire –CH_3 served instead for building N^6-methyladenine and 5-methylcytosine on bacterial DNA and 5-methylcytosine alone on human DNA. In humans, although a slight extra-S asymmetric methylation appeared *de novo* yielding on parental DNA 5'-m^5CpC-3'/3'-GpG-5', 5'-m^5CpT-3'/3'-GpA-5' and 5'-m^5CpA-3'/3'-GpT-5' monomethylated dinucleotide pairs, a heavy symmetric methylation involved in S semiconservatively newly made DNA to guarantee genetic *maintenance* of –CH_3 in 5'-m^5CpG-3'/3'-Gpm^5C-5' dimethylated dinucleotide pairs. In this framework, an inverse correlation was found between bulk genomic DNA methylation occurring in S and bulk polyA-containing pre-mRNA transcription taking place in G_1 and G_2. Thus, *probes* of 1 × 10^6 Daltons (constructed using sheared by sonication newly made methylated DNA filaments) revealed a modular organization in genes: after the hypermethylated promoter, they exhibited an alternation of unmethylated coding and methylated uncoding sequences. This encouraged the search for a language that genes regulated by methylation should have in common. An initial deciphering of *restriction minimaps* with hypomethylatable exons *vs.* hypermethylatable promoters and introns was improved when the bisulfite technique allowed a direct sequencing of m^5C. In *lymphocytes*, where the transglutaminase gene is inactive, its promoter exhibited two fully methylated CpG-rich domains at 5' and one fully unmethylated CpG-rich domain at 3', including the site +1 and a 5'-UTR. At variance, in *HUVEC* cells, where the transglutaminase gene is active, in the first CpG-rich domain of promoter few doublets lost their –CH_3. Such an inverse correlation suggested new hypotheses especially in connection with *repair-modification*: UV radiation would cause de-methylation in given *loci* of a promoter by chance, whilst even a partial de-methylation in this promoter would be able to resume a previously silent pre-mRNA transcription.

Keywords

Coding *vs.* Non-Coding Pre-Messenger RNA Regions, Exons and Introns, Multigenic and Monogenic Transcriptional Units, Regulation of Gene Expression, Repair-Modification

1. Introduction

Initial contributions established that, while the *Sea urchin* DNA contains the sole 5-methylcytosine (m^5C) [1], the *E. coli* DNA contains N^6-methyladenine (m^6A) and m^5C [2] [3].

The finding of these two classes of post-synthetically modified nitrogen bases along the bacterial heteroduplex was accompanied by basic observations dealing with the *restriction-modification* enzyme system [4]-[6]: m^6A was shown to be due in part to the activity of the deoxyadenylate (dam) DNA methyltransferases (DNAmets) [7]-[9] and in part to the activity of type I, type II and type III methylating restriction-modification endonucleases involved in the digestion of the infecting m^6A-free phage DNA [10]-[13]; m^5C was found to be due in part to the activity of the deoxycytidylate (dcm) DNAmets [9] [14] [15] and in part to the methylase activity of the sole type II restriction-modification system involved in the digestion of the infecting m^5C-free phage DNA [11].

The presence of m^5C along the animal double helix was known earlier [16] [17], and a small amount of m^6A was soon found in DNA of protozoa [18] [19], fungi [20], algae [21] and higher vegetables [22]. The analysis of base composition in DNA of *Drosophila* was contradictory: one technique suggested an absolute absence of m^5C [23]; another showed that, in some of its sequences, C was still modified [24] [25]. Later a modest presence of m^5C in the genome of insects was correlated with lethality [26]-[28]. In any case, it was established that, if in animal genomes, 1 to 5 out of every 100 Cs are modified [29] [30], in plant genomes, this amount could increase up to 10 or even 30 per cent [31] [32].

Our laboratory at the International Institute of Genetics and Biophysics (IIGB) of the National Research Council (CNR), in Naples, got interested in the topic because of studies on structuristics of rare nucleotides found in ox brain [33] [34], *plus* and *minus* strand poliovirus RNAs synthesized at hourly intervals through the *HeLa* mitotic cycle [35] [36] and the nature of a high-molecular weight RNA extracted with hot phenol from nerve cells [37] [38] pertaining to the origin of mRNA. On the other hand, at variance with the polygenic structure of the *operon* regulating protein biosynthesis in bacteria [39], while tentatively an operon organization was thought to act in higher organisms too [40], the results collected by us at the Albert Einstein College of Medicine, in New York, on the amphi-directional regulation of a reversible reaction common to two inverted enzyme sequences [41] [42] accounted for a monogenic function of the transcriptional unit acting in mammals.

These points—apparently distant from each other—converged into an intriguing macromolecular manipulation, shedding light on the internal structure of the eukaryotic gene [46]. At the beginning of the 70s, in fact, while attention was paid to methylation of specific DNA sequences [47], a problematic discussion focused not only on the origin and size of the eukaryotic pre-mRNAs and mRNAs [37] [73] [74] but consequently on the design of the transcriptional unit in higher cells [75] [76]. The background experiment, performed to elucidate both these aspects, originated from previous knowledge regarding the timing of DNA methylation during the cellular cycle [47] [67] and from the non-random genetic scattering of m^5C along a semiconservatively replicating DNA chain [46]. This experiment showed a preferential methylation of gene promoter [46] (as reviewed in [88]) and of all regulatory and signal sequences that do not code for mRNAs [46] (as reviewed in [89]). In brief, in *HeLa* cells, if half of the population of hybrids between genomic DNA fragments of about 1×10^6 Daltons (used as "probes") and pre-processed high molecular weight mRNAs (purified from nuclei) contained a large number of m^5Cs, the whole population of hybrids between genomic DNA fragments of about 1×10^6 Daltons (also used as "probes") and processed low molecular weight mRNAs (purified from polysomes) contained few, if any, m^5Cs [46]. Since polysomal mRNAs were known to be much shorter than nuclear pre-mRNAs [75] [76] [111], these results suggested for the first time that, at variance with the structure of bacterial genes (in an operon the cistrons were known to be constituted of coding sequences [39]), the eukaryotic gene had to be thought of as a repetition, after the promoter, of intermittent "coding" and "uncoding" regions [46] [93]. For the sake of clarity, the DNA tracts containing methylated bases essentially were left out by mRNA [46], namely methylation did not significantly involve the coding regions: methylation preferentially involved the uncoding ones, complementary to parts of pre-mRNAs to be removed during processing [46]. Three years later, Mandel and Chambon supported this finding [116]. Their electron microscopic pictures showed how a cDNA intermittently excludes segments of the corresponding pre-mRNA from hybridization: the sequences of pre-mRNA, non-hybridized with cDNA [116], were nothing but the uncoding sequences [46] (Crick [87] and Gilbert [86] proposed to name them "introns"); the sequences of pre-mRNA, hybridized with cDNA [116], were nothing but the coding sequences [46] (Crick [87] and Gilbert [86] suggested to call them "exons").

The objective of this essay is to celebrate the eukaryotic gene fragmentation discovery and to emphasize how complex was the early 70s basic biochemical research promoting the modern Epigenetics and Genetic Engineering (as recognized in [150]-[160]). Our next investigation will concentrate on the physiological evolution of coding *vs*. non-coding pre-messenger RNA regions. The exon and intron methylation differences will be evaluated as a function of the vertebrate evolution, while the perturbed regulation of transcription will be investigated in mammals in relation to the radiation-induced neoplastic consequences of the "repair-modification" activity.

2. The Pathway Using Methionine to Synthesize DNA Remained Unchanged in Phylogeny, While the Pathway Using Methionine to Modify DNA Evolved from Bacteria to Humans

Let us begin with description of the biochemical reactions lying on the basis of DNA polymerization and modification. The study was initiated in *HeLa* cells [46] [47] and completed in *E. coli* [48] [49]. However, coherently with evolution, it seemed appropriate to first treat the results regarding prokaryotes and, then, those regarding eukaryotes.

2.1. Synthesis and Methylation of DNA in Prokaryotes

One of the main purposes of this part of analysis was to verify if *E. coli* would be able to employ Met as a sole common donor of atoms for DNA synthesis and methylation, as *HeLa* cells do [47] [50]; another purpose was to verify whether, on bacterial DNA, the timing for optimal yield of m^5C would coincide with that for optimal yield of m^6A [48] [49].

It emerged that, when percolated through reverse phase HPLC [51], control nitrogen bases, mixed in equimolar proportions, were eluted in the following time-order: C, m^5C, G, T, A and m^6A (**Figure 1(a)**). But there was a striking quantitative difference between the eluted unmethylated *vs*. methylated bases, when these were constituents of an *E. coli* DNA hydrolyzate: C, G, T, and A resulted in equimolar proportions, as expected; m^5C and m^6A were present in very small amounts (**Figure 1(b)**). **Figure 1(c)** illustrated the profile of an *E. coli* DNA hydrolyzate labelled with [^{14}C]methyl-L-methionine ([^{14}C]Met) in mid-culture growth cycle (CGC): C was the only base not incorporating the labelled carbon coming from the precursor; by contrast, m^5C was labelled to a notable extent comparable to that of m^6A. The radioactivity found in the two modified bases meant that post-synthetic DNA modification took place, as for *HeLa* cells [47] [50], by virtue of transfer, *via* S-adenosyl-L-methionine (AdoMet), of the whole radioactive $-CH_3$ from [^{14}C]Met to the DNA Cs and As. In addition, in **Figure 1(c)**, the radioactivity found in the A, G and T bases revealed that, once again, as for humans [47] [50], DNA synthesis emerged because there was, *via* C_1-chain oxidation, an insertion of the labelled carbon coming from the $-CH_3$ of [^{14}C]Met into the purine heterocycle of A and G and into the $-CH_3$ of T.

The quantitative analysis of the experiments performed through the CGC of *E. coli* (in the same conditions corresponding to those of **Figure 1(c)**), indicated that, while the curve of bacterial proliferation was S-shaped [48], the synthetic and methylase pathways of DNA developed in a differential fashion (**Figure 2(a)**).

i. The specific labelling accounting for synthesis of A, G and T appeared to be much lower than that accounting for construction of m^5C and m^6A (this meant that the methylase pathway evolved faster than the synthetic pathway).

ii. The specific labelling of T was always lower than that of G and A since, one thought, the exit of the C_1-chain, oriented towards synthesis of G and A, was highly facilitated in comparison with that oriented towards synthesis of T.

iii. The specific labelling of m^5C was much lower than that of m^6A, probably because the dam enzyme system worked faster than the dcm enzyme system.

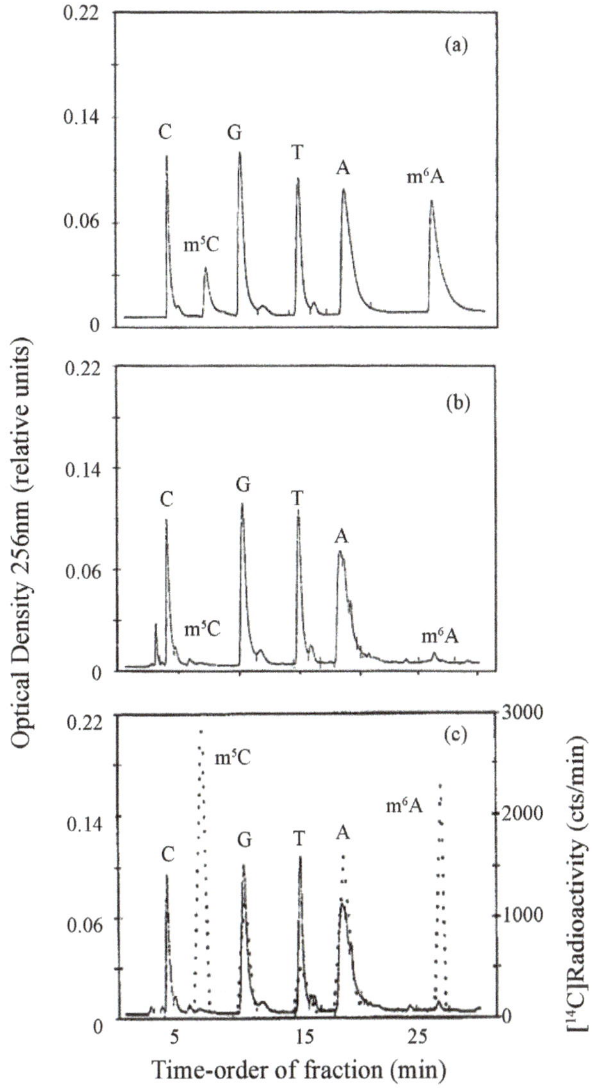

Figure 1. Chromatographic analysis of a bacterial DNA hydrolyzate. Cultures of *E. coli MRE* 600 were grown on Petri dishes by inoculating 0.5 - 1 ml of minimal medium (containing 1×10^{11} cells/ml) into 100 ml of Luria-Bertani medium. At intervals of time during their CGC (roughly lasting 10 hrs), DNA was labelled for 40 min at 37°C with 2 microCi/ml [^{14}C]Met [48]. The ^{14}C-labelled DNA was extracted with minor modification of the method described by Marmur [145]. A 30-micrograms aliquot of purified DNA was introduced into a glass vial containing 300 microliters of 88% formic acid (75% final concentration). The vial was closed with flame and left at 160°C for 1 hr [50]. The hydrolyzed residues were dried, suspended in 30 microliters of 20 mM trichloroacetic acid at pH 2.2 and run at 13,000 rev./min for 5 - 10 min in an Eppendorf centrifuge. A 20-microliters aliquot was percolated through a 131051 L ODS-55 Biorad Bio-Sil HPLC column (250 × 4 mm) in reverse phase condition [51]. Fractions of 0.4 ml were collected every 30 sec, while their content was resuspended in 50 microliters of water and tested for radioactivity in a Packard Tri-Carb TR beta-counter [48] [49]. (a) Free bases from a standard solution; (b) Bases present in 30 micrograms of an unlabelled DNA hydrolyzate extracted from a 2-ml suspension of 1×10^8 - 1×10^{10} bacteria maintained in minimal medium; (c) Bases present in 30 micrograms of a labelled DNA hydrolyzate obtained from a 2-ml suspension of 1×10^8 - 1×10^{10} bacteria maintained in Luria-Bertani medium. The left ordinate showed the concentration of bases detected in arbitrary units by a computerized instrument (continuous line); the right ordinate showed their radioactivity (dotted line). The analysis was repeated a dozen of times.

iv. In harmony with the development of the specific labelling of all nitrogen bases taken together and of the specific labelling of single A, G and T, the specific labelling of m^{6}A gradually decreased during the CGC, when that of m^{5}C rapidly decreased after the inoculum to rise again, with a sharp peak, in mid-CGC (an interpretation of these patterns was that the dam and dcm DNAmet activities are uncoupled) [48] [49].

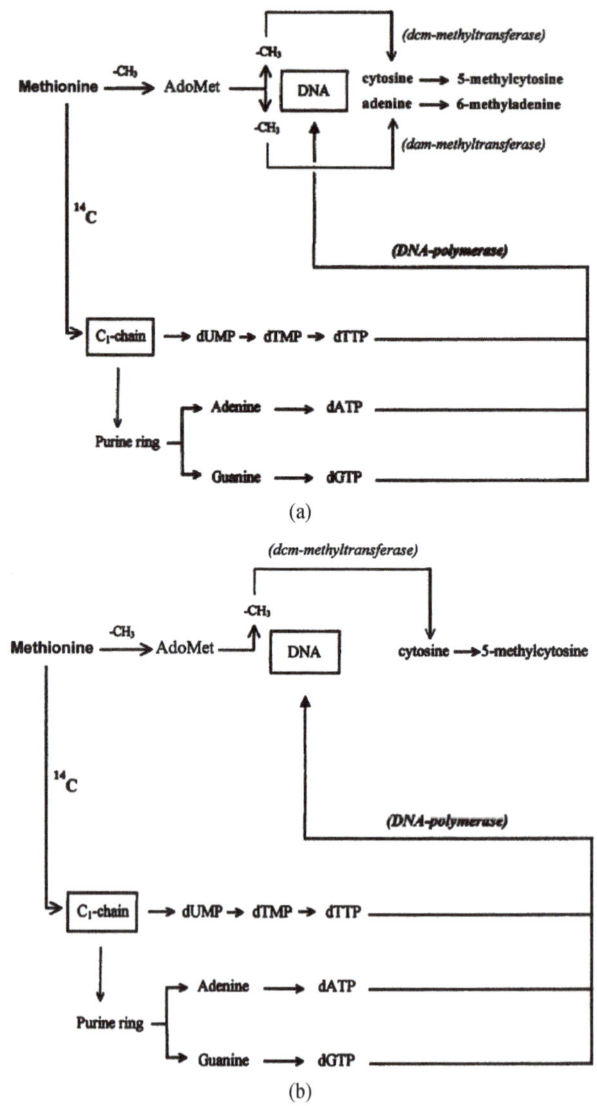

Figure 2. Pathways leading to synthesis and post-synthetic modification of DNA. In *E. coli* [48] [49], as in *HeLa* cells [47] [50], to accomplish replication, the DNApol system exploited the dATP, dGTP and dTTP resulting from the C_1-chain oxidation which used, *via* formic acid, the carbon coming from the $-CH_3$ of Met to build the purine heterocycle of A and G and the $-CH_3$ of T (this carbon did not enter the pyrimidine ring of C). (a) In bacteria, two parallel pathways led *via* AdoMet to post-synthetic DNA modification [48] [49]: on the one hand, the dam DNAmet system (including the methylation exerted by the type I, II and III RM endonucleases) constructed m^6A on new DNA circles; on the other, the dcm DNAmet system (including the methylation exerted by type II restriction-modification endonucleases) constructed m^5C on the same new DNA circles. (b) In Humans, Met also served as a common donor of atoms [46] [47] [138]: in the cytosol, the carbon coming from its $-CH_3$ through the C_1-chain entered the purine ring of A and G and the $-CH_3$ of T; after formation of the corresponding dATP, dGTP, and dTTP and, after release of a PP_i from each, in the nucleus the DNApol *alfa* [112] introduced, during S, the dAMP, dGMP, and dTMP into semiconservatively newly synthesized DNA chains, while Met catapulted *via* AdoMet its entire $-CH_3$ on given C residues located along these chains.

2.2. Synthesis and Methylation of DNA from the World of Bacteria to the World of Humans

The data reported above demonstrated that the two metabolic pathways, leading to synthesis and methylation of DNA, originated in the prokaryotic world by exploiting Met as donor of carbon atoms and $-CH_3$ groups (**Figure 2(a)**). This crucial property was preserved fairly well by the mammalian world (**Figure 3(a)** and **Figure 3(b)**), even if at the level of the DNA modifying enzyme bifurcation one of the two branches—the one yielding m^6A on DNA (**Figure 2(a)**)—was erased in *HeLa* cell DNA (**Figure 2(b)**).

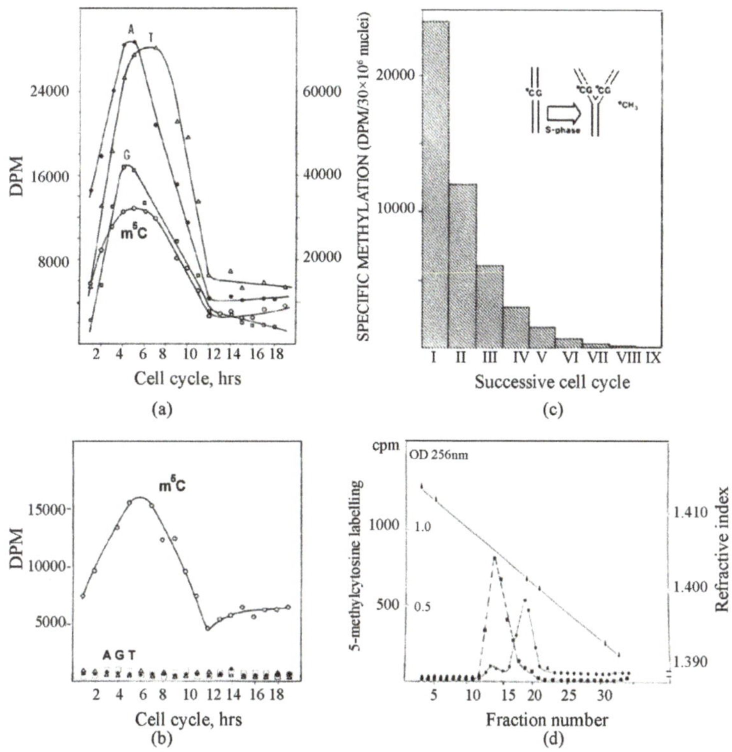

Figure 3. Semiconservative transmission of m^5C in human DNA. (a) Through an entire *HeLa* cell cycle [67], the genomic DNA (labelled with [^{14}C]Met) was hydrolyzed to bases and chromatographed: radioactivity in A, G and T (right ordinate) showed synthesis, while radioactivity in m^5C (left ordinate) showed methylation [46] [47]. (b) During an entire mitotic cycle, nuclei isolated from *HeLa* cells were labelled with [^{3}H-methyl]AdoMet and their DNA was hydrolyzed to bases and chromatographed: radioactivity in A, G, and T, accounting for synthesis, was negligible, while radioactivity in m^5C showed methylation [47]; (c) During S, *HeLa* cells were labelled with [^{14}C]Met; then, after 14 hrs of growth in fresh medium (in the absence of radioisotope), their genomic DNA was hydrolyzed to bases and chromatographed: the measurement of radioactivity in m^5C was repeated in the course of twelve CGCs. The insert suggested how, in the replication fork, methylation semiconservatively followed synthesis, since the labelled m^5C per cell always decreased by half [109]. (d) Ultracentrifugation of genomic *HeLa* cell DNA in alkaline CsCl gradient [50]: the dashed line showed the m^5C radioactivity (originating in [^{14}C]Met) detected along the semiconservatively newly replicating chains, made heavier by a previous incorporation of BrdUrd; the solid line showed the OD at 256 nm of the separated lighter parental chains.

The abandonment of m^6A by DNA of the majority of the highly evoluted cells was a consequence of the substitution, in them, of the obsolete bacterial antiphage restriction-modification defense [6] with the innovative immune reaction. Actually, with the extinction of the system of dam and dcm DNAmets [9] and of the system of restriction-modification endonucleases [11]-[13], both characterizing the world of bacteria [52], molecular selection led in mammals to the loss of m^6A not only in nuclear [46] but also in mitochondrial [53] [54] and chloroplastic [55] DNAs. In these different nucleic acids, the m^5C base was conserved by a novel DNAmet family that likely evolved through an *alternate splicing* [56]-[58], as in mammals the fundamental defense against antigens was exerted by antibodies.

In such a structure, the question concerning the speed of reactions in the pathway leading to DNA synthesis *vs.* the speed of reactions in the pathway leading to post-synthetic DNA modification assumed great interest since, during the bacterial CGC, the rate of formation of T was always lower than that of A and G. Namely, the time required by the carbon, on the way from the $-CH_3$ of Met to the $-CH_3$ of T, was longer when compared to the time required for its insertion into the purine heterocycle of A and G [48] [49]. This differential rate was in harmony with the situation seen in *HeLa* cells: quantitatively, the biosynthetic rate of T was of the same order of that of A, even though a notable delay of its optimum with respect to the optimum of synthesis of A and G was observed (**Figure 3(a)**).

This kind of correspondence was expected, since either in bacteria (**Figure 1**) or mammals (**Figure 3(a)**) the process leading to synthesis of T had to join into a pyrimidine ring the carbamylphosphate and aspartate molecules before accepting, at position 5 of the newly formed U, the $-CH_3$ constructed with the carbon coming from formic acid. This carbon entered directly the A and G purine heterocycle (**Figure 2(a)**).

Biosynthesis of A, G and T transpired in comparable loci (in liquid matrix of a bacterial cell or in cytosol of a mammalian cell), whereas post-synthetic modification of DNA occurred in different compartments: in *E. coli*, it yielded m^5C and m^6A in the same liquid matrix (**Figure 1(b)** and **Figure 1(c)**); in *HeLa* cells, it yielded the sole m^5C in the nucleus (**Figure 3(b)**). As follows, the longer time required by the nucleoside AdoMet to enter the nucleus would justify the advantage of the bacterial cell over the mammalian cell in exploiting the $-CH_3$ of Met: if in *E. coli* the specific building of A, G and T was much lower than that of m^5C and m^6A [48] [49], in *HeLa* cells the specific building of A, G and T turned out to be either similar or only a little lower than that of m^5C [47] [50].

Summarizing, the large percentage of DNA methylation, necessary for antiphage defense in bacteria, became unnecessary in mammals.

Was this a non-turning point in evolution?

To make a long reasoning short: the prokaryotic world used Met to promote both the pathways leading to synthesis and methylation of DNA (**Figure 1(b) Figure 1(c)** and **Figure 2(a)**); this capacity was inherited almost completely by the mammalian world, with the exception that, in *HeLa* cells, the reactions yielding m^6A on DNA were suppressed (**Figure 3(a)** and **Figure 3(b)**); the diversification of the speed of single reactions (in the pathways leading to synthesis and methylation of DNA) depended on the evolution of cellular organization, of course, since the eukaryotic nucleus concentrated in itself important genetic functions that were diffused earlier in the bacterial liquid matrix [47]-[49].

3. In Humans the Cell-Cycle Methodology Served to Reveal Two Basic Laws Regulating DNA Methylation

Notwithstanding the exhaustive investigation on the evolution of the pathways leading to DNA synthesis and methylation, there were doubts in elucidating the biocatalytic processes dealing with the complex phenomenon of DNA methylation in eukaryotes: the existence of DNA demethylases, deaminases and reaminases was debated [59]-[61]; the heterogeneity of the DNAmet system was revealed much later [56] [57] [62]-[64]. For this reason, in the early 70s, we preferred to investigate the timing of DNA methylation and consequently the precise macromolecular acceptors of the $-CH_3$ groups, namely the old or the new DNA chains: the experiments were made feasible by a method set up to determine the lengths of an entire mitotic cycle and its single G_1, S, G_2 and M phases [65]-[67].

3.1. DNA Methylation Followed DNA Synthesis

It was suggested that DNA methylation would continue for some hours after DNA synthesis is completed [68] [69]. In this research, by incubating synchronized *HeLa* cells with [^{14}C]Met used as common tracer for both DNA synthesis and methylation [47], it was possible to verify that DNA methylation accompanied DNA synthesis: the labelled carbon atom, coming from the $-CH_3$ of Met, did not enter the pyrimidine ring but pierced *via* C_1-chain the purine ring of A and G and the $-CH_3$ of T, while the entire $-CH_3$ of [^{14}C]Met was transferred *via* AdoMet to DNA C (**Figure 2(b)**). Truly, the maximal labelling emerging from the ^{14}C constituting the $-CH_3$ of [^{14}C]Met was always found in S in four hydrolyzed bases: in A, G and T, implying synthesis, and in m^5C, signifying methylation (**Figure 3(a)**).

3.2. Newly Replicating DNA Chains Were Semiconservatively Methylated

It was known that in isolated nuclei, in the absence of triphosphonucleosides (supplied by cytosol), DNA could not be synthesized. On this basis, by employing nuclei labelled with [^{14}C-methyl]AdoMet, one could verify whether or not the absence of DNA replication would influence DNA methylation. The result was unexpected: in nuclei isolated from *HeLa* cells during S, among other DNA bases, only C continued to be methylated in the absence of DNA synthesis (**Figure 3(b)**). This showed that the two pathways of DNA replication and DNA methylation were separated from one another (**Figure 2(b)**): in whole cells, DNA methylation followed DNA replication during S (**Figure 3(a)**); in isolated nuclei, DNA methylation proceeded during S in the absence of DNA synthesis (**Figure 3(b)**). By itself the occurrence of DNA methylation in S-phase nuclei was not an absolute demonstration that it involved new chains formed just before their isolation; but the correlations between *in vivo*

DNA synthesis and *in vivo* and *in vitro* DNA methylation strongly suggested this possibility [47]:

$$\left[\Delta DNA - CH_3/\Delta t\right]_{in\ vivo} = K_1\left[\Delta DNA\right]_{in\ vivo}$$

$$\left[\Delta DNA - CH_3/\Delta t\right]_{in\ vitro} = K_2\left[\Delta DNA\right]_{in\ vivo}$$

$$\left[\Delta DNA - CH_3/\Delta t\right]_{in\ vitro} = K_3\left[\Delta DNA - CH_3/\Delta t\right]_{in\ vivo}$$

These equations established that the cells entering S would carry old chains inherited by parental cells, while targets of DNA met both *in vivo* and *in vitro* should be nothing but nascent chains: several of them still remained in the nuclei to be isolated [47]. A similar conclusion was reached thinking that, if old and new DNAs were methylated in isolated nuclei [70], new DNAs would be formed without m^5C in whole cells [71].

Whatever the case would be, the semiconservativity of DNA methylation (**Figure 3(c)**), revealed by treating synchronized cells with the [^{14}C]Met tracer [47], was confirmed not only by the analysis that used restriction endonucleases [72] but also by the direct separation in alkaline CsCl of heavy bromodeoxyuridine (BrdUrd)-containing new chains and light non-containing BrdUrd old chains (**Figure 3(d)**).

4. Hybridizations of Pre-Processed Messenger RNA or Processed Messenger RNA with DNA "Probes" (Constructed during the S Phase by Shearing Newly Methylated DNA Filaments) Revealed an Alternation of Coding and Non-Coding Regions in the Human Gene

The hypotheses regarding the organization of the eukaryotic transcriptional unit had some analogy with those concerning the role of DNA methylation in higher organisms. From this, one deduced that the easiest way to directly characterize the eukaryotic gene design would be to exploit pure biochemical tools.

4.1. From Finding of Pre-mRNA and mRNA to Construction of DNA Probes

It is worth re-visiting in some more details the basic dispute into which we get involved. Georgyev and Mantieva found the polyribonucleotide called *giant nuclear* RNA in mouse [73], as also evidenced by us in nerve cells [37] [38]. Penman, Vesco and Penman claimed that their polydisperse RNA had nothing to do with the giant nuclear RNA [74]. In return, Georgyev assumed that the polydisperse RNA would likely correspond to a product of maturation of the giant nuclear RNA [75].

Accepting this assumption, Darnell, Jelinek and Molloy considered the giant nuclear RNA as *pre-processed* mRNA (measuring the size of a gene) and the polydisperse RNA as *processed* mRNA (much shorter than a gene): an agreement stating that a transcriptase would copy the whole eukaryotic gene to form a large pre-processed pre-mRNA, namely the *heterogeneous nuclear* RNA [76].

This conclusion helped indeed to elucidate the structure of the monogenic transcriptional unit in mammals, since we judged such a monogenic unit as an extraordinary innovation exerted by evolution with respect to the complex bacterial polygenic transcriptional unit known as operon.

Therefore, the information reported above on semiconservative methylation of newly replicating DNA chains (**Figure 3(c)** and **Figure 3(d)**) served as a key for a crucial biochemical analysis [46]: segments of these chains, obtained through shearing, were employed as *probes*, labelled in m^5C, in order to look for particular sequences located inside the eukaryotic gene (**Figure 4(A)**).

4.2. Our So-called "Pioneering Experiment on Gene Fragmentation" Demonstrated the Existence in the Human Transcriptional Unit of Coding *vs.* Non-Coding Regions

Filaments of semiconservatively newly made methylated DNA, previously separated during the S-phase in alkaline CsCl gradients (**Figure 3(d)**), were shortened through sonication into fragments of about 1×10^6 Daltons (**Figure 4(A)**, left top) and hybridized with pre-mRNA (**Figure 4(Ab)**) extracted from *HeLa* cell nuclei [46] [66] [67] or with mRNA (**Figure 4(Ac)**) extracted from polysomes of these same cells [46] [77]: in the first case, the modified Cs were found in about 50 per cent of the hybrids "pre-mRNA/DNA probe" (signed with "plus" *vs.* "minus" in **Figure 4(A)**); in the second, the modified Cs were essentially absent in 100 percent of the hybrids "mRNA/DNA probe" (marked with "minus" in **Figure 4(A)**).

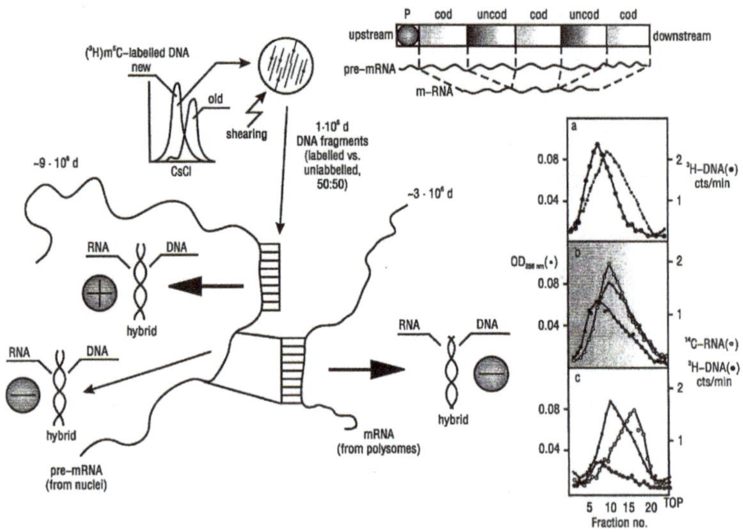

(A)

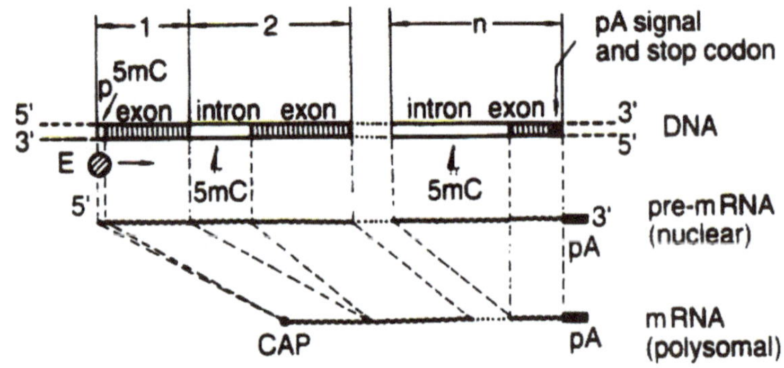

(B)

Figure 4. Gene design. (A) Employing *HeLa* cells, semiconservatively newly made DNA chains, separated through CsCl in mid S-phase (a), served to construct by shearing methylated probes of 1×10^6 Daltons; their association with larger polyA-pre-processed messenger RNA (from nuclei) yielded methylated (50%) and unmethylated (50%) DNA/RNA hybrids (b); the same probes, associated with shorter polyA-containing processed messenger RNA (from polysomes), essentially yielded 100% unmethylated DNA/RNA hybrids (c). Actually, after the promoter, an alternation of coding and uncoding regions appeared to take place, since a preferential methylation characterized the same promoter and the sequences that did not code for mRNAs [46] [138] [147]. (B) The monogenic transcriptional unit was thought to imply an adjacency of 1-, 2-, n-blocs, each containing an initial methylated part and a final unmethylated part [46] [93]. The promoter (p) would correspond to the methylated part of the first bloc, followed by the first unmethylated coding (exon) part. Each following bloc should contain a methylated non-coding (intron) part and an unmethylated coding (exon) part. The last bloc should carry, at the end of its coding part (exon), the polyA signal and the stop codon(s). The RNApol II would synthesize pre-mRNA along the whole gene, while the pre-mRNA tracts—complementary to the methylated non-coding DNA sequences (from 2- to n-blocs)—would remain out of processed mRNA. (CAP) 5'-Methylated mRNA end; (pA) 3'-post-transcriptional polyadenylated pre-mRNA end.

For clarity's sake, the result of the former experiment signified that the m^5Cs were located along the promoter and along the non-coding regions inside the gene (all uncoding sequences, including the promoter, being complementary to regions to be removed from the unprocessed pre-mRNA); by contrast, the result of the latter experiment suggested that the joined translatable sequences of processed mRNA were complementary to the unmethylated coding sequences scattered in the gene.

Such a design of human transcriptional unit-structured in alternating non-informative and informative sequences, starting from the promoter (**Figure 4(A)**, right top)—was illustrated at the 9*th Meeting of the Federation of the European Biochemical Societies* held in Budapest in 1974: the oral presentation was preceded by an invited paper that opened the ritual Special Issue of *FEBS-Letters* [46].

Three years later, in 1977, the novelty brought by this paper was confirmed in France by Breathnach, Mandel and Chambon [78]: in their electron microscopic observations, the ovalbumin mRNA repeatedly did not hybridize sequences of the ovalbumin gene, forming characteristic *loops* (these were nothing else but the uncoding regions found earlier by us); differently, the sequences of the ovalbumin gene, that did become hybridized to the ovalbumin mRNA, were nothing else but our coding regions.

Soon, a number of other laboratories, mostly in USA, observed that the eukaryotic gene looked *split*: the pre-mRNA coding sequences were interrupted by uncoding ones [79]-[82]; uncoding sequences were seen even in the *adenovirus* mRNA [83] [84].

Then, when an intervening sequence was also described in the mouse *beta*-globin gene [85], Gilbert [86] and Crick [87] proposed to call *exons* and *introns*, respectively, the coding and non-coding regions found in Naples [46].

Nevertheless, while two years earlier the proposal of these new appellations Maclean and Hilder [88], from United Kingdom, in the *International Review of Cytology*, attributed to us the priority of discovery of the "*preferential methylation of bases in the promoter proximal region of the transcriptional unit*", two years later the introduction of these new appellations, Tentravhi, Guntaka, Erlanger and Miller [89], from United States, in the *Proceedings of the National Academy of Sciences of USA*, attributed to us also the priority of discovery of the "*preferential methylation of the early replicating regulatory sequences that do not code for mRNAs*".

In sum, if in *HeLa* cells half of the population of hybrids between genomic DNA fragments (used as probes) and pre-processed mRNAs (purified from nuclei) contained a large number of m^5Cs (**Figure 4(Ab)**), the whole population of hybrids between genomic DNA fragments (used as probes) and processed mRNAs (purified from polysomes) contained few, if any, m^5Cs (**Figure 4(Ac)**). Since polysomal mRNAs were known to be much shorter than nuclear pre-mRNAs, these results suggested that, in comparison with the bacterial operon structure (containing *cistrons* essentially representing informative sequences), the eukaryotic gene had to be thought of as a repetition, after the promoter, of alternate *coding* and *uncoding* regions (**Figure 4(A)**, top right): methylation did not significantly involve the coding regions but involved the non-coding ones, complementary to parts of pre-mRNAs to be removed during processing [46].

5. Evolution Combined a Variable Number of *Coding-Uncoding Modules* in Eukaryotic Genes Depending on the Lengths of the Polypeptides to Be Expressed

Following the demonstration that the infection by a lytic virus, like *poliovirus*, was found to be rigorously cell-cycle dependent [35] [36], the observation of the design of the eukaryotic transcriptional unit [46] induced to investigate the site of recombination between the genome of an oncovirus, like *SV40*, and that of a host-cell, like *mouse fibroblasts* (the oncogenome integration needing at least an *uncoding-coding module* as specific target).

The working hypothesis was based on the assumption of two possible mechanisms of virus-induced cell transformation: one involving a complete loss of cell regulation and the other involving a partial loss of cell regulation. In the first instance, the virus DNA would be integrated between the uncoding and coding regions of the module while, in the second, the virus DNA would be integrated in the uncoding region only [90].

The genetic engineering experiments spoke in favor of the former mechanism, being facilitated during the S-phase [91] [92]. This mechanism—in addition to the information carried by the *SV*40 early gene—exploited the information carried by an adjacent short piece of host-cell DNA to yield the so-called *non-truncated*, *i.e.* entire (*fused*), T-antigen: the *truncated* T-antigen, encoded by the *SV40* early gene, did not contain the signals for polyadenylation and stop codons; these signals were brought by the joined host-cell DNA piece [93].

These new acquisitions led us to carefully consider the question, posed by Gilbert [86], on why the genes are broken down into pieces.

The origin and evolution of the uncoding-coding structures seemed problematic to him, not only because both the "exon theory of genes" and the "insertional theory of introns" debated the presence or absence of uncoding sequences in "primordial genes", but also because the statistical analysis assumed that an "exon shuffling" played an important role in the origin of either *ancient* or *modern* genes.

Despite the fact that we did agree with such a viewpoint, our attention was focused, rather, on the order of the two classes of regions in a module. Why, we questioned, an uncoding region should precede a coding one, and not *vice versa*?

Independently from the concept of *exon shuffling*, one would answer that transcription starts in any case from a promoter.

A *modular gene organization* should be in good harmony with this need: in the initial module, the promoter should play the role of an uncoding region, preceding the first exon; in the last module, after the intron, an exon should necessarily carry an end with signals for polyadenylation and stop codons (**Figure 4(B)**). Thus, inside the gene, the promoter and the various non-coding regions should exert a regulatory function, while the various coding regions should carry information for a given protein [46].

But were the $-CH_3$ groups (preferentially carried by the promoter and non-coding regions along the gene) correlable indeed with the genetic regulation of transcription?

A circumstantial attention to this correlation will be paid below.

Allow us emphasize the fact that the association of non-coding regions with methylation *vs.* the practical absence of methylation in coding regions (**Figure 4(B)**) represented ideal properties useful to evaluate the "age" of the regions themselves, *i.e.* if unmethylated coding sequences should be considered phylogenetically older while methylated non-coding sequences should be considered phylogenetically younger.

From the viewpoint of evolution, we assumed that the unmethylated coding regions probably became 'conservative' because the conventional excision-repair system always reconstructed them by maintaining any possibility of error at a minimum (e. g. the high conservation of genes coding for histones), as the continuous repetition of possible incomplete repair of the methylated uncoding sequences—through a long chain of cell cycles [90]—probably caused their wide variability.

On the basis of this variability selection would have operated by creating (through *ad hoc* transpositions) new transcriptional units containing non-coding regulatory and signal sequences [91] [93]-[96].

An insight of this kind might also explain why the amount of the uncoding portion in many eukaryotic genes was found to be so large when compared with that of their coding part, being the proportion dealing with the lengths of the proteins to be made, of course [53]. After all, the chain of biocatalytical reactions able to join in sequence various uncoding-coding modules to form the genes expressing these proteins had not yet been understood.

6. The Alternation of Coding and Non-Coding Regions in Eukaryotic Genes Was Convalidated by Their Restriction Minimaps Showing a Scattered Distribution of the Methylatable Sites

At the *7th International Symposium on Macromolecules in the Functioning Cell*, held in Taormina in 1990, my talk entitled "*Is 5-methylcytosine a regulatory signal in eukaryotic gene expression*?" emphasized the concept of split eukaryotic gene [46] analyzing its methylatable sites recognized by restriction-modification endonucleases [97]. Then, the split design of the gene, observed in *HeLa* cells, that I extensively elucidated in research seminars given at the King College of London, in 1991, and at the Cancer Center of the University of Texas, Houston, in 1992, contributed to motivate at the "Incubator of Ideas" of Stony Brook, USA, in 1993, the project of a long-term biotechnological co-operation between the Italian CNR and the State of New York [98], while a review on "*5-methylcytosine in genes with methylation-dependent regulation*" opened an issue of *FEBS-Letters* [99].

Complementing this, it is not superfluous to further make the subject clearer through a careful reading of *restriction minimaps* constructed for a large number of human genes regulated by m^5C: their genetically discontinuous "geometry" (**Figure 4(A)**) exhibited, after the promoter, a *sui generis* wavy chain of hypomethylatable coding and hypermethylatable uncoding regions (**Figure 4(B)** and **Figure 5**). Both in *housekeeping* (HK) and *tissuespecific* (TS) transcriptional units, they were roughly contained beneath the limits of a "scalene triangle".

For example, the *calcitonin* (CALC) transcriptional unit–representative of about thirty fully A, T, G, C-sequenced and m^5C-regulated TSs—was heavily methylatable upstream and slightly methylatable downstream (**Figure 5(b)**).

In such a framework, at least four pairs of dinucleotides behaved as targets recognizable by restriction-modification endonucleases [98] [99].

In exons, two of them were monomethylatable:

5'-Tpm^5C-3'	5'-Cpm^5C-3'
3'-ApG-5'	3'-GpG-5'

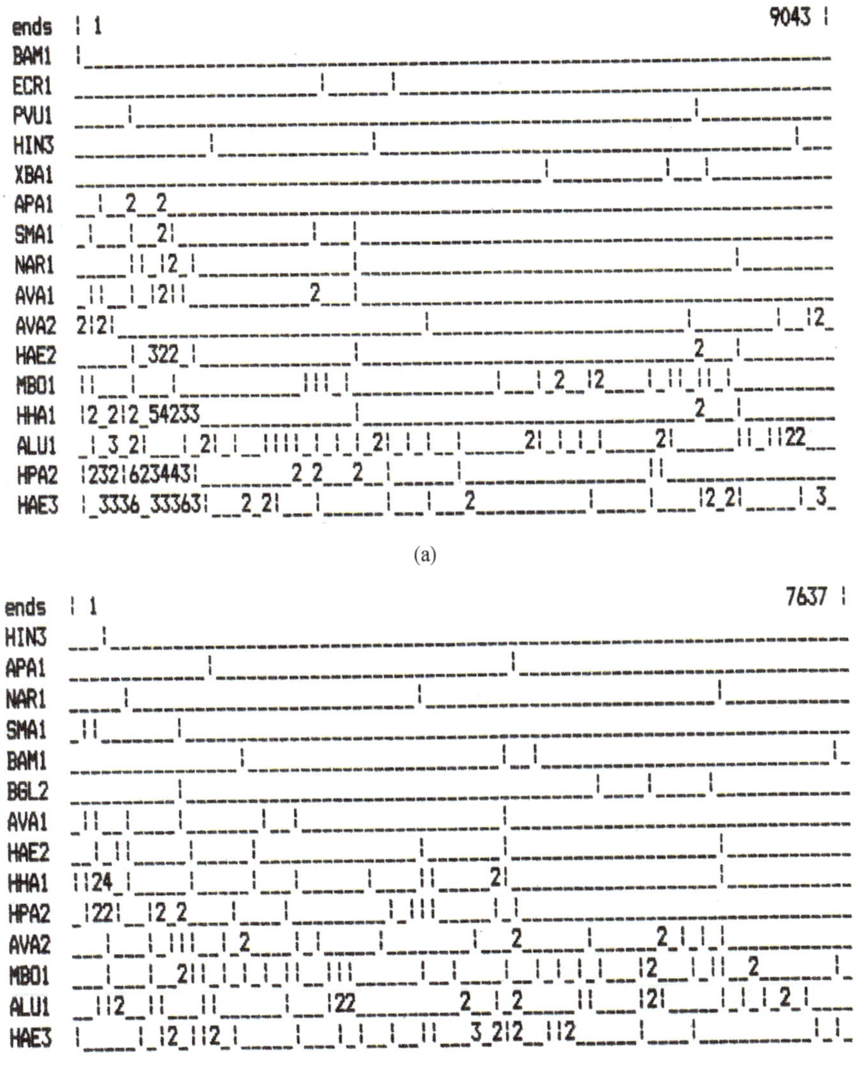

Figure 5. Scattered methylatable cites along the HK and TS genes in mammals. The restriction minimaps were drawn using the Mapdrow DNAstar Program. At the 5'-end of a gene, the promoter region showed the highest number of dcm-sites, accounting for a significant hypermethylation [98] [99] [146]. Going from 5' to 3', while the alternate coding (exon) sequences carried a relatively low number of dcm-sites, the alternate uncoding (intron) sequences carried a relatively high number of dcm-sites. However, this level steadily decreased going to the 3'-end. (a) The HK *ODC* gene contained 9.043 nucleotide pairs; (b) The TS *CALC* gene contained 7.637 nucleotide pairs. (Ends) dcm-inhibited restriction endonucleases; (1) single dcm-sites; (2-6) multiple dcm-sites.

In promoter and introns, the other two were dimethylatable:

5'-m^{5}CpG-3'   5'-Gpm^{5}C-3'
3'-Gpm^{5}C-5'   3'-m^{5}CpG-5'

At any rate, the average quantitative distribution of these pairs–which corresponded to those isolated from the DNAse I digests [100]-[102]—was similar in HKs (**Figure 5(a)**) and TSs (**Figure 5(b)**), despite the fact that in the case of TSs (going in the direction from 5' to 3') the monomethylatable dinucleotide pairs proportionally increased while the dimethylatable dinucleotide pairs proportionally decreased [98] [99].

Thanks to identification of all these methylatable "words", the promising 1993 seemed to anticipate, therefore, that the biomolecular studies were moving into a challenging new cycle, whilst new resources such as *restriction gene minimaps* were enabling genome-wide investigations that could potentially identify most common genetic determinants of human health, disease or drug design.

7. The "Repair-Modification" System Played a Role in Either Gene Evolution or Neoplastic Cell Transformation

If we look back for a moment at the investigations regarding the cell-cycle dependence of DNA methylation [47] *vs.* the non-random scattering of m^5C along semiconservatively newly replicating DNA chains [46], we can appreciate how they not only paved the way to the discovery of the hypermethylation of promoter and all regulatory and signal sequences that, in the eukaryotic gene, did not code for mRNAs [46] but also considered the corollary experiments which, albeit at the beginning oriented towards other goals (particularly towards deciphering of the $-CH_3$ language), at the end helped to shed some new light on the fine control of post-synthetic DNA modification.

7.1. A Differential Methylation of Euchromatic *vs.* Heterochromatic DNAs Took Place

Evidence indicated that the S-phase should be subdivided into two parts with respect to the characteristics of DNA, since early replicating euchromatic DNA tended to be CG-rich, while late replicating heterochromatic DNA tended to be AT-rich [103].

Furthermore, the DNA fraction extracted from *Chinese hamster* cells in early S was methylated to a greater extent when compared to the DNA fraction extracted from the same cells in late S [104].

This difference was reproduced in *HeLa* cells [46]. On its basis, we supposed that, while in newly replicating DNA chains there would be sequences not uniformly methylated in the course of S, the CG-rich sequences would be preferentially methylated [46].

7.2. A Biological Clock Controlled Methylation of Specific Replicating Sequences

Since DNA did not appear to be uniformly methylated in S, investigations continued to focus on nitrogen bases specifically adjacent to m^5C [46] rather than considering the rough subdivision of hypermethylated euchromatic *vs.* hypomethylated heterochromatic replicons [104].

With this purpose in mind, methylated DNA coming from *HeLa* cells was fractioned in alkaline Ag^+/Cs_2SO_4 gradients: a small heavy CG-rich fraction and a large light AT-rich fraction were obtained (the first—containing genes for rRNA—was mainly expressed in early S and, the second, was mainly expressed in late S); as for the m^5C concentration, if in early S it was found to increase on the heavier peak, in late S it was found to increase on the lighter peak [105]-[107].

All this suggested that, at the different stages of S, one dealt with CG- or AT-rich regions probably polymerized in correspondence of specific hypo- or hypermethylated templates: diversely, it seemed that the genes would be methylated according to a given order and intensity along the newly replicating chains [105].

This idea was definitely demonstrated when we observed that the foldback DNAs were subjected to hypermethylation: if in *HeLa* cells—through the S-phase—a methylation wave chronologically involved palindromic, highly repeated and moderately repeated sequences, the unique sequences were characterized by a minimal late methylation [108].

7.3. Both CG- and AT-Rich Sequences Were Targets for Methylation

A circumstantial analysis of the two genomic DNA fractions, separated in alkaline Ag^+/Cs_2SO_4 gradients during S, confirmed that m^5C had a differential distribution in them: the highest concentration of methylated sequences was found in the denser side of the heavy fraction; in the light DNA the highest concentration of methylated sequences was found in the lighter side [105]-[107].

This implied two generalizations: hypermethylation on CG-rich sequences would offer no surprise because of its rather statistical character (the CG-rich rRNA genes would be hypermethylated according to the size of their CG targets); hypermethylation on AT-rich sequences would seem to work, instead, against the principle of CG target size, since a decrease in C residues would correspond to an increase in their methylation. Hence, when the extra-S time was characterized by a minimal, probably *de novo* DNA methylation, S exhibited maintenance DNA methylation [47].

For this reason, one assumed, the mechanisms discussed above should function in S [90].

But it was obvious that, with consecutive cell cycles (**Figure 6(a)**), the methylated sequences could not increase without limits [109], also because changes in the amount of methylated DNA during development [100]

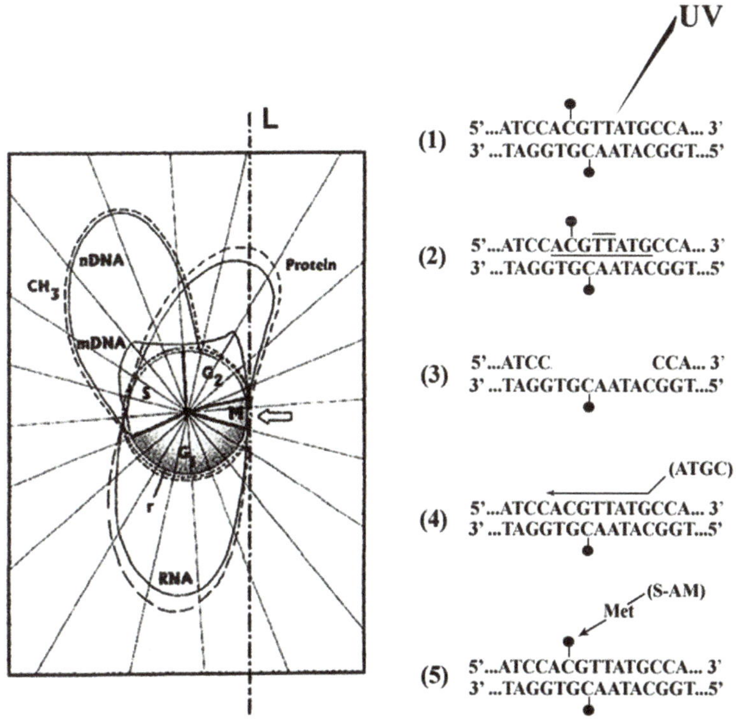

(a) (b)

Figure 6. Inverse correlation between DNA methylation and gene expression as a function of cell cycle or repair-modification. (a) DNA methylation [47] *vs.* biosynthesis of DNA [101], RNA [110], and protein [111]: the hours were signed by points where spokes crossed the circular "abscissa"; the ordinates were represented by the same spokes; the white arrow, against M, referred to a line of cell-cycle asymmetry L [65] [67]. Following appropriate labelling, at hourly intervals, macromolecules were extracted, purified and analyzed on gradients [50] [101] [110] [111]. Replicative synthesis of genomic DNA (–) and its methylation (--) showed apogees in S, as in **Figures 3(a)-(b)**, while repair synthesis of genomic DNA (--) was rather constant around the cycle [50] [90] [138]. Replication of mtDNA (–) showed maxima in S and G_2, whereas its methylation also occurred in S [54] [101]. Differently, transcription of RNA (–) and its translation into protein (- -) exhibited almost overlapped apogees in G_1 and G_2 [90] [138]. In M, all macromolecular biosyntheses were rather repressed [35] [36] [90] [138]. (b) Repair-modification mechanism [53] [95] [138] [142]-[144] [148]: UV at random would induce the building of TT-dimers in vicinity of symmetrically dimethylated 5'-m^5CpG-3'/3'-Gpm^5C-5' dinucleotide pairs, for instance (2); in this case, after digestion of damaged regions (3), which previously included a m^5CpG dinucleotide (1), the cell cycle-independent excision-repair would replace the m^5C nucleotide with a simple C nucleotide, since repair DNApol would not find any methylated dCTP in the soluble pool of triphosphonucleosides; the resultant hemi-methylated CpG/Gpm^5C dinucleotide pairs, namely the incompletely reconstructed RPs (4), should gain the capability of resuming transcription; in S, the cell cycle-dependent DNAmet would add to C the lost –CH_3 (5), providing completely reconstructed RPs able to silence the transcription again.

and in differentiated tissues [29] [30] were not conspicuous.

So, whatever the mechanism of DNA methylation was in S, statistical or specific, it was reasonable to imagine that, mostly during the extra-S time, each small wave of *de novo* DNA methylation should be followed by a corresponding small wave of DNA de-methylating repair [53].

7.4. After a Gene Radiodamage, the Complete Rebuilding of Unmethylated Coding Regions Was Guaranteed by "Excision-Repair", While the Complete Rebuilding of Methylated Uncoding Regions Was Guaranteed by "Repair-Modification"

In *HeLa* cells, replication of genomic DNA [65] [67] [101] was followed by its methylation [47] in S, whereas transcription [110] and translation [111] mainly resulted in G_1 and G_2. This was the first concrete suggestion as to a macroscopic inverse correlation observable between gene methylation and expression (**Figure 6(a)**). As for the enzyme components: DNA polymerase (DNApol) *alpha*, responsible for replication of genomic DNA, was induced in S; DNApol *gamma*, responsible for replication of mtDNA, was made in S and G_2; DNApol *beta*, involved in rebuilding, was produced at a rather low level through the whole mitotic cycle with some increase in early S (the increased DNApol *beta* repair activity in early S signifying an effort by the cell to highly reconstruct the genomic DNA just before its replication) [121]. Moreover, in conformity with the sharp peak of maintenance genomic DNA methylation in S (**Figure 6(a)**), the DNAmet system [63] was induced in S as well (**Figure 3(b)**), while a moderate presence of DNAmet characterized the rest of the cell cycle mainly accounting for *de novo* genomic DNA methylation (**Figure 3(b)**). During S, a *coupling*—definable as *repair-modification*—took place between the activities of DNApol *beta* (involved in *excision-repair*) and DNAmet (making *re-methylation*), whereas during the extra-S time an *uncoupling* occurred between these two enzyme systems [53] [64].

Thus, using synchronized *HeLa* cells [65] [67], we took a look at the differential *reconstruction* of the UV-damaged unmethylated coding and methylated uncoding gene sequences (**Figure 6(b)**): in a *repair patch* (RP) located along given promoters or introns (1), it could happen that a symmetrically dimethylated 5'-m^{5}CG-3'/3'-Gm^{5}C-5' dinucleotide pair would flank a UV-induced TT-dimer (2); after digestion of the damaged nucleotide sequence (3), the *excision-repair* (being the methylated dCTP absent in the soluble pool of triphosphonucleosides) would replace—in the "incompletely" reconstructed sequence—the previous m^{5}C nucleotide with a normal unmethylated cytosine nucleotide (4). Even if the existence of DNA de-methylating proteins should not be excluded in general [113], this "non-enzymatic de-methylation" certainly would occur through the whole cell cycle. Anyhow, during S—*via* AdoMet—DNAmet would add to the incompletely reconstructed (unmethylated) DNA sequence (4) the lost –CH_3 group, rebuilding the previous symmetrically di-methylated 5'-m^{5}CG-3'/3'-Gm^{5}C-5' dinucleotide pair (5). This "complete" rebuilding of a previously methylated RP would require, in any situation, a "re-methylation" catalyzed by DNAmet [53] [95] [114], as demonstrated by the fact that the labelled carbon atom coming from the –CH_3 of [^{14}C]Met (**Figure 3** and **Figure 2(b)**) was found in the T, G, A and m^{5}C nitrogen bases carried by the re-methylated RPs (**Figure 7(c)**) distributed along the light parental strands (**Figure 7(b)**). Yet, to what kind of regions—along the eukaryotic transcriptional unit—did such RPs belong?

Sedimentation of the semiconservatively newly replicating chains, made heavier in alkaline CsCl by BrdUrd incorporation (**Figure 7(a)**), was confronted with that of the labelled RPs appearing in the light parental chains (**Figure 7(b)**). These light parental chains, after an extensive shearing by sonication [46], were hybridized with polyA-containing pre-mRNA extracted from nuclei (**Figure 7(c)**) or with polyA-containing mRNA extracted

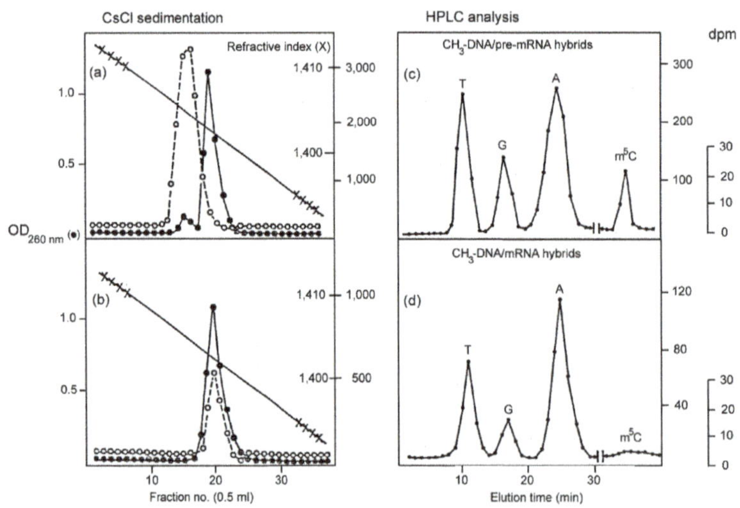

Figure 7. Differential repair-modification of introns and exons following a DNA radiodamage [96]. (a) *HeLa* cells were labelled, in mid S-phase, for 1.5 hrs with 80 Ci/mmol [^{14}C]Met in the presence of 10 mM BrdUrd; (b) *HeLa* cells were UV-irradiated for 45 sec and labelled with 80.0 Ci/mmol [^{14}C]Met in the presence of 10 mM BrdUrd plus 5 mM hydroxyurea. In both cases, the CsCl gradients, brought to pH 8.5, were run in the rotor SW55Ti for 60 hrs at 32,000 rev./min. The [^{14}C]-repaired parental DNA (b), sheared by sonication to fragments of about 1 × 10^6 Daltons, was hybridized with the [^{3}H]-labelled pre-processed polyA-containing mRNA fraction, purified from nuclei (c), or with the [^{3}H]-labelled processed polyA-containing mRNA fraction, purified from polysomes (d). Both pre-processed and processed messenger RNA fractions were labelled in unsynchronized cells with [^{3}H]uridine for 30 min in the presence of 0.04 micrograms/ml Actinomycin D and purified through oligo-dT cellulose. The hybrids were hydrolyzed to bases and chromatographed by HPLC. The reduced scale at the right-hand side (^{14}C-label) concerned m^{5}C.

from polysomes (**Figure 7(d)**). In the first combination, the hybrids "DNA/pre-mRNA" showed an appreciable amount of m^5Cs; in the second, the hybrids "DNA/mRNA" did not appear to be methylated. These results signified that re-methylation involved the uncoding parts of the repaired genes and that the coding parts of these genes only underwent a rebuilding of their primary base sequence, without modified bases.

In summary, we considered one of the possible mechanisms—*repair-modification*—through which the presence of m^5C in eukaryotic genes might have exerted a key role in the evolution of their unmethylated coding *vs.* methylated uncoding regions: after radioinduced damage, the RPs formed in exons (where the modified bases were practically absent) would be completely reconstructed by the sole excision-repair, whereas the RPs formed in promoters and introns (where the modified bases were normally present) would be incompletely reconstructed by excision-repair. Only later, in S, the unmethylated RPs would become completely reconstructed by re-methylation.

We assumed, therefore, that the repair-modification machinery exerts a continuous differential pressure on evolution of given gene pieces, when methylated to a different extent (**Figure 7(c)** and **Figure 7(d)**).

In particular, according to Darwinian concepts, *repair-modification* would lead to a very high variability of *introns* (if hypermethylated), whereas on the basis of this wide variability selection would favor their transposition from one gene to another. In them, the conservative nature of *exons* (if unmethylated) would be ensured by *excision-repair*.

Besides, it is worth stressing the possible influence that the incomplete repair of RPs (see the de-methylation shown in **Figure 6(b4)**) could exert on resumption of transcription by virtue of the cell cycle-dependent inverse correlation seen between DNA methylation and RNA synthesis (**Figure 6(a)**).

In this case, in harmony with the *azacytidine effect* [94] observed in *HeLa* cells (**Figure 8**), certain previously silent (**Figure 6(b1)**) and then re-activated (**Figure 6(b4)**) genes would lead even to neoplasy.

After all, the idea suggesting that the radio-induced DNA de-methylation could be considered as a feasible molecular basis for cancerogenesis [53] conformed with the known fact denouncing that at least 30 per cent of tumors are really induced in humans by radiations [148] [149].

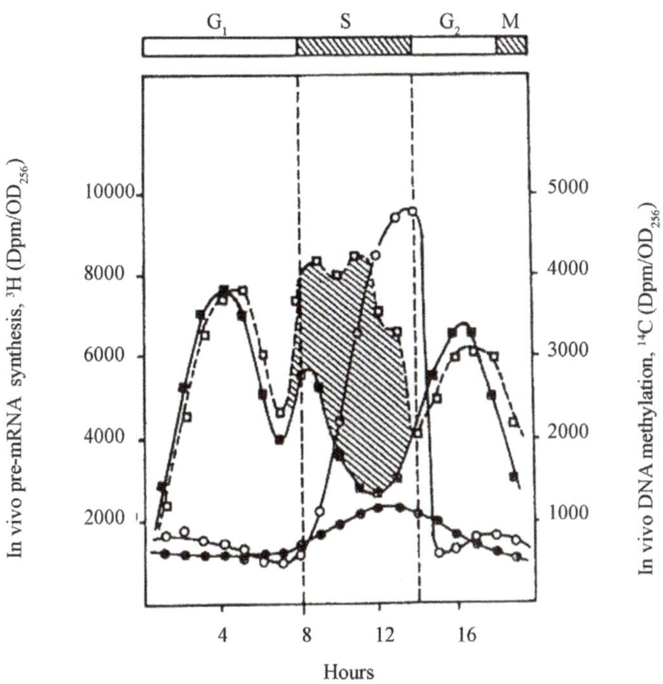

Figure 8. Inverse correlation in S phase between inhibition of DNA methylation by 5-azacytidine and polyA-containing pre-mRNA transcription. The experiments were performed in synchronized *HeLa* cells, as for **Figure 3(a)** and **Figure 6(a)**: following the inhibition of genomic DNA methylation by 5-azacytidine [94], some resumption of transcription was evidenced (dashed area). The left ordinate showed the rates of pre-mRNA synthesis in control (black squares) and 5-azacytidine-treated (white squares) cells labelled with [^{3}H]uridine; the right ordinate showed the rates of genomic DNA methylation in control (open circles) and 5-azacytidine-treated (closed circles) cells labelled with [^{14}C]Met.

8. The Arduous Biochemical Approach to Decipher the Language of Gene Methylation in Eukaryotes Was Crowned with Success by Its Direct Bisulfite Reading

The finding of an alternation, after the promoter, of coding and uncoding regions in the eukaryotic transcriptional unit and the consequent demonstration that these regions are differentially methylated, or re-methylated, inevitably stimulated curiosity to decipher the language marked by their m^5Cs.

8.1. At Variance with the Systems of Regulation of the Operon Activity in Bacteria, an Inverse Correlation between Gene Methylation and Transcription Was Shown in Humans

In bacterial DNA the m^6A and m^5C modified bases were related not only to the restriction-modification anti-phage defense [6]: complementing the induction or repression operon functions [39], they were further associated with an auxiliary mechanism capable of partially regulating transcription in the same phage-infected microorganisms [115]. This "embryonic" stage of control evolved in the major part of eukaryotes: in them the higher concentration of m^5C, present in the promoter and introns, was suggested to be responsible for modulation of gene expression [46], as supported by others [60] [116] [117] and reviewed [89] [118]. Such an involvement found its first clear demonstration in the fact that the maximal rate of DNA methylation [47] [119] followed the maximal rate of DNA replication [101] during S, while the maximal rates of transcription [110] and translation [77] [111] appeared during G_1 and G_2 (**Figure 6(a)**). Further results supporting the idea about the role of m^5C in regulation of gene expression were soon detected [120]-[123], afresh to those on the methylation-dependent regulation of integrated viral genes [92]-[109] [117] [124] [125]. Moreover, *maintenance* DNAmet [47] [72] was shown to recognize hemimethylated 5'-CpG-3'/3'-Gpm^{5}C-5' dinucleotide pairs to yield dimethylated 5'-m^{5}CpG-3'/3'-Gpm^{5}C-5' dinucleotide pairs (**Figure 3(c)**), whilst an inverse correlation between methylation and expression was established for a large number of sequenced HK and TS genes [98] [99]. In fact, an inverse correlation between gene methylation and transcription did occur in animals signifying, at the same time, that it could be caused most likely by the higher concentration of m^5C in promoter and intron sequences: one theory justifies the other, and *vice versa*.

8.2. A Study Focused on Isolation of Specific Targets for Methylation in Native Human DNA

After establishing that in eukaryotes the $-CH_3$ groups are preferentially attached to the CpG dinucleotides [106] [107] [126], as in bacteria [127], clearcut evidence about the existence of two classes of targets for DNAmet emerged, as mentioned already, from analysis of *HeLa* cell double-helical DNA in alkaline Ag^+/Cs_2SO_4 [106].

This type of gradient separated a heavier fraction (representing about 20% of the total) from a lighter one (representing about 80% of the total) [128].

However, the consequent ultracentrifugation of these fractions in CsCl [106] further showed that the heavier, banding at 1.715 g/cm^3, contained 53% CG (10% of the total CG), whereas the lighter, banding at 1.703 g/cm^3, contained 40% CG (32% of the total CG).

Generalizing, four possible triplets as sites for DNAmet recognition were suggested: GCG and CCG, in CG-rich sequences; ACA and ACG, in AT-rich sequences [107].

The evidence that the specific methylation of CG-rich sequences (as those for rRNAs) was maximal in early S and that the specific methylation of AT-rich sequences was maximal in late S not only demonstrated that the DNA sequences were replicated and methylated following an order during S [107] [108] but also pointed that the DNAmet activity exerted the role of *maintenance* modification along CG-rich sequences and the role of *de novo* modification along AT-rich sequences [107].

This important deduction was in accordance with the fact that the extra-S methylation (about 10% of the total) involved some AT-rich sequences either along the old or the new chains, when the S phase methylation (about 90% of the total) almost exclusively involved newly replicating chains [47] [50].

In S, the *maintenance* DNAmet function [63] followed at a distance of about 30 min [47] the DNApol *alpha* activity [112].

The reasoning made sense because, in S, the *maintenance* modification—involving CpG targets in newly replicating chains—would be induced by the $-CH_3$ of the m^5C present in the complementary strand as a genetically encoded signal.

Otherwise, a CpG dinucleotide (**Figure 3(c)**) would be methylated on the new chain in the presence of a complementary already methylated GpC dinucleotide on the old chain [47] [72]; differently, during the extra-S time, a *de novo* modification occurring on an ApC dinucleotide, for instance, could not be induced by a signal coming from the complementary TpG dinucleotide since this cannot be methylated at all.

8.3. Both *De Novo* Monomethylated and *Maintenance* Dimethylated Dinucleotide Pairs Were Found in Humans

For the sake of clarity, the above-established sedimentation of CG- and AT-rich sequences in alkaline Ag^+/ Cs_2SO_4 suggested the existence of at least four targets for methylation [106]:

GCG, CCG, ACA, and ACG.

In one of the trinucleotides there was a methylatable CpA; in three of them, the methylatable CpG was repeated. For this reason, a direct isolation of such restricted targets was required (**Figure 9**).

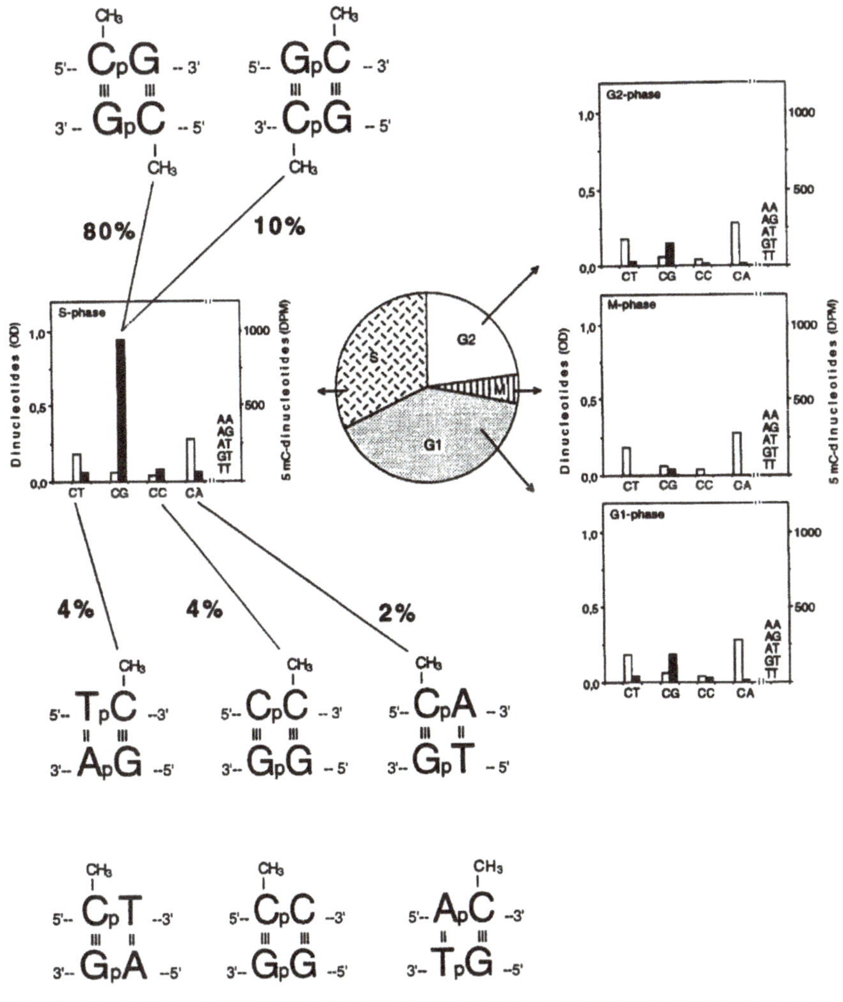

Figure 9. Methylated "words" in eukaryotic genomic DNA. Experiments with synchronized *HeLa* cells [65] [67] demonstrated that, in double helix (characterized by antiparallelism of complementary chains) two symmetrically dimethylated palindromic dinucleotide pairs and six asymmetrically monomethylated non-palindromic dinucleotide pairs were detected [102] [138]. DNA was labelled in isolated nuclei with [^{14}C-methyl]AdoMet, as for **Figure 3(b)**: after extraction, in the middle of each cell-cycle phase (G_1, S, G_2 and M), it was digested with pancreatic DNAse I [129] to chromatografically separate various dinucleotides [102]. The white columns show the dinucleotide molar concentration, expressed in OD at 256 nm; the black columns, expressed in DPM, show the radioactivity of their m^5C.

This was achieved by exploiting DNAase I digests [126] [129] of DNA samples methylated in isolated nuclei of previously synchronized cells with [^{3}H-methyl]AdoMet [47] [109]. Chromatographic analysis of the material contained in these digests [102] yielded a number of obviously unmethylated dinucleotides

ApA, ApG, ApT, CpT, and TpT

and, for the rest, four dinucleotides methylated to a different extent:

m^{5}CpT, Cpm^{5}C, m^{5}CpA, and m^{5}CpG.

Their methylation level was cell-cycle dependent, in agreement with [47]: in M, few CpGs were methylated; in G_1 and G_2, methylation of CpGs slightly increased, in comparison with that seen in M, and there appeared some methylation on CpTs, CpCs, and CpAs; in S, again, a methylation of four dinucleotides

m^{5}CpTs, Cpm^{5}Cs, m^{5}CpAs, and m^{5}CpGs

was detected. In this phase, methylation of CpGs became intensive (90% of the total), accounting for its *maintenance* character along newly made chains (**Figure 3(c)**). The occurrence of methylation on CpTs, CpCs, and CpAs accounted instead for a *de novo* phenomenon, which characterized especially the extra-S part of the interphase [47]. Considering the two possible directions of the separated dinucleotides, due to the double helix antiparallelism (the 5'-3' and 3'-5' directions could not be distinguished chromatographically), the analysis actually regarded the identification of eight *words* written in the genome in terms of dinucleotide pairs:

5'-m^{5}CpG-3'	5'-Tpm^{5}C-3'	5'-Cpm^{5}C-3'	5'-m^{5}CpA-3'
5'-Gpm^{5}C-3'	3'-ApG-5'	3'-GpG-5'	3'-GpT-5'
+	+	+	+
3'-m^{5}CpG-5'	5'-m^{5}CpT-3'	5'-m^{5}CpC-3'	5'-Apm^{5}C-3'
3'-Gpm^{5}C-5'	3'-GpA-5'	3'-GpG-5'	3'-TpG-5'

Briefly, if the CpG target only allowed a final dimethylation of dinucleotide pairs, as expected from semiconservative methylation [47] [72] [102] [130], the other combinations always led to a monomethylation of dinucleotide pairs [102]. Still, what could be demonstrated about the diversity of the methylation code at the level of the promoter and at that of the intervening uncoding sequences?

8.4. In the Promoter of the Transglutaminase Gene the Bisulfite Sequencing of 5-Methylcytosine Showed Two Fully Methylated CpG-Rich Domains Close to the 5'-End and One Fully Unmethylated CpG-Rich Domain Close to the 3'-End

Using the model of human *transglutaminase* (hTGc) gene [131] and the method based on bisulfite conversion of C to T residues along a DNA filament [132], the following analysis showed that m^{5}C, behaving as bisulfite-independent base [133], could be sequenced directly despite its low molar proportion [134].

The hTGc gene was chosen principally because it was among those that are regulated by methylation [135]. In its 1665 bp long promoter [131], three CpG-rich domains were characterized (**Figure 10**): the first domain (330 bp), close to 5', contained 12 CpGs corresponding to an average frequency of 3.63% with respect to the total number of nucleotides; the second (227 bp), roughly located in the middle of the promoter, contained eight CpGs corresponding to an average frequency of 3.52%; the third (264 bp), close to 3' (including 70 bp of the 73 bp long 5'-UTR), contained 31 CpGs corresponding to an average frequency of 11.74% [136].

In leukocytes and lymphocytes, where hTGc is silent [135], out of the three CpG-rich domains, only the first two were found to be methylated, while the third, on the 3'-side, did not present any m^{5}Cs: the 11 CpGs of domain 1 and the 7 CpGs of domain 2 were methylated 100%; the 31 CpGs of domain 3 were on the contrary unmethylated 100% (in this domain the 5'-UTR, also resulting 100% unmethylated, was almost entirely included) [136].

The lack of m^{5}C in domain 3 was expected, in agreement with the conventional definition of unmethylated CpG-rich "island" (where the intrafilament CpG/GpC ratio has to be higher than 0.6); but the occurrence of methylation in all CpGs of domains 1 and 2 was unexpected. Domain 1 contained 79 Cs and 11 CpGs out of 330 bases; domain 2 contained 44 Cs and 7 CpGs out of 227 bases; domain 3 contained 116 Cs and 31 CpGs out of 264 bases. Compared with the average 5%-methylation of human genomic DNA [137], methylation of domains 1 and 2 corresponded to 15.18% and 18.18%, respectively [136]. This confirmed the idea that the promoter of an inactive gene should be characterized by hypermethylation [46] [98] [99].

Certainly, *HUVEC* cells joined their hTGc gene activity with a loss of m^{5}C along the hTGc promoter domain 1, at least at the −1380, −1349, −1338, and −1320 CpG sites [138]. This is an appropriate example of direct

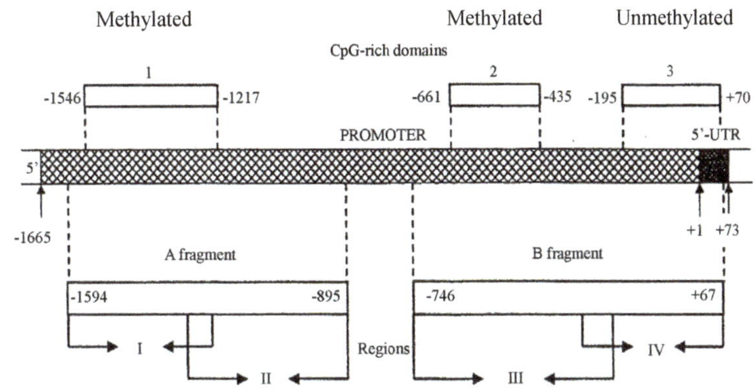

Figure 10. Scattered methylation along the repressed hTGc gene promoter in lymphocytes. The sequenced bisulfite-converted fragments A and B were investigated [136]: all CpGs found in domain 1 were methylated; all CpGs found in domain 2 also were methylated; in contrast, all CpGs found in domain 3 were unmethylated (starting from site +1, the fragment B, with its domain 3, overlapped a 5'-UTR, also unmethylated). In *HUVEC* cells, carrying an active hTGC gene, several CpGs of domain 1 were unmethylated [136] [138].

base-sequence analysis showing the inverse correlation between the activity of a gene and the methylation of its promoter (of course, in harmony with this correlation, the on-going research is aimed at clarifying the regulatory role of m^{5}Cs along the non-coding sequences).

9. Interpretations of the Mechanisms Switching-on and -off Transcription in Higher Cells: Future Perspectives

From the picture described herewith what hypothesis could one deduce on the role of methylation in regulating gene activity?

Supporting the message coming from **Figure 8** and **Figure 10** and from the long list of genes regulated by methylation [98] [99], the analysis based on the use of methylation-sensitive restriction endonucleases further showed that in human lymphocytes and monocytes an inverse correlation between the hTGc gene expression and methylation of its promoter took place [135].

Furthermore, the study based on the direct sequencing of m^{5}C showed that in the same cell species, in the absence of hTGc activity, the hTGc promoter was hypermethylated in its CpG-rich domains 1 and 2 and unmethylated in its CpG-rich domain 3 [136].

This neat division of hTGc promoter in two parts—one methylated and the other not—deserved particular attention in correlating DNA heterochromatization with A or B m^{5}CpG-binding proteins [123] [139]-[141] [139]-[141]: one could assume that they might transform domains 1 and 2 into heterochromatic structures and that, in turn, these structures would be sufficient to prevent transcription (methylation would be relevant to the repressor complex able to bind the promoter 5'-end).

Then, one could ask why, in hTGc promoter, the unmethylated domain 3, remaining in normal relaxed state, would be unable to interfere with the basal transcription mechanism at the promoter 3'-end [138].

Combined with the inactivity of the hTGc gene in leukocytes [136], the postulated heterochromatization of domains 1 and 2, in its promoter, could acquire great interest if considered within the framework of the repair-modification scheme (**Figure 6(b)**). Causing an *at random* DNA de-methylation [95] [142], it would lead, in general, to switch-on of previously silent genes, as in the case of genes for *alpha* and *beta* chains of hemoglobin in *Friend erythroleukemia* cells [99] [143].

In this situation, the switch-on and -off of transcription—implying conformational changes of the two hemoglobin gene promoters—would correspond, first, to their de-methylation and, second, to their re-methylation [99] [143] [144].

In other words, following the damages caused to double helix by ionizing radiations [95] [142], excision-repair would be sufficient to guarantee a complete reconstruction of previously unmethylated regions; nevertheless, to complete the reconstruction of previously methylated regions, a coupling would arise [53] between excision-repair (re-establishing the basic code in A, G, T, and C) and DNAmet (re-establishing the position of m^{5}C among A, G, T, and C). Actually, a coupling (properly meaning *repair-modification*) took place in S [64]: a specific DNApol, crucial for excision-repair, was active during the whole cell cycle [112]; DNAmet was highly active in S and almost inactive during the major part of the extra-S time [63]. Once de-methylated through excision-repair, given genes, silent when methylated, could be expressed if their transcription was inversely correlated with their methylation.

10. Conclusion

In conclusion, our basic hypothesis was that m^5C might function as a signal for genetic regulation of transcription through the hypermethylated promoter and the intermittent methylated non-coding regions. After a possible association with m^5C-binding proteins, they would physically interfere with the slip of RNA polymerase (RNApol) on the transcriptional unit. Probably, the block of transcription starts from the methylated promoter (initially recognized by a type A m^5C-binding protein able to compete with RNApol) and then is potentiated, inside the gene, by an intermittent association of uncoding methylated regions with a type B m^5C-binding protein.

Acknowledgements

The re-visited "*pioneering experiment on gene fragmentation*" is dedicated to F.H.C. Crick, who inspired me with his enthusiastic lectures on the structure of DNA and the genetic code at the International Institute of Genetics and Biophysics (IIGB) of CNR in Naples in 1964 and 1965, and to S. Ochoa, who affectionately referred to me as his "*nephew*" at the end of a research seminar on the language of DNA methylation in normal and virus-transformed cells that I took in 1981 at the Roche Institute of Molecular Biology of Nutley, USA. The investigation was encouraged by the geneticist A. Buzzati Traverso, founder of the IIGB in the 60s, who suggested I study nucleic acids in relation to gene expression, and by the biochemist A. Ruffo, director of the IIGB in the early 70s, who identified the significance of the discovery of non-coding eukaryotic gene sequences to the nascent *Genetic Engineering*. An appreciation is sincerely expressed to T. Eremenko, whose dedication ensured the success of the research, initiated at the IIGB in Naples and continued for many years both at the Albert Einstein College of Medicine in New York and at the Institute of Experimental Medicine of CNR in Rome. The competence of a large number of coworkers is gratefully acknowledged: R. Sawamura and H.J. Strecker, of the Albert Einstein College of Medicine, New York; E.P. Whitehead, of the Department of Biology of the Commission of the European Communities, Bruxelles; M.Y. Timofeeva, of the Institute of Molecular Biology of the USSR Academy of Sciences, Moscow; R. Cocchiara, D. Geraci, A. Giuditta, A. Granieri, T. Menna, F. Morelli, G. Pagano, and E. Scarano, of the IIGB, Naples; E. Alfani, A.L. Crema, D. Delfini, V. De Venezia, G. Hoberholtzer, G. Sapora, and C. Tomasello, of the High Institute of Health, Rome; G. Elia, C. Esposito, P. Iacovacci, T. Parasassi, and G. Ravagnan, of the Institute of Experimental Medicine of CNR, Rome; R.H. Butler, of the Institute of Cell Biology of CNR, Rome; Benedetto A., of the Laboratory of Virology of the United Hospitals, Rome; Palitti F., of the Department of Biochemical Sciences of the "La Sapienza" University of Rome, Rome; O. Cascio, A. Fazzio, and M.G. Sarpietro, of the Institute of Biochemical and Pharmacological Sciences of the University of Catania, Catania; E. Bertoli, T. Cacciamani, M. Centurelli, T. Duranti, A. La Teana, and S. Virgili, of the Institute of Biochemistry of the University of Ancona, Ancona.

References

[1] Scarano, E., Iaccarino, M., Grippo, P. and Winckelmans, D. (1965) On Methylation of DNA during Development of the Sea Urchin Embryo. *Journal of Molecular Biology*, **14**, 603-607. http://dx.doi.org/10.1016/S0022-2836(65)80211-X

[2] Vanyushin, B.F., Belozersky, A.N., Kokurina, N.A. and Kadirova, D.X. (1968) 5-Methylcytosine and 6-Methylaminopurine Inbacterial DNA. *Nature*, **218**, 1066-1067. http://dx.doi.org/10.1038/2181066a0

[3] Lark, C. (1968) Studies on the *in Vivo* Methylation of DNA in *Escherichia coli* 15T. *Journal of Molecular Biology*, **31**, 389-399. http://dx.doi.org/10.1016/0022-2836(68)90416-6

[4] Arber, W. (1965) Host Specificity of DNA Produced by *Escherichia coli*. V. The Role of Methionine in the Production of Host Specificity. *Journal of Molecular Biology*, **11**, 247-256. http://dx.doi.org/10.1016/S0022-2836(65)80055-9

[5] Arber, W. and Linn, S. (1969) DNA Modification and Restriction. *Annual Review of Biochemistry*, **38**, 467-500. http://dx.doi.org/10.1146/annurev.bi.38.070169.002343

[6] Arber, W. (1974) DNA Modification and Restriction. *Progress in Nucleic Acids Research and Molecular Biology*, **14**, 1-37. http://dx.doi.org/10.1016/S0079-6603(08)60204-4

[7] Hattman, S., Brooks, J.E. and Masurekar, M. (1978) Sequence Specificity of the PI Modification Methylase (M·*Eco P*1) and DNA Methylase (M·*Eco dam*) Controlled by *Escherichia coli dam* Gene. *Journal of Molecular Biology*, **126**, 367-380. http://dx.doi.org/10.1016/0022-2836(78)90046-3

[8] Messer, W. and Noyer-Weidnetr, M. (1988) Timing and Targeting: The Biological Functions of *dam* Methylation in *E. coli*. *Cell*, **54**, 735-737. http://dx.doi.org/10.1016/S0092-8674(88)90911-7

[9] Palmer, B.R. and Marinus, M.G. (1994) The *dam* and *dcm* Strains of *Escherichia coli*. *Gene*, **143**, 1-12. http://dx.doi.org/10.1016/0378-1119(94)90597-5

[10] Marinus, M.G. and Morris, N.R. (1974) Biological Function for 6-Methyladenine Residues in the DNA of *Escherichia coli* K12. *Journal of Molecular Biology*, **85**, 309-322. http://dx.doi.org/10.1016/0022-2836(74)90366-0

[11] Raleigh, E.A. and Wilson, G. (1986) *Escherichia coli* K-12 Restricts DNA-Containing 5-Methylcytosine. *Proceedings of the National Academy of Sciences of the United States of America*, **83**, 9070-9074. http://dx.doi.org/10.1073/pnas.83.23.9070

[12] Heitman, J. and Model, P. (1987) Site-Specific Methylases Induce the SOS DNA Repair Response in *Escherichia coli*. *Journal of Bacteriology*, **169**, 3243-3250.

[13] Raleigh, E.A., Murray, N.E., Revel, H., Blumenthal, R.M., Wastaway, D., Reith, A.D., Rigby, P.W.J., Elhai, J. and Hanahan, D. (1988) *McrA* and *McrB* Restriction Phenotypes of Some *E. coli* Strains and Implications for Gene Cloning. *Nucleic Acids Research*, **16**, 1563-1575. http://dx.doi.org/10.1093/nar/16.4.1563

[14] Marinus, M.G. and Morris, N.R. (1973) Isolation of Deoxyribonucleic Acid Methylase Mutants of *Escherichia coli K*12. *Journal of Bacteriology*, **114**, 1143-1150.

[15] May, M. and Hattman, S. (1975) Analysis of Bacteriophage Deoxyribonucleic Acid Sequences Methylated by Host- and R-Factor-Controlled Enzymes. *Journal of Bacteriology*, **123**, 768-770.

[16] Wyatt, G.R. (1951) The Purine and Pyrimidine Composition of Deoxypentose Nucleic Acids. *Biochemical Journal*, **48**, 584-590.

[17] Vanyushin, B.F., Tkacheva, S. and Belozersky, A.N. (1970) Rare Bases in Animal DNA. *Nature*, **225**, 948-994. http://dx.doi.org/10.1038/225948a0

[18] Gorovsky, M.A., Hattman, S. and Plager, G.L. (1973) [^{6}N]methyl-adenine in the Nuclear DNA of a Eucaryote, *Tetrahymena pyriformis*. *Journal of Cell Biology*, **56**, 697-701. http://dx.doi.org/10.1083/jcb.56.3.697

[19] Cummings, D.J., Tait, A. and Goddard, J.M. (1974) Methylated Bases in DNA from *Paramecium aurelia*. *Biochimica et Biophysica Acta*, **374**, 1-11. http://dx.doi.org/10.1016/0005-2787(74)90194-4

[20] Buryanov, Y.I. and Skryabin, G.K. (1970) Detection of 6-Methylaminopurine in DNA of the Fungus *Mucor hiemalis*. *Doklady Akademii Nauk SSSR*, **195**, 728-730.

[21] Pakhomova, M.V., Zaitseva, G.N. and Belozersky, A.N. (1968) The Presence of 5-Methylcytosine and 6-Methylaminopurine in the Composition of DNA in Some Algae. *Doklady Akademii Nauk SSSR*, **182**, 712-715.

[22] Buryanov, Y.I., Eroshina, N.V., Vagabova, L.M. and Ilyin, A.V. (1972) On the Detection of 6-Methylaminopurine in DNA of Higher Plant Pollen. *Doklady Akademii Nauk SSSR*, **206**, 992-994.

[23] Urieli-Shove, S., Gruenbaum, Y., Sedat, J. and Razin, A. (1982) The Absence of Detectable Methylated Bases in *Drosophila melanogaster* DNA. *FEBS Letters*, **146**, 148-152. http://dx.doi.org/10.1016/0014-5793(82)80723-0

[24] Adams, R.L.P., MacKay, E.L., Graig, L.M. and Burdon, R.H. (1979) Methylation of Mosquito DNA. *Biochimica et Biophysica Acta*, **563**, 72-81. http://dx.doi.org/10.1016/0005-2787(79)90008-X

[25] Achwal, C.W., Ganguly, P. and Chandra, H.S. (1984) Estimation of the Amount of 5-Methylcytosine in *Drosophila melanogaster* DNA by Amplified ELISA and Photoacoustic Spectroscopy. *EMBO Journal*, **3**, 263-266.

[26] Lyko, F., Ramsahoye, B.H., Kashevsky, H., Tudor, M., Mastrangelo, M.A., Orr-Weaver, T.L. and Jaenish, R. (1999) Mammalian (Cytosine-5) Methyltransferases Cause Genomic DNA Methylation and Lethality in *Drosophila*. *Nature Genetics*, **23**, 363-366. http://dx.doi.org/10.1038/15551

[27] Lyko, F., Ramsahoye, B.H. and Jaenisch, R. (2000) DNA Methylation in *Drosophila melanogaster*. *Nature*, **408**, 538-540. http://dx.doi.org/10.1038/35046205

[28] Gowher, H., Leismann, O. and Jeftsch, A. (2000) DNA of *Drosophila melanogaster* Contains 5-Methylcytosine. *EMBO Journal*, **19**, 6918-6923. http://dx.doi.org/10.1093/emboj/19.24.6918

[29] Vanyushin, B.F., Mazin, A.L., Vasilyev, V.K. and Belozersky, A.N. (1973) The Content of 5-Methylcytosine in Animal DNA: The Species and Tissue Specificity. *Biochimica et Biophysica Acta*, **299**, 397-403. http://dx.doi.org/10.1016/0005-2787(73)90264-5

[30] Vanyushin, B.F., Nemirovsky, L.E., Klimenko, V.V., Vasilyev, V.K. and Belozersky, A.N. (1973) The 5-Methylcytosine in DNA of Rats: Tissue and Age Specificity and the Changes Induced by Hydrocortisone and Other Agents.

Gerontology, **19**, 138-152. http://dx.doi.org/10.1159/000211967

[31] Vanyushin, B.F. and Belozersky, A.N. (1959) Base Composition of Deoxyribonucleic Acids of Higher Plants. *Doklady Akademii Nauk SSSR*, **129**, 944-946.

[32] Vanyushin, B.F. (1984) Replicative DNA Methylation in Animals and Higher Plants. *Current Topics in Microbiology and Immunology*, **108**, 99-114. http://dx.doi.org/10.1007/978-3-642-69370-0_7

[33] Giuditta, A. and Volpe, P. (1967) Studies on an Unknown Compound from Ox Brain. I. Isolation and Analysis. *Journal of Neurochemistry*, **14**, 626-636. http://dx.doi.org/10.1111/j.1471-4159.1967.tb09566.x

[34] Volpe, P. and Giuditta, A. (1967) Studies on an Unknown Compound from Ox Brain: II. Separation into Several Components. *Journal of Neurochemistry*, **14**, 637-647. http://dx.doi.org/10.1111/j.1471-4159.1967.tb09567.x

[35] Eremenko, T., Benedetto, A. and Volpe, P. (1972) Poliovirus Replication during the *HeLa* Cell Life Cycle. *Nature New Biology*, **237**, 114-116. http://dx.doi.org/10.1038/newbio237114a0

[36] Eremenko, T., Benedetto, A. and Volpe, P. (1972) Virus Infection as a Function of the Host Cell Cycle: Replication of Poliovirus RNA. *Journal of General Virology*, **16**, 61-68. http://dx.doi.org/10.1099/0022-1317-16-1-61

[37] Volpe, P. and Giuditta, A. (1967) Kinetics of RNA Labelling in Fractions Enriched with Neuroglia and Neurons. *Nature*, **216**, 154-155. http://dx.doi.org/10.1038/216154a0

[38] Volpe, P. and Giuditta, A. (1967) Biosynthesis of RNA in Neuron- and Glia-Enriched Fractions. *Brain Research*, **6**, 228-240. http://dx.doi.org/10.1016/0006-8993(67)90193-X

[39] Jacob, F. and Monod, J. (1961) Genetic Regulatory Mechanisms in the Synthesis of Proteins. *Journal of Molecular Biology*, **3**, 318-356. http://dx.doi.org/10.1016/S0022-2836(61)80072-7

[40] Tomkins, G.M. and Ames, B.N. (1966) The Operon Concept in Bacteria and in Higher Organisms. *National Cancer Institute Monographs*, **27**, 221-234.

[41] Volpe, P. and Strecker, H.J. (1968) Amphi-Directional Control of a Reversible Reaction Common to Two Enzyme Sequences. *Biochemical and Biophysical Research Communications*, **32**, 240-245. http://dx.doi.org/10.1016/0006-291X(68)90375-6

[42] Volpe, P., Sawamura, R. and Strecker, H.J. (1969) Control of Ornithine-Delta-Ransaminase in Rat Liver and Kidney. *Journal of Biological Chemistry*, **244**, 719-726.

[43] Volpe, P. and Eremenko, T. (1971) Sructural Enzyme Heterogeneity as Expressed by Differential Regulation during the Phases of the Cell Cycle. *Advances in Cytopharmacology*, **1**, 257-261.

[44] Volpe, P. and Eremenko, T. (1971) Biochemical Differentiation and Cancerogenesis. In: Neskovic, B., Ed., *Proceedings of the International Symposium on Experimental Oncology*, Ognjon Prica Publishers, Zagreb, 200-203.

[45] Volpe, P., Menna, T. and Pagano, G. (1974) Ornithine-Delta-Transaminase Heterogeneity and Regulation: Sequential Expression of the "Liver" and "Kidney" Enzyme Forms during the *HeLa* Cell Cycle. *European Journal of Biochemistry*, **44**, 455-458. http://dx.doi.org/10.1111/j.1432-1033.1974.tb03503.x

[46] Volpe, P. and Eremenko, T. (1974) Preferential Methylation of Regulatory Gene Sequences in *HeLa* Cells. *FEBS Letters*, **44**, 121-126.

[47] Geraci, D., Eremenko, T., Cocchiara, R., Granieri, A., Scarano, E. and Volpe, P. (1974) Correlation between Synthesis and Methylation of DNA in *HeLa* Cells. *Biochemical and Biophysical Research Communications*, **57**, 353-361. http://dx.doi.org/10.1016/0006-291X(74)90937-1

[48] Duranti, T., La Teana, A., Cacciamani, T. and Volpe, P. (2006) The Prokaryotic Origin of the Pathways for Synthesis and Post-Synthetic Modification of DNA. *RNA Biology*, **3**, 49-53. http://dx.doi.org/10.4161/rna.3.1.2794

[49] Duranti, T., La Teana, A., Cacciamani, T. and Volpe, P. (2006) Evolution of the Pathways Leading to Synthesis and Modification of DNA. *Journal of Biological Research*, **81**, 47-50.

[50] Eremenko, T., Palitti, F., Morelli, F., Whitehead, E.P. and Volpe, P. (1985) Hypomethylation of Repair Patches in *HeLa* Cells. *Molecular Biology Reports*. **10**, 177-182. http://dx.doi.org/10.1007/BF00778526

[51] Gehrke, C.W., Kuo, K.C., Davis, G.E., Suits, R.D., Waalkesb, R.D. and Borek, E. (1978) Quantitative High-Performance Liquid Chromatography of Nucleosides in Biological Materials. *Journal of Chromatography*, **150**, 455-476. http://dx.doi.org/10.1016/S0021-9673(00)88205-9

[52] Marinus, M.G. (1987) DNA Methylation in *Escherichia coli*. *Annual Review of Biochemistry*, **21**, 113-131. http://dx.doi.org/10.1146/annurev.genet.21.1.113

[53] Volpe, P. and Eremenko, T. (1989) Repair-Modification and Evolution of the Eukaryotic Genome Organization. *Cell Biophysics*, **15**, 41-60. http://dx.doi.org/10.1007/BF02991578

[54] Volpe, P. (2007) Post-Synthetic Methylation of Macromolecules in Synchronized Mammalian Cells. *Progress in DNA Methylation Research*, **8**, 203-224.

[55] Feng, T.Y. and Chang, K.S. (1984) Persistence of Maternal Inheritance in *Chlamydomonas* Despite Hypomethylation of Chloroplast DNA Induced by Inhibitors. *Proceedings of the National Academy of Sciences of the United States of America*, **81**, 3438-3442. http://dx.doi.org/10.1073/pnas.81.11.3438

[56] Volpe, P. and Cascio, O. (1993) Detection of a Mammalian DNA Methylase Protein Family. *Rendiconti Lincei*, **4**, 345-357. http://dx.doi.org/10.1007/BF03001194

[57] Volpe, P. and Cascio, O. (1994) Evolution of the DNA Methylase Protein Family in Vertebrates. *Rendiconti Lincei*, **5**, 79-87. http://dx.doi.org/10.1007/BF03001439

[58] Franchina, M., Hooper, J. and Kay, P.H. (2001) Five Novel Alternatively Spliced Transcripts of DNA (Cytosine-5) Methyltransferase 2 in Human Peripheral Blood Leukocytes. *International Journal of Biochemistry and Cell Biology*, **33**, 1104-1115. http://dx.doi.org/10.1016/S1357-2725(01)00074-7

[59] Scarano, E., Iaccarino, M., Grippo, P. and Parisi, E. (1967) Heterogeneity of Thymine Methyl Group Origin in DNA Pyrimidine Isostichs of Developing Sea Urchin Embryos. *Proceedings of the National Academy of Sciences of the United States of America*, **57**, 1394-1400. http://dx.doi.org/10.1073/pnas.57.5.1394

[60] Holliday, R. and Pugh, J.E. (1975) DNA Modification Mechanisms and Gene Activity during Development. *Science*, **187**, 226-232. http://dx.doi.org/10.1126/science.1111098

[61] Borek, E. (1975) Model for Differentiation Based on DNA Modifying Enzymes. *Science*, **190**, 591-593. http://dx.doi.org/10.1126/science.1188359

[62] Bestor, T.H. and Ingram, V.M. (1983) Two Species of DNA Methyltransferase from Murine Erythroleukemia Cells: Purification, Sequence Specificity, and Mode of Interaction with DNA. *Proceedings of the National Academy of Sciences of the United States of America*, **80**, 5559-5563. http://dx.doi.org/10.1073/pnas.80.18.5559

[63] Delfini, C., Crema, A.L., Alfani, E., Lo Presti, E., Eremenko, T. and Volpe, P. (1987) DNA-Methylases Separated through the *HeLa* Cell Cycle Methodology Show Allosteric Properties. *FEBS Letters*, **210**, 17-21.

[64] Eremenko, T., Delfini, C., Crema, A.L., Alfani, E. and Volpe, P. (1988) Uncoupling of the DNA Polymerase and Methylase Systems Leads to Hypomethylation of Repair Patches. *Macromolecules in the Functioning Cell*, **5**, 37-41.

[65] Volpe, P. and Eremenko, T. (1970) A Method to Measure the Length of Each Phase of the Cell Cycle in Spinner Cultures. *Experimental Cell Research*, **60**, 456-458. http://dx.doi.org/10.1016/0014-4827(70)90542-2

[66] Volpe, P. and Eremenko, T. (1970) Quantitative Studies on Cell Proteins in Suspension Cultures. *European Journal of Biochemistry*, **12**, 195-200. http://dx.doi.org/10.1111/j.1432-1033.1970.tb00837.x

[67] Volpe, P. and Eremenko, T. (1973) A Method for Measuring Cell Cycle Phases in Suspension Cultures. *Methods in Cell Biology*, **6**, 113-126. http://dx.doi.org/10.1016/S0091-679X(08)60049-7

[68] Kappler, J.A. (1970) The Kinetics of DNA Methylation in Cultures of a Mouse Adrenal Cell Line. *Journal of Cell Physiology*, **75**, 21-34. http://dx.doi.org/10.1002/jcp.1040750104

[69] Evans, H.H., Evans, T.E. and Littman, S. (1973) Methylation of Parental and Progeny DNA Strands in *Physarum polycephalum*. *Journal of Molecular Biology*, **74**, 563-574. http://dx.doi.org/10.1016/0022-2836(73)90047-8

[70] Adams, R.L.P. and Hogarth, C. (1973) DNA Methylation in Isolated Nuclei: Old and New DNAs Are Methylated. *Biochimica et Biophysica Acta*, **331**, 214-220. http://dx.doi.org/10.1016/0005-2787(73)90434-6

[71] Adams, R.L.P. (1974) Newly Synthesized DNA Is Not Methylated. *Biochimica et Biophysica Acta*, **335**, 365-373. http://dx.doi.org/10.1016/0005-2787(74)90159-2

[72] Bird, A.P. (1978) Use of Restriction Enzymes to Study Eucaryotic DNA Methylation: The Symmetry of Methylated Sites Supports Semi-Conservative Copying of the Methylation Pattern. *Journal of Molecular Biology*, **118**, 49-60. http://dx.doi.org/10.1016/0022-2836(78)90243-7

[73] Georgyev, G.P. and Mantieva, V.I. (1962) The Isolation of DNA-Like RNA and Ribosomal RNA from the Nucleolo-Chromosomal Apparatus of Mammalian Cells. *Biochimica et Biophysica Acta*, **61**, 153-162.

[74] Penman, S., Vesco, C. and Penman, M. (1968) Localization and Kinetics of Formation of Nuclear Heterodisperse RNA, Cytoplasmic Heterodisperse RNA and Polyribosome Associated Messenger RNA in *HeLa* Cells. *Journal of Molecular Biology*, **34**, 49-62. http://dx.doi.org/10.1016/0022-2836(68)90234-9

[75] Georgyev, G.P. (1969) On the Structural Organization of Operon and on the Regulation of RNA Synthesis in Animal Cells. *Journal of Theoretical Biology*, **25**, 227-231. http://dx.doi.org/10.1016/S0022-5193(69)80034-2

[76] Darnell, J.E., Jelinek, W.R. and Molloy, G.R. (1973) Biogenesis of mRNA: Genetic Regulation in Mammalian Cells. *Science*, **181**, 1215-1221. http://dx.doi.org/10.1126/science.181.4106.1215

[77] Volpe, P. and Eremenko, T. (1973) Polysomes during the *HeLa* Cell Life Cycle. *9th International Congress of Biochemistry*, Stockholm, Abstracts, 195.

[78] Breathnach, R., Mandel, J.L. and Chambon, P. (1977) Ovalbumin Gene Is Split in Chicken DNA. *Nature*, **270**, 314-

319. http://dx.doi.org/10.1038/270314a0

[79] Brack, C. and Tonegawa, S. (1977) Variable and Constant Parts of the Immunoglobulin Light Chain Gene of a Mouse Myeloma Cell Are 1250 Nontranslated Bases Apart. *Proceedings of the National Academy of Sciences of the United States of America*, **74**, 5652-5656. http://dx.doi.org/10.1073/pnas.74.12.5652

[80] Doel, M.T., Houghton, M., Cook, E.A. and Carey, N.H. (1977) The Presence of Ovalbumin mRNA Coding Sequences in Multiple Restriction Fragments of Chicken DNA. *Nucleic Acids Research*, **4**, 3701-3713. http://dx.doi.org/10.1093/nar/4.11.3701

[81] Jeffreys, A.J. and Flavell, R.A. (1977) A Rabbit Beta-Globin Gene Contains a Large Insert in the Coding Sequence. *Cell*, **12**, 1097-1108. http://dx.doi.org/10.1016/0092-8674(77)90172-6

[82] Leder, P., Tilghman, S.M., Tiemeier, D.C., Polsky, F.I., Seidman, J.G., Edgell, M.H., Enquist, L.W., Leder, A. and Norman, B. (1977) Cloning of Mouse Globin and Surrounding Gene Sequences in *Bacteriophage lambda*. *Cold Spring Harbor Symposium on Quantitative Biology*, **42**, 915-920. http://dx.doi.org/10.1101/SQB.1978.042.01.093

[83] Berget, S.M., Moore, C. and Sharp, P.A. (1977) Spliced Segments at the 5' Terminus of Adenovirus 2 Late mRNA. *Proceedings of the National Academy of Sciences of the United States of America*, **74**, 3171-3175. http://dx.doi.org/10.1073/pnas.74.8.3171

[84] Chow, L.T.R., Gelinas, R.E., Broker, T.R. and Roberts, R.J. (1977) An Amazing Sequence Arrangement at the 5' Ends of *Adenovirus* 2 Messenger RNA. *Cell*, **12**, 1-8. http://dx.doi.org/10.1016/0092-8674(77)90180-5

[85] Tilghman, S.M., Tiemeier, D.C., Seldman, J.G., Peterlin, B.M., Sullivan, M., Maizel, J.V. and Leder, P. (1978) Intervening Sequence of DNA Identified in the Structural Portion of a Mouse Beta-Globin Gene. *Proceedings of the National Academy of Sciences of the United States of America*, **75**, 725-729. http://dx.doi.org/10.1073/pnas.75.2.725

[86] Gilbert, W. (1978) Why Genes in Pieces. *Nature*, **271**, 501. http://dx.doi.org/10.1038/271501a0

[87] Crick, F.H.C. (1979) Split Genes and RNA Splicing. *Science*, **204**, 264-271. http://dx.doi.org/10.1126/science.373120

[88] Maclean, M. and Hilder, V.A. (1977) Mechanisms of Chromatin Activation and Repression. *International Review of Cytology*, **48**, 54-97. http://dx.doi.org/10.1016/S0074-7696(08)61742-8

[89] Tentravahi, V., Guntaka, R.V., Erlanger, B.F. and Miller, O.J. (1981) Amplified Ribosomal RNA Genes in a Rat Hepathoma Cell Line Are Enriched in 5-Methylcytosine. *Proceedings of the National Academy of Sciences of the United States of America*, **78**, 489-493. http://dx.doi.org/10.1073/pnas.78.1.489

[90] Volpe, P. (1976) The Gene Expression during the Cell Life Cycle. *Horizons in Biochemistry and Biophysics*, **2**, 285-340.

[91] Eremenko, T., Menna, T. and Volpe, P. (1979) Differential Gene Expression during the Cell Life Cycle. *Macromolecules in the Functioning Cell*, **1**, 79-108. http://dx.doi.org/10.1007/978-1-4684-3465-1_7

[92] Eremenko, T. and Volpe, P. (1984) Virus Genome Integration as a Function of the Host Cell Genome Replicative Cycle. *FEBS Letters*, **169**, 211-214.

[93] Volpe, P. and Eremenko, T. (1984) Position of the Integration Site of SV40 Virus Genome at the Level of the Transcriptional Unit. *Voprosy Virusologii*, **5**, 610-619.

[94] Volpe, P. and Eremenko, T. (1984) Resumption of Transcription Following Inhibition of DNA Methylation. *Macromolecules in the Functioning Cell*, **3**, 59-69.

[95] Volpe, P. and Eremenko, T. (1985) The Gene Repair-Modification System in Eukaryotes. In: Ovcinnikov, Y.A., Ed., *Proceedings of the* 16*th FEBS Congress*, Vol. 3, Science Press, Utrecht, 123-129.

[96] Volpe, P. and Eremenko, T. (1986) Differential Evolution of Coding and Non-Coding Eukaryotic DNAs through Repair-Modification. *Macromolecules in the Functioning Cell*, **4**, 8-20.

[97] Eremenko, T., Esposito, C., Iacovacci, P., Benedetto, A., Elia, G., Delfini, C., Alfani, E., Cascio, O., Sarpietro, M.G., Fazzio, A. and Volpe, P. (1990) Is 5-Methylcytosine a Regulatory Signal in Eukaryotic Gene Expression? 7*th International Symposium on Macromolecules in the Functioning Cell*, Taormina, Abstracts, 22.

[98] Volpe, P., Esposito, C., Iacovacci, P., Butler, R.H. and Eremenko, T. (1993) Language of Genes with Inverse Correlation between Methylation and Transcription. *Macromolecules in the Functioning Cell*, **7**, 59-71.

[99] Volpe, P., Iacovacci, P., Butler, R.H. and Eremenko, T. (1993) 5-Methylcytosine in Genes with Methylation-Dependent Regulation. *FEBS Letters*, **329**, 233-237. http://dx.doi.org/10.1016/0014-5793(93)80228-M

[100] Scarano, E. (1969) Enzymatic Modifications of DNA and Embryonic Differentiation. *Annals in Embryology and Morphology*, Suppl. 1, 51-61.

[101] Volpe, P. and Eremenko, T. (1973) Nuclear and Cytoplasmic DNA Synthesis during the Mitotic Cycle of *HeLa* Cells. *European Journal of Biochemistry*, **32**, 227-232. http://dx.doi.org/10.1111/j.1432-1033.1973.tb02600.x

[102] Volpe, P. and Eremenko, T. (1978) A Language of DNA Modification during Replication. 14*th International Congress*

of Genetics, Moscow, Abstracts, 1, 194.

[103] Lima de Faria, A. (1969) Handbook of Molecular Cytology. Wiley, New York, 277-283.

[104] Comings, D.E. (1972) Methylation of Euchromatic and Heterochromatic DNA. *Experimental Cell Research*, **74**, 383-391. http://dx.doi.org/10.1016/0014-4827(72)90391-6

[105] Volpe, P., Granieri, A. and Eremenko, T. (1975) Negli eucarioti i geni si metilano ordinatamente durante la replica. *12th Meeting of the Italian Society of Biophysics and Molecular Biology*, Pavia, Abstracts, 49.

[106] Eremenko, T., Granieri, A. and Volpe, P. (1978) Organization, Replication and Modification of the Human Genome: I. Differential Methylation of Two Classes of *HeLa* Nuclear DNA Separated on Ag^+/Cs_2SO_4 Gradients. *Molecular Biology Reports*, **4**, 163-170. http://dx.doi.org/10.1007/BF00777518

[107] Eremenko, T., Granieri, A. and Volpe, P. (1978) Organization, Replication and Modification of the Human Genome: II. Temporal Order of Synthesis and Methylation of Two Classes of *HeLa* nDNA Separated in Ag^+/Cs_2SO_4 Gradients. *Molecular Biology Reports*, **4**, 237-240. http://dx.doi.org/10.1007/BF00777561

[108] Eremenko, T., Timofeeva, M.Y. and Volpe, P. (1980) Organization, Replication and Modification of the Human Genome: Synthesis and Methylation of Palindromic, Repeated and Unique *Hela* nDNA Sequences during the S-Phase. *Molecular Biology Reports*, **6**, 131-136. http://dx.doi.org/10.1007/BF00775405

[109] Volpe, P. and Eremenko, T. (1982) Cell and Virus Macromolecular Modification during the Cell Life Cycle. *Macromolecules in the Functioning Cell*, **2**, 179-195.

[110] Volpe, P., Menna, T. and Eremenko, T. (1976) RNA Transcription and Processing as a Function of the Cell Cycle. *Bulletin of Molecular Biology and Medicine*, **1**, 18-28.

[111] Eremenko, T. and Volpe, P. (1975) Polysome Translational State during the Cell Cycle. *European Journal of Biochemistry*, **52**, 203-210. http://dx.doi.org/10.1111/j.1432-1033.1975.tb03988.x

[112] Delfini, C., Alfani, E., De Venezia, V., Oberholtzer, G., Tomasello, C., Eremenko, T. and Volpe, P. (1985) Cell-Cycle Dependence and Properties of the *HeLa* Cell DNA Polymerase System. *Proceedings of the National Academy of Sciences of the United States of America*, **82**, 2220-2224. http://dx.doi.org/10.1073/pnas.82.8.2220

[113] Battacharya, S.K., Ramchandani, S., Cervoni, N. and Szyf, M. (1999) A Mammalian Protein with Specific Demethylase Activity for mCpG DNA. *Nature*, **397**, 579-583. http://dx.doi.org/10.1038/17533

[114] Volpe, P. and Eremenko, T. (1994) Complete and Incomplete Repair of Radiodamaged Gene Sequences. *UNESCO Technical Reports*, **19**, 27-43.

[115] Kruger, D., Schroeder, C., Reuter, M., Bogdarina, I., Buryanov, Y. and Bickle, T. (1985) DNA Methylation of Bacterial Viruses T3 and T7 by Different DNA Methylases in *Escherichia coli* K12 Cells. *European Journal of Biochemistry*, **150**, 323-330. http://dx.doi.org/10.1111/j.1432-1033.1985.tb09024.x

[116] Mandel, J.L. and Chambon, P. (1979) DNA Methylation: Organ Specific Variations in the Methylation Pattern within and Oround Ovalbumin and Other Chicken Genes. *Nucleic Acids Research*, **7**, 2081-2092. http://dx.doi.org/10.1093/nar/7.8.2081

[117] Sutter, D. and Doerfler, W. (1980) Methylation of Integrated Adenovirus Type 12 DNA Sequences in Trasformed Cells Is Inversely Correlated with Viral Gene Expression. *Proceedings of the National Academy of Sciences of the United States of America*, **77**, 253-256. http://dx.doi.org/10.1073/pnas.77.1.253

[118] Liau, M.C., Chang, C.F., Saunders, G.F. and Tsai, Y.H. (1981) S-Adenosylhomocysteine Hydrolases as the Primary Target Enzymes in Androgen Regulation of Methylated Complexes. *Archives of Biochemistry and Biophysics*, **208**, 261-272. http://dx.doi.org/10.1016/0003-9861(81)90148-X

[119] Geraci, D., Eremenko, T., Granieri, A., Scarano, E. and Volpe, P. (1973) The *in Vivo* and *in Vitro* Methylation of DNA in Synchronized *HeLa* Cells. *9th International Congress of Biochemistry*, Stockholm, Abstracts, 3m64, 190.

[120] Razin, A. and Riggs, A.D. (1980) DNA Methylation and Gene Function. *Science*, **210**, 604-610. http://dx.doi.org/10.1126/science.6254144

[121] Eden, S. and Cedar, H. (1994) Role of DNA Methylation in the Regulation of Transcription. *Current Opinions in Genetics and Development*, **4**, 255-259. http://dx.doi.org/10.1016/S0959-437X(05)80052-8

[122] Razin, A. (1998) CpG Methylation, Chromatin Structure and Gene Silencing—A Three-Way Connection. *EMBO Journal*, **17**, 4905-4908. http://dx.doi.org/10.1093/emboj/17.17.4905

[123] Bird, A.P. and Wolffe, A.P. (1999) Methylation-Induced Repression: Belts, Braces and Chromatin. *Cell*, **99**, 451-454. http://dx.doi.org/10.1016/S0092-8674(00)81532-9

[124] Volpe, P., Menna, T. and Eremenko, T. (1978) Supermethylation of Eukaryotic nDNA after Oncovirus Genome Integration. *11th European Tumor Virus Meeting*, Balatonfured, Abstracts, 4, 194.

[125] Eremenko, T. and Volpe, P. (1984) *In Vitro* Methylation of Total and Foldback DNAs in Normal and Virus-Trans-

formed Cells. *FEBS Letters*, **173**, 233-237.

[126] Grippo, P., Iaccarino, M., Parisi, E. and Scarano, E. (1968) Methylation of DNA in Developing Sea Urchin Embryos. *Journal of Molecular Biology*, **36**, 195-208. http://dx.doi.org/10.1016/0022-2836(68)90375-6

[127] Doskocil, J. and Sormova, Z. (1965) The Occurrence of 5-Methylcytosine in Bacterial Deoxyribonucleic Acid. *Biochimica et Biophysica Acta*, **95**, 513-515. http://dx.doi.org/10.1016/0005-2787(65)90199-1

[128] Eremenko, T., Granieri, A., Scarano, E. and Volpe, P. (1975) Fractionation of Methylated Nuclear DNA in *HeLa* Cells. 10*th FEBS Meeting*, Paris, Abstracts, 130.

[129] Augusti-Tocco, G., Carestia, C., Grippo, P., Parisi, E. and Scarano, E. (1968) Thin-Layer Electrophoresis of Nucleic Acid Derivatives on Microcrystalline Cellulose. *Biochimca et Biophysica Acta*, **155**, 8-18. http://dx.doi.org/10.1016/0005-2787(68)90329-8

[130] Eremenko, T. and Volpe, P. (1976) Replica semiconservativa del DNA metilato. 13*th Meeting of the Italian Society of Biophysics and Molecular Biology*, Albano Laziale, Abstracts, 23.

[131] Lu, S., Saydak, M., Gentile, V., Steins, J.P. and Davies, J.A. (1995) Isolation and Characterization of the Human Tissue Transglutaminase Gene Promoter. *Journal of Biological Chemistry*, **270**, 9748-9756. http://dx.doi.org/10.1074/jbc.270.17.9748

[132] Clark, S.J., Harrison, J., Paul, C.L. and Frommer, M. (1994) High Sensitivity Mapping of Methylated Cytosines. *Nucleic Acids Research*, **22**, 2990-2997. http://dx.doi.org/10.1093/nar/22.15.2990

[133] Frommer, M., McDonald, L.E., Millar, D.S., Collins, C., Watt, F., Grigg, G.W., Molloy, P.L. and Paul, C.L. (1992) A Genomic Sequencing Protocol That Yields a Positive Display of 5-Methylcytosine Residues in Individual DNA Strands. *Proceedings of the National Academy of Sciences of the United States of America*, **89**, 1827-1831. http://dx.doi.org/10.1073/pnas.89.5.1827

[134] Clark, S.J., Harrison, J. and Frommer, M. (1995) CpNpG Methylation in Mammalian Cells. *Nature Genetics*, **10**, 20-27. http://dx.doi.org/10.1038/ng0595-20

[135] Lu, S. and Davies, P.J.A. (1997) Regulation of the Expression of the Tissue Transglutaminase Gene by DNA Methylation. *Proceedings of the National Academy of Sciences of the United States of America*, **94**, 4692-4697. http://dx.doi.org/10.1073/pnas.94.9.4692

[136] Cacciamani, T., Virgili, S., Centurelli, M., Bertoli, E., Eremenko, T. and Volpe, P. (2002) Specific Methylation of the CpG-Rich Domains in the Promoter of the Human Tissue Transglutaminase Gene. *Gene*, **297**, 103-112. http://dx.doi.org/10.1016/S0378-1119(02)00874-0

[137] Razin, A. and Cedar, H. (1977) Distribution of 5-Methylcytosine in Chromatin. *Proceedings of the National Academy of Sciences of the United States of America*, **74**, 2725-2728. http://dx.doi.org/10.1073/pnas.74.7.2725

[138] Volpe, P. (2005) The Language of Methylation in Genomics of Eukaryotes. *Biochemistry* (*Moscow*), **70**, 584-595. http://dx.doi.org/10.1007/s10541-005-0152-x

[139] Meehan, R., Antequera, F., Lewis, J., McLeod, D., McKay, S., Kleiner, E. and Bird, A.P. (1990) A Nuclear Protein That Binds Preferentially to Methylated DNA *in Vitro* May Play a Role in the Inaccessibility of Methylated CpGs in Mammalian Nuclei. *Philosophical Transactions of the Royal Society of London*, **B326**, 199-205. http://dx.doi.org/10.1098/rstb.1990.0004

[140] Nan, X., Ng, H.-H., Johnson, C.A., Laherty, C.D., Turner, B.M., Eisenman, R.N. and Bird, A.P. (1998) Transcriptional Repression by the Methyl-CpG-Binding Protein MeCP2 Involves a Histone Deacetylase Complex. *Nature*, **393**, 386-389. http://dx.doi.org/10.1038/30764

[141] Wade, P.A., Gegonne, A., Jones, P.L., Ballestar, E., Aubry, F. and Wolffe, A.P. (1999) Mi-2 Complex Couples DNA Methylation to Chromatin Remodelling and Histone Deacetylation. *Nature Genetics*, **23**, 62-66. http://dx.doi.org/10.1038/12664

[142] Volpe, P. and Eremenko, T. (1995) Repair-Modification of Radiodamaged Genes. *Radiation Protection Dosimetry*, **62**, 19-22.

[143] Volpe, P., Parasassi, T., Sapora, O., Ravagnan, G. and Eremenko, T. (1999) Influence of the Low Doses of Radiation on the DNA Double Helix, Gene Expression and Membranal State. *International Journal of Radiation Medicine*, **1**, 78-89.

[144] Volpe, P. (2001) The Gene "Repair-Modification" Theory in Biophysics of Radiations. *International Journal of Radiation Medicine*, **3**, 123-137.

[145] Marmur, J. (1961) A Procedure for the Isolation of Deoxiribonucleic Acid from Microorganisms. *Journal of Molecular Biology*, **3**, 208-218. http://dx.doi.org/10.1016/S0022-2836(61)80047-8

[146] Volpe, P., Iacovacci, P., Esposito, C. and Eremenko, T. (1992) A Common Language in Eukaryotic Genes with Methylation-Dependent Regulation. *Rendiconti Lincei*, **3**, 383-394.

[147] Volpe, P. (1996) Dall'origine del codice all'evoluzione del gene eucariotico (*Lectio Doctoralis* in Occasion of the *Laurea Honoris Causa* in *Biological Sciences with Biotechnological Specialization*). Università degli Studi, Ancona, 1-35.

[148] Volpe, P. (1999) Introduzione alla Biofisica delle Radiazioni. UNESCO Publishers, Venice, 1-256.

[149] Volpe, P. (2014) A Theoretical Overview of Bioresponse to Magnetic Fields on the Earth's Surface. *International Journal of Geosciences*, **5**, 1149-1162. http://dx.doi.org/10.4236/ijg.2014.510097

[150] Kruger, D.H., Shroeder, C., Reuter, M., Bogdarina, I.G., Buryanov, Y.I. and Bickle, T.A. (1985) DNA Methylation of Bacterial Viruses T3 and T7 by Different DNA Methylases in *Escherichia coli* K12 Cells. *European Journal of Biochemistry*, **150**, 323-330. http://dx.doi.org/10.1111/j.1432-1033.1985.tb09024.x

[151] Gorelick, R. (2003) Transposable Elements Suppress Recombination in All Meiotic Eukaryotes, Including Automictic Ancient Asexuals: A Reply to Schön and Martens. *Journal of Natural History*, **37**, 903-909. http://dx.doi.org/10.1080/0022293021000007705

[152] Gorelick, R. (2005) Theory for Why Dioecious Plants Have Equal Length Sex Chromosomes. *American Journal of Botany*, **92**, 979-984. http://dx.doi.org/10.3732/ajb.92.6.979

[153] Lingua, G., Trotta, A., Prigione, V., Ugoccioni, R. and Berta, G. (2005) Nucleus Size in the Host Cells of an Arbuscular Mycorrhizal System: A Mathematical Approach to Estimate the Role of Ploidy and Chromatin Condensation. *Karyologia*, **58**, 112-121.

[154] Zemskov, E.A., Janiak, A., Hang, J., Waghray, A. and Belkin, A.M. (2006) The Role of Tissue Transglutaminase in Cell-Matrix Interactions. *Frontiers in Bioscience*, **11**, 1057-1076. http://dx.doi.org/10.2741/1863

[155] Lai, T.S., Liu, Y., Li, W. and Greenberg, C.S. (2007) Identification of Two GTP-Independent Alternatively Spliced Forms of Tissue Transglutaminase in Human Leukocytes, Vascular Smooth Muscle, and Endothelial Cells. *FASEB Journal*, **21**, 4131-4143. http://dx.doi.org/10.1096/fj.06-7598com

[156] Malousi, A., Maglaveras, N. and Kouidou, S. (2008) Intronic CpG Content and Alternative Splicing in Human Genes Containing a Single Cassette Exon. *Epigenetics*, **3**, 69-73. http://dx.doi.org/10.4161/epi.3.2.6066

[157] Łukasik, M., Karmalska, J., Szutowski, M.M. and Łukaszkiewicz, J. (2009) Wplyw metylacji DNA na funkcjonowanie genomu. *Biuletyn Wydziału Farmaceutycznego Warszawskiego Uniwersytetu Medycznego*, **2**, 13-18.

[158] Shleikin, A.G. and Danilov, N.P. (2011) Evolutionary-Biological Peculiarities of Transglutaminase. Structure, Physiological Functions, Application. *Journal of Evolutionary Biochemistry and Physiology*, **47**, 1-14. http://dx.doi.org/10.1134/S0022093011010014

[159] Lytvynenko, O.O. and Bugaicov, S.G. (2013) Ionizing Radiation in the Process of Initiation and Development of Cancer Pathology. *Archiv Klinichnoj ta Eksperimentalnoj Medizini*, **22**, 120-123.

[160] Lytvynenko, O. and Bugaitsov, S. (2013) Ionizing Radiation in Origin and Development Oncology Pathology. *Actual Problems of Transport Medicine*, **4**, 18-23.

Abbreviations

N^6-methyladenine (m^6A), 5-methylcytosine (m^5C), deoxyadenylate (dam), deoxycytidylate (dcm), methionine (Met), [^{14}C]methyl-L-methionine ([^{14}C]Met), S-adenosyl-L-methionine (AdoMet), bromo-deoxyuridine (BrdUrid), DNA polymerase (DNApol), RNA polymerase (RNApol), methyltransferase (met), transglutaminase (hTGc), housekeeping (HK) and tissuespecific (TS) genes, calcitonin (CALC), mitochondrial DNA (mtDNA), culture growth cycle (CGC), Simian virus 40 (SV40), ultraviolet (UV), repair patch (RP).

12

Bioinformatics Analysis of NprR-NprX Quorum-Sensing System of *Bacillus thuringiensis* Isolates from the Papaloapan Region, Oaxaca-Mexico

Humberto Rafael Bravo-D[1], Alain Cruz-Nolasco[1], Luis Raúl Gutiérrez-Lucas[2], Ana Karin Navarro-Mtz[2*]

[1]División de Estudios de Posgrado, Universidad del Papaloapan, Circuito Central 200, Parque Industrial, Tuxtepec, Oaxaca, México

[2]Instituto de Biotecnología, Universidad del Papaloapan, Circuito Central 200, Parque Industrial, Tuxtepec, Oaxaca, México

Email: [*]anavarro@unpa.edu.mx

Abstract

Quorum sensing is a chemical communication process that bacteria use to regulate collective behaviors. In Gram-positive bacteria, oligopeptides (called autoinducers) are the signaling molecules to elicit quorum sensing. In *Bacillus thuringiensis*, NprR is a transcriptional regulator whose activity depends on the NprX signalling peptide. *Bacillus thuringiensis* is closely related to *Bacillus cereus* and *Bacillus anthracis*. The principal difference between them is that *Bacillus thuringiensis* is the only one that produced Cry protein. The aim of this study is to explore the relation of *nprR* and 16S rRNA genes in *Bacillus thuringiensis.* Phylogenetic trees of nucleotide sequences of *nprR* and 16S rRNA genes were built. Sequences of fourteen new isolates from Papaloapan region were included in those phylogenetic trees. In order to identify the isolates, a simple and fast methodology considering the Cry protein formation was used. The 16S rRNA phylogenetic tree allows identify eight isolates as *Bacillus thuringiensis* and the others as *Bacillus spp*. The *nprR* phylogenetic tree does not match with the 16S rRNA phylogenetic tree. This confirms that *nprR* is not a molecular marker for evolution. Most of the new isolates have the same NprR sequence (WTSDIVG). However, the SKPDIVG is the most common NprR sequence in *thuringiensis* species.

[*]Corresponding author.

Keywords

Quorum-Sensing, *Bacillus thuringiensis*, NprR-NprX, Phylogenetic Tree

1. Introduction

Several bacterial species use the Quorum Sensing (QS, cell-cell communication) to coordinate their behavior as a whole community [1]. In Gram-positive bacteria, the signaling molecules are mostly small secreted peptides that are actively released into the extracellular environment [2]. The QS regulators of soil bacteria have been grouped in a new protein family called RNPP (Rap, NprR, PlcR and PrgX). Genome analysis indicated that genes encoding a putative regulator (NprR) and a putative signaling peptide (NprX or NprRB) were found upstream from *nprA* in the bacteria of the *Bacillus cereus* (*B. cereus*) group [3] [4]. These genes encode for a receptor and for a small protein. The small protein has a putative signal sequence used in the export pathway and a secreted domain [5]. The signaling peptide exported is processed by extracellular proteases and it is internalized by an oligopeptide permease [6]. The secreted QS factor allowing the activation of *nprA* expression corresponds to the central part of NprX and it is at least seven amino acids long [3]. The NprR-NprX system was found in all the species of *B. cereus* group, where 31 different NprR polypeptide sequences were identified [3].

The *B. cereus* group comprises a number of closely related pathogenic species (*Bacillus thuringiensis*, *B. thuringiensis*; *Bacillus anthracis*, *B. anthracis*; and *B. cereus*) [7]. *B. thuringiensis* is a Gram-positive endospore-forming bacterium which is the only one in *B. cereus* group that synthesizes a crystalline delta-endotoxin protein (named Cry) [8]. Cry protein is formed during sporulation process and it is an insoluble and crystalline protein. The main application of the Cry protein is in the biological control and it has been also reported that some Cry proteins are highly cytotoxic to a wide range of human cancer cells [9].

The aim of the present study is to determine if the genes encoding the signaling peptide (*nprX*) and the 16S rRNA genes are related in *B. thuringiensis*. New isolates from the Papaloapan region were used in the study. Also, a simple and rapid isolation method for *B. thuringiensis* is proposed.

2. Material and Methods

2.1. Soil Samples

The new *B. thuringiensis* isolates were obtain from uncultivated and cultivated soil samples from the Papaloapan region. The cultivated soil samples were from sugarcane, coffee and banana crops. The geographic location and coordinates of the soil sampling is show in the Electronic Supplementary Material, **Table S1**. Papaloapan region is a humid tropical region with annual average conditions of 70% of humidity, 33˚C and 3000 mm of precipitation. Five samples of 500 g at 15 cm deep of soil per crop (four corners and the center) were collected. The samples were stored at 4˚C for 24 h.

2.2. Isolation and Identification Method

The isolation method was design considering the principal characteristics of *B. thuringiensis* which is forming an insoluble and crystalline protein (Cry protein). The method consist them in two steps: 1) colony and microscopically morphology and 2) production of an insoluble protein at the end of the submerged culture. In order to compare the isolates *B. thuringiensis* var. *kurstaki* HD-73 (ATCC-35866), which produces a 133.3 kDa Cry1A(c) insecticidal crystal protein, was used as a reference. The isolates were phylogenetically identified comparing the molecular chronometer 16S rRNA from the isolates with 16S rRNA from *B. thuringiensis* strains reported in databases.

Steps 1: *Colony and Microscopically Morphology*. From the collected sample serial dilutions were done with sterile distilled water. The 1×10^{-3} dilutions were inoculated by triplicate in nutrient agar and were incubated at 30˚C for 24 hours. Colonies were selected by their morphology looking for white to off-white color, opaque, slightly raised elevation, and regular outlined colonies [10]. The selected colonies were inoculated in serial subcultures of nutrient agar until pure cultures were obtained. After that, the selected isolates were inoculated in 250 ml of nutrient broth by duplicate at 30˚C and 180 rpm during 48 h. At eight hours of culture (during the exponential growth) samples were collected for Gram staining, looking for gram positive bacteria. At 24 hours of culture (during the sporulation phase) samples were collected for spore formation staining. The stains were analyzed with a microscope under dark field illumination, with a 40× objective and 10× ocular lenses. It is important for a proper Gram staining of *B. thuringiensis* to take the sample during the exponential growth; otherwise, according to our experiments, the bacteria could be stained as a gram negative. At the end of the cultures the broth were collected for Cry protein detection.

Step 2: *Detection of Insoluble Protein.* The Cry proteins are insoluble at neutral pH, they are soluble just under alkaline conditions [9] [11]. The detection of Cry protein was done by sodium dodecyl sulfate polyacrylamide gel electrophoresis (SDS-PAGE). Bovine serum albumin (BSA) was used as protein standard. Considering the insolubility of Cry protein, the culture broth were centrifuged in 50 ml corning tubes at 5580 g during 30 min. The pellets were resuspended in 2 ml of distilled water, concentrated in a 15 ml corning tube and centrifuged at 5580 g during 30 min. The samples preparation and the solubilization process were done using the methodology described by Navarro *et al.* [12]. Four replicates were done for each sample.

2.3. Bioinformatic Analysis

The selected isolates were inoculated in Gerry Rowe culture medium, and incubated in orbital shaker (Thermo Scientific) at 30˚C and 200 rpm during 48 h. At eight hours of culture, 5 ml samples were centrifuged at 11,180 g during 15 min at 4˚C. Genomic DNA was extracted using the Ultra Clean microbial DNA isolation kit (MOBIO) following the manufacturer's recommendations. DNA was quantified using a nano-spectrophotometer (Nanodrop 2000, Thermo Scientific).

The PCR amplification of 16S DNA gene was done using 100 ng of genomic DNA, 10 pM fD1 primer (CCGAATTCGTCGACAACAGAGTTTGATCCTGGCTCAG) and rD1 primer (CCCGGGATCCAAGCTTAA GGAGGTGATCCAGCC) [13]. 16S DNA PCR was performed in a MaxyGene (Axygen Scientific®) thermal cycler under the following conditions: 95˚C 5 min; 35 cycles of 95˚C for 1 min, 55˚C for 40 s and 72˚C for 30 s; finally, 72˚C for 10 min. The PCR amplification of *nprX* gene was done using the primers NprR-1 (GGGCAT TTGTTCTGTCTC) and NprR-2 (GCTAACACTAACGCTAAAC) [4] for NprR-NprX in 50 μL reaction (DreamTaq kit ™ 2X Green PCR Master Mix, Fermentas). *nprX* gene PCR was performed in a MaxyGene (Axygen Scientific®) thermal cycler under the following conditions: 95˚C 5 min; 35 cycles of 95˚C for 1 min, Ta 1 min and 72˚C for 1 min; finally, 72˚C for 10 min. The amplifications were observed with an imaging and analyzing system INGENIUS SYNGENNE®. The PCR products were purified using Thermo Scientific GeneJET PCR Purification Kit. The sequencing service was performed in Macrogen Inc. Korea. The nucleotide sequences obtained in this study were compared with sequences retrieved from databases in a pairwise mode, with the BLAST2 sequences tool (http://www.ncbi.nlm.nih.gov/BLAST/bl2seq/wblast2.cgi). The identification criteria proposed by Pothathil and Lazzazzera [6], and Perchat *et al.* [3] were used to identify the isolates putative signal pentapeptide sequence.

Likelihood trees of 16S rRNA gene and *nprX* gene nucleotide sequences were built for isolates bioinformatic analysis. For phylogenetic tree based on 16S rRNA, the sequences from 29 isolates and 42 sequences from *B. cereus* group strains were included. For phylogenetic tree based on *nprX* gene, 24 isolates and 33 sequences from *B. cereus* group strains were included (Electronic Supplementary Material **Table S2** and **Table S3**). The *B. cereus* group strains sequences were obtained from the GenBank database of the NCBI (http://www.ncbi.nlm.nih.gov/genbank) and IMG database (Integrated Microbial genomes, http://img.jgi.doe.gov). The phylogenetic tree based on 16S rRNA gene was built under the Statistical Method Maximum Likelihood. The Substitution Model used was Tamura Nei [14]. The phylogenetic tree based on 16S rRNA gene topology was optimized with the nearest neighbor interchanges (NNI). The phylogenetic tree based on *nprX* gene was built under the general time reversible (GTR) substitution model using Phyml [15], while the topology was optimized with the nearest neighbor interchanges and sub-tree pruning and regrafting approaches. Statistical support was determined by 10,000 bootstrap replicates. The phylogenetic trees based on 16S rRNA and *nprX* gene were built using MEGA 6.0 software and CLUSTALW analysis.

3. Results

3.1. Isolation and Identification of Strains

The results of the isolation method are shown in **Table 1**. From colony and microscopically morphology 45 isolates were selected as *B. thuringiensis*; from those, 41 isolates produce an insoluble protein between 60 and 130 kDa (91%). Considering just colony morphology, approximately 63% of the isolates produce an insoluble protein. Therefore, the combination of colony and microscopically morphology increase the possibility to isolate correctly *B. thuringiensis* strains.

3.2. Molecular Characterization and Phylogenetic Analysis

For isolates identification three consecutive phylogenetic trees were built. The first one was built using 25 reported sequences of 16S rRNA gene: 19 of *B. cereus* group strains (10 for *B. thuringiensis*, 6 for *B. cereus*, 3 for *B. anthracis*), 4 of *Bacillus subtilis* (*B. subtilis*), 1 of *Bacillus licheniformis* (*B. licheniformis*) and 1 of *Escherichia coli* (*E. coli*), **Figure 1**. This phylogenetic tree shows three groups, one with all the strains of *B. cereus*

Table 1. Results from the isolation method.

	Isolates of soil crop			
	Uncultivated	Banana	Coffee	Sugarcane
Colony morphology	2	6	5	52
Microscopically morphology	2	4	3	36
Insoluble protein detection	2	4	3	32
16S rRNA gene amplification	2	4	3	32
Forward and reverse positive sequencing	2	4	2	21
Phylogenetic identification	2	2	-	20

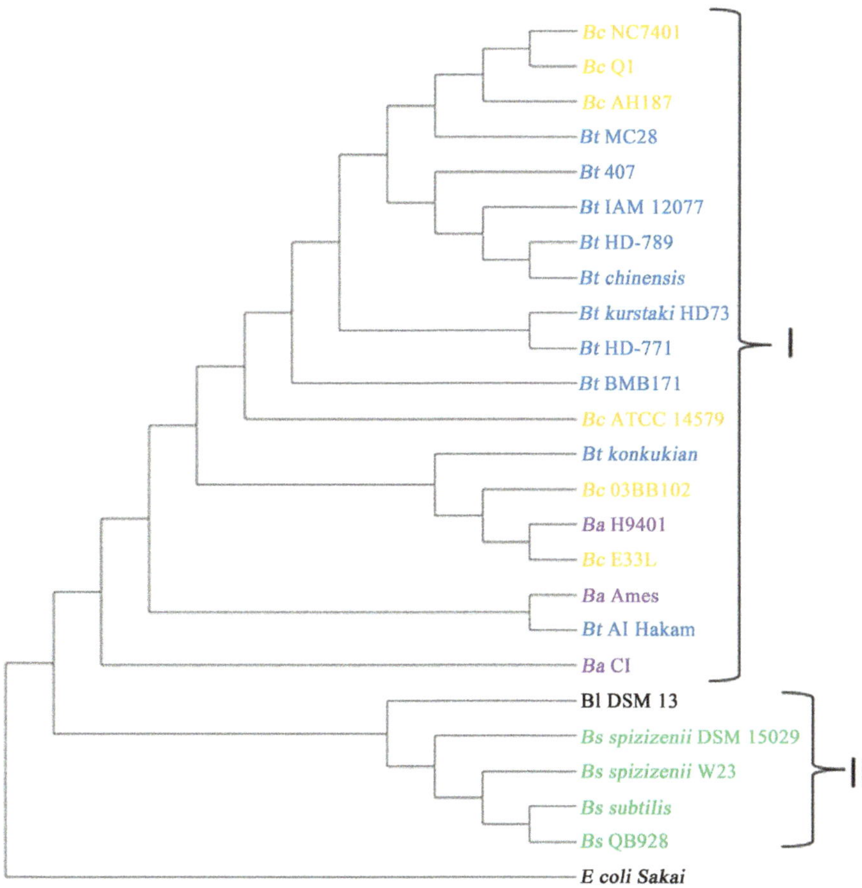

Figure 1. Rectangular phylogenetic tree built with the 16S rRNA genes sequences (first tree) for several strains of *B. cereus* group (obtained from the GenBank and IMG). Abbreviations: Bt: *Bacillus thuringiensis* (blue), Bc: *B. cereus* (orange), Ba: *B. anthracis* (purple), Bs: *B. subtilis* (green), Bl DSM 13: *B. licheniformis* DMS 13 (black). Group I corresponds to *B. cereus* group strains and group II corresponds to *B. subtilis* strains.

group, the second one with *B. subtilis* and *B. licheniformis* and the last one with *E. coli.*

A ClustalW analysis [16] using the isolates 16S rRNA gene sequences and the sequences used for the first tree with the MEGA 6.0 was done. With this alignment the second phylogenetic tree was built using the Tamura Nei model [14] (**Figure 2**). The second phylogenetic tree shows two groups, one with the *B. cereus* strains and

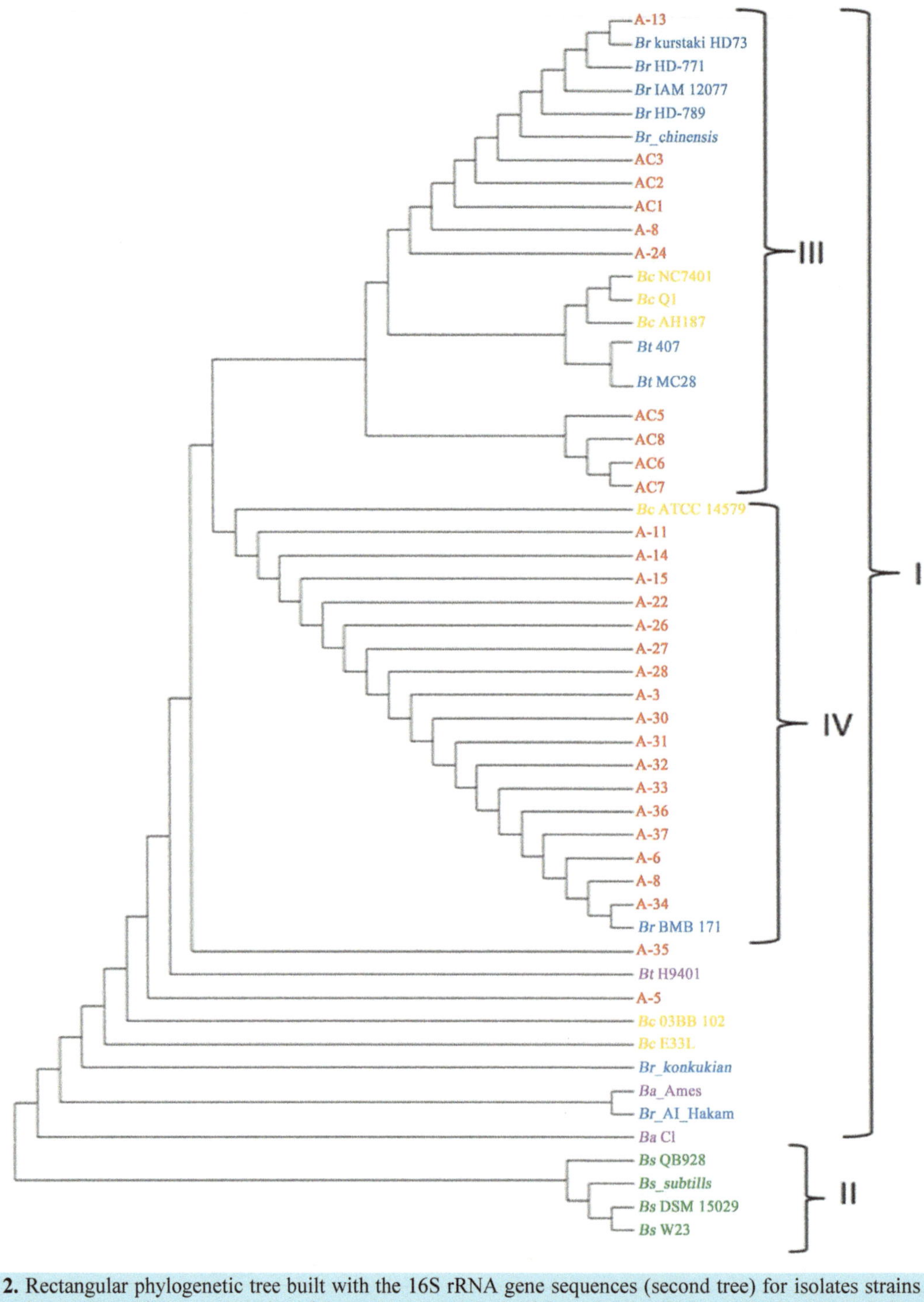

Figure 2. Rectangular phylogenetic tree built with the 16S rRNA gene sequences (second tree) for isolates strains of Papaloapan region (red) and for several strains of *B. cereus* group (obtained from the GenBank and IMG). Abbreviations: Bt: *Bacillus thuringiensis* (blue), Bc: *B. cereus* (orange), Ba: *B. anthracis* (purple), Bs: *B. subtilis* (green). Group I corresponds to *B. cereus* group strains, group II corresponds to *B. subtilis* strains, group III corresponds to *B. thuringiensis* strains and group IV corresponds to A-X isolates group.

the other one with *B. subtilis*. All the isolates sequences are closer to the *B. cereus* group (**Figure 2**). The 16S rRNA gene sequences of 16 isolates (A-8, A-6, A-37, A-36, A-33, A-32, A-31, A-30, A-3, A-28, A-27, A-26, A-22, A-15, A-14 and A-11) have a comb-like configuration. Through the alignment of these sequences, it was observed that all the sequences are identical. Therefore, they were considered as one group (A-X group).

Considering the results of the first and the second phylogenetic tree, the third one was built (**Figure 3**). For the third phylogenetic tree 22 sequences of *B. thuringiensis* strains were added (Electronic Supplementary Material **Table S2**). In this third phylogenetic tree (**Figure 3**) it can be observed that isolated AC1 presents similarity with *B. thuringiensis alesti*; as well as AC2 with *B. thuringiensis* HD-771, AC3 with *B. thuringiensis higo* and A-13 with *B. thuringiensis* IAM 12077. Besides, A-9 and A-24 show similarity not only with *B. thuringiensis* IAM 12077 but also with different varieties of *B. thuringiensis* such as *chinensis*, *oswaldocruzi*, *kenyae* and *thompsoni*. Isolates AC5, AC6, AC7 and AC8 show similarity with *B. thuringiensis* and *B. cereus*. The A-X group share information with *B. thuringiensis* (*ostriniae*, *seoulensis* and BMB) and with *B. cereus* (ATCC 14579). The isolated A-34 shows similarity with *B. thuringiensis tolworthi*. The isolated A-35 and A-5 show similarity with the group of *B. thuringiensis*, *B. cereus* and *B. anthracis* species. **Table 2** shows the results from

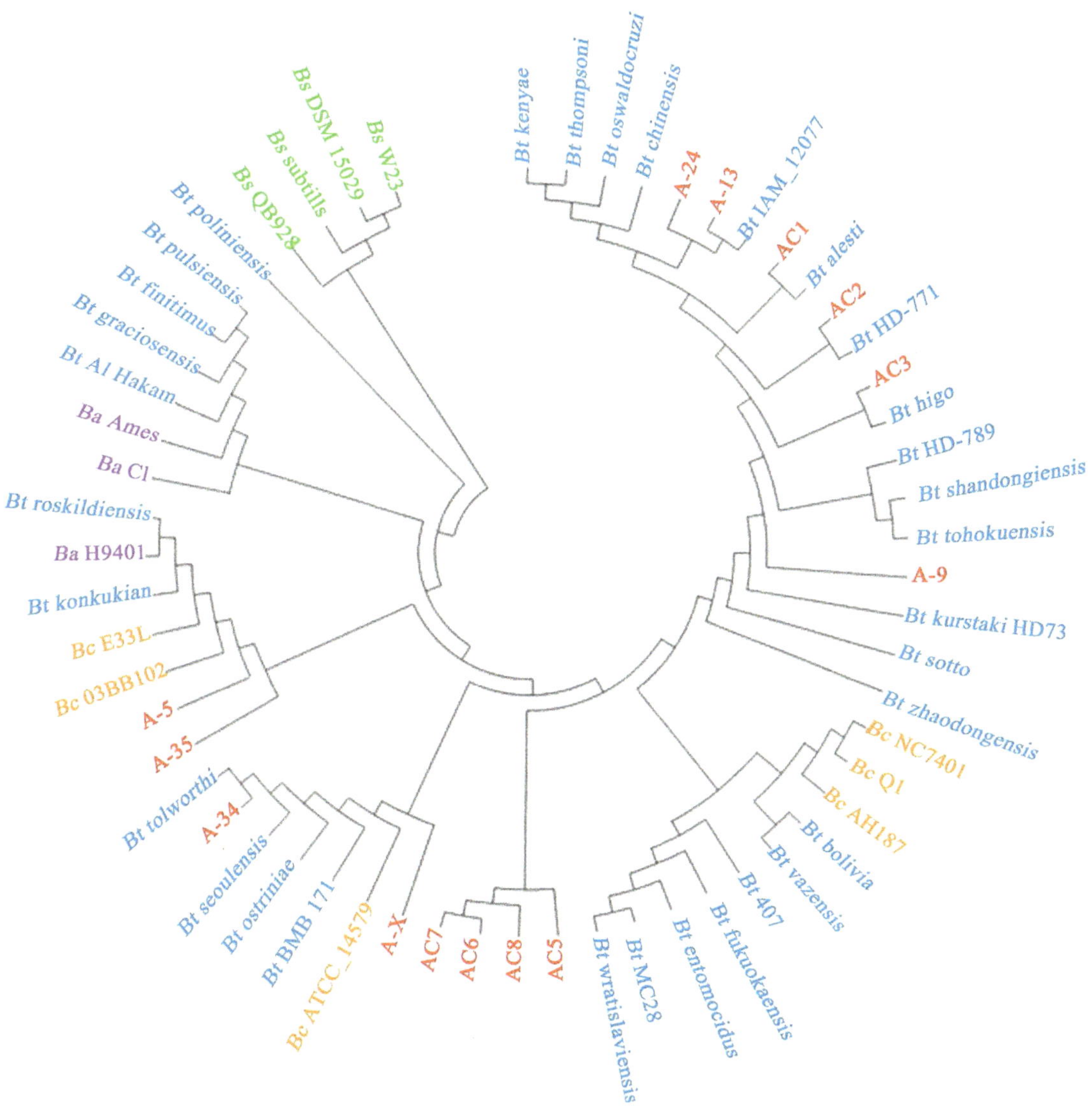

Figure 3. Circular phylogenetic tree built with the 16S rRNA gene sequences (third tree) for isolates strains of Papaloapan region (red) and several strains of *B. cereus* group (obtained from the GenBank and IMG). Abbreviations: Bt: *Bacillus thuringiensis* (blue), Bc: *B. cereus* (orange), Ba: *B. anthracis* (purple), Bs: *B. subtilis* (green). A-X group represent A-8, A-6, A-37, A-36, A-33, A-32, A-31, A-30, A-3, A-28, A-27, A-26, A-22, A-15, A-14 and A-11 isolates.

Table 2. Phylogenetic analysis results of 16S rRNA gene sequences of isolates strains of Papaloapan region.

Isolated	Soil type	Species	Heptapeptide type
AC-1	Uncultivated	*thuringiensis* var *alesti*	WTSDIVG
AC-2	Uncultivated	*thuringiensis* HD-771	WTSDIVG
AC-3	Banana	*thuringiensis* var *higo*	Unidentified
AC-5	Coffe	*spp*	WTSDIVG
AC-6	Banana	*spp*	WTSDIVG
AC-7	Coffe	*spp*	WTSDIVG
AC-8	Banana	*spp*	WTSDIVG
A-X	Sugarcane	*thuringiensis*	WTSDIVG
A-5	Sugarcane	*spp*	SRPDVLT
A-9	Banana	*thuringiensis*	SKPDIVG
A-13	Sugarcane	*thuringiensis* IAM 12077	SKPDIVG
A-24	Sugarcane	*thuringiensis*	SKPDIVG
A-34	Sugarcane	*thuringiensis* var *tolworthi*	Unidentified
A-35	Sugarcane	*spp*	SRPDVLT

the phylogenetic identification of the isolates from the Papaloapan region. For AC-5, AC-6, AC-7, AC-8, A-5 and A-35 isolates the phylogenetic analysis is not conclusive, therefore, they were identified as *Bacillus spp.* However, all of these isolates produce an insoluble protein at the end of the culture.

3.3. NprX Bioinformatic Analysis

The phylogenetic tree based on *nprR* gene nucleotide sequence built on 29 strains isolated and 32 strains of *B. cereus* group from the literature (**Figure 4**, Electronic Supplementary Material **Table S3**) show a perfect correlation of seven putative mature NprX heptapeptide sequences [3]. Analysis of the NprX peptide was performed using the software SignalP 3 [15], which incorporates a prediction of cleavage sites and a signal peptide/non-signal peptide prediction based on a combination of several artificial neural networks and hidden Markov models. Alignment of the secreted region was done using ClustalX to identify putative mature signaling peptides. The putative mature NprX for isolated strains are show in **Table 2**.

In phylogenetic tree based on *nprR* gene (**Figure 4**) different putative mature NprX can be observed: SKPDIVG for 15 *B. thuringiensis* strains, 3 *B. anthracis* strains and one *B. cereus* strain; WTSDIVG for 5 *B. thuringiensis* strains and 4 *Bacillus spp.* strains; SRPDVLT** for 2 strains of *B. thuringiensis*, one of *B. cereus* and two of *Bacillus spp.*; WKPDVLG* for 2 *B. thuringiensis* strains; WKPDTLG* for 2 *B. thuringiensis* strains and one *B. cereus* strain; SNPDIY** for one *B. thuringiensis* strain and one *B. cereus* strain; SKPDTYG for 2 strains of *B. cereus*.

The phylogenetic tree based on *nprR* gene also shows the type of Cry protein produced for the reported strains (**Figure 4**, Electronic Supplementary Material **Table S3**). No relationship was found between the Cry protein type produced and the putative mature peptides of all strain studied.

4. Discussion

The isolation method results show that the phase contrast microscopy strategy is not enough for *B. thuringiensis* isolation. Just 63% of the isolates produce an insoluble protein (Cry protein) at the end of the culture. In general, phase contrast microscopy is the backbone of the isolation strategy [17]. Another disadvantage of the phase contrast microscopy strategy is that a large proportion of isolates contain *cry*1 gene (bipyramidal shaped crystal) and relatively few novel isolates [17].

According to the isolation method results (**Table 1**), the critical step in the method is to obtain forward and reverse sequencing. With both sequences, around 83% of the strains were identified by phylogenetic analysis. In this study, the phylogenic analysis of the 16S rRNA gene from 30 isolates from the Papalopan region 24 isolates was identified as *B. thuringiensis* while 6 of them were unidentified (**Table 2**). However, *B. thuringiensis* is the

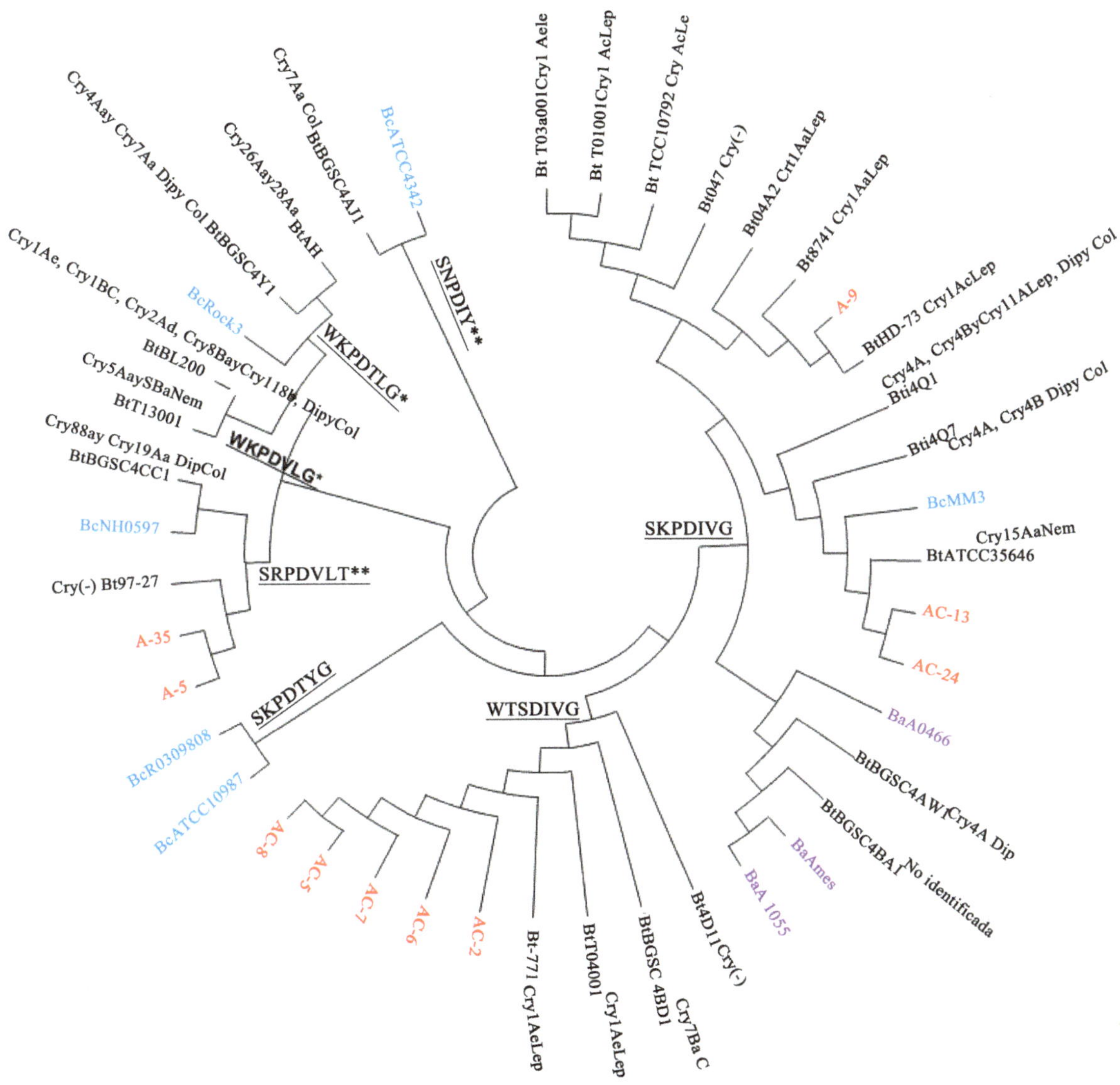

Figure 4. Phylogenetic tree based on *nprX* gene sequence. 34 sequences of the *B. cereus* group were used under the model of amino acid substitution. One asterisk represents sequences of NprX that are lowly predicted to be exported from the cell. Two asterisks represent NprX sequences that are not predicted to be secreted by the cell.

only strain in *B. cereus* group that produces an insoluble protein at the end of the culture, named Cry protein. The 6 unidentified isolates produce this insoluble protein; therefore, they cannot be discarded as *B. thuringiensis*.

Several methods have been used in order to classify the *B. cereus* group. However, the classification results depend on the method, the strains, and the molecular marker used for each study. For example, multi-locus enzyme electrophoresis and the analysis of nine chromosomal genes showed that *B. cereus*, *B. anthracis*, and *B. thuringiensis* are the same species [18]. On the other hand, analysis of 16S rRNA, 23S, *rpoB* and *gyrB* genes from *B. cereus* group strains indicated that *B. anthracis* can be distinguished from *B. cereus* and *B. thuringiensis* strains [17] [19].

The phylogenetic tree based on *nprR* gene does not match with the phylogenetic tree based on 16S rRNA gene indicating that the *nprR* is not specific for the species but it is specific for the strain. Several species of *Bacillus* could have the same heptapeptide sequence. For *B. thuringiensis* five different heptapeptide sequences were found. Although the SKPDIVG is the most common heptapeptide in *thuringiensis* species. For the isolates strain, the most common peptide sequence is WTSDIVG and there is no relationship with the soil type. Considering A-X as the same strain, for *thuringiensis* species SKPDIVG is the most common peptide. It has been reported that the SKPDIVG and the SKPDI are the most common peptide in *thuringiensis* species [3] [4]. The NprR-NprX quorum-sensing system was found to be strain-specific with a possible cross-talk between some pherotypes. The phylogenic relationship between NprR and NprX suggests a coevolution of the regulatory protein and its signalling peptide [3]. The incongruence between the phylogenetic trees based on *nprR* and 16S rRNA genes indicates that NprR is not a molecular marker for evolution. Therefore, the *nprR* gene is not useful for strain identification. Although phylogenetic tree based on *nprR* gene provides significant biological information, as it constitutes a system that controls functions like extracellular enzyme productions and plays a role in sporulation and *cry* transcription [3].

Acknowledgements

This work had been supported by Consejo Nacional de Ciencia y Tecnología (2008-105057) and by Programa de Mejoramiento del Profesorado (103.5/10/0246). Alain Cruz-Nolasco acknowledges the fellowship provided by Programa de Mejoramiento del Profesorado. Humberto Rafael Bravo-D acknowledges the fellowship provided by Consejo Nacional de Ciencia y Tecnología Project 2008-105057.

References

[1] Waters, C.M. and Bassler, B.L. (2005) Quorum Sensing: Cell-to-Cell Communication in Bacteria. *Annual Review of Cell and Developmental Biology*, **21**, 319-346. http://dx.doi.org/10.1146/annurev.cellbio.21.012704.131001

[2] Monnet, V. and Gardan, R. (2015) Quorum-Sensing Regulators in Gram-Positive Bacteria: Cherchez le Peptide. *Molecular Microbiology*, **97**, 181-184. http://dx.doi.org/10.1111/mmi.13060

[3] Perchat, S., Dubois, T., Zouhir, S., Gominet, M., Poncet, S., Lemy, C., Aumont-Nicaise, M., Deutscher, J., Gohar, M., Nessler, S. and Lereclus, D. (2011) A Cell-Cell Communication System Regulates Protease Production during Sporulation in Bacteria of the *Bacillus cereus* Group. *Molecular Microbiology*, **82**, 619-633. http://dx.doi.org/10.1111/j.1365-2958.2011.07839.x

[4] Rocha, J., Flores, V., Cabrera, R., Soto-Guzman, A., Granados, G., Juaristi, E., Guarneros, G. and de la Torre, M. (2012) Evolution and Some Functions of the NprR-NprRB Quorum-Sensing System in the *Bacillus cereus* Group. *Applied Microbiology and Biotechnology*, **94**, 1069-1078. http://dx.doi.org/10.1007/s00253-011-3775-4

[5] Lazazzera, B.A. (2001) The Intracellular Function of Extracellular Signaling Peptides. *Peptides*, **22**, 1519-1527. http://dx.doi.org/10.1016/S0196-9781(01)00488-0

[6] Pottathil, M. and Lazazzera, B.A. (2003) The Extracellular Phr Peptide-Rap Phosphatase Signaling Circuit of *Bacillus subtilis*. *Frontiers in Bioscience*, **8**, 32-45. http://dx.doi.org/10.2741/913

[7] Raymond, B., Johnston, P.R., Nielsen-LeRoux, C., Lereclus, D. and Crickmore, N. (2010) *Bacillus thuringiensis*: An Impotent Pathogen? *Trends in Microbiology*, **18**, 189-194. http://dx.doi.org/10.1016/j.tim.2010.02.006

[8] Rowe, G.E. and Margaritis, A. (1987) Bioprocess Development in the Production of Bioinsecticides by *Bacillus thuringiensis*. *Critical Reviews in Biotechnology*, **6**, 87-127. http://dx.doi.org/10.3109/07388558709086986

[9] Ito, A., Sasaguri, Y., Kitada, S., Kusaka, Y., Kuwano, K., Masutomi, K., Mizuki, E., Akao, T. and Ohba, M. (2004) A *Bacillus thuringiensis* Crystal Protein with Selective Cytocidal Action to Human Cells. *Journal of Biological Chemistry*, **279**, 21282-21286. http://dx.doi.org/10.1074/jbc.M401881200

[10] Muniady, S., Rathinam, X. and Subramaniam, S. (2011) Quick Isolation and Characterization for the Confirmation of a Novel *Bacillus thuringiensis* Strains from Chicken Manure Samples. *African Journal of Microbiology Research*, **5**, 3131-3137.

[11] Wabiko, H. and Yasuda, E. (1995) *Bacillus thuringiensis* Protoxin: Location of Toxic Border and Requirement of Non-Toxic Domain for High-Level *in Vivo* Production of Active Toxin. *Microbiology*, **141**, 629-639. http://dx.doi.org/10.1099/13500872-141-3-629

[12] Navarro, A.K., Farrera, R.R., Lopez, R. and Perez-Guevara, F. (2006) Relationship between Poly-β-Hydroxybutirate Production and δ-Endotoxin for *Bacillus thuringiensis* var. *kurstaki*. *Biotechnology Letters*, **28**, 641-644. http://dx.doi.org/10.1007/s10529-006-0029-0

[13] Weisburg, W., Barns, S., Pelletier, D. and Lane, D. (1991) 16S Ribosomal DNA Amplification for Phylogenetic Study. *Journal of Bacteriology*, **173**, 697-703.

[14] Tamura, K., Peterson, D., Peterson, N., Stecher, G., Nei, M. and Kumar, S. (2011) MEGA5: Molecular Evolutionary Genetics Analysis Using Maximum Likelihood, Evolutionary Distance, and Maximum Parsimony Methods. *Molecular Biology and Evolution*, **28**, 2731-2739. http://dx.doi.org/10.1093/molbev/msr121

[15] Emanuelsson, O., Brunak, S., von Heijne, G. and Nielsen, H. (2007) Locating Proteins in the Cell Using Target P, Signal P, and Related Tools. *Nature Protocols*, **2**, 953-971. http://dx.doi.org/10.1038/nprot.2007.131

[16] Thompson, J., Higgins, D. and Gibson, T. (1994) Clustal W: Improving the Sensitivity of Progressive Multiple Sequence Alignment through Sequence Weighting, Position-Specific Gap Penalties and Weight Matrix Choice. *Nucleic Acids Research*, **22**, 4673-4680. http://dx.doi.org/10.1093/nar/22.22.4673

[17] Bavykin, S., Lysov, Y., Zakhariev, V., Kelly, J., Jackman, J., Stahl, D. and Cherni, A. (2004) Use of 16S rRNA, 23S rRNA, and *gyrB* Gene Sequence Analysis to Determine Phylogenetic Relationships of *Bacillus cereus* Group Microorganisms. *Journal of Clinical Microbiology*, **42**, 3711-3730. http://dx.doi.org/10.1128/JCM.42.8.3711-3730.2004

[18] Helgason, E., Okstad, O.E., Caugant, D.A., Johansen, H.A., Fouet, A., Mock, M., Hegna, I. and Kolsto, A.B. (2000) *Bacillus anthracis*, *Bacillus cereus*, and *Bacillus thuringiensis* One Species on the Basis of Genetic Evidence. *Applied and Environmental Microbiology*, **66**, 2627-2630. http://dx.doi.org/10.1128/AEM.66.6.2627-2630.2000

[19] Zasada, A. and Gierczyński, R. (2006) Some *Bacillus thuringiensis* Strains Share *rpoB* Nucleotide Polymorphisms Also Present in *Bacillus anthracis*. *Journal of Clinical Microbiology*, **44**, 1606-1607. http://dx.doi.org/10.1128/JCM.44.4.1606-1607.2006

Supplementary Tables

Table S1. Soil type and location of sampling in the Papaloapan region in the state of Oaxaca.

Soil type	Geographic location	Geographic coordinates	Meters above sea level
Uncultivated	San Bartolo Tuxtepec, Oaxaca	N18°00'744"; WO96°10'146"	55
Cane	Higway Tuxtepec-Palomares	N31°37'514"; W131°21'293"	89
		N45°03'5044"; W115°52'022"	92
		N17°59'597"; W96°06'895"	45
		N17°59'597"; W96°06'882"	53
		N17°59'597"; W96°06'884"	49
		N18°06'342"; W96°15'249"	50
		N18°06'349"; W96°15'264"	59
		N18°06'357"; W96°15'291"	61
		N18°06'366"; W96°15'241"	66
		N18°06'402"; W96°15'234"	35
Coffe	Armadillo Chico, Valle Nacional, Oaxaca	N17°50'764"; W96°18'992"	605
Banana	San Bartolo Tuxtepec, Oaxaca	N18°05'924"; W96°06'767"	53

Table S2. Strains used for phylogenetic bioinformatics study of the 16S rRNA gene.

Species	Access Number	Species	Access Number
Bacillus anthracis CI	NC_014335	*Bacillus thuringiensis aralesti*	EF210300
Bacillus cereus E33L	NC_006274	*Bacillus thuringiensis fukuokaensis*	EF210301
Bacillus cereus NC7401	NC_016771	*Bacillus thuringiensis sotto*	EF210309
Bacillus subtilis QB928	CP003783	*Bacillus thuringiensis kenyae*	EF210315
Bacillus subtilis spizizenii DSM 15029	NC_016047	*Bacillus thuringiensis entomocidus*	EF210312
Bacillus subtilis subsp. *spizizenii* str. W23	NC_014479	*Bacillus thuringiensis ostriniae*	EF210314
Bacillus thuringiensis BMB171	NC_014171	*Bacillus thuringiensis tolworthi*	EF210288
Bacillus thuringiensis Bt407	CP003889	*Bacillus thuringiensis thompsoni*	EF210310
Bacillus thuringiensis HD-771	CP003752	*Bacillus thuringiensis tohokuensis*	EF210294
Bacillus thuringiensis HD-789	CP003763	*Bacillus thuringiensis shandongiensis*	EF210287
Bacillus thuringiensis MC28	CP003687	*Bacillus thuringiensis seoulensis*	EF210286
Bacillus thuringiensis var. *chinensis* CT-43	CP001907	*Bacillus thuringiensis oswaldocruzi*	EF210284
Bacillus thuringiensis var *konkukian* str. 97-27	NC_005957	*Bacillus thuringiensis higo*	EF210308
*Bacillus thuringiensis*str. *Al Hakam*	NC_008600	*Bacillus thuringiensis roskildiensis*	EF210304
Bacillus anthracis str. *Ames*	NR_074453	*Bacillus thuringiensis wratislaviensis*	EF210311
Bacillus cereus ATCC 14579	NC_004722	*Bacillus thuringiensis poloniensis*	EF210305
Bacillus licheniformis DSM 13 ATCC 14580	NC_006322	*Bacillus thuringiensis zhaodongensis*	EF210289
Bacillus subtilis subsp. *subtilis* 6051-HGW	NC_020507	*Bacillus thuringiensis Bolivia*	EF210299
Bacillus thuringiensis var. *kurstaki* str. HD73	NC_020238	*Bacillus thuringiensis pulsiensis*	EF210285
Bacillus thuringiensis strain IAM 12077	NR_043403	*Bacillus thuringiensis graciosensis*	EF210302
Bacillus thuringiensis finitimus	EF210290	*Bacillus thuringiensis vazensis*	EF210293

Table S3. Strains used for phylogenetic study of the NprRB gene.

Strain	Abbreviation	GenBank Access Number	Cry protein	Target groups	Isolation location	Heptapeptide type
Bacillus thuringiensis strain 407	Bt407	ACMZ01000019.1	Cry1Ab	Lepidoptera	-	SKPDIVG
Bacillus thuringiensis var. *andalousiensis* BGSC 4AW1	BtBGSC4AW1	ACNG01000020.1	Cry 4Aa	Diptera	Spain	SKPDIVG
Bacillus thuringiensis var. *berliner* ATCC 10792	BtATCC10792	NZ_ACNF01000019.1	Cry1Ac	Lepidoptera	-	SKPDIVG
Bacillus thuringiensis var. *kurstaki* str. T03a001	BtT03a001	NZ_ACND01000019.1	Cry1Ae	Lepidoptera	Japan	SKPDIVG
Bacillus thuringiensis var. *pondicheriensis* BGSC 4BA1	BtBGSC4BA1	NZ_ACNH01000016.1	-	-	India	SKPDIVG
Bacillus thuringiensis var. *thuringiensis* str. T01001	BtT01001	NZ_ACNA01000020.1	Cry1Ac	Lepidoptera	-	SKPDIVG
Bacillus thuringiensis var. *israelensis* 4Q1	Bti4Q1	JN637469	Cry4A, Cry4B and Cry11A	Diptera, Lepidoptera and Coleoptera	Israel	SKPDIVG
Bacillus thuringiensis var. *israelensis* 4Q7	Bti4Q7	JN637470	Cry4A and Cry4B	Diptera	-	SKPDIVG
Bacillus thuringiensis var. *thuringiensis* 4A2	Bt4A2	JN637471	Cry1Aa	Lepidoptera	-	SKPDIVG
Bacillus thuringiensis var. *thuringiensis* 8741	Bt8741	JN637473	Cry1Aa	Lepidoptera	-	SKPDIVG
Bacillus thuringiensis var. *kurstaki* 4D11	Bt4D11	JN637468	Cry(-)	-	-	SKPDIVG
Bacillus thuringiensis var. *huazhongensis* BGSC 4BD1	BtBGSC 4BD1	NZ_ACNI01000018.1	Cry7Ba	Coleoptera	China	WTSDIYG
Bacillus thuringiensis var. *monterrey* BGSC 4AJ1	BtBGSC4AJ1	NZ_ACNE01000016.1	Cry7Aa	Coleoptera	México	SNPDIYG
Bacillus thuringiensis var. *pakistani* str. T13001	BtT13001	NZ_ACNC01000023.1	Cry5Ba and Cry5Aa	Nematode	-	WKPDVLG
Bacillus thuringiensis var. *pulsiensis* BGSC 4CC1	BtBGSC4CC1	NZ_ACNJ01000017.1	Cry8Ba and Cry19Aa	Coleoptera and Diptera	Pakistan	SRPDVLT
Bacillus thuringiensis var. *sotto* str. T40001	BtT04001	NZ_ACNB01000029.1	Cry1Ae	Lepidoptera	China	WTSDIYG
Bacillus thuringiensis var. *tochigiensis* BGSC 4Y1	BtBGSC4Y1	NZ_ACMY01000020.1	Cy4Aa and Cry7Aa	Diptera and Coleoptera	Japan	WKPDTLG
Bacillus thuringiensis var. *israelensis* ATCC 35646	BtATCC35646	AAJM01000002	Cry15Aa	Nematode	-	SKPDIVG
Bacillus thuringiensis str. *Al Hakam*	BtAH	CP000485	Cry26Aa and 28Aa	Nonpathogenic	-	WKPDTLG
Bacillus thuringiensis IBL 200	BtIBL200	CM000758	Cry1Ae, Cry1Bc, Cry2Ad, Cry8Ba and Cry11Bb	Lepidoptera, Coleoptera and Diptera	-	WKPDVLG
Bacillus thuringiensis var. *konkukian* str. 97-27	Bt97-27	NC_005957	Cry(-)	-	-	SRPDVLT
Bacillus thuringiensis BMB171	BtBMB171	ADH05341.1	Cry1Ac	Lepidoptera	-	SKPDIVG
Bacillus thuringiensis HD-73	BtHD-73		Cry1Ac	Lepidoptera	-	SKPDIVG
Bacillus thuringiensis HD-771	Bt-771	CP003752	Cry1Aa	Lepidoptera	-	WTSDIYG

13

Somatolactin Transcription during Oogenesis in Female Blue Gourami (*Trichogaster trichopterus*)

Gad Degani[1,2]

[1]School of Science and Technology, Tel-Hai Academic College, Tel Hai, Israel
[2]MIGAL—Galilee Technology Center, Kiryat Shmona, Israel
Email: gad@migal.org.il

Abstract

Somatolactin (SL), a specific pituitary hormone belonging to the prolactin (PRL) super family, is involved in background adaptation, osmoregulation, reproduction and fatty acid metabolism. The goal of this study was to examine the gene transcription of SL changes in the ovary of blue gourami females (*Trichogaster trichopterus*) during oogenesis by quantitative real-time polymerase chain reaction (Real-Time PCR). Somatolactin in the pituitary was higher in females at low vitellogenesis compared to females with oocytes in maturation, and the difference was significant ($p < 0.05$). No significant differences were found in mRNA levels between low and high vitellogenesis, and high vitellogenesis and maturation. The findings of this and previous studies demonstrate that SL, growth hormone (GH) and PRL are involved in oogenesis in blue gourami; however, considerably more studies are required in order to separate the functions of these hormones.

Keywords

Female Blue Gourami, Oogenesis, RNA, Somatolactin

1. Introduction

Blue gourami (*T. trichopterus*) is a tropical fish. It serves as a useful model for studying the role of endocrine regulation in reproduction [1]. Blue gourami (*T. trichopterus*) belongs to the Anabantidae family, which contains 16 genera and about 50 species distributed throughout most of southern Asia, India and Central Africa. This family belongs to the Labyrinthici suborder, which is characterized by the presence of an air-filled breathing cavity (labyrinth) located above the gills under the operculum [1]. Blue gourami is multi-spawning and male-dependent, with asynchronic ovary development [2]. The gonadotropin-releasing hormone (GnRH) plays a central role in the control of blue gourami reproduction by affecting various hormones, e.g., gonadotropins (GtHs), which in turn regulate gametogenesis and steroidogenesis. In addition, final oocyte maturation occurs in females characterized by a large percentage of high vitellogenic oocytes in the ovary, only in the presence of a mature male [3]. Sexual glucosteroids 17α-hydroxyprogesterone(17-P), 17α, 20β-Dihydroxy-4-pregnen-3-one (17, 20P) and 17β-estradiol (E_2) are secreted into the water; they might serve as chemical pheromones during reproductive behavior and be involved in preparing females to spawn via the promotion of oogenesis E_2 and 17, 20P plasma levels [4] [5]. Thus, gonadal stages in females can be controlled and examined separately in the laboratory [2] [6]. SL, GH and PRL are pituitary hormones closely related to hormones belonging to the growth hormone family [7]. All these hormones are involved in many biological functions, including growth and reproduction [8].

Blue gourami GH is a 22 kDa single-chain polypeptide [9]-[11]. The GnRH and pituitary adenylate cyclase-activating polypeptide (PACAP) hormones family controls the GH axis and the GTH axis, enabling them to stimulate differential biological functions in oogenesis regulation [12]-[14]. In some species, including salmonids, there are two forms, SLα and SLβ [15]. The SL receptor (SLR) has recently been cloned, and phylogenetic analysis shows that it is similar to previously cloned GH receptors (GHRs) of non-salmonids. The ligand-specificity of GHR/SLR is unclear. Little is known about the role of SLs in sexual maturation of fish [15]. The aim of this study is to increase our knowledge about the regulatory role of SLs during oogenesis in fish, focusing on the female blue gourami model of the Labyrinthici suborder.

2. Methods

2.1. Fish and Sampling

T. trichopterus (blue gourami) females and males purchased from local pet shops were separated into different containers and grown for about a month. The fish were grown in aquaria at 27˚C under a light regime of 12 h light: 12 h darkness, fed twice a day with commercial tropical fish food (Tetra Bits 47.5% protein, 6.5% fat) and supplemented once a day with frozen live food (*Artemia salina*) [16]. Males were divided into two groups: 1) mature males (isolated from females); and 2) individual mature males under reproductive conditions that include plants adapted for nest-building (**Figure 1**). Females were divided into three groups: 1) non-mature females (BW = 1.13 ± 0.38 g) (number = 12)—the ovary of these fish contains mostly oocytes at their pre-vitellogenic stage; 2) mature females that were kept in a group without males (6.85 ± 1.17 g) (number = 12); and 3) mature females that were kept with nest-building males for 24 hours (7.20 ± 1.20 g) (number = 8) (**Figure 1**). The latter

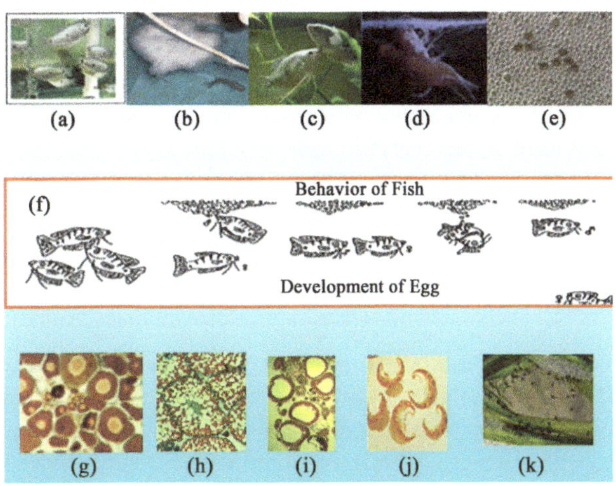

Figure 1. (a) Mature fish found in the group with no sexual behavior; (b) Nest-building male found in a separate aquarium; (c) Female and male in one aquarium, sexual behavior under the nest; (d) Mating, the female lays the eggs and the male fertilizes them, the fertilized eggs in the nest; (f) Different stages of sexual behavior; (g) Oocytes in a juvenile female (no vitellogenesis); (h) Oocytes in advanced vitellogenes is in a female found in groups; (i) Oocytes in maturation in females during sexual behavior of the male; (j) Oocytes in ovulation; (k) Larvae on the day of hatching in the nest.

showed oocytes at their final maturation stage. Fish were sampled and the gonads (small pieces of the ovary) were dissected [17]. The pituitaries were collected from females at various stages of gonadal development from non-mature females, from mature females that were kept in a group without males, and from mature females that were kept in a group without males [18] [19].

2.2. Histological Analysis

Gonad samples were fixed in Bouin and subsequently processed for light microscopy. Paraffin sections of 6 µm were stained with hematoxylin and eosin, as previously described [17] (**Figure 1**).

2.3. SL mRNA Extraction

SL mRNA levels were determined in the groups of female blue gourami: 1) non-mature females, the ovary of these fish contains mostly oocytes at their pre-vitellogenic stage (low vitellogenesis); 2) mature females that were kept in a group without males; and 3) mature females that were kept with nest-building males for 24 hours. For sampling, each fish was anesthetized in a clove oil bath (0.25 mg/l), and weight and length were recorded. Tissues (pituitary for mRNA SL expression studies) were removed and stored in RNA Later buffer (Ambion Inc., Austin, TX). The gonads were removed and weighed, and a portion was taken for histology. Total RNA was extracted from RNA Later stored tissues by means of the RNeasy® Total RNA Kit (QIAGEN, Alameda, CA) according to manufacturer's recommendations [8]. First-strand cDNA was synthesized from 2 µg of total RNA by the Superscript System (Invitrogen, Carlsbad, CA). The single-strand cDNA was used to amplify a cDNA internal fragment using gene-specific primers [15].

2.4. Real-Time PCR

The mRNA levels of SL in the gourami pituitary were determined [20]. The SL in the pituitary and the relative abundance of mRNA were normalized with the mRNA of the endogenous reference gene, 18 S subunit of rRNA (18 S rRNA), using the comparative threshold cycle (CT) method according to [21]. The relative amount of each gene was calculated by the formula 2_DCT, where DCT corresponds to the difference between the CT measured for each target gene and that determined for 18S rRNA. To validate this method, serial dilutions were prepared from a brain cDNA sample, and the efficiencies of gene amplifications were compared by plotting _DCT versus log (template) according to the method of [22]. Linear regressions of the plots showed R 2 values of 0.99 and efficiency values of >95%. Gene-specific primers for real-time PCR were designed using Primer 3 Software [12]. Syber Green Master Mix (ABgene) was added to each of the above PCR mixtures, and amplification was carried out in a Rotor Gene 3000 Sequence Detection System (Corbett Research, Sydney, Australia) under the following conditions: for GnRH1 and GnRH2 (Accession numbers KF113107 and KF113108, respectively), initial denaturation at 95˚C for 15 min, followed by 40 cycles of 95˚C for 10 sec, 52˚C for 20 sec, 72˚C for 20 sec, and 82˚C for 15 sec; for detecting SL and r18S CT values, the reactions were carried out as previously reported [23]. Amplifications of each target gene and reference gene (18 S rRNA) cDNAs were performed in separate tubes in duplicate, and the results were analyzed using Q-Gene software (Bio Techniques Software Library at: www.BioTechniques.com). Dissociation-curve analysis was run after each real-time experiment to ensure that there was only one product. To control for false positives, a non-template negative control was run for each primer pair.

2.5. Analysis

The phylogenetic tree showing the relationship between SL amino acid sequences was generated by maximum Clustal W using DNA Star WI Megalign software. All sequences were obtained from the NCBI Gene bank according to the accession numbers. Data of the transcription of each gene are presented as mean ± SEM. The significance of the differences between group means of hormone mRNA levels was determined by one-way analysis of variance (ANOVA). In order to compare between non-mature females and the ovary of these fish containing mostly oocytes at their pre-vitellogenic stage, mature females that were kept in a group without males, and mature females, student's t-test data analysis was performed using the Statistical Package for the Social Sciences 17.0 software (Chicago: SPSS Inc.). The comparison differences were considered to be statistically significant at $p < 0.05$.

3. Results

In order to determine whether SL is expressed in brains and pituitaries, total RNA from these organs of gourami fish was reverse-transcribed. Reactions using cDNA derived from brain RNA and a set of primers specific to a segment of the gourami SL gene produced a fragment (**Figure 2**) of *T. trichopterus*. The partial length of cDNA sequence of somatolactin compared to other fish species is shown in **Figure 2**. A comparison between the amino acid sequences of SL sequences was made using the Clustal W method. The phylogenetic tree showing the relationship between somatolactin amino acid sequences compared to differences inteloses is presented in **Figure 3**.

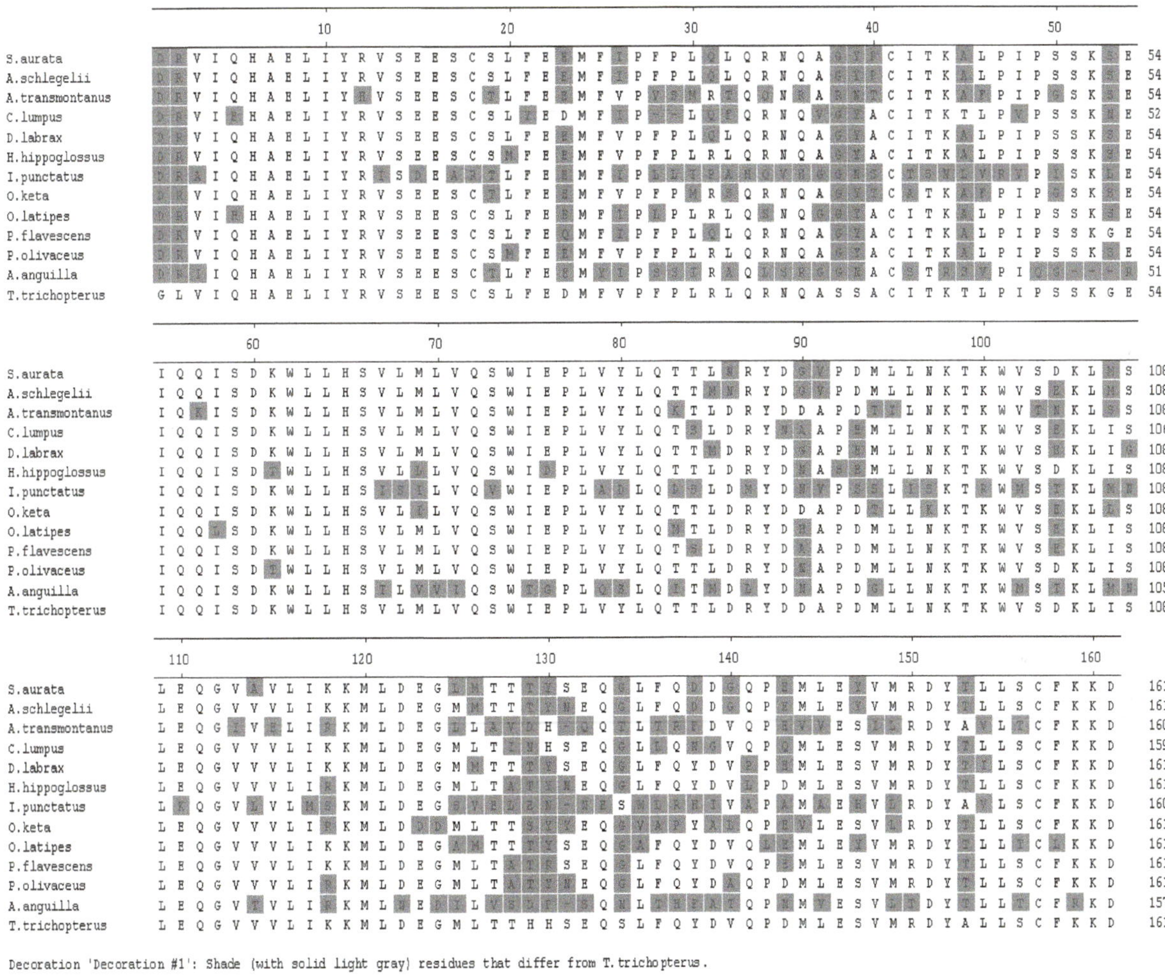

Figure 2. Comparison of cDNA sequences of *Trichogaster trichopterus* somatolactin to various fish species.

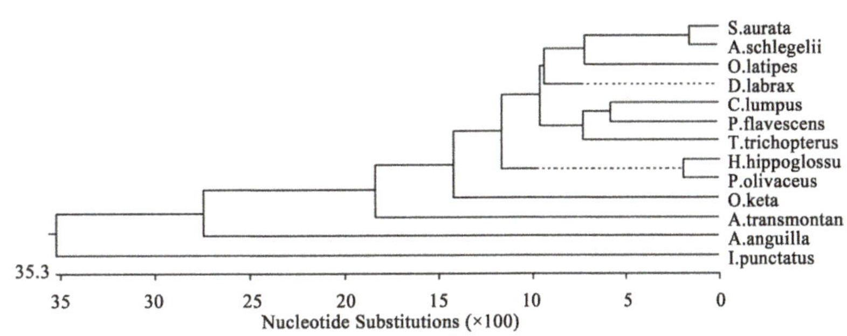

Figure 3. Phylogenetic tree showing the relationship between SL amino acid sequences. The tree was generated by maximum Clustal W using DNA Star WI Megalign software. All sequences were obtained from NCBI Gene bank according to the accession numbers provided in **Figure 2**.

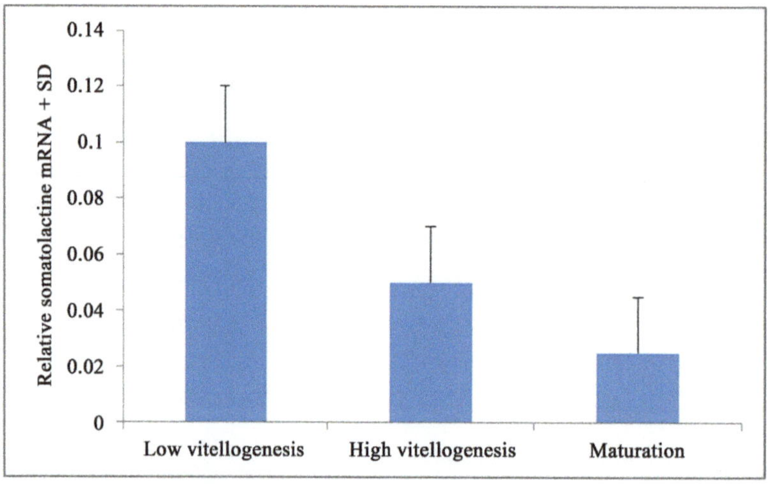

Figure 4. Relative mRNA levels of somatolactin in the pituitary of blue gourami females at different stages of oogenesis: low vitellogenesis, high vitellogenesis and maturation.

The *T. trichopterus* somatolactin sequences are very similar to *Cyclopterus lumpus* and *P. flavescens* (**Figure 3**). *T. trichopterus* somatolactin is evolutionally very distant from *Ictalurus punctatus* and *Anguilla andguilla*. In all the other fish species, the somatolactin sequence is relatively similar to *T. trichopterus*.

The mRNA of somatolactin during oogenesis is shown in **Figure 4**.

The transcription of somatolactin in the pituitary was higher in females at low vitellogenesis compared to females where oocytes were in maturation, and the difference was significant ($p < 0.05$) (**Figure 4**). However, no significant differences were found in mRNA levels between low and high vitellogenesis, and high vitellogenesis and maturation (**Figure 4**).

4. Discussion

SL has been found in all fish species studied and its structure is conserved. In the present study on partial SL of blue gourami (*T. trichopterus*) belonging to the Anabantidae family, it was compared to other fish species for the first time and the mRNA level was measured during oogenesis. The two SL forms are produced by different cells in the pituitary gland of fish [24]. Like GH and PRL, SL is thought to have multiple functions [15]. In the present study, it was found that SL might be involved in the vitellogenesis of female blue gourami and not in maturation.

There is apparently a small difference in mRNA level of PRL in blue gourami. The average level of PRL mRNA in juvenile and low vitellogenetic females was lower than in mature females (at high vitellogenesis and maturation), but the differences were not significant [7]. The GH in blue gourami seem to be involved in a similar way as are the SL and PRL hormones. The expression of GH and mRNA was measured during various stages of oogenesis: immature, low vitellogenesis, high vitellogenesis and final oocyte maturation. The highest levels of GH mRNA were found in immature females during high vitellogenesis and maturation. These results suggest that GH may play a role in the gonadal cycle of female blue gourami. High levels of GH mRNA found during the late stages of the gonadal cycle validate the concept that GH participates in reproduction [9]-[11].

Medaka (*Oryzias latipes*), a small freshwater fish, is a powerful model organism. This mutant exhibits a remarkably pale body color due to constitutively increased white pigment cells. Transcription of medaka SL in the brain (pituitary) is dramatically enhanced or suppressed when the fish is kept in a black or white tank. These results strongly suggest a conserved and major role of SLs in chromatophore regulation [25]. In blue gourami, it is well known that during sexual behavior, the oocytes reach maturation (**Figure 1**) and their color changes consider ably in males but less in females [1]. In this study, it was found that the SL mRNA level is relatively low in females during oocyte maturation.

5. Conclusion

In conclusion, the findings of this and previous studies demonstrate that SL, GH and PRL are involved in oogenesis in blue gourami; however, considerably more studies are required in order to separate the functions of these hormones.

References

[1] Degani, G. (2001) Blue Gourami (*Trichogaster trichopterus*) Model for Labyrinth Fish. Laser Pages Publishing, Israel.

[2] Degani, G. (1993) The Effect of Sexual Behaviour on Oocyte Development and Steroid Changes in *Trichogaster trichopterus*. *Copeia*, **4**, 1091-1096. http://dx.doi.org/10.2307/1447089

[3] Jackson, K., Degani, G. and Goldberg, D. (2004) Transcription of FSH-b, LH-b and Growth Hormone in Female *Trichogaster trichopterus* (Pallas 1770) Treated with GnRH Analog. *IV International Congress on the Biology of Fish*, Manaus, 45-57.

[4] Becker, D., Galili, M. and Degani, G. (1992) GCNS-Identified Steroids and Steroids Goucoronides in Gonads and Holding Waterin Gonads and Holding Water of *Trichogaster tricopterus* (Anabantidea, Pallas 1770). *Comparative Biochemistry and Physiology*, **103B**, 15-19.

[5] Degani, G. and Schreibman, M.P. (1993) Pheromone of Male Blue Gourami and Its Effect on Vitellogenesis and Gonadotropin Cells in Ptuitary of Females. *Journal of Fish Biology*, **43**, 475-485.

[6] Jackson, K., Goldberg, D., Ofir, M., Abraham, M. and Degani, G. (1999) Blue Gourami (*Trichogaster trichopterus*) Gonadotropic Beta Subunits (I and II) cDNA Sequences and Expression during Oogenesis. *Journal of Molecular Endocrinology*, **23**, 177-187. http://dx.doi.org/10.1677/jme.0.0230177

[7] Degani, G., Yom-Din, S., Goldberg, D. and Jackson, K. (2010) cDNA Cloning of Blue Gourami (*Trichogaster trichopterus*) Prolactin and Its Expression during the Gonadal Cycles of Males and Females. *Journal of Endocrinological Investigation*, **33**, 7-12. http://dx.doi.org/10.1007/BF03346543

[8] Degani, G. (2014) Expression of the Growth Hormone and Insulin-Like Growth Factor 1 Genes in the Male and Female Blue Gourami (*Trichogaster trichopterus*) at Different Temperatures. *Journal of Asian Scientific Research*, **4**, 413-427.

[9] Degani, G., Jackson, K. and Goldberg, G. (2000) Blue Gourami b GTH-I, b GTH-II and GH Gene Expression Affected by GnRH Analog. *Comparative Biochemistry and Physiology—Part A*, **126**, 42. http://dx.doi.org/10.1016/s1095-6433(00)80081-1

[10] Goldberg, D., Jackson, K., Yom-Din, S. and Degani, G. (2004) Growth Hormone of *Trichogaster trichopterus*: cDNA Cloning, Sequencing and Analysis of mRNA Expression during Oogenesis. *Journal of Aquaculture in the Tropics*, **19**, 215-229.

[11] Degani, G., Jackson, K., Yom-Din, S. and Goldberg, D. (2006) CDNA Cloning and mRNA Expression of Growth Hormone in Belontiidae (Anabantoidei Suborder). *The Israeli Journal of Aquaculture*, **58**, 124-136.

[12] Levy, G. and Degani, G. (2012) Involvement of GnRH, PACAP and PRP in the Reproduction of Blue Gourami Females (*Trichogaster trichopterus*). *Journal of Molecular Neuroscience*, **48**, 603-616. http://dx.doi.org/10.1007/s12031-012-9730-8

[13] Levy, G. and Degani, G. (2011) Evidence of a Reproduction-Related Function for Pituitary Adenylate Cyclase-Activating Polypeptide-Related Peptide in an Anabantidae Fish. *Journal of Molecular Endocrinology*, **46**, 101-110. http://dx.doi.org/10.1530/JME-10-0065

[14] Levy, G. and Degani, G. (2013) The Role of Brain Peptides in the Reproduction of Blue Gourami Males (*Trichogaster trichopterus*). *Journal of Experimental Zoology Part A*: *Ecological Genetics and Physiology*, **319**, 461-470.

[15] Perea, S.G. (2008) Growth Hormone and Somatolactin Function during Sexual Maturation of Female Atlantic Salmon. Department of Zoology/Zoophysiology University of Gothenburg, Sweden, 1-79.

[16] Degani, G. (2014) Involvement of GnRH and Gonadotropin Genes in Oocyte Development of Blue Gourami Females (*Trichogaster trichopterus*). *Advances in Biological Chemistry*, **4**, 197-202. http://dx.doi.org/10.4236/abc.2014.43024

[17] Jackson, K., Abraham, M. and Degani, G. (1994) Oocyte Maturation Triggered by the Presence of Male in the Blue Gourami, *Trichogaster trichopterus*. *Journal of Morphology*, **220**, 1-9. http://dx.doi.org/10.1002/jmor.1052200102

[18] Degani, G., Gal, E. and Vaya, J. (1994) *In Vitro* Biosynthesis of Steroids in Ovary of Asynchronic *Trichogaster trichopterus* (Pallus, 1770). *Comparative Biochemistry and Physiology*, **109**, 715-723.

[19] Jackson, K., Abraham, M. and Degani, G. (2005) Cellular Events in *Trichogaster trichopterus* Adenohypophysis during Final Oocyte Maturation. *Indian Journal of Fisheries*, **52**, 1-13.

[20] Degani, G. (2013) Underground Water Affects Sexual Behavior and Gene Expression of Hormones Related to Reproduction in Blue Gourami Males. *Advances in Biological Chemistry*, **3**, 133-140. http://dx.doi.org/10.4236/abc.2013.31016

[21] Pfaffl, M.W. (2001) A New Mathematical Model for Relative Quantification in Real-Time RT-PCR. *Nucleic Acids Research*, **29**, e45. http://dx.doi.org/10.1093/nar/29.9.e45

[22] Muller, P.Y., Janovjak, H., Miserez, A.R. and Dobbie, Z. (2002) Processing of Gene Expression Data Generated by Quantitative Real-Time RT-PCR. *Biotechniques*, **32**, 1372-1374, 1376, 1378-1379.

[23] Levy, G., Gothilf, Y. and Degani, G. (2009) Brain Gonadotropin Releasing Hormone3 Expression Variation during Oogenesis and Sexual Behavior and Its Effect on Pituitary Hormonal Expression in the Blue Gourami. *Comparative Biochemistry and Physiology Part A*: *Molecular & Integrative Physiology*, **154**, 241-248. http://dx.doi.org/10.1016/j.cbpa.2009.06.010

[24] Björnsson, B.T., Benedet, S., Taranger, G.L. and Andersson, E. (2008) Cloning of Somatolactin Alpha, Beta Forms and the Somatolactin Receptor in Atlantic Salmon: Aeasonal Expression Profile in Pituitary and Ovary of Maturing Female Broodstock. *Reproductive Biology and Endocrinology*, **6**, 42.

[25] Fukamachi, S., Yada, T. and Mitani, H. (2005) Medaka Receptors for Somatolactin and Growth Hormone: Phylogenetic Paradox among Fish Growth Hormone Receptors. *Genetics*, **171**, 1875-1883. http://dx.doi.org/10.1534/genetics.105.048819

14

Chemical Composition and Antimicrobial Activity of *Pituranthos chloranthus* (Benth.) Hook and *Pituranthos tortuosus* (Coss.) Maire Essential Oils from Southern Tunisia

Hedi Mighri*, Khawla Sabri, Hajer Eljeni, Mohamed Neffati, Ahmed Akrout

Range Ecology Laboratory, Arid Lands Institute, University of Gabès, Medenine, Tunisia
Email: *mighrih@yahoo.fr

Abstract

Essential oils (EO) from fresh and dry aerial parts of *Pituranthos chloranthus* (Benth.) Hook and *Pituranthos tortuosus* (Coss.) Maire were isolated by hydrodistillation and analyzed by GC/MS. The main constituents of the EO obtained from fresh herb of *P. chloranthus* were found to be α-pinene, sabinene, *cis*-ocimene and myrcene. In dry biomass, a significant increase of the content of some compounds such as α-phellandrene, Δ,3-carene and β-phellandrene characterized the oil. Minor changes in the chemical composition of the *P. tortuosus* EOs obtained from fresh or dry herbs and the major constituents were found to be sabinene and myrcene with equilibrate amounts of α-pinene, *p*-cymene, *cis*-ocimene, limonene, *trans*-β-ocimene, γ-terpinene and *cis*-verbenol. The paper disc diffusion method was used to evaluate the antibacterial activity and results showed an important inhibitory effect of oils obtained from fresh herb against most tested bacteria.

Keywords

***Pituranthos*, *Chloranthus*, *Tortuosus*, Essential Oil, South Tunisia**

1. Introduction

Pituranthos chloranthus (Benth.) Hook (Apiaceae) is an endemic and aromatic plant, locally named "Aljène" which grows naturally in North Africa and is widespread in central and southern Tunisia. Stems of this plant have been traditionally used as straw for farmers to dry figs and grapes. This plant has a double advantage: first,

*Corresponding author.

it has been used for its aroma and distinctive taste that adhere to the dry fruits; second, it has an insecticidal effect. In the southern Tunisian, the tuft of *P. chloranthus* was, traditionally, suspended to the surface of the water to disinfect the underground cisterns of the rain water storage used for the human drink. Furthermore, *Pituranthos* species are used in traditional medicine for the treatment of asthma, rheumatism, postpartum care, spasms, pains, fevers, diabetes, lice (head and pubis), hepatitis, digestive difficulties, urinary infections and scorpion's stings [1].

Previous studies have shown an interest changes in the chemical composition of EOs of *P. chloranthus*. Indeed, from distinct geographical areas in Tunisia and at various development stages of this species, the aerial part contains various oils types mainly composed of α-pinene, β-pinene, α-phellandrene, β-phellandrene, β-myrcene, *p*-cymene, 8-methyldecanal, exo-2-hydroxycineole acetate, carvacrol, geraniol and β-damascenone [2]. These oil types were found to exhibit antibacterial, antioxydant and anti-genotoxic activities [2] [3]. Yanghui *et al.* [1] demonstrated that *P. chloranthus* EO collected at Sfax (Tunisia) was mainly composed of terpinen-4-ol, 8-hydroxy-*p*-cymene, myrtenol, p-menth-2-en-1-ol and α-terpineol and exhibited antioxidant, antifungal and insecticidal activities suggesting its potential use as an alternative natural agent for disinfection.

Pituranthos tortuosus (Coss.) Maire (Apiaceae), locally named as "Guezzah", is a small shrub without leaves. This aromatic plant, which grows naturally in North Africa, is also, widespread in central and southern Tunisia. The plant is used by the Egyptians for the preparation of carminative drink and is occasionally eaten by grazing animals [4] [5]. It is also used for relief of stomach pains, against intestinal parasites, when blood is excreted in the urine or when coughing blood, and for the regulation of menstruation [4] [5]. In Tunisia, *P. tortuosus* is used traditionally as an anti-asthmatic and against scorpion's stings [5]. The composition and biological activities of *P. tortuosus* EO have been previously reported [4]-[8]. Abdewahed *et al.* [6] show that the chemical composition and the bacterial activity of EO isolated from fresh aerial parts collected in Monastir (center of Tunisia) depend on the period of collect: the November EO mainly composed of myrtenol is more effective than that of April mainly composed of terpinen-4-ol against the Gram-positive bacteria *Enterococcus faecalis* and *Staphylococcus aureus*. These oils also showmutagenic, antimutagenic, cytotoxic and apoptotic activities [7]. EO is isolated from dried aerial parts of this species collected at Sousse showed strong insecticidal, antifungal and allelopathic activities [5]. In Egypt, Hossam *et al.* [8] report that the composition and anticancer activity of EO isolated from aerial parts depend on the method of extraction: the simultaneous hydrodistillation-solvent (*n*-pentane) extraction method shows the most potent activity against the three human cancer cell lines tested (liver, colon and breast cancer cell lines). The EO from southern Sinai, contained 32 components with camphene (31.0%) as the major constituent, is ineffective against Gram-positive bacteria [4].

Never previously studies have been compared these two type of oils isolated from fresh and dried aerial parts of these endemic plants from southern region of Tunisia. The purpose of this present study is to compare the oil yields, chemical composition and the antibacterial activity of these oils which attract the interest of the local farmers and/or EO producers in arid zones of Southern Tunisia.

2. Materials and Methods

2.1. Plant Material

Aerial parts of *P. tortuosus* were collected from Beni-khedech (Medenine) and *P. chloranthus* from Douiret (Tatouine) just before the flowering stage. These two areas belong to Matmata's mountainous chain in southern Tunisia. The collected biomass were divided into two groups, one used as fresh herb and the other air-dried during 20 days in shade at ambient temperature for the EO extraction with a modified Clevenger-type apparatus for 4 h. The EOs were collected, dried by anhydrous sodium sulfate and stored at 4˚C in tight vials until analysis.

2.2. GC and GC-MS Identification

GC analysis was carried out using an Agilent 6890N Network GC system gas chromatograph fitted with flame ionization detector (FID) and an electronic integrator, using a HP-5 fused silica capillary column (30 m × 0.32 mm i.d., film thickness 0.25 mm). The oven temperature was programmed from 50˚C - 280˚C at 7˚C/min; injector temperature: 220˚C; detector temperature: 240˚C; carrier gas: nitrogen (1.0 mL/min); sample manually injected: 0.2 mL. Retention indices (RIs) were determined relative to the retention times of a series of n-alcanes (C_6 - C_{22}). The relative amount of components in the oil was calculated by electronic integration of FID peak areas and normalized without the use of response factor correction.

EOs constituents were also analyzed by GC-MS using the Agilent 6890N Network GC system combined with Agilent 5975 B Inert MSD detector (quadrupole) with electron impact ionization (70 eV). A HP-5-MS fused silica capillary column (30 m × 0.25 mm i.d., film thickness 0.25 mm) was used. The column temperature was programmed to rise from 50˚C - 280˚C at rate at 7˚C/min. The carrier gas was helium adjusted to a linear velocity of 34 cm/s. Scan time and mass range were 2.2 s and 50 - 550 *m/z*, respectively. Samples (0.1 mL) were injected with a split ratio of 1:100.

Identification of the components was based: 1) on comparison of their GC RIs on apolar column (HP-5) with those of literature data [6]-[9]. 2) by comparison of their recorded mass spectra with those of a computer library (Wiley 275 library and NIST98 database/Chem Station data system) provided by the instrument software and MS literature data [9] [10]; 3) identities of some other components were further confirmed by co-injection of pure standards available in the laboratory under the same GC/MS conditions as above.

2.3. Antibacterial Testing

The antibacterial activity of the different EOs was evaluated by the paper-disc agar diffusion method [11] against the *Streptococcus pyogenes* (ATCC 19615), *Staphylococcus aureus* (ATCC 25923), *Enterobacter aerogenes* (ATCC 13048), *Escherichia coli* (ATCC 25922) and *Klebsielle pneumoniae* (ATCC 3583). These clinical strains were obtained from Microbiology and Immunology Laboratory (EPS Habib Bourguiba, Medenine, Tunisia). Microorganisms were maintained on Muller-Hinton agar (MH) (BIORAD) medium. Inocula were prepared by diluting overnight (24 h at 37˚C) cultures in Muller Hinton Broth medium to approximately 106 CFU/mL. Absorbent discs (Whatman N˚3 discs, 6 mm diameter) were impregnated with 10 µL of oil and then placed on the surface of inoculated plates (90 mm). Positive control discs of gentamicin (10 µg/disc) were included in each assay. Diameters of growth inhibition zones were measured after incubation at 37˚C for 24 h.

3. Results and Discussion

The distilled EOs from fresh and dry biomass of *P. chloranthus* and *P. tortuosus* species collected in two localities from the Matmata's mountainous chain in southern Tunisia, were yellow. Air-drying the plant material in shade at ambient temperature (17˚C - 20˚C) not only resulted in a decrease of the *P. chloranthus* and *P. tortuosus* EO yields (1.6 and 2.0 from the fresh herb and 0.9 and 1.1 from the dry herb, respectively) but affected the density of these oils (0.8922 and 0.9216 of the oils obtained from the fresh herb and 0.7123 and 0.8931 from the dry herb, respectively). The refractive indexes of oils were similar ranging from 1.4987 to 1.4840 and 1.5000 to 1.4826, respectively (**Table 1**).

To determine the difference between analyzed oil samples, only sixteen major volatile compounds (amount > 1%) accounting for 93.3% to 97.2% of different oil types were retained as shown in **Table 1**. Qualitatively, the chemical composition remains stable between the two species belonging to the same genus regardless the biomass status. The main constituents of the EO obtained from fresh herb of *P. chloranthus* were found to be α-pinene (47.4%), sabinene (15.0%), *cis*-ocimene (6.6%) and myrcene (6.6%). In dry biomass, the amount of α-pinene decreases to 32.5% and *cis*-ocimene is present as trace but a significant increase of the content of some compounds such as α-phellandrene (3.8% to 7.8%), Δ,3-carene (trace to 5.7%) and β-phellandrene (3.4% to 13.9%) characterize this oil. Minor changes in the chemical composition of the *P. tortuosus* EO obtained from fresh or dry herbs and the major constituents were found to be sabinene (35.8% - 38.9%) and myrcene (8.9% - 12.5%) with equilibrate amounts of α-pinene, *p*-cymene, *cis*-ocimene, limonene, *trans*-β-ocimene, γ-terpinene and *cis*-verbenol (from 3.9% to 8.8%). The only compound that its content has been modified after drying the biomass is 3-n-butyl phthalide from 5.0% to 1.5%.

The influence of drying the aboveground biomass on the chemical composition has been previously reported by several authors [12]-[17]. A differential response of the aromatic species is attributed generally, to the loss of some compounds during the storage of the biomass after deteriorating oil glands and/or due to some physiological process that continue even after harvesting. The results of this work are in agreement with those obtained with others species confirming these qualitative and quantitative changes in the EO isolated from fresh and dried plant materials.

The composition of these EO isolated from our samples seems to be different from other regions in Tunisia and other countries. In fact, Yanghui *et al.* [1] demonstrated that the EO of *P. chloranthus* collected from Sfax (Tunisia) was mainly composed of terpinen-4-ol (30.3%), 8-hydroxy-*p*-cymene (4.2%), myrtenol (4.1%),

Table 1. Major compounds of the essential oils of *Pituranthos choloranthus* and *Pituranthos tortuosus.*

				Pituranthos chloranthus		*Pituranthos tortuosus*		
				Fresh herb	Dry herb	Fresh herb	Dry herb	
			Yield %	1.64	0.85	2.03	1.06	
Major			Refractive index	1.4987	1.4840	1.5000	1.4826	Identification
Compounds	Ri_{lit}	Ri	Density	0.8922	0.7123	0.9216	0.8931	Methods
α-pinene	937	942		47.4	32.5	6.2	5.1	A, B
Sabinene	976	964		15.0	12.6	38.9	35.8	A, B
β-pinene	980	989		2.0	1.7	1.8	2.1	A, B
Myrcene	986	993		6.6	4.6	8.9	12.5	A, B
α-phelandrene	1008	996		3.6	7.7	0.8	1.8	A
Δ-3-carene	1010	1005		tr.	5.7	2.9	1.9	A, B
p-cymene	1023	1011		7.5	6.4	4.9	8.8	A, B
Limonene	1031	1018		tr.	tr.	4.3	5.0	A, B
β-phellandrene	1032	1026		3.4	13.9	1.3	1.0	A
cis-β-ocimene	1037	1037		6.6	tr.	5.6	6.6	A, B
trans-β-ocimene	1045	1041		0.4	0.9	5.0	3.9	A, B
γ-terpinene	1058	1070		0.5	0.9	4.3	5.5	A, B
cis-verbenol	1140	1149		0.7	0.7	6.1	5.1	A
t-cadinol	1640	1625		1.5	1.8	tr.	0.7	A, B
β-eudesmol	1649	1667		tr.	1.4	tr.	tr.	A
3-n-butyl phthalide	1720	1723		2.0	2.7	5.0	1.5	A

Ri_{lit}: Retention indexes on HP-5 column according to literature data; *Ri*: Retention index on a HP-5 column; tr.: trace (less than 1%); A: GC-GC/MS; B: co-injection with authentic standard.

p-menth-2-en-1-ol (4.0%) and α-terpineol (3.5%). EOs isolated from *P. chloranthus* harvested at the vegetative, flower budding, flowering and fruiting stages from three different areas of southern Tunisia (Gabès, Médenine and Benguerdane) were found to be mainly composed by α-pinene, β-pinene, α-phellandrene, β-myrcene, β-phellandrene, *p*-cymene, 8-methyldecanal, exo-2-hydroxycineole acetate and carvacrol (more than 10% for each compound) [2]. However, this composition varied with respect to both the geographical area and the season: *p*-cymenene was only detected at the floral budding stage (February), whereas high amounts of exo-2-hydroxy-cineole and exo-2-hydroxycineole acetate were specific for the flowering period (April). Carvacrol was shown to be characteristic of the fruiting period (August), whereas the vegetative stage (November) could be distinguished by the presence of α/β-pinene, limonene, camphene, geraniol and β-damascenone [2]. This difference on composition between oils suggests that different chemotypes of *P. chloranthus* exist in Tunisia. *P. tortuosus* oil types extracted from the fresh herb collected in Monastir (central Tunisia) depends on the period of the collect of vegetal samples [6]: EO of samples collected during November was mainly composed of myrtenol (26.2%), sabinene (11.0%), limonene (10.9%), α-pinene (5.5%) and 3-*n*-butyl phthalide (5.9%) whereas the oil of samples harvested during April was mainly composed of terpinen-4-ol (39.6%), 3-*n*-butyl phthalide (11.4%), butylidene phthalide (4.1%), limonene (4.1%) and sabinene (3.2%). The yield and the composition of EOs isolated from aerial parts of *P. tortuosus* growing wild in Egypt vary with the method of extraction [8]: the major components of the oil prepared by hydrodistillation were β-myrcene (18.8%), sabinene (18.5%), *trans-iso*-elemicin (12.9%), and terpinen-4-ol (8.1%); those predominant in the oil extracted by the simultaneous hydrodistillation-solvent (*n*-pentane) extraction were terpinen-4-ol (29.7%), sabinene (7.4%), γ-terpinene (7.3%) and β-myrcene (5.5%); while the prominent ones in the oil isolated by the conventional volatile solvent extraction sample were terpinen-4-ol (15.4%), dillapiol (7.9%), and *allo*-ocimene (4*E*, 6*Z*) (6.0%). The volatile oil of *P. tortuous* from Southern Sinai was found to have a different composition from other regions in Egypt with camphene (31.0%) as the major constituent [4]. This difference on composition between oils suggests that different chemotypes of *P. tortuosus* exist in Tunisia and around the world.

Table 2. Antimicrobial activity of the investigated essential oils and the standard antibiotic (gentamicin) against five bacteria (Inhibition zone diameters: mm).

	Gentamicin	*Pituranthos chloranthus*		*Pituranthos tortuosus*	
		Fresh herb	Dry herb	Fresh herb	Dry herb
Streptococcus pyogenes	40	30	12	27	20
Staphylococcus aureus	30	35	14	15	15
Enterobacter aerogenes	20	10	0	30	9
Escherchia coli	20	9	0	8	10
Klebsielle pneumoniae	15	7	8	10	6

The antimicrobial activities of *P. chloranthus* and *P. tortuosus* EOs collected from southern Tunisia were evaluated by a paper disc diffusion method against some bacteria. As shown in **Table 2**, results revealed that oils obtained from fresh herb (especially *P. chloranthus* oils) exhibited higher antibacterial activity than those of dried herbs. Indeed, concerning *P. chloranthus* oils, all tested bacteria were found to be more sensitive against the oil isolated from fresh herbs than that extracted from dried herbs except *K. pneumoniae* which showed the same weak activity with both type of oils (from fresh and dried herbs). Some strains, such as *E. aerogenes* and *E. coli*, which were resistant against dried herbs oil, showed a weak activity against fresh herbs oil. The highest activity has been observed for *S. pyogenes* and *S. aureus* with fresh herb oil (30 and 35 mm respectively) while dried herbs oil revealed a weak activity against these two strains (12 and 14 mm). These results demonstrated that the EO isolated from fresh *P. chloranthus* exhibited a potent antibacterial activity against *S. pyogenes* and *S. aureus* bacteria which was similar as 10 μg of gentamicin (standard antibiotic). This important activity could be attributed to the high amount of α-pinene (47.4%), sabinene (15.0%), myrcene (6.6%) and *cis*-β-ocimene (6.6%) known to have exhibit a potent activity against these stains. Minor components such as γ-terpinene and *cis*-verbenol could also contribute to this activity in the synergism with major components [18]-[21].

Concerning *P. tortuosus* EOs, except *E. aerogenes* which exhibited a significant difference on sensitivity against EO isolated from fresh and dried herbs (30 and 9 mm respectively), other strains were not affected by the type of the oil. *P. tortuosus* oils exhibited a weak activity against *E. coli* and *K. pneumoniae*, moderate activity against S. aureus and important activity against *S. pyogenes* and *E. aerogenes* (especially fresh herb oil for this last strain). These results showed that EO isolated from fresh *P. tortuosus* could be a potent antibacterial natural product, similar as 10 μg of gentamicin, against *E. aerogenes* bacteria. The important activity oft the fresh *P. tortuosus* oil against *S. pyogenes* and *E. aerogenes* could be attributed to its high content of sabinene (38.87%), *cis*-verbinol (6.09%) and 3-n-butyl phthalide (5.00%) (16 - 23). Some other components such as *cis*-β-ocimene, *trans*-β-ocimene, limonene and γ-terpinene could be also contribute to the antibacterial activity of this EO [18]-[21].

4. Conclusion

Our results have shown that air-drying aboveground biomass of *P. choloranthus* and *P. tortuosus* in shade at ambient temperature results in a decrease of the yield of EO but conversely, the drying time of the plant material has no effect on the qualitative oil composition. For both species, the oil isolated from fresh aerial parts could be revealed a potent antibacterial activity against *S. pyogenes*, *S. aureus* and *E. aerogenes* and have potential to be used as natural antibacterial agents for these types of bacteria. For this purpose, further studies should be undertaken on these EOs, especially those extracted from dried herbs, which could exhibited other biological activities due to their chemical composition which is different from fresh herbs.

References

[1] Yangui, T., Bouaziz, M., Dhouib, A. and Sayadi, S. (2009) Potential use of Tunisian *Pituranthos chloranthus* Essential Oils as a Natural Disinfectant. *Letters in Applied Microbiology*, **48**, 112-117. http://dx.doi.org/10.1111/j.1472-765X.2008.02499.x

[2] Neffati, A., Hennequin, D., Basset, B., Chekir-Ghedira, L., Ghedira, K., Barillier, D. and Ledauphin, J. (2009) Influence of Growth phase And Geographic Origin on the Essential Oil Composition of *Pituranthos chloranthus* from Tu-

nisia. *Natural Product Communications*, **4**, 1585-1594.

[3] Neffati, A., Bouhlel, I., Ben Sghaier, M., Boubaker, J., Limem, I., Kilani, S., Skandrani, I, Bhouri, W., Le Dauphin, J., Barillier, D., Mosrati, R., Chekir-Ghedira, L. and Ghedira, K. (2009b) Antigenotoxic and antioxidant activities of *Pituranthos chloranthus* essential oils. *Environmental Toxicology and Pharmacology*, **27**, 187-194. http://dx.doi.org/10.1016/j.etap.2008.10.010

[4] Al-Gaby, A.M. and Allam, R.F. (2009) Chemical Analysis, Antimicrobial Activity of the Essential Oil from Some Wild Herbs in Egypt. *Journal of Herbs Spices and Medicinal Plants*, **7**, 15-23. http://dx.doi.org/10.1300/J044v07n01_03

[5] Krifa, M., Gharada, T. and Haoualab, R. (2011) Biological Activities of Essential Oil, Aqueous and Organic Extracts of *Pituranthos tortuosus* (Coss.) Maire. *Scientia Horticulturae*, **128**, 61-67. http://dx.doi.org/10.1016/j.scienta.2010.12.016

[6] Abdelwahed, A., Hayder, N., Kilani, S., Mahmoud, A., Chibani, J., Hammami, M., Chekir-Ghedira, L. and Ghedira, K. (2006) Chemical Composition and Antimicrobial Activity of Essential Oils from Tunisian *Pituranthos tortuosus* (Coss.) Maire. *Flavour and Fragrence Journal*, **21**, 129-133. http://dx.doi.org/10.1002/ffj.1542

[7] Abdelwahed, A., Skandrani, I., Kilani, S., Neffati, A., Ben Sghaier, M., Bouhlel, I., Boubaker, J., Ben Ammar, R., Mahmoud, A., Ghedira, K. and Chekir-Ghedira, L. (2008) Mutagenic, Antimutagenic, Cytotoxic, and Apoptotic Activities of Extracts from *Pituranthos tortuosus. Drug and Chemical Toxicology*, **31**, 37-60. http://dx.doi.org/10.1080/01480540701688634

[8] Hossam, M.A. and Shahira, M.E. (2011) Effect of the Method of Preparation on the Composition and Cytotoxic Activity of the Essential Oil of *Pituranthos tortuosus. Zeitschrift fur Naturforschung*, **66**,143-148. http://dx.doi.org/10.5560/ZNC.2011.66c0143

[9] Joulain, D., Knig, W.A. and Hochmuth, D.H. (2001) Terpenoids and Related Constituents of Essential Oils. Library of Mass Finder, Hamburg.

[10] Adams, R.P. (2001) Identification of Essential Oil Components by Gas Chromatography/Mass Spectroscopy. Allured Publishing Corporation, Carol Stream.

[11] Yadegarinia, D., Gachkar, L., Rezaei, M.B., Taghizadeh, M., Astaneh, S.A. and Rasooli, I. (2006) Biochemical Activities of Iranian *Mentha piperita* L. and *Myrtus communis* L. Essential Oils. *Phytochemistry*, **67**, 1249-1255. http://dx.doi.org/10.1016/j.phytochem.2006.04.025

[12] Mejri, J., Abderrabba, M. and Mejri, M. (2010) Chemical Composition of the Essential Oil of *Ruta chalepensis* L: Influence of Drying, Hydro-Distillation Duration and Plant Parts. *Industrial Crops and Products*, **32**, 671-673. http://dx.doi.org/10.1016/j.indcrop.2010.05.002

[13] Ashafa, A.O.T., Grierson, D.S. and Afolayan, A.J. (2008) Effects of Drying Methods on the Chemical Composition of Essential Oil from *Felicia muricata* Leaves. *Asian Journal of Plant Sciences*, **7**, 603-606. http://dx.doi.org/10.3923/ajps.2008.603.606

[14] Venskutonis, P. (1997) Effect of Drying on the Volatile Constituents of Thym (*Thymus vulgaris* L.) and Sage (*Salvia officinalis* L.). *Food Chemistry*, **59**, 219-227. http://dx.doi.org/10.1016/S0308-8146(96)00242-7

[15] Sefidkon, F., Abbasi, K. and Khaniki, G.B. (2006) Influence of Drying and Extraction Methods on Yield and Chemical Composition of the Essential Oil of *Satureja hortensis. Food Chemistry*, **99**, 19-23. http://dx.doi.org/10.1016/j.foodchem.2005.07.026

[16] Okoh, O.O., Sadimenko, A.P., Asekun, O.T. and Afolayan, A.J. (2008) The Effects of Drying on the Chemical Components of Essential Oils of *Calendula officinalis* L. *African Journal of Biotechnology*, **7**, 1500-1502.

[17] Mighri, H., Akrout, A., Tomi, F., Casanova, J. and Neffati, M. (2009) Influence of Drying Time and Process on *Artemisia herba-alba* Asso Essential Oil Yield and Composition. *Journal of Essential Oil Bearing Plants*, **12**, 358-364. http://dx.doi.org/10.1080/0972060X.2009.10643731

[18] Elaissi, A., Rouis, Z., Mabrouk, S., Bel-Haj Salah, K., Aouni, M., Khouja, M.L., Farhat, F., Chemli, R. and Harzallah-Skhiri, F. (2012) Correlation between Chemical Composition and Antibacterial Activity of Essential Oils from Fifteen Eucalyptus Species Growing in the Korbous and Jbel Abderrahman Arboreta (North East Tunisia). *Molecules*, **17**, 3044-3057. http://dx.doi.org/10.3390/molecules17033044

[19] Dorman, H.J.D. and Deans, S.G. (2000) Antimicrobial Agents from Plants: Antibacterial Activity of Plant Volatile Oils. *Journal of Applied Microbiology*, **28**, 308-316. http://dx.doi.org/10.1046/j.1365-2672.2000.00969.x

[20] Lis-Balchin, M., Deans, S.G. and Eaglesham, E. (1998) Relationship between Bioactivity and Chemical Composition of Commercial Essential Oils. *Flavour and Fragrance Journal*, **13**, 98-104. http://dx.doi.org/10.1002/(SICI)1099-1026(199803/04)13:2<98::AID-FFJ705>3.0.CO;2-B

[21] Solórzano-Santos, F. and Miranda-Novales, M.G. (2012) Essential Oils from Aromatic Herbs as Antimicrobial Agents. *Current Opinion in Biotechnology*, **23**, 136-141. http://dx.doi.org/10.1016/j.copbio.2011.08.005

15

In Vitro Model Systems to Investigate Drug Resistance Mechanisms in Pancreatic Cancer Cells

Eric Romney[1], Vinay J. Nagaraj[2*]

[1]Arizona College of Osteopathic Medicine, Midwestern University, Glendale, Arizona
[2]Department of Biochemistry, Midwestern University, Glendale, Arizona
Email: eromney92@midwestern.edu, [*]nvinay@midwestern.edu

Abstract

With a 5-year survival rate of less than 6%, the diagnosis of pancreatic cancer is devastating news for any patient. Gemcitabine, the most commonly used chemotherapy drug, only improves survival by approximately 1.5 months. A major obstacle to the treatment of pancreatic cancer with gemcitabine is the development of drug resistance. To better understand the precise mechanisms by which patient tumor cells gain resistance to gemcitabine, a cell culture model system that more accurately reflects the development of drug resistance *in vivo* is required. In this study, cultured pancreatic adenocarcinoma BxPC-3 cells were subjected to two different treatment regimens. The first method—termed pulse method—involves periodically treating separate cultures of BxPC-3 cells with constant predetermined doses of gemcitabine. The second treatment regimen—termed incremental method—consists of treating BxPC-3 cells with increasing doses of gemcitabine from 10 to 100 nM. While all treated cells showed enhanced resistance to gemcitabine, low-dose pulse treatments were sufficient to produce highly drug-resistant cells as evidenced by higher IC_{50} measurements. Pulse treatments also resulted in slower growth rates and increased doubling time of the drug-resistant cells. Morphological changes indicate cellular abnormalities linked to likely epithelial-to-mesenchymal transition and drug resistance. Our preliminary results indicate that the pulse method may better simulate resistance observed in patients undergoing chemotherapy and may serve as a superior model to investigate drug-resistance. This model can also help with identification of appropriate markers that determine the presence of drug-resistant cells and help clinicians adjust treatment strategies to improve outcomes for patients suffering from pancreatic cancer.

Keywords

Pancreatic Cancer, Gemcitabine, Drug-Resistance, Pulse Method, Incremental Method

[*]Corresponding author.

1. Introduction

In the United States, pancreatic cancer is the 5th most deadly cancer with an estimated 40,560 deaths out of the 48,960 people that were diagnosed with this disease in 2015 [1]. In spite of the great amount of research on pancreatic cancer over the last few decades, there has only been a marginal increase in the 5-year survival rate from 2% for 1975-1977 to 6% for 2002-2008 [2]. The exceptionally poor prognosis for pancreatic cancer can be attributed to several factors. Among them, late diagnosis of the diseases seems to be an important factor [2] [3]. A study of patients from 2002-2008, shows that as many as 90% of the pancreatic cancers that could be staged had either regional or distant metastases at the time of diagnosis [2]. While early detection of pancreatic cancer is critical to improve patient outcomes [4], there is an urgent need to develop more effective chemotherapies [5]. In 1997 gemcitabine (GEM), a nucleoside analog, replaced fluorouracil (5-FU) as the standard of care for patients with metastatic pancreatic cancer and continues to be the chemotherapeutic of choice to date [5] [6]. Since that time, different chemotherapeutic combination strategies have been used clinically with marginal effects on survivability [5]. Most of these therapies have included gemcitabine as a chemotherapeutic backbone combined with another drug to enhance its effects [5] [7].

Development of acquired drug resistance is a significant factor for the poor prognosis of patients with pancreatic cancer during gemcitabine chemotherapy [8]. There are indications that the development of resistance to gemcitabine and other chemotherapeutics, in part, is also responsible for the development of pancreatic cancer metastasis by encouraging an epithelial-mesenchymal transition (EMT) in cancer cells [9]-[11]. This is important because it is the widespread metastases that are blamed for the rapid progression of the clinical disease leading to mortality. In addition, there have been several studies regarding pancreatic cancer stem cells and the possible role these cell subpopulations may play in the development of drug resistance [12]-[18]. Interestingly, as few as 100 pancreatic cancer stem cells appear to be enough to form a new tumor in a mouse model [17]. Based on this new understanding of pancreatic adenocarcinoma tumor biology, we sought to explore *in vitro* the mechanisms by which patient tumor cells gain resistance to gemcitabine during their chemotherapy. This knowledge is key to identifying novel biomarkers as well as developing therapies that can prevent resistance from ever occurring.

In vitro techniques to create drug resistant cancer cells in the lab vary considerably. Previous studies on gemcitabine resistance most commonly use an "incremental method" where cultured pancreatic cancer cells receive increasing doses of gemcitabine in a step-wise manner. Typically, time is allowed for cell recovery in between treatments, and gemcitabine doses provided to cells can reach up to 50 times the initial IC_{50} [19] [20]. A lesser used method that has not been previously used on pancreatic cancer is the "pulse method" [21]. In this method, cells are given a metronomic dose of the chemotherapeutic agent with time to recover in between treatments. Preliminary results from ovarian cancer cells indicate that this method is more similar to the standard dosing regimen used on patients undergoing treatment [21]. In this short report, we compared the incremental and pulse methods using cultured human pancreatic adenocarcinoma cell line BxPC-3 with the goal of better modelling the resistance process that occurs *in vivo* during gemcitabine treatment.

2. Results and Discussion

2.1. Development of Drug-Resistance in Response to Different Drug Treatment Methods

Currently, there are a wide variety of methods used by researchers to induce drug resistance in pancreatic cancer cells but a variation of the incremental method described by Shah *et al.* [18] [20] is the most frequently used. This method involves choosing an initial low dose and exposing the cells to that dose for a given length of time. After one or more treatments at that initial dose, the drug is increased by some factor and the cycle is repeated. This is in contrast to a lesser used method known as the "pulse method" [21]. The pulse method is designed to closely resemble the weekly intravenous pulse dosing regimen of gemcitabine administered to pancreatic cancer patients. Specifically, cells are periodically exposed to an unchanged dose without increasing its initial concentration. We compared the two different treatment protocols with the end goal of determining if there is a better clinical model to determine pharmacodynamic biomarkers of response and resistance to gemcitabine during pancreatic cancer therapy.

For this study, cultured pancreatic adenocarcinoma BxPC-3 cells were subjected to the two different treatment regimens. The first treatment regimen consists of treating BxPC-3 cells with increasing doses of gemcitabine from 10 to 100 nM (**Table 1**). The second method involves periodically treating separate cultures of BxPC-3

Table 1. Treatment schedule for the development of gemcitabine resistance in the pancreatic adenocarcinoma cell line BxPC-3. Gemcitabine treatments (indicated in nM) were administered when the cells reached 50% confluence in the tissue culture flask. After incubation of the cells with the drug for two days, the cells were allowed to recover in drug-free media until they reached 70% confluence. At this stage, cells were passaged and the next treatment was commenced at 50% confluency.

	Gemcitabine treatment schedule									
Treatment number	1	2	3	4	5	6	7	8	9	10
Control BxPC-3 (0 nM)	0	0	0	0	0	0	0	0	0	0
Incremental (10 - 100 nM)	10	20	30	40	50	60	70	80	90	100
Pulse (60 nM)	60	60	60	60	60	60	60	60	60	60
Pulse (30 nM)	30	30	30	30	30	30	30	30	30	30
Pulse (15 nM)	15	15	15	15	15	15	15	15	15	15

cells with constant predetermined doses of gemcitabine (**Table 1**). Using these different protocols we developed 4 different gemcitabine-resistant subcultures of BxPC-3 cells and compared their characteristics (**Figure 1**).

Our results indicate that as few as 10 treatments of gemcitabine are sufficient to create highly drug-resistant BxPC-3 subcultures (**Figure 1**). All of the subcultures showed a two-fold or higher increase in resistance to gemcitabine despite large differences in maximum treatment dose (15 nM to 100 nM) and treatment regimen. Gemcitabine IC_{50} measurement increased significantly from about 30 nM for the parental BxPC-3 cells to about 55 - 80 nM for the gemcitabine treatments. **Figure 1(A)** through **Figure 1(E)** show representative dose response curves of 3 independent experiments for gemcitabine which were used for calculating IC_{50} measurements.

It is very interesting to note that pulse treatment at the lowest dose (15 nM) leads to the greatest drug resistance (IC_{50} of ~80 nM). We hypothesize that exposure to the pulse treatment, especially at sub-lethal doses, may allow cancer cells to better activate molecular pathways for drug resistance. These may include drug-efflux mechanisms, drug degradation pathways, intracellular drug sequestration mechanisms, alterations to the drug target, activation of repair mechanisms to undo drug damage, etc. A molecular analysis of each of these mechanisms is necessary to understand their contribution to the development of drug resistance in the BxPC-3 subcultures.

2.2. Effects of Incremental and Pulse Treatments on the Growth of BxPC-3 Cells

We compared the growth rates of BxPC-3 subcultures derived by pulse and incremental methods to identify if there are any differences. Previous studies with ovarian cancer cells have found that drug-resistant subcultures have a decreased rate of proliferation [21]. We report similar findings with BxPC-3 pancreatic adenocarcinoma cells. The pulse-treated cells, in particular, showed a greater decrease in cell proliferation (**Figure 2(A)**). The incremental method yielded only a 19.4% increase in doubling time whereas the 15 nM pulse and 60 nM pulse showed an increase of 45.4% and 60.2% respectively (**Figure 2(B)**). We believe these results may show clinically relevant differences between the incremental and pulse methods. Since chemotherapeutic agents typically target fast-growing cells in the body, a reduction in the doubling time is likely to help the drug-resistant cells evade some of the effects of gemcitabine and contribute to drug-resistance.

2.3. Drug-Resistance-Induced Changes in the Morphology of Cells

Throughout the gemcitabine treatment protocols, cells were observed and imaged to document any morphological changes. During that time, we specifically looked for any morphological changes related to metastasis and signs of possible epithelial-mesenchymal transitions (EMT) [9] [10] [20] [21]. Gemcitabine-sensitive (parental) BxPC-3 cells grow in colonies with a typical epithelial cobblestone appearance (**Figure 3(A)**). During gemcitabine treatment, two new morphologies were observed in both the incremental and pulse dosing regimen. The first morphology was very large isolated cells after gemcitabine exposure (**Figure 3(B)**). These cells were more prominent in the pulse treatments, appeared to increase in number at later treatments, and persisted for at least 10 cell culture passages after the last treatment. The significance, if any, of these cells is unknown as these cells have not been previously described in the context of drug-resistance. Studies with ovarian cancer cells hypothesize that these enlarged cells may contain vesicles for the intracellular sequestration of the chemotherapeutic agent [21]. Since the drug resistant cells show a slower growth rate—another possibility is that the accumulation of cellular components in preparation for cell division, with a delay in the progression of the cell cycle, may also lead to large sized cells.

The second phenotype is the spindle-shaped cells with long neuron-like morphology (**Figure 3(C)**). This morphology seemed to appear after 3 treatments and were observed in all the different drug treatments. The proportion of these cells were higher when the cell cultures were at lower confluency and decreased significantly

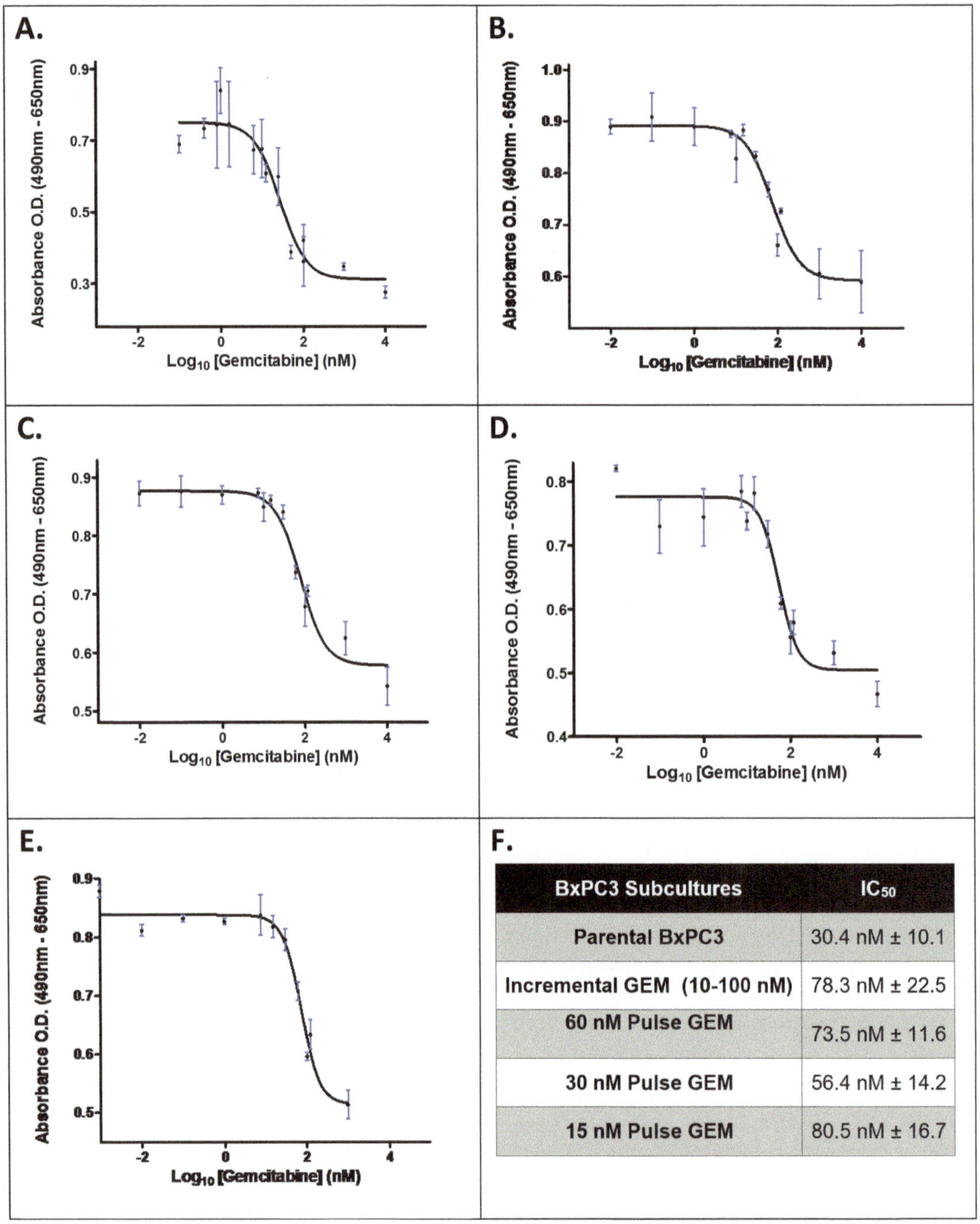

BxPC3 Subcultures	IC_{50}
Parental BxPC3	30.4 nM ± 10.1
Incremental GEM (10-100 nM)	78.3 nM ± 22.5
60 nM Pulse GEM	73.5 nM ± 11.6
30 nM Pulse GEM	56.4 nM ± 14.2
15 nM Pulse GEM	80.5 nM ± 16.7

Figure 1. (A) Shows the dose response curves of the parental (untreated) BxPC-3 cells to gemcitabine and that of the BxPC-3 cells exposed to gemcitabine by 10 - 100 nM incremental treatment (B), 15 nM pulse treatment (C), 30 nM pulse treatment (D) and 60 nM pulse treatment (E). The IC_{50} for each BxPC-3 subculture following treatments with gemcitabine is the shown in the table in (F).

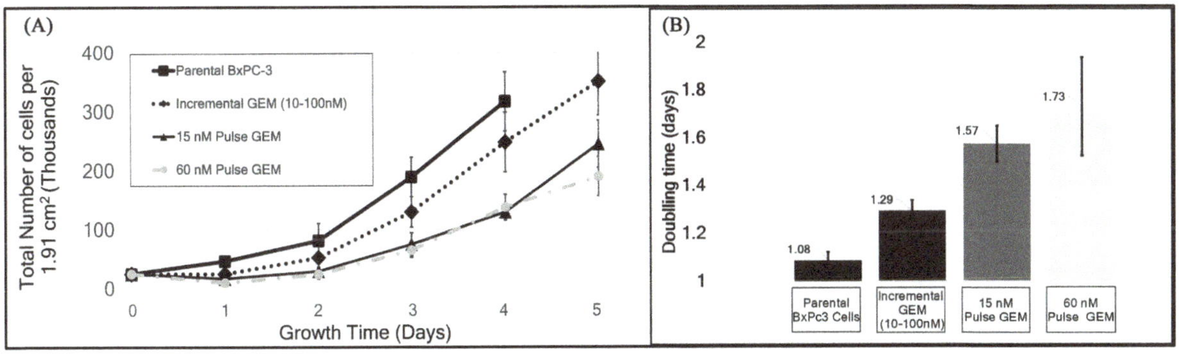

Figure 2. (A) Growth curves for the parental BxPC-3 cell line and gemcitabine-resistant BxPC-3 subcultures; (B) Doubling times for the parental BxPC-3 cell line and gemcitabine-resistant BxPC-3 subcultures.

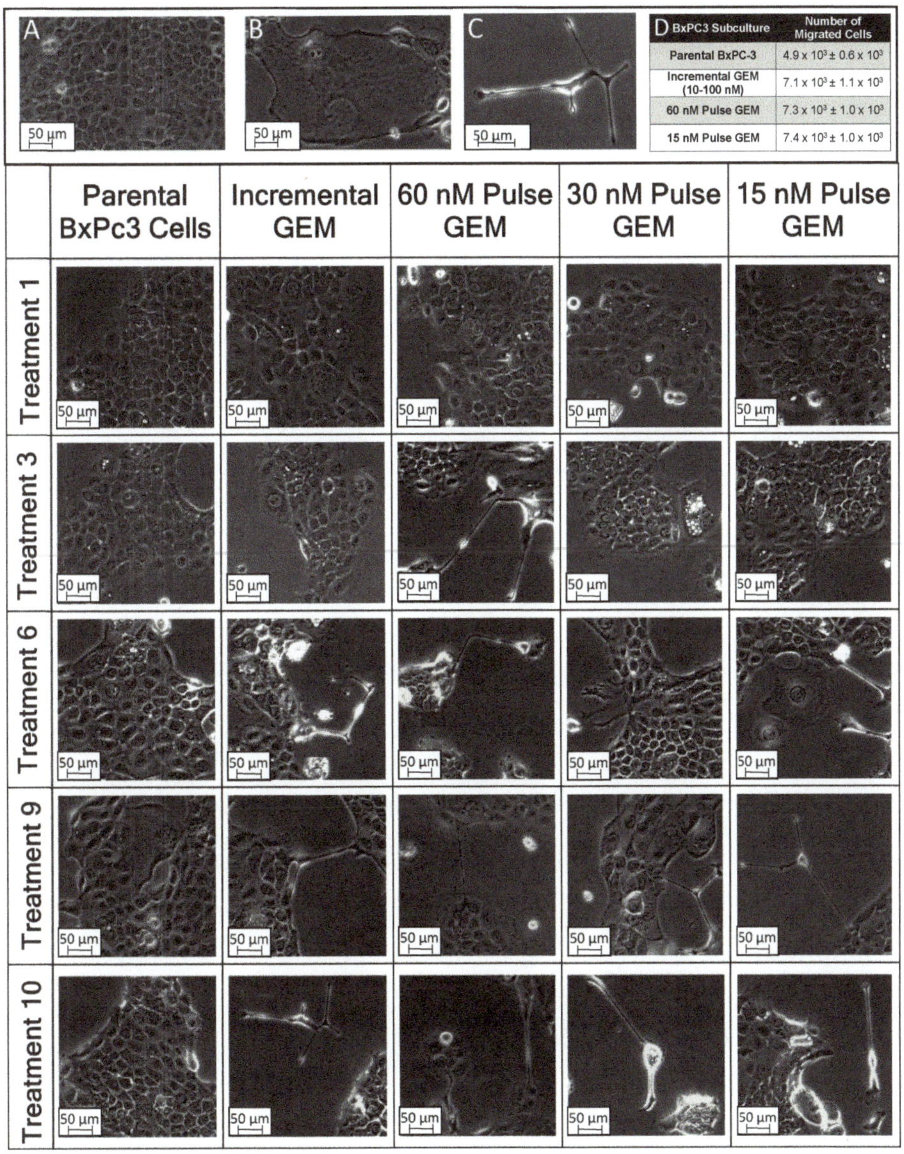

Figure 3. (A), (B), (C) and (E) Morphological changes observed in pancreatic cancer BxPC-3 cells during the development of resistance to gemcitabine; (D) Table shows the number of migratory cells in a Boyden chamber-based cell migration assay that was performed to assess the metastatic potential of BxPC-3 subcultures.

at higher confluences. We believe that these cells may be responsible for the increase in migration and metastasis observed *in vivo* in patients undergoing chemotherapy. Moreover, their transient nature fits with the current "seed and soil" model of metastasis which is said to require an epithelial-mesenchymal transition (EMT) to leave the dense stroma of the pancreas. However a reversal of this phenotype *i.e.* mesenchymal-epithelial transition (MET) is required to complete metastasis [8]-[11] [22]. In order to determine the metastatic potential of the drug-treated cells, a Boyden chamber-based migration assay was performed. All drug-resistant subcultures showed a significant increase in the number of cells able to migrate through the 8 μm pores when compared with the parental BxPC-3 cell line used as a control (**Table 1**, **Figure 3**). If we take into account the relatively slower growth rate of the resistant subcultures, actual migratory potential of the pulse-treated cells may be even higher than what has been experimentally determined. It appears that drug-resistance is strongly associated with enhanced metastatic potential of the cancer cells.

3. Conclusions

Pancreatic cancer is a deadly disease with a very high rate of mortality. Drug resistance is a major reason for the failure of chemotherapy and results in numerous deaths from pancreatic cancer. Previous research on the mechanisms of drug-resistance in cancer cells in the lab uses the incremental dose method to create drug-resistant cells. Although popular for studying resistance, these studies by design may not appropriately reflect the development of drug-resistance *in vivo*. In the incremental method, drug-resistant cells are selected by treatment with extremely high concentrations of the drug. A consequence to this approach may be the elimination of subpopulation of cells that may display other clinically relevant properties and phenotypes important for drug resistance, metastasis and occurrence of cancer stem cells.

Previous studies have indicated that chemoresistant cells are more metastatic [8] [19] [21]. Gemcitabine-resistant cells in the current study displayed strong migratory capabilities even with a low-dose pulse treatment. Future studies investigating changes in E- and N-Cadherins and other EMT/metastases markers will shed more light on this subpopulation of cells.

Our preliminary results revealed significant differences in the growth rates between the pulse and incrementally treated cells. This may suggest—in part—a greater presence of cancer stem cell population in the pulse treatment subcultures versus the traditional incremental method [12]. Cells triple positive for CD44, CD24 and epithelial cell adhesion molecule (ESA, CD326, EpCAM) will confirm the presence of pancreatic cancer stem cells [17]. Other markers which indicate pancreatic cancer stem cell-like properties that need to be examined include CD184 (CXCR4) and CD133, and the more recently described CD24, and CD338 (ABCG2) [18]. Drug-resistant cells derived from this study can also be useful to identify other novel determinants of pancreatic cancer stem cells.

In conclusion, we have shown that, the chemo-resistance process can occur *in vitro* with periodical exposure to a minimal amount of drug. In addition to drug-resistance, gemcitabine treatment may induce metastases while also enriching the tumor with cancer stem cells. Indeed, our observations seem to match the actual clinical course for patients treated with gemcitabine. Patients experience an initial decrease in tumor burden only to succumb months later to metastasis. Further research using the pulse method on cell lines may provide more clinically relevant information on these cellular/molecular processes and identify appropriate markers that determine drug-resistance. This knowledge will help clinicians adjust treatment strategies to improve outcomes for patients suffering from pancreatic cancer.

Acknowledgements

We thank Bernardo Chavira, Whitney Wilson, Zachary Emmons and Omar Jawhar for technical help. Eric Romeny was supported by the Kenneth A Suarez research fellowship.

References

[1] Siegel, R.L., Miller, K.D. and Jemal, A. (2015) Cancer Statistics, 2015. *CA: A Cancer Journal for Clinicians*, **65**, 5-29. http://dx.doi.org/10.3322/caac.21254

[2] Siegel, R., Naishadham, D. and Jemal, A. (2013) Cancer Statistics, 2013. *CA: A Cancer Journal for Clinicians*, **63**, 11-30. http://dx.doi.org/10.3322/caac.21166

[3] Jemal, A., Bray, F., Center, M.M., Ferlay, J., Ward, E. and Forman, D. (2011) Global Cancer Statistics. *CA*: *A Cancer Journal for Clinicians*, **61**, 69-90. http://dx.doi.org/10.3322/caac.20107

[4] Paulson, A.S., Tran Cao, H.S., Tempero, M.A. and Lowy, A.M. (2013) Therapeutic Advances in Pancreatic Cancer. *Gastroenterology*, **144**, 1316-1326. http://dx.doi.org/10.1053/j.gastro.2013.01.078

[5] Michl, P. and Gress, T.M. (2013) Current Concepts and Novel Targets in Advanced Pancreatic Cancer. *Gut*, **62**, 317-326. http://dx.doi.org/10.1136/gutjnl-2012-303588

[6] Burris 3rd, H.A., Moore, M.J., Andersen, J., *et al*. (1997) Improvements in Survival and Clinical Benefit with Gemcitabine as First-Line Therapy for Patients with Advanced Pancreas Cancer: A Randomized Trial. *Journal of Clinical Oncology*: *Official Journal of the American Society of Clinical Oncology*, **15**, 2403-2413.

[7] Kothari, N., Saif, M.W. and Kim, R. (2013) First-Line Treatment for Advanced Pancreatic Cancer. *JOP*: *Journal of the Pancreas*, **14**, 129-132.

[8] Long, J., Zhang, Y., Yu, X., *et al*. (2011) Overcoming Drug Resistance in Pancreatic Cancer. *Expert Opinion on Therapeutic Targets*, **15**, 817-828. http://dx.doi.org/10.1517/14728222.2011.566216

[9] Arumugam, T., Ramachandran, V., Fournier, K.F., *et al*. (2009) Epithelial to Mesenchymal Transition Contributes to Drug Resistance in Pancreatic Cancer. *Cancer Research*, **69**, 5820-5828. http://dx.doi.org/10.1158/0008-5472.CAN-08-2819

[10] Wu, Q., Miele, L., Sarkar, F.H. and Wang, Z. (2012) The Role of EMT in Pancreatic Cancer Progression. *Pancreatic Disorders & Therapy*, **2**. http://dx.doi.org/10.4172/2165-7092.1000e121

[11] Sarkar, F.H., Li, Y., Wang, Z. and Kong, D. (2009) Pancreatic Cancer Stem Cells and EMT in Drug Resistance and Metastasis. *Minerva Chirurgica*, **64**, 489-500.

[12] Hermann, P.C., Huber, S.L., Herrler, T., *et al*. (2007) Distinct Populations of Cancer Stem Cells Determine Tumor Growth and Metastatic Activity in Human Pancreatic Cancer. *Cell Stem Cell*, **1**, 313-323. http://dx.doi.org/10.1016/j.stem.2007.06.002

[13] Hage, C., Rausch, V., Giese, N., *et al*. (2013) The Novel C-Met Inhibitor Cabozantinib Overcomes Gemcitabine Resistance and Stem Cell Signaling in Pancreatic Cancer. *Cell Death & Disease*, **4**, e627. http://dx.doi.org/10.1038/cddis.2013.158

[14] Vaz, A.P., Ponnusamy, M.P., Rachagani, S., Dey, P., Ganti, A.K. and Batra, S.K. (2014) Novel Role of Pancreatic Differentiation 2 in Facilitating Self-Renewal and Drug Resistance of Pancreatic Cancer Stem Cells. *British Journal of Cancer*, **111**, 486-496.

[15] Hong, S.P., Wen, J., Bang, S., Park, S. and Song, S.Y. (2009) CD44-Positive Cells Are Responsible for Gemcitabine Resistance in Pancreatic Cancer Cells. *International Journal of Cancer*, **125**, 2323-2331.

[16] Ischenko, I., Seeliger, H., Jauch, K.W. and Bruns, C.J. (2009) Metastatic Activity and Chemotherapy Resistance in Human Pancreatic Cancer—Influence of Cancer Stem Cells. *Surgery*, **146**, 430-434. http://dx.doi.org/10.1016/j.surg.2009.05.003

[17] Li, C., Lee, C.J. and Simeone, D.M. (2009) Identification of Human Pancreatic Cancer Stem Cells. *Methods in Molecular Biology*, **568**, 161-173. http://dx.doi.org/10.1007/978-1-59745-280-9_10

[18] Niess, H., Camaj, P., Renner, A., *et al*. (2014) Side Population Cells of Pancreatic Cancer Show Characteristics of Cancer Stem Cells Responsible for Resistance and Metastasis. *Targeted Oncology*, **10**, 215-227.

[19] Hung, S.W., Mody, H.R. and Govindarajan, R. (2012) Overcoming Nucleoside Analog Chemoresistance of Pancreatic Cancer: A Therapeutic Challenge. *Cancer Letters*, **320**, 138-149. http://dx.doi.org/10.1016/j.canlet.2012.03.007

[20] Shah, A.N., Summy, J.M., Zhang, J., Park, S.I., Parikh, N.U. and Gallick, G.E. (2007) Development and Characterization of Gemcitabine-Resistant Pancreatic Tumor Cells. *Annals of Surgical Oncology*, **14**, 3629-3637. http://dx.doi.org/10.1245/s10434-007-9583-5

[21] Yan, X.D., Li, M., Yuan, Y., Mao, N. and Pan, L.Y. (2007) Biological Comparison of Ovarian Cancer Resistant Cell Lines to Cisplatin and Taxol by Two Different Administrations. *Oncology Reports*, **17**, 1163-1169. http://dx.doi.org/10.3892/or.17.5.1163

[22] Castellanos, J.A., Merchant, N.B. and Nagathihalli, N.S. (2013) Emerging Targets in Pancreatic Cancer: Epithelial-Mesenchymal Transition and Cancer Stem Cells. *OncoTargets and Therapy*, **6**, 1261-1267.

16

The Polymorphisms in the Vitamin D Receptor Gene and Disease Severity in Sickle Cell Disease

E. Leila Jerome Clay[1*], Alison Motsinger-Reif[2], Janelle Hoskins[3], Lindsay Veit[4], Ali Calikoglu[5], Rupa Redding-Lallinger[6]

[1]Departments of Pediatrics and Internal Medicine/Division of Hematology and Oncology, Georgia Regents University, Augusta, GA, USA
[2]Department of Statistics, Bioinformatics Research Center, North Carolina State University, Raleigh, NC, USA
[3]Gentris Corporation, Morrisville, NC, USA
[4]Orthopedic Surgery Resident, University of Toledo Medical Center, Toledo, OH, USA
[5]Department of Pediatrics/Division of Endocrinology, University of North Carolina School of Medicine, Chapel Hill, NC, USA
[6]Departments of Pediatrics and Internal Medicine/Division of Hematology and Oncology, University of North Carolina School of Medicine, Chapel Hill, NC, USA
Email: [*]leila.clay@gru.edu

Abstract

Vitamin D is important in multiple aspects of health and its effects are mediated through the Vitamin D Receptor (VDR). We wanted to test the hypothesis that specific haplotypes of the VDR gene are associated with markers of disease severity, inflammation and bone health in Sickle Cell Disease (SCD). Genotyping was performed on DNA specimens from 1141 study participants in the NIH-funded Silent Infarct Transfusion (SIT) trial. We used the clinical and laboratory data to create separate endothelial dysfunction, vaso-occlusive severity scores and phenotype variables. Seventy-nine Single Nucleotide Polymorphisms (SNP) in the VDR gene and three associated genes—CYP27B1, VD binding protein, retinoid X receptor, were evaluated. The discovery cohort individuals had VDR haplotype information from a prior Genome-Wide Association Study (GWAS). The validation cohort was analyzed for SNPs that were significant in the discovery cohort. The phenotype data were obtained from the demographic and clinical information of the participants, and were used to create the severity scores, vaso-occlusive score, endothelial dysfunction severity, and overall severity score. Potential gene-gene interactions were analyzed for prediction of disease severity within each severity score. Two SNPs were associated with the overall severity score, 3

[*]Corresponding author.

SNPs with the endothelial dysfunction severity score and 4 SNPs with the vaso-occlusive severity score. After permutation testing to correct for multiple comparisons, only one of the associations remained significant. SNP rs7965281 was found to be associated with the endothelial dysfunction severity score and remained significant after correcting for multiple comparisons using permutation testing. In the validation cohort, that SNP was again tested for association with each of the severity scores. There was no association with the endothelial or the overall severity score but a trend towards association with the vaso-occlusive severity score (p = 0.02). None of the known functional polymorphisms in the VDR gene were found to have an association with severity in sickle cell disease. Further work analyzing for gene-gene interaction using the same significant SNPs remains to be done in association with inflammatory markers and measure of bone health. Those studies may provide insight on the contribution of VDR polymorphisms to sickle cell disease severity.

Keywords

Vitamin D (VD), Sickle Cell Disease (SCD)

1. Introduction

Sickle Cell Disease (SCD) is a hereditary disorder primarily affecting people of African ancestry [1]. SCD is a single gene disorder and the complications of this disease affect every organ of the body including the vascular and immune systems along with bone metabolism. Vitamin D deficiency is commonly seen in patients with sickle cell disease especially those with SCD-SS [2] [3]. Vitamin D and calcium are required for optimal bone health with close to 90% of required vitamin D synthesis coming from exposure to sunlight. Dark-skinned individuals usually require 5 - 10 times more exposure to sunlight to produce the same amount of vitamin D3 in their skin [4]. As common as vitamin D deficiency is in SCD, its contribution to disease manifestations is not yet known. Vitamin D is important in multiple aspects of health, including cardiovascular, immune and skeletal systems. The active metabolite of the pro-hormone vitamin D exerts its effect through interaction with the Vitamin D Receptor (VDR). A Single Nucleotide Polymorphism (SNP) is DNA sequence variation occurring commonly within a population in which a single nucleotide in the genome differs between members of a biological species or paired chromosome. The VDR is expressed in multiple cell types and there are specific VDR polymorphisms associated with diseases such as diabetes, hyperparathyroidism, and renal disease (to name a few) [5]. Some of the polymorphisms such as bsmI, fokI, taqI have been found to be associated with bone health, inflammation and vascular disease, respectively [6] [7]. Although vitamin D deficiency has been studied in sickle cell disease to some extent, the polymorphisms in the VDR gene have not been explored in sickle cell disease or in Africans or African Americans [8]. Because of the presence of the VDR on cells in endothelial, inflammatory and bone systems, polymorphisms in this gene may be associated with disease manifestations in these organ systems in SCD. Our hypothesis is that specific VDR polymorphisms are associated with disease severity in sickle cell disease. Potentially some of the variation in the severity of sickle cell disease can be explained by differing polymorphisms in the VDR gene.

2. Materials and Methods

After approval of the study was obtained from the Biomedical Institutional Review Board of the University of North Carolina, genotyping was performed on DNA specimens from 1141 study participants in the NIH-funded Silent Infarct Transfusion (SIT) trial (WU-04-60/PO29892B). In this multi-center international trial, the participants were children ages 4 through 13 with SCD who were screened for the presence of silent cerebral infarction and had demographic and clinical data collected, as well as samples for a biologic repository which included DNA. We used the clinical and laboratory data to create separate endothelial dysfunction and vaso-occlusive severity scores for phenotype variables, following the current thinking on the subtypes of sickle cell disease expression [9]. The initial 570 participants served as our discovery cohort, used to detect potential associations from all the variants tested. The subsequently enrolled 530 individuals formed our validation cohort to try to replicate/validate statistically significant findings from the discovery cohort. We evaluated 79 Single Nucleotide Polymorphisms (SNPs) in the VDR gene, three associated genes: CYP27B1, VD binding protein, retinoid X receptor, and tagging SNPs from the African American population from Hap map (**Appendix 1**). The discovery cohort individuals had VDR haplotype information from a prior Genome-Wide Association Study (GWAS) study, and analysis for additional VDR-related SNPs was performed using a specifically designed Sequenom assay.

The validation cohort was analyzed for SNPs that were significant in the discovery cohort.

The phenotype data was obtained from the demographic and clinical information of the participants, and was used to create severity scores. The vaso-occlusive severity score includes: number of hospitalizations for pain, number of hospitalizations for acute chest syndrome and avascular necrosis. The endothelial dysfunction severity score includes priapism, Transient Ischemic Attacks (TIA), silent cerebral infarct, systolic and diastolic blood pressure, transcranial doppler velocity, white count and baseline hemoglobin. The overall severity score includes all of the endothelial dysfunction and vaso-occlusive severity variables (**Figures 1(a)-(c)**). To derive the scores, the variables were transformed into quartiles. Each individual subject was assigned values of 1, 2, 3, or 4 for each variable with 1 representing lowest severity and 4, the highest. In addition, in concert with prior analyses of the SIT data, the variable for number of hospitalizations for pain was used alone as a severity measure.

Before association analysis, genotypes underwent quality control filters. First, markers with missing data rates greater that 5% were removed. Second, individuals with missing data rates greater than 5% were removed. Next, markers were tested for deviation from Hardy-Weinberg proportions using a Fisher's exact test and a Bonferroni corrected alpha =0.05 for the number of SNPs tested and markers with significant deviations were removed prior to analysis. This QC resulted in no markers of individuals needing to be removed.

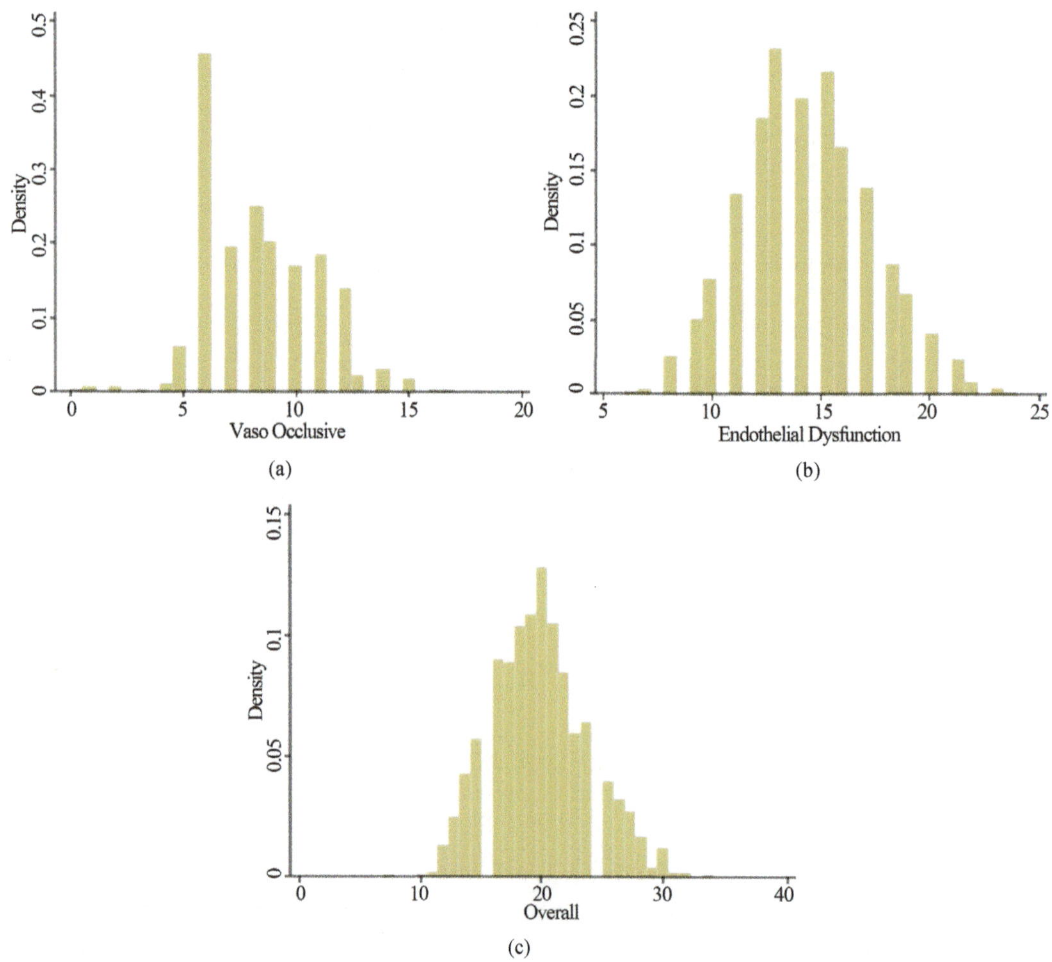

Figure 1. Histogram of severity score distributions in the discovery cohort. The relative density of the validation population for the (a) vaso-occlusive, (b) endothelial dysfunction and (c) overall severity score are shown.

After QC, genetic markers were tested for association with each of the severity scores using a two stage approach. Initially, univariate tests for association were performed. Because severity scores were not normally distributed and were not totally continuous distributions, the Kruskal-Wallis test was used in association analysis (treating each genotype as a category). To correct for multiple comparisons, permutation testing was performed. One thousand permuted datasets (permuting the risk score outcomes while preserving the genotype matrix) were created and the Kruskal-Wallis tests were performed for each of the variants. The lowest p-value from the analysis of each permuted dataset was used to build an empirical distribution of p-values expected by chance, and the p-values for the original analyses were determined by their hypothetical percentile rank in the empirical distribution. Such a permutation approach is a well validated approach for multiple testing correction that is less conservative than a Bonferroni correction when there is correlation between predictor variables [10] [11].

To look for complex genetic models, including potential gene-gene interactions for prediction of disease severity, the Generalized Multifactor Dimensionality Reduction (GMDR) method was utilized, with repeat analyses performed for each severity score. GMDR introduces the concept of a score statistic into the Multifactor Dimensionality Reduction (MDR) framework to obtain an appropriate statistic to classify multifactor contingency table cells into two different groups. The score value is based on a generalized linear model of the phenotype variable, and allows for the use of covariates in the score approximation. Details of GMDR have been previously described in detail [12].

In step one, data was divided into a training set and an independent testing set for cross validation. Five-fold cross-validation was used, with 4/5 of the data used for training and 1/5 for testing. A set of n genetic factors were then selected. These factors and their multiple classes were divided in n-dimensional space. In the traditional MDR approach, the ratios of cases to controls were then calculated within each multifactor class. Each multifactor cell class was labeled "high risk" if the ratio exceeded 1.0, or "low risk" if less than 1.0, thus reducing n-dimensional space to one dimension with two levels. In GMDR, the ratio of cases to controls in each cell is replaced by the score values. The null hypothesis assumes there are no effects of the putative factors or their interactions, so the score values are the same for all different factor classifications. In the third step of GMDR, the cumulative score value was calculated within each multifactor cell and in the fourth step, each multifactor cell was labeled either as high-risk if the average score meets or exceeded a pre-assigned threshold T (e.g., 0), or as low-risk if the threshold was not exceeded. The collection of these multifactor classes comprised the GMDR model. The result is a set of models, one for each model size considered. The final model was chosen that minimizes prediction error while maximizing Cross Validation Consistency (CVC). The statistical significance of the final best model was determined through permutation testing, which involved creating 1000 permuted datasets by randomizing the value of the phenotype variable. The entire procedure was repeated for each, generating a distribution of 1000 prediction errors that could be expected by chance alone. The significance of the final model was determined by comparing the prediction error of the final model to the distribution. A p-value was extracted for the model by its theoretical location in the permutation distribution.

In the present study, GMDR analysis was performed with the same outcome variables listed above as the phenotypes, using stratification when necessary and all other genetic markers as potential predictor variables. Analysis was performed with 5-fold cross-validation, and single-variable through four-variable interactions were evaluated. The permutation testing was used to empirically define the prediction accuracies that are significant with a family-wise type I error rate of 0.05.

The study was designed to have at least 80% power to detect odds ratios of 2.15 for the univariant association of variants with minor allele frequencies of at least 0.25; 58% of our candidate genes fell within this range. These power calculations were performed assuming an F distribution with a Bonferroni corrected alpha of 0.05. SNPs that were significantly associated with severity in the discovery cohort were tested for association in the validation cohort. Each SNP was directly tested for association using a Kruskal-Wallis test after QC checks. Since this replication analysis represents a single statistical hypothesis, no correction for multiple testing was performed.

3. Results

By univariate testing in the discovery cohort, 2 SNPs (rs1491710, rs11574114) were nominally associated with the overall severity score, 3 SNPs (rs11829917, rs2853563, rs11574138) with the endothelial dysfunction severity score and 4 (rs7855881, rs12348547, rs11574114, rs7965281) with the vaso-occlusive severity score. One of the SNPs, rs11574114, was associated with both the overall and the vaso-occlusive severity scores.

Our severity scores, although based on current thinking about sickle cell disease, have not been validated in other work. The measure that has been validated as a measure severity in the SIT population (the source of the data for this study) is number of hospitalization for pain. When tested by univariate analysis, there was a nominal association with 2 of the SNIPS (rs7855881, rs34312136) with this variable. However, after permutation testing to correct for multiple comparisons, none of the associations remained significant.

Using MDR to test for significant gene-gene interactions in the discovery cohort, one single locus association between SNP rs7965281 and the endothelial dysfunction score was significant, and remained significant after permutation testing to correct for multiple comparisons. The mean endothelial dysfunction severity score for those homozygous for A and for G (the variant allele) was 15, and it is 13 for the heterozygotes (AG). The range for the endothelial dysfunction score is 7-23, with higher numbers corresponding to greater severity.

In the validation cohort, there was no univariate association with the one significant SNP from the discovery cohort, rs7965281. However, that SNP was significantly associated with the vaso-occlusive severity score in the validation cohort by MDR testing (p value = 0.02).

4. Discussion

Overall there does not appear to be a significant association between any of the currently described functional VDR genes, such as bsml, fokI, taqI and severity in SCD. With the current study's sample size and power, only 58% of our candidate genes had a minor allele frequency sufficient for testing for an association with disease severity. The association between the VDR and the SCD severity could be further explored in a larger SCD population. In addition, a sickle cell disease population for which there are data on bone health, disease and inflammatory markers should be studied in relation to VDR haplotypes.

Concerning the single SNP that was found to be significantly associated with SCD severity, not much is known currently. In the literature, SNP rs7965281 is a tagging SNP whose significance is not fully clear. It is located in the regulatory region UTR + 3713: Chr position 45262658 as per dbSNP (HuRef). It was found to be associated with reduced risk for cutaneous melanoma in a large population [13]. In a marker report, a trend analysis was done with this SNP and an association with systolic blood pressure was found in a British population, with an adjusted p value of 0.029. More work is being currently done on this particular SNP (personal communication with Dr. Orlow [13]).

While our study did not show a relationship between the well-described VDR gene polymorphisms and the SCD severity, one SNP was found which may be of importance. In this study with a two-stage approach to association analysis, the SNP found was associated with disease severity in both the discovery and the validation cohorts, but it was not associated with the same severity scores in the two cohorts, making the results somewhat less precise that one would like. However, in the current conception of the clinical expression of sickle cell disease, the vaso-occlusive and the endothelial dysfunction disease manifestations do overlap, so the association of SNP rs7965281 with both of the scores does not invalidate its significance. Clearly, an understanding of the biologic function of SNP rs7965281 is necessary if the results of the current study are to be built upon, and these results should be evaluated in other studies of the VDR and sickle cell disease severity to confirm this association.

Acknowledgements

This project is an ancillary study of the SIT trial (NIH WU-04-60/PO29892B).

This manuscript was prepared during the corresponding author's training and was supported by the T32 NIH grant *PHS GRANT 5T32 HL* 7149-35.

This project was supported by a grant from North Carolina Translational and Clinical Sciences Institute.

We thank the staff, clinicians and patients for their participation in the Silent Infarct Transfusion (SIT) Trial study. We would like to thank the members of the SIT Executive Committee for their comments and suggestions during the planning and conduct of this study. The SIT study (PI: M. DeBaun, MD) is supported by the National Institute of Neurological Disorders and Stroke (NINDS) (NIH-NINDS 5U01-NS042804). We also thank the SIT Trial Biologic Repository (J. F. Casella, MD, E. Barron-Casella, PhD and K. Jones) for providing plating f genomic DNA. Genotyping services were provided by the Center for Inherited Disease Research (CIDR) and the Centers for Disease Control and Prevention (CDC). CIDR is fully funded through a federal contract from the

National Institutes of Health to The Johns Hopkins University (PI: J.F. Casella, MD) (CIDR contract #: HHSN268200782096C). Additional assistance with the GWAS data was provided by Drs. D.E. Arking and P. Bhatnagar (JHU) and W.C. Hooper and C.J. Bean (CDC).

We also would like to acknowledge David Barrow, DDS, for his contribution in the early stages and planning of this project.

References

[1] Fixler, J. and Styles, L. (2002) Sickle Cell Disease. *Pediatric Clinics of North America*, **49**, 1193-210, vi. http://dx.doi.org/10.1016/S0031-3955(02)00089-5

[2] Buison, A.M., Kawchak, D.A., Schall, J., Ohene-Frempong, K., *et al.* (2004) Low Vitamin D Status in Children with Sickle Cell Disease; Osteopenia and Vitamin D Deficiency in Children with Sickle Cell Disease. *Journal of Pediatrics*, **145**, 622-627. http://dx.doi.org/10.1016/j.jpeds.2004.06.055

[3] Reed, J.D., Redding-Lallinger, R. and Orringer, E.P. (1987) Nutrition and Sickle Cell Disease. *American Journal of Hematology*, **24**, 441-455. http://dx.doi.org/10.1002/ajh.2830240416

[4] Adewoye, A.H., Chen, T.C., Ma, Q., McMahon, L., Mathieu, J., Malabanan, A., *et al.* (2008) Sickle Cell Bone Disease: Response to Vitamin D and Calcium. *American Journal of Hematology*, **83**, 271-274. http://dx.doi.org/10.1002/ajh.21085

[5] Valdivielso, J.M. and Fernandez, E. (2006) Vitamin D Receptor Polymorphisms and Diseases. *Clinica Chimica Acta*, **371**, 1-12. http://dx.doi.org/10.1016/j.cca.2006.02.016

[6] Bouillon, R., Norman, A.W. and Lips, P. (2007) Vitamin D Deficiency. *New England Journal of Medicine*, **357**, 1980-1981. http://dx.doi.org/10.1056/NEJMc072359

[7] Rovner, A.J., Stallings, V.A., Kawchak, D.A., Schall, J.I., Ohene-Frempong, K. and Zemel, B.S. (2008) High Risk of Vitamin D Deficiency in Children with Sickle Cell Disease. *Journal of the American Dietetic Association*, **108**, 1512-1516. http://dx.doi.org/10.1016/j.jada.2008.06.433

[8] Uitterlinden, A.G., Fang, Y., Van Meurs, J.B., Pols, H.A. and Van Leeuwen, J.P. (2004) Genetics and Biology of Vitamin D Receptor Polymorphisms. *Gene*, **338**, 143-156. http://dx.doi.org/10.1016/j.gene.2004.05.014

[9] Kato, G.J., Gladwin, M.T. and Steinberg, M.H. (2007) Deconstruction Sickle Cell Disease: Reappraisal of the Role of Hemolysis in the Development of Clinical Subphenotypes. *Blood Reviews*, **21**, 37-47. http://dx.doi.org/10.1016/j.blre.2006.07.001

[10] Lou, X.Y., Chen, G.B., Yan, L., Ma, J.Z., Zhu, J., Elston, R.C., *et al.* (2007) A Generalized Combinatorial Approach for Detecting Gene-by-Gene and Gene-by-Environment Interactions with Application to Nicotine Dependence. *American Journal of Human Genetics*, **80**, 1125-1137. http://dx.doi.org/10.1086/518312

[11] Good, P. (2001) Extensions of the Concept of Exchangeability and Their Applications. *Journal of Modern Applied Statistical Methods*, **1**, 243-247.

[12] Lou, X.Y., Chen, G.B., Yan, L., Ma, J.Z., Zhu, J., Elston, R.C. and Li, M.D. (2007) A Generalized Combinatorial Approach for Detecting Gene by Gene and Gene by environment Interactions with Application to Nicotine Dependence. *American Journal of Human Genetics*, **80**, 1125-1137. http://dx.doi.org/10.1086/518312

[13] Orlow, I., Roy, P., Reiner, A.S., Yoo, S., Patel, H., Paine, S., *et al.* (2011) Vitamin D Receptor Polymorphisms in Patients with Cutaneous Melanoma. *International Journal of Cancer*, **130**, 405-418. http://dx.doi.org/10.1002/ijc.26023

Appendix 1

Original SNPs	Gene	Chromosome	Position	Reference allele	Other allele	African ancestry in South-west USA (HapMap III) other allele frequency	Yoruba in Ibadan, Nigeria (HapMap III) other allele frequency	Source	Genotyped SNPs	r^2	Genotype platform	Comment/ Alt. snp name
rs1048691	CYP27B1	12	58152948	C	T	0.33	0.394	Haploview Tagger	rs1048691		650Y	
rs10488854	GC	4	72624135	C	T	0.198	0.155	Haploview Tagger	rs10488854		Sequenom	
rs10735810 /rs2228570	VDR	12	48272895	A	G	0.802	0.808	PharmGKB	rs2228570		Sequenom	FokI
rs10877012	CYP27B1	12	58162085	G	T	0.16	0.088	Haploview Tagger	rs10877012		Sequenom	
rs10881578	RXRA	9	137232535	A	G	0.377	0.394	Haploview Tagger	rs10881578		Sequenom	
rs10881582	RXRA	9	137256078	G	A	0.623	0.73	Haploview Tagger	rs10881582		650Y	
rs11168266	VDR	12	48251533	C	T	0.528	0.673	Haploview Tagger	rs11168266		Sequenom	
rs11168268	VDR	12	48251812	G	A	0.566	0.708	Haploview Tagger	rs11168268		Sequenom	
rs11172327	CYP27B1	12	58161660	G	A	0.066	0.071	Haploview Tagger	rs8176353		Sequenom	
rs11185649	RXRA	9	137228370	T	C	0.585	0.643	Haploview Tagger	rs11185649		Sequenom	
rs11185659	RXRA	9	137243383	C	T	0.283	0.319	Haploview Tagger	rs11185659		650Y	
rs1155563	GC	4	72643488	T	C	0.057	0.049	Haploview Tagger	rs1155563		650Y	Ahn *et al.* 2010
rs11568820	VDR	12	48302545	C	T	0.717	0.978	PharmGKB	rs11568820		Sequenom	Cdx2
rs11574114	VDR	12	48238883	C	T	0.151	0.226	Haploview Tagger	rs11574114		Sequenom	
rs11574138	VDR	12	48235702	T	C	0.066	0.124	Haploview Tagger	rs11574138		Sequenom	
rs11574143	VDR	12	48234917	C	T	0.132	0.08	Haploview Tagger	rs11574143		650Y	
rs11829917	CYP27B1	12	58164455	G	A	0.123	0.124	Haploview Tagger	rs11829917		Sequenom	
rs12308082	VDR	12	48252139	G	A	0.066	0.054	Haploview Tagger	rs12308082		Sequenom	
rs12314197	VDR	12	48242722	A	G	0.245	0.217	Haploview Tagger	rs12314197		Sequenom	
rs12339187	RXRA	9	137229327	A	G	0.206	0.221	Haploview Tagger	rs12339187		650Y	
rs12348547	RXRA	9	137246262	T	C	0.123	0.168	Haploview Tagger	rs12348547		Sequenom	
rs12351315	RXRA	9	137225103	A	T	0.689	0.788	Haploview Tagger	rs11185647	0.965	650Y	LD with SNP on original list
rs12640179	GC	4	72612687	C	G	0.085	0.081	Haploview Tagger	rs12640179		Sequenom	

Continued

rs1352841	GC	4	72605170	T	C	0.16	0.124	Haploview Tagger	rs1352841		Sequenom	
rs1352844	GC	4	72647749	C	T	0.113	0.125	Haploview Tagger	rs1352844		650Y	
rs1352845	GC	4	72647762	A	G	0.34	0.403	Haploview Tagger	rs2298849	1	650Y	LD with SNP on original list
rs1491709	GC	4	72613566	G	A	0.075	0.08	Haploview Tagger	rs1491709		650Y	
rs1491710	GC	4	72609094	A	C	0.415	0.433	Haploview Tagger	rs1491710		Sequenom	
rs1491711	GC	4	72602252	C	G	0.358	0.354	Haploview Tagger	rs1491711		Sequenom	
rs1540339	VDR	12	48257326	C	T	0.245	0.204	PharmGKB	rs1540339		650Y	
rs1544410	VDR	12	48239835	C	T	0.217	0.279	PharmGKB	rs1544410		650Y	BsmI
rs16847024	GC	4	72650679	C	T	0.066	0.102	Haploview Tagger	rs16847024		Sequenom	
rs1873590	GC	4	72653480	A	G	0.142	0.15	Haploview Tagger	rs1873590		Sequenom	
rs188812	GC	4	72627684	A	T	0.189	0.137	Haploview Tagger	rs188812		Sequenom	
rs2107301	VDR	12	48255570	G	A	0.179	0.137	Haploview Tagger	rs2107301		650Y	
rs2189480	VDR	12	48263828	G	T	0.387	0.389	Haploview Tagger	rs2189480		650Y	
rs222014	GC	4	72632931	C	T	0.066	0.064	Haploview Tagger	rs222014		650Y	
rs222016	GC	4	72634975	G	A	0.462	0.389	Haploview Tagger	rs222016		650Y	
rs222047	GC	4	72610208	C	A	0.802	0.888	Haploview Tagger	rs222047		650Y	
rs222049	GC	4	72605955	G	C	0.154	0.177	Haploview Tagger	rs222046	0.869	650Y	LD with SNP on original list
rs2239179	VDR	12	48257766	T	C	0.311	0.292	PharmGKB	rs2239179		650Y	
rs2239182	VDR	12	48255411	T	C	0.519	0.611	Haploview Tagger	rs2239182		650Y	
rs2239184	VDR	12	48244583	G	A	0.594	0.634	Haploview Tagger	rs2239184		650Y	
rs2239185	VDR	12	48244559	G	A		0.542	PharmGKB	rs2239185		Sequenom	
rs2248098	VDR	12	48253356	A	G	0.423	0.593	Haploview Tagger	rs2248098		Sequenom	
rs2282678	GC	4	72608792	A	G	0.264	0.312	Haploview Tagger	rs16846912 and rs4752	0.959	650Y	LD with SNP on original list
rs2525045	VDR	12	48233020	G	A	0.132	0.168	Haploview Tagger	rs2525045		650Y	
rs2853563	VDR	12	48235738	C	T	0.17	0.177	Haploview Tagger	rs2853563		Sequenom	
rs34312136	RXRA	9	137269456					Haploview Tagger	rs34312136		Sequenom	
rs3733359	GC	4	72649774	G	A	0.217	0.283	Haploview Tagger	rs3733359		650Y, Sequenom	Failed 650Y, passed on sequenom

Continued

rs3737549	GC	4	72631668	G	A	0.2080.219	Haploview Tagger	rs3737549		650Y	
rs3755967	GC	4	72609398	C	T	0.094 0.04	Haploview Tagger	rs3755967		Sequenom	In complete LD with rs2282679 (Wang *et al.* 2010)
rs3782905	VDR	12	48266167	G	C	0.217	PharmGKB	rs3782905		Sequenom	Failed Sequenom
rs3819545	VDR	12	48265006	A	G	0.3020.221	Haploview Tagger	rs3819545		650Y	
rs41400444	RXRA	9	137231651	G	T	0.0750.098	Haploview Tagger	No snps in LD			Failed Sequenom assay design
rs4364228	GC	4	72623347	A	G	0.3960.468	Haploview Tagger	rs4364228		650Y	
rs4646536	CYP27B1	12	58157988	A	G	0.368 0.27	Haploview Tagger	rs703842	0.876	650Y	LD with SNP on original list
rs4646537	CYP27B1	12	58157281	T	G	0.0750.097	Haploview Tagger	rs4646537		650Y	
rs4917352	RXRA	9	137232030	T	C	0.528 0.58	Haploview Tagger	rs4917352		Sequenom	
rs4917354	RXRA	9	137237661	C	T	0.453 0.5	Haploview Tagger	rs4917354		650Y	
rs6817912	GC	4	72653705	C	T	0.0940.121	Haploview Tagger	rs6817912		Sequenom	
rs7039190	RXRA	9	137266704	A	C	0.104 0.17	Haploview Tagger	rs7039190		650Y	
rs7041	GC	4	72618334	A	C	0.1790.088	Haploview Tagger	rs7041		650Y	Ahn *et al.* 2010
rs7048602	RXRA	9	137253447	G	A	0.3870.527	Haploview Tagger	rs7048602		Sequenom	
rs705117	GC	4	72608115	C	T	0.2980.155	Haploview Tagger	rs705117		650Y	
rs705119	GC	4	72613036	A	C	0.1790.097	Haploview Tagger	rs705119		Sequenom	
rs705120	GC	4	72614140	A	C	0.5470.496	Haploview Tagger	rs705120		Sequenom	
rs731236	VDR	12	48238757	A	G	0.2360.288	PharmGKB	rs731236		Sequenom	TaqI
rs739837	VDR	12	48238221	G	T	0.5570.588	Haploview Tagger	rs739837		650Y	
rs757343	VDR	12	48239675	C	T	0.1130.062	Haploview Tagger	rs757343		650Y	
rs7853934	RXRA	9	137252760	C	T	0.2920.288	Haploview Tagger	rs7853934		Sequenom	
rs7855881	RXRA	9	137220543	G	T	0.1790.221	Haploview Tagger	rs7855881		Sequenom	
rs7954412	VDR	12	48237287	T	C	0.0850.071	Haploview Tagger	rs7967673	1	650Y	LD with SNP on original list
rs7962898	VDR	12	48242837	C	T	0.34 0.336	Haploview Tagger	rs7962898		Sequenom	
rs7965281	VDR	12	48231610	A	G	0.3580.385	Haploview Tagger	rs7965281		Sequenom	
rs7967152	VDR	12	48244184	A	C	0.6230.615	Haploview Tagger	rs7967152		650Y	

Continued

rs7968585	VDR	12	48232093	C	T	0.6040.628	Haploview Tagger	rs7968585		650Y	
rs7975232	VDR	12	48238837	C	A	0.6320.628	PharmGKB	rs7967152	0.815	650Y	ApaI, LD with SNP on original list
rs8176348	CYP27B1	12	58157133	G	A	0.1470.146	Haploview Tagger	rs8176348		Sequenom	
rs872298	RXRA	9	137222137	G	A	0.3490.381	Haploview Tagger	rs872298		650Y	
rs886441	VDR	12	48262964	G	A	0.5850.606	Haploview Tagger	rs886441		650Y	

17

Identification and Characterization of Human Genomic Binding Sites for Retinoic Acid Receptor/Retinoid X Receptor Heterodimers

Kenta Hosoda, Masashi Sato, Kazuyuki Yanai*

Department of Biomolecular Science, Faculty of Science, Toho University, Chiba, Japan
Email: *yanai@biomol.sci.toho-u.ac.jp

Abstract

All-trans retinoic acid (ATRA) triggers a wide range of effects on vertebrate development by regulating cell proliferation, differentiation, and apoptosis. ATRA activates retinoic acid receptors (RARs) which heterodimerize with retinoid X receptors (RXRs). RAR/RXR heterodimers function as ATRA-dependent transcriptional regulators by binding to retinoic acid response elements (RAREs). To identify RAR/RXR heterodimer-binding sites in the human genome, we performed a modified yeast one-hybrid assays and identified 193 RAR/RXR heterodimer-binding fragments in the human genome. The putative target genes included genes involved in development process and cell differentiation. Gel mobility shift assays indicated that 160 putative RAREs could directly interact with the RAR/RXR heterodimer. Moreover, 19 functional regulatory single nucleotide polymorphisms (rSNPs) on the RAR/RXR-binding sequences were identified by analyzing the difference in the DNA-binding affinities. These results provide insights into the molecular mechanisms underlying the physiological and pathological actions of RAR/RXR heterodimers.

Keywords

All-Trans Retinoic Acid, Retinoic Acid Receptor, Retinoid X Receptor, Yeast One-Hybrid System, Polymorphism, Regulatory SNP

1. Introduction

All-trans retinoic acid (ATRA) which is a vitamin A derivative affects physiological processes ranging from

*Corresponding author.

embryonic development to homeostasis of adult tissues and organs [1]. The actions of ATRA are mediated through the retinoic acid receptors (RARs), which is a member of the nuclear receptor super family. RARs heterodimerize with retinoid X receptors (RXRs) and function as ATRA dependent transcriptional regulators by binding to retinoic acid response elements (RAREs). Their target genes involve cell proliferation, differentiation, apoptosis, and homeostasis [2] [3]. As synthetic retinoic acids have been widely used and are known to have therapeutic effects in cancers and other various types of disease [4] [5], identification of target genes may lead to the development of more specific and effective treatments.

Methods to identify DNA-binding sites for transcription factors such as electrophoretic mobility shift assays (EMSAs), SELEX, DNase I foot printing assays, chromatin immunoprecipitation (ChIP) analysis, and reporter assays can be used to identify transcription factor binding sites, but these processes tend to be laborious and not well matched for screening large numbers of DNA elements [6]. Recently, the dramatic rising of the data for RAR target genes is generated by high-throughput technologies, such as microarrays, genome-wide ChIP analyses, and computational methods. However microarrays could not discriminate between primary and secondary target genes. Furthermore, the results of ChIP analyses actually include indirect transcription factor-DNA interactions [7]. In addition, these approaches have obstacles including high cost, reproducibility, high false positive rates, and cell-specific context. Therefore, it was important to develop alternative methods to assess regulatory regions in the genome.

Recently, we developed a modified yeast one-hybrid system which enables rapid and efficient identification of genomic targets for DNA-binding proteins [8]. Using this system, we reported here functional screening for RAR/RXR heterodimer-binding sites from human genome. As a result, we identified 193 genomic fragments including RAR/RXR heterodimer-binding motifs. At least 160 genomic fragments were confirmed as direct binding sites for the RAR/RXR heterodimer. In addition, we identified 19 functional regulatory single nucleotide polymorphisms (rSNPs) on the identified RAR/RXR-binding sequences.

2. Materials and Methods

2.1. Plasmid Constructions

Human RXR alpha and human RAR alpha was amplified by the polymerase chain reaction (PCR) from Uterus cDNA (PCR Ready-cDNA, Maxim Biotech, Inc., USA) and pituitary Cap site cDNA (Wako Pure Chemical Industries, Ltd., Japan.), respectively. These cDNA fragments were cloned into pGADT7 (CLONTECH, USA) and reamplified by PCR with primers (**Table 1**, RXR-5', RXR-3', RAR-5) to generate the restriction sites for sub-cloning. The Hind III fragment including nuclear localization signal (NLS) and GAL4 activation domain (GAL4 AD) was removed from pGADT7andthe cDNAs was inserted into the plasmid, respectively. The resulting plasmid was named pADH1_RXR and pADH1_RAR. The fragment including NLS and GAL4AD was amplified from pGADT7 by PCR and cloned. The fragment was inserted *Bgl* II sites at N-terminal end of RAR in pADH1_RAR and confirmed by sequencing. Resulting plasmid was named pADH1_NLS_GAL4AD_RAR. Foot-and-mouth disease virus (FMDV) 2A peptide sequence was amplified by annealing synthetic oligonucleotides (**Table 1**, FMDV_2A-F and FMDV_2A-R) and reamplified by PCR with primers (**Table 1**, FMDV_2A-5' and FMDV_2A-3') to generate the restriction sites. Then FMDV 2A fragment and NLS_GAL4AD_RAR was tandemly inserted at C-terminal end of RXR in pADH1_RXR. The resulting plasmid was named pADH1_RXR_2A_NLS_GAL4AD_RAR.pSUR (Gene Bank AB425277) was constructed previously and used as a reporter in the modified yeast one-hybrid system [8]. The positive control of a modified one-hybrid assay were constructed by inserting eight copies of the mouse cellular retinoic acid binding protein II (mCRABPII) RARE (**Table 1**, mCRABP2_F and mCRABP2_R) [9] at the upstream of the SPO13 promoter of pSUR. To generate the restriction sites and Kozak sequence for sub-cloning, N-terminal end side of RXR was amplified by PCR. The resulting PCR product was digested with *Hind* III and *Sma* I. C-terminal end side of RXR was obtainedby digesting with *Sma* I and *Xho*I. Both fragments were inserted into *Hind* III and *Xho*I sites pSP64 Poly (A) (Promega, USA) for *in vitro* transcription/translation. The cDNA for human RAR alpha was also inserted into the *Sal* I and the *Xba* I sites pSP64 Poly (A).

2.2. A Modified Yeast One-Hybrid Assay

The human genomic library for a modified yeast one-hybrid assay was generated as previously described [8].

Table 1. Primer sequences.

Primer name	Sequence (5'→3')
RXRA-5'	aaaAAGCTTACGCGTGCCGCCACCATGGACACC
RXRA-3'	tttAAGCTTTCTAGACTACTCGAGAGTCATTTGGTGCGGCGC
RARA-5'	aaaGTCGACAGATCTgccgccaccATGGCCAGCAACAGCAGC
RARA-3'	tttTCTAGACTACTCGAGCGGGGAGTGGGTGGCCGGG
FMDV_2A-F	aaaAGATCTTAAAATTGTCGCTCCTGTCAAACAAACTCTTAACTTTGATTTACT-CAAACTGGCTG
FMDV_2A-R	aaaTCTAGAGGATCCtttACTAGTTGGACCTGGATTGCTTTCTACATCCCCAGCCAGTTTGAG-TAAATCA
FMDV_2A-5'	aaaGTCGACAAAATTGTCGCTCCTGTCAA
FMDV_2A-3'	tttTCTAGAGAATTCCCGCGGCTCGAGACGCGTTGGACCTGGATTGCTTTC
mCRABP2_F	GATCCGAGGCTAGAAGGGCAGAGGTCACAGCAGATCTG
mCRABP2_R	AATTCAGATCTGCTGTGACCTCTGCCCTTCTAGCCTCG
pSRU-K1-F	TCGCGTTGCATTTTTGTTCTACAAAATGAAGCAC
pSRU-K1-R	ACTTCCTTTTTGTCGGCGGCTATTTCTCAATATAC
pSRU-K2-F	TGTTCTACAAAATGAAGCACAG
pSRU-K2-R	CGGCTATTTCTCAATATACTCC
rCRBPII-F	GTCCCACTCTGCTGTCACAGGTCACAGGTCACAGGTCACAGGTCATTTT
rCRBPII-R	AAAATGACCTGTGACCTGTGACCTGTGACCTGTGACAGCAGAGTGGGAC
RARE-F	TAGGGTTCACCGAAAGTTCACTC
RARE-R	GAGTGAACTTTCGGTGAACCCTA
random_F	CGCGTTGTGTGTGTTTTATTCC
random_R	GGAATAAAACACACACAACGCG

The expression vector by placing FMDV 2A peptide sequence between RXR and RAR (pADH1_RXR_2A_NLS-GAL4_RAR) was transformed into the 5FOA-selected yeast cells. The obtained transformants were grown on synthetic complete media lacking leucine, tryptophan, and uracil but containing 25 μg/ml 6-azauracil, 10 μM LG100754 (Wako Pure Chemical Industries, Ltd., Japan.), and 1 μM TTNPB (Funakoshi, Tokyo, Japan) for 3 weeks. Human genomic fragments were recovered from the positive colonies by colony-direct PCR with primers corresponding to the vector sequences (**Table 1**, pSRU-K1-F and pSRU-K1-R). The PCR fragments were directly sequenced with the forward and reverse primers (**Table 1**, pSRU-K2-F and pSRU-K2-R).

2.3. EMSAs

The RARE of rat rat cellular retinol binding protein type II (rCRBPII) were labeled with Cy5 and used as a probe (**Table 1**, rCRBPII-F and rCRBPII-R) [10]. Synthesis of human RAR and RXR proteins was performed with the TNT SP6 High Yield System (Promega, USA). *In vitro* synthesized RAR and RXR proteins were mixed with 500 ng of Calf thymus DNA (Invitrogen, USA) and 0.25 pmol of labeled oligonucleotide at 4°C. In competition experiments, a 25- and 200-fold molar excess of unlabeled oligonucleotide was added to the reaction mixture. Used competitor sequences were listed in **Table 2** and **Table 3**. The RARE of human RAR beta (**Table 1**, RARE-F and RARE-R) [11] and random sequences (**Table 1**, random_F and random_R) were used as positive and negative controls, respectively. The binding reaction was carried out in the EMSA binding buffer

Table 2. The validated RAR/RXR heterodimer-binding sites in the human genome.

Position	RARE on genome sequence	Motif	Symbol	Distance	EMSA competition %	
					×25	×200
chr1: 4974378-4974396	ctcAGGTCAaAGGTGAgga	DR1	AJAP1	259,282	46 ± 2.2	81.8 ± 2.7
chr1: 12160796-12160815	ttgAGGTCAagAGTTCAaga	DR2	TNFRSF8	37,372	63.8 ± 1.5	95.1 ± 3.1
chr1: 13537350-13537369	acgAGGTCAggAGTTCAaga	DR2	LRRC38	302,882	63.1 ± 1.9	96.8 ± 0.3
chr1: 25551801-25551820	ttgAGGTCAggAGTTCAaga	DR2	SYF2	7202	57.5 ± 1	88.9 ± 1.2
chr1: 28932727-28932746	tctTGAACTccTGACCTcgt	DR2	TAF12	36,867	63.1 ± 1.9	96.8 ± 0.3
chr1: 30925056-30925085	ataTGACCCccgtatcacttaTGACCTctg	DR12	MATN1	271,361	52.3 ± 1.8	92.5 ± 0.3
chr1: 42478144-42478163	ttgAGGTCAagAGTTCAaga	DR2	HIVEP3	−93,658	67.1 ± 0.3	95.2 ± 6
chr1: 45330672-45330691	ttgAGGTCAagAGTTCAaga	DR2	PTCH2	−22,066	67.1 ± 0.3	95.2 ± 6
chr1: 51547811-51547830	tctTGAACTccTGACCTcaa	DR2	C1orf185	−20,085	57.5 ± 1	88.9 ± 1.2
chr1: 77460865-77460884	tagAGGTCAggAGTTCAaga	DR2	ST6GALNAC5	127,689	69.4 ± 0.6	100 ± 0.5
chr1: 84134340-84134359	tctTGAACTccTGACCTcgt	DR2	TTLL7	330,483	63.1 ± 1.9	96.8 ± 0.3
chr1: 144530379-144530398	tggAGGTCAagAGGTCAagg	DR2	PPIAL4A	−166,143	66.5 ± 0.5	106.5 ± 0.9
chr1: 154106408-154106427	ctgAGGTCAggAGTTCAagt	DR2	NUP210L	21,174	66 ± 1.6	92.4 ± 1.1
chr1: 182413124-182413143	tctTGAACTccTGACCTcat	DR2	RGSL1	−6122	54.6 ± 2.2	93.7 ± 0.4
chr1: 194536260-194536279	tctTGAACTttTGACCTcag	DR2	NONE	NONE	65 ± 1.5	97 ± 0.8
chr1: 202259783-202259802	ctgAGGTCAggAGTTCAaga	DR2	UBE2T	51,301	62.3 ± 3.2	98.6 ± 0.3
chr1: 211421924-211421943	acaAGGTCAggAGTTCAaga	DR2	RCOR3	−10,774	53.3 ± 1.5	92.1 ± 0.5
chr1: 246944336-246944358	agaGGGTCAgagatAGGATAatg	DR5	SCCPDH	56,969	16.1 ± 0.6	56.9 ± 2.7
chr2: 20711074-20711099	ttcTGACCTctcacctcTGACCTtat	DR8	RHOB	64,252	58.2 ± 1	94.5 ± 0.4
chr2: 48263266-48263285	ttgAGGTCAaaAGTTCAaga	DR2	FBXO11	−130,344	64 ± 0.2	99.4 ± 1.1
chr2: 49177869-49177888	tctTGACCTctTGACCTttt	DR2	LHCGR	−194,999	76 ± 0.5	108 ± 1.4
chr2: 99943751-99943770	tctTGAACTccTGACCTcaa	DR2	TXNDC9	9099	57.5 ± 1	88.9 ± 1.2
chr2: 99971483-99971501	tctTGAACTcTGACCTcaa	DR1	EIF5B	17,658	64.9 ± 2.1	97.2 ± 0.9
chr2: 102169950-102169969	tctTGAACTccTGACCTcag	DR2	RFX8	−78,795	62.3 ± 3.2	98.6 ± 0.3
chr2: 114517066-114517085	acgAGGTCAagAGTTCAaga	DR2	SLC35F5	−2676	75.6 ± 0.2	109.5 ± 1.1
chr2: 130041252-130041271	ctgAGGTCAggAGTTCAaga	DR2	RAB6C	−695,973	62.3 ± 3.2	98.6 ± 0.3
chr2: 130228881-130228900	ttgAGGTCAggAGTTCAaga	DR2	RAB6C	−508,344	57.5 ± 1	88.9 ± 1.2
chr2: 176862952-176862971	acaAGGTCAagAGTTCAaga	DR2	KIAA1715	4056	65.8 ± 0.2	114.4 ± 7.8
chr2: 205912271-205912289	tttTGACCTcTCACCTctc	DR1	PARD3B	501,764	52.3 ± 0.7	87.7 ± 0.6
chr3: 3949000-3949021	caaTGAACTctggTGACCTggt	DR4	LRRN1	107,890	60.6 ± 2.4	98.3 ± 0.9
chr3: 24598921-24598946	catTGACCTtttttccaAGGTCAatt	ER8	THRB	−62,621	62.6 ± 0.6	96.5 ± 1.2
chr3: 25637996-25638015	acaTGAACCctTGACCCcaa	DR2	TOP2B	67,825	66.9 ± 0.9	97.3 ± 1.3
chr3: 75150440-75150462	cagTGACCTccactTCACCCcag	DR5	FRG2C	−563,030	63.5 ± 1	102.1 ± 1.3
chr3: 94076622-94076641	tctTGAACTccTGACCTcag	DR2	DHFRL1	−294,565	62.3 ± 3.2	98.6 ± 0.3

Continued

chr3: 118151024-118151046	caaTGTCCTaaacaTGACTGaga	DR5	IGSF11	602,641	2.6 ± 3	21.3 ± 0.6
chr3: 121266626-121266645	tctCGAACTccTGACCTcaa	DR2	POLQ	−1783	50.2 ± 0.9	86.9 ± 1
chr3: 145586120-145586139	tctTGAACTccTGACCTcaa	DR2	PLOD2	293,152	57.5 ± 1	88.9 ± 1.2
chr3: 149024261-149024280	ttgAGGTCAggAGTTCAaga	DR2	CP	−84,439	57.5 ± 1	88.9 ± 1.2
chr3: 151488824-151488850	aaaAGGTCAactggtaaaAGGTCAagg	DR9	AADACL2	37,133	64 ± 1.3	101.1 ± 1.4
chr3: 186241436-186241456	ctgTGACCCcagTGACCCcaa	DR3	CRYGS	20,721	70.3 ± 2.4	106.6 ± 0.5
chr4: 49258702-49258721	tctTGAACTccTGACCTcaa	DR2	CWH43	270,447	57.5 ± 1	88.9 ± 1.2
chr4: 74427504-74427523	tctTGAACTccTGACCTcag	DR2	RASSF6	58,826	62.3 ± 3.2	98.6 ± 0.3
chr4: 103004436-103004455	ttgAGGTCAggAGTTCAgga	DR2	SLC39A8	262,209	54.4 ± 1.9	98 ± 4.2
chr4: 112083722-112083741	ccaTGACCCaaACACCTtcc	DR2	PITX2	−520,453	−1.5 ± 0.6	27.8 ± 1.4
chr4: 169802530-169802550	caaAGGTCAtggATGTCAgtg	DR3	CBR4	128,928	42.2 ± 1.7	87.1 ± 1.2
chr5: 16950445-16950464	ttgAGGTCAggAGTTCAaaa	DR2	MYO10	−14,070	66.3 ± 0.3	99.6 ± 1.5
chr5: 21398739-21398756	attAGGTCATGACCTttt	IR0	GUSBP1	−60,841	70 ± 0.8	105.8 ± 1.3
chr5: 23515290-23515309	tcaTGAACTccTGACCTcaa	DR2	PRDM9	7576	66.8 ± 0.9	100.7 ± 0.6
chr5: 69214462-69214481	tctTGAACTttTGACCTcaa	DR2	SERF1B	−106,600	64 ± 0.2	99.4 ± 1.1
chr5: 72858058-72858077	tctTGAACTccTGACCTcaa	DR2	ANKRA2	3443	57.5 ± 1	88.9 ± 1.2
chr5: 73742058-73742077	ttgAGGTCAggAGTTCAaca	DR2	ENC1	195,181	61.8 ± 0.4	106.3 ± 6.5
chr5: 80685562-80685581	tctTGAACTtcTGACCTcag	DR2	ACOT12	4416	51 ± 2.7	88.8 ± 3.2
chr5: 138039418-138039437	ttgAGGTCAggAGTTCAaga	DR2	CTNNA1	−49,679	57.5 ± 1	88.9 ± 1.2
chr5: 143214556-143214575	tctTGAACTccTGACCTagt	DR2	YIPF5	335,712	52.8 ± 1.1	91.9 ± 0.2
chr5: 145898007-145898026	ccgAGGTCAggAGTTCAaga	DR2	GPR151	−2341	63.2 ± 0.6	97.1 ± 0.7
chr5: 166602931-166602950	tctTGAACTccTGACCTcaa	DR2	ODZ2	−108,902	57.5 ± 1	88.9 ± 1.2
chr5: 180015655-180015674	cggTGTCCCcaTGACCTtgt	DR2	SCGB3A1	2822	26.8 ± 0.9	80.1 ± 3.3
chr6: 22196558-22196577	ctgAGGTCAggAGTTCAaga	DR2	PRL	106,514	62.3 ± 3.2	98.6 ± 0.3
chr6: 31915297-31915315	aggTGACCTtTGACCTgta	DR1	SKIV2L	−11,275	72.6 ± 1.4	103.2 ± 0.3
chr6: 38272104-38272123	acgAGGTCAagAGGTCAaga	DR2	BTBD9	335,810	75.8 ± 3.3	104.5 ± 0.7
chr6: 42310079-42310097	ctgGGGTCAaAGGTCAcca	DR1	TRERF1	109,695	72.1 ± 2.9	104.2 ± 0.3
chr6: 42437740-42437759	ggtGGATCAcaAGGTCAgga	DR2	TRERF1	−17,967	21.9 ± 2.3	66.5 ± 2.2
chr6: 93902272-93902291	tctTGAACTccTGACCTcaa	DR2	EPHA7	227,018	57.5 ± 1	88.9 ± 1.2
chr6: 107377360-107377379	atgAGGTCAagAGGTCAaga	DR2	C6orf203	27,963	73 ± 0.6	99.9 ± 2.1
chr6: 127695151-127695170	tctTGAACTccTGACCTcat	DR2	ECHDC1	−31,609	54.6 ± 2.2	93.7 ± 0.4
chr6: 137349337-137349356	ctgAGGTCAggAGTTCAaga	DR2	IL20RA	16951	62.3 ± 3.2	98.6 ± 0.3
chr7: 2263677-2263699	cggAGACCCtacaaTGACCCctc	DR5	MAD1L1	8895	52.3 ± 1.7	100.7 ± 0.8
chr7: 18015916-18015935	tctTGAACTccTGACCTcat	DR2	SNX13	−35,795	54.6 ± 2.2	93.7 ± 0.4

Continued

chr7: 28744762-28744781	acgAGGTCAggAGTTCAaga	DR2	TRIL	253,257	63.1 ± 1.9	96.8 ± 0.3
chr7: 41717021-41717040	ttgAGGTCAggAGTTCAaga	DR2	INHBA	25,675	57.5 ± 1	88.9 ± 1.2
chr7: 56952727-56952746	tcaTGAACTccTGACCTcaa	DR2	ZNF479	254,834	66.8 ± 0.9	100.7 ± 0.6
chr7: 67465473-67465492	ctgAGGTCAagAGTTCAaga	DR2	NONE	NONE	65.7 ± 1.3	98.9 ± 0.9
chr7: 88072665-88072684	gcaTGACCTccTGACCTctg	DR2	STEAP4	−136,447	67.9 ± 1.6	100.2 ± 0.3
chr7: 92803857-92803876	ttaAGGTCAggAGTTCAaga	DR2	HEPACAM2	51,915	63.3 ± 1.5	91.4 ± 1.9
chr7: 152384707-152384726	acaAGGTCAggAGTTCAaga	DR2	XRCC2	−11,467	53.3 ± 1.5	92.1 ± 0.5
chr8: 53147041-53147060	tctTGAACTccTGACCTcat	DR2	ST18	175,388	54.6 ± 2.2	93.7 ± 0.4
chr8: 56752302-56752321	ttgAGGTCAtaAGTTCAaga	DR2	LYN	−40,074	68.1 ± 5.6	102.8 ± 0.8
chr8: 104877729-104877748	tctTGAACTccTGACCTcag	DR2	RIMS2	364,763	62.3 ± 3.2	98.6 ± 0.3
chr8: 104984015-104984034	tctTGAACTccTGACCTcaa	DR2	TM7SF4	−368,029	57.5 ± 1	88.9 ± 1.2
chr8: 141528665-141528684	tctTGAACTccTGACCTcaa	DR2	CHRAC1	7278	57.5 ± 1	88.9 ± 1.2
chr9: 10621119-10621138	tctTGAACTccTGACCTcgt	DR2	PTPRD	−8406	63.1 ± 1.9	96.8 ± 0.3
chr9: 25217250-25217269	ctgAGGTCAagAGTTCAaga	DR2	LOC100506422	−849,413	65.7 ± 1.3	98.9 ± 0.9
chr9: 83921356-83921375	ttgAGGTCAagAGTTCAaga	DR2	TLE1	382,230	67.1 ± 0.3	95.2 ± 6
chr9: 92908690-92908709	tctTGAACTccTGACCTcag	DR2	DIRAS2	496,408	62.3 ± 3.2	98.6 ± 0.3
chr9: 128762284-128762303	ttgAGGTCAggAGTTCAaga	DR2	PBX3	252,677	57.5 ± 1	88.9 ± 1.2
chr10: 101670478-101670496	gatAGGTCAaAGGGCAcag	DR1	DNMBP	99,189	51.3 ± 1.5	93.6 ± 1.8
chr10: 120477590-120477612	tggTGACCCttcttTGACCTtag	DR5	PRLHR	−122,441	67.5 ± 0.5	103 ± 0.7
chr10: 122751598-122751617	atgAGGTCAggAGTTCAaga	DR2	WDR11	140,921	54.6 ± 2.2	93.7 ± 0.4
chr10: 127772676-127772694	ttgTGAACTtTGACCTctg	DR1	FANK1	187,577	67.3 ± 1	104.5 ± 1.2
chr11: 9987853-9987872	tctTGAACTccTGACCTcaa	DR2	SWAP70	302,235	62.3 ± 3.2	98.6 ± 0.3
chr11: 10151333-10151352	ttgAGGCCAcgAGTTCAaga	DR2	SBF2	164,411	37.8 ± 2.1	78.1 ± 0.6
chr11: 73263784-73263803	ttgAGGTCAgaAGTTCAaga	DR2	PLEKHB1	−94,800	58.6 ± 1.3	94.4 ± 0.8
chr11: 84351765-84351784	atgAGGTCAggAGTTCAaga	DR2	DLG2	986,539	54.6 ± 2.2	93.7 ± 0.4
chr11: 102121325-102121344	tccTGACCTcaTGATCTgcc	DR2	BIRC3	−66,846	20.9 ± 1.1	62.9 ± 1.8
chr11: 113168930-113168949	ctgAGGTCAggAGTTCAaga	DR2	TTC12	−16,311	62.3 ± 3.2	98.6 ± 0.3
chr11: 133310404-133310423	ctgAGGTCAagAGTTCAaga	DR2	SPATA19	404,978	65.7 ± 1.3	98.9 ± 0.9
chr12: 1757331-1757350	tggAGGTCAggAGTTCAaga	DR2	WNT5B	18,929	60.7 ± 1.4	93.7 ± 1.2
chr12: 5137518-5137541	actTGACCCcactgtTGACCTctc	DR6	KCNA5	−15,555	71.2 ± 1.1	105.4 ± 0.4
chr12: 13637881-13637900	tctTGAACTccTGACCTcaa	DR2	EMP1	288,289	57.5 ± 1	88.9 ± 1.2
chr12: 53367382-53367401	aagAGGTCAggAGTTCAaga	DR2	KRT18	24,549	59.3 ± 1.6	91.8 ± 0.7
chr12: 111147348-111147367	agaAGGCCAtgGGGTCAaat	DR2	HVCN1	−20,412	56.6 ± 0.7	88.3 ± 0.5
chr12: 122916626-122916645	ctgAGGTCAggAGTTCAaga	DR2	CLIP1	−9520	62.3 ± 3.2	98.6 ± 0.3
chr13: 64476525-64476544	ctgAGGTCGggAGTTCAaga	DR2	NONE	NONE	66.9 ± 2	103 ± 0.8
chr13: 77978240-77978259	actTGAACTccTGACCTcaa	DR2	MYCBP2	−77,073	62.3 ± 0.4	99.4 ± 1.6

Continued

chr14: 70604077-70604096	tctTGAACTccTGACCTcgt	DR2	SLC8A3	51,700	63.1 ± 1.9	96.8 ± 0.3
chr14: 74027852-74027871	ctgAGGTCAaaAGTTCAaga	DR2	HEATR4	−2211	65 ± 1.5	97 ± 0.8
chr14: 85281297-85281316	tctTGAACTctTGACCTttt	DR2	FLRT2	−715,181	68.9 ± 3.1	101.4 ± 1.8
chr14: 98241445-98241464	ctgAGGTCAggAGTTCAaga	DR2	VRK1	977,771	62.3 ± 3.2	98.6 ± 0.3
chr15: 25002465-25002484	ttgAGGTCAggAGTTCAaga	DR2	C15orf2	81,934	57.5 ± 1	88.9 ± 1.2
chr15: 45889090-45889109	tctTGAACTtcTGACCTcag	DR2	PLDN	9683	51 ± 2.7	88.8 ± 3.2
chr15: 63680120-63680140	tggAGGACAgtgTGACCTgga	IR3	CA12	−6055	5.5 ± 1.9	35 ± 1.7
chr15: 74836867-74836889	tgaAGGCCAgagagGGGTCAtgg	DR5	ARID3B	3330	39.3 ± 1.2	84.5 ± 0.5
chr16: 2926654-2926673	ctgAGGTCAagAGTTCAaga	DR2	FLYWCH2	−6532	65.7 ± 1.3	98.9 ± 0.9
chr16: 8744664-8744683	ctgAGGTCAggAGTTCAaga	DR2	ABAT	−23,770	62.3 ± 3.2	98.6 ± 0.3
chr16: 21194228-21194247	ttgGGGTCAggAGTTCAaga	DR2	DNAH3	−23,476	69.3 ± 1.2	97.1 ± 1
chr16: 24396387-24396406	tctTGAACTctTGACCTcag	DR2	CACNG3	129,523	65.7 ± 1.3	98.9 ± 0.9
chr16: 28365690-28365709	tctCCAACTcgTGACCTctt	DR2	NPIPL1	2275	67.1 ± 0.3	95.2 ± 6
chr16: 56152959-56152978	tctTGAACTccTGACCTcaa	DR2	GNAO1	−72,282	57.5 ± 1	88.9 ± 1.2
chr16: 57488795-57488816	gaaAGGTCAgtcgAGGTCAaat	DR4	CIAPIN1	−7437	75.5 ± 1.7	104.1 ± 1.8
chr16: 65859695-65859714	aagAGGTGAagAGGTCAaag	DR2	CDH5	−540,820	67 ± 0.7	102.5 ± 0.2
chr16: 79019910-79019932	ggaGGGGCAtcccaGGGTCAgaa	DR5	MAF	614,701	23.1 ± 1.1	62.8 ± 0.8
chr16: 79567620-79567653	cagTGACCTctccgccacctcccag TGACCTctg	DR16	MAF	66,985	64.5 ± 2.2	97.6 ± 1.4
chr17: 35224339-35224358	ctgAGGTCAggAGTTCAaga	DR2	LHX1	−70,423	62.3 ± 3.2	98.6 ± 0.3
chr17: 40470361-40470380	tctTGAACTccTGACCTcag	DR2	STAT5A	30,806	62.3 ± 3.2	98.6 ± 0.3
chr17: 43654747-43654766	ttgAGGTCAgaAGTTCAaga	DR2	PLEKHM1	−86,611	58.6 ± 1.3	94.4 ± 0.8
chr17: 55839651-55839679	cgaTGACCTtacaaaggctcTCACCTaaa	DR11	MSI2	505,734	27.3 ± 1.7	70.4 ± 1.5
chr17: 60825539-60825558	acgAGGTCAggAGTTCAagc	DR2	10-Mar	60,156	58.1 ± 1.6	92.2 ± 2.7
chr17: 64280560-64280579	ctgGGGTCAggAGTTCAaga	DR2	PRKCA	−18,356	65.1 ± 2.3	96.3 ± 0.8
chr17: 65785502-65785521	tctTGAACTccTGACCTcaa	DR2	BPTF	−36,268	57.5 ± 1	88.9 ± 1.2
chr17: 65810594-65810613	ctgAGGTCAggAGTTCAaga	DR2	BPTF	−11,176	62.3 ± 3.2	98.6 ± 0.3
chr17: 72730765-72730784	ctaAGGTCAggAGTTCAaga	DR2	RAB37	−2581	53.2 ± 1.7	95.3 ± 0.7
chr17: 73619194-73619213	ctgAGGTAGgaGTTGCAgac	DR2	C17orf109	−10,310	62.3 ± 3.2	98.6 ± 0.3
chr18: 8161336-8161361	ctgAGTTCAgatcctaaAGGTCAtag	DR8	RAB12	−448,094	35.1 ± 0.7	78 ± 2.5
chr18: 12322082-12322101	ttgAGGTCAggAGTTCAaga	DR2	TUBB6	13,835	57.5 ± 1	88.9 ± 1.2
chr18: 20703035-20703054	ttgAGGTCAggAGTTCAagg	DR2	CABLES1	−12,682	58.8 ± 0.8	96.2 ± 0.5
chr18: 55414467-55414486	ctaAGGTCAggAGTTCAaga	DR2	ATP8B1	55,850	53.2 ± 1.7	95.3 ± 0.7
chr19: 14416278-14416297	tctTGAACTctTGACCTtag	DR2	CD97	−75,668	65.9 ± 1.6	106.7 ± 1.4
chr19: 21290261-21290280	tctTGAACTccTGACCTcag	DR2	ZNF714	25,318	62.3 ± 3.2	98.6 ± 0.3

Continued

chr19: 33629174-33629193	tctTGAACTccTGACCTcag	DR2	LRP3	−56,415	62.3 ± 3.2	98.6 ± 0.3
chr19: 36456308-36456326	cggAGGTCAgAGTTCGaga	DR1	LRFN3	28,295	45.1 ± 2.8	83.3 ± 1
chr19: 37396556-37396575	tctTGAACTccTGACCCcag	DR2	ZNF829	10,357	65.1 ± 2.3	96.3 ± 0.8
chr19: 45539547-45539566	tctTGAACTtcTGACCTcag	DR2	CLASRP	−2741	51 ± 2.7	88.8 ± 3.2
chr20: 16197032-16197054	gagTCACCTggtcgTGACCTttg	DR5	KIF16B	357,036	66.3 ± 1.2	102.6 ± 0.3
chr20: 24050829-24050848	tctTGAACTccTGACCTcgt	DR2	GGTLC1	−81,423	63.1 ± 1.9	96.8 ± 0.3
chr20: 29578657-29578676	tctTGAACTccTGACCTcag	DR2	DEFB115	−266,800	62.3 ± 3.2	98.6 ± 0.3
chr20: 32549760-32549779	ttgAGGTCAggAGTTCAaga	DR2	RALY	−31,688	57.5 ± 1	88.9 ± 1.2
chr20: 35935942-35935961	ttgAGGTCAggAGTTCAaga	DR2	MANBAL	17,901	57.5 ± 1	88.9 ± 1.2
chr20: 49728437-49728456	tctCGAACTccTGACCTcaa	DR2	KCNG1	−88,772	57.5 ± 1	88.9 ± 1.2
chr21: 11186672-11186691	acaAGGTCAagAGATCAaga	DR2	BAGE	−87,745	57.1 ± 1.8	96.4 ± 0.5
chr21: 32961765-32961784	ctgAGGTCAggAGTTCAaca	DR2	TIAM1	−30,485	64.7 ± 0.7	101.6 ± 2
chr21: 33535792-33535813	ggcAGGTCAacccAGGCCAagg	DR4	MIS18A	115,573	16.1 ± 2.7	49.5 ± 6.3
chr21: 37807490-37807509	ttgAGGTCAggAGTTCAaga	DR2	CLDN14	31,225	57.5 ± 1	88.9 ± 1.2
chr22: 31732573-31732592	ctgAGGTCAggAGTTCAaga	DR2	PATZ1	9666	62.3 ± 3.2	98.6 ± 0.3
chr22: 33943562-33943581	atgAGGTCAagAGTTCAaga	DR2	LARGE	372,844	66.7 ± 1.5	106.1 ± 0.2
chr22: 35512814-35512833	tctTGAACTtcTGAACTcct	DR2	ISX	50,694	57.6 ± 3.4	91.4 ± 0.9
chr22: 37370457-37370476	tctTGATCTccTGACCTcgt	DR2	TST	45,024	42.5 ± 1.7	90.7 ± 0.3
chr22: 45947191-45947210	ctgTGGACTtgTGACCTctc	DR2	FBLN1	48,482	51.5 ± 1.8	90.3 ± 0.2
chr22: 48865816-48865835	tctTGAACTctTGACCTcaa	DR2	FAM19A5	−19,462	67.1 ± 0.3	95.2 ± 6
chrUn_gl000225:106816-106836	acaTAACCTacgTGACCTgtg	DR3	NONE	NONE	35.8 ± 1.5	84.1 ± 0.3
chrY: 13899010-13899029	tctTGAACTccTGACCTcag	DR2	USP9Y	−914,140	62.3 ± 3.2	98.6 ± 0.3

containing 12 mM HEPES (pH 7.9), 60 mM KCl, 4 mM $MgCl_2$, 1 mM EDTA, 12% glycerol, and 0.5% Nonidet P-40. The reaction mixtures were directly loaded onto a 4% nondenaturing polyacrylamide gels made in 0.5xTBE. After electrophoresis was performed at 4˚C, the gels were analyzed by using a bio-imaging analyzer (FLA-7000 FUJIFILM).

2.4. Bioinformatics

To map the obtained sequences on the human genomic assembly (GRCh37), these sequences were analyzed using NCBI's BLAST and were searched for AGGTCA motifs. For the stringency of the search, we allowed up to 2-bp mismatches. For each predicted RAR/RXR binding site, the nearest gene and the distance from the center of the binding site to the transcriptional start site of the gene within 1000 kb was identified with GREAT (http://bejerano.stanford.edu/great/public/html/).

3. Results

3.1. Identification of Human Genomic Binding Sites for the RAR/RXR Heterodimer

To identify RAR/RXR heterodimer—but not RXR homodimer-binding sites by a modified yeast one-hybrid system, the simultaneous expression of two transcription factors in yeast was required. Internal ribosome entry site (IRES) has been widely used for this purpose, but it has a major limitation. Namely, translation efficiency of a gene placed after IRES is much lower than that of a gene located before IRES [12]. The limitation can be over

Table 3. Identified functional rSNPs in RAR/RXR binding sites.

dbSNP ID	Gene (positon)	Sequence	EMSA (×25) Competition %	p-Value
rs191541	HIVEP3	ttgAGGTCAagAGTTCAaga	80.7 ± 2.0	9.00E−05
	(−93658)	ttgAGGTCAagACTTCAaga	21.4 ± 3.1	
rs3781025	FANK1	cagAGGTCAaAGTTCAcaa	77.9 ± 1.6	1.10E−05
	(+187577)	cagAGATCAaAGTTCAcaa	33.1 ± 0.6	
rs6907105	EPHA7	ttgAGGTCAggAGTTCAaga	69.2 ± 1.0	2.50E−04
	(+227018)	ttgAAGTCAggAGTTCAaga	43.3 ± 1.8	
rs12164613	CWH43	ttgAGGTCAggAGTTCAaga	63.5 ± 0.9	5.70E−05
	(+270447)	ttgAGGCCAggAGTTCAaga	38.2 ± 1.1	
rs36129646	unplaced	cacAGGTCAcgtAGGTTAtgt	34.4 ± 1.9	1.70E−03
		cacAGGTCAcgtACGTTAtgt	11.7 ± 2.3	
rs36129646	unplaced	cacAGGTCAcgtAGGTTAtgt	34.4 ± 1.9	2.40E−02
		cacAGGTCAcgtATGTTAtgt	18.3 ± 4.1	
rs72422438	GNAO1	ttgAGGTCAggAGTTCAaga	69.2 ± 1.0	3.70E−05
	(−72282)	ttgAGGTCAggAGTCAAgac	41.5 ± 1.0	
rs112102213	TOP2B	ttgGGGTCAagGGTTCAtgt	74.4 ± 0.5	5.60E−04
	(+67825)	ttgGGGCCAagGGTTCAtgt	43.7 ± 3.0	
rs113359665	SKIV2L	tacAGGTCAaAGGTCAcct	73.4 ± 1.6	4.80E−04
	(−11275)	tacAGGTCAaAGGTCTcct	45.7 ± 2.1	
rs114889780	RALY	ttgAGGTCAggAGTTCAaga	69.2 ± 1.0	1.20E−05
	(−31688)	ttgAGGTCAggATTTCAaga	27.0 ± 1.2	
rs138010912	INHBA	ttgAGGTCAggAGTTCAaga	69.2 ± 1.0	1.20E−02
	(+25675)	ttgATGTCAggAGTTCAaga	42.2 ± 6.1	
rs139841157	SCGB3A1	acaAGGTCAtgGGGACAccg	36.7 ± 2.6	2.80E−04
	(+2822)	acaAGATCAtgGGGACAccg	0.3 ± 1.7	
rs145976536	SCGB3A1	acaAGGTCAtgGGGACAccg	36.7 ± 2.6	3.80E−03
	(+2822)	acaAGGTCTtgGGGACAccg	9.2 ± 3.8	
rs140387758	RAB12	ctgAGTTCAgatcctaaAGGTCAtag	47.1 ± 5.9	7.40E−03
	(−448094)	ctgAGTTCAgatcctaaAGGTTAtag	14.1 ± 2.8	
rs147157947	RAB12	ctgAGTTCAgatcctaaAGGTCAtag	47.1 ± 5.9	4.40E−02
	(−448094)	ctgAGTTCTgatcctaaAGGTCAtag	27.4 ± 3.3	
rs148188266	TM7SF4	ttgAGGTCAggAGTTCAaga	63.5 ± 0.9	1.40E−03
	(−368029)	ttgAGGTTAggAGTTCAaga	45.2 ± 2.1	
rs149916363	TTC12	ctgAGGTCAggAGTTCAaga	68.7 ± 0.3	5.10E−04
	(−16311)	ctgAGATCAggAGTTCAaga	38.9 ± 2.9	
rs191800593	CDH5	aagAGGTGAagAGGTCAaag	63.5 ± 1.7	2.10E−02
	(−540820)	aagTGGTGAagAGGTCAaag	47.9 ± 3.9	
rs192782031	MANBAL	ttgAGGTCAggAGTTCAaga	69.2 ± 1.0	1.90E−05
	(+17901)	ttgAGGTCAggAGTTAAaga	41.8 ± 0.6	

come by a 2A peptide, a "self-cleaving" small peptide, which was identified in the FMDV [13]. As 2A-me-diated cleavage is a universal phenomenon in eukaryotic cells [14] [15], we constructed a single expression vector by placing FMDV 2A segment between RXR and RAR. To minimize the effect of RXR homodimer on the reporter activity, RXR is expressed as the native protein, although RAR is expressed as a fusion to the GAL4 AD (**Figure 1**). To evaluate the function of RAR/RXR heterodimer, yeast cells were transformed with these effectors and the indicated reporters (**Figure 2**). The transformants were grown on synthetic complete media includ-

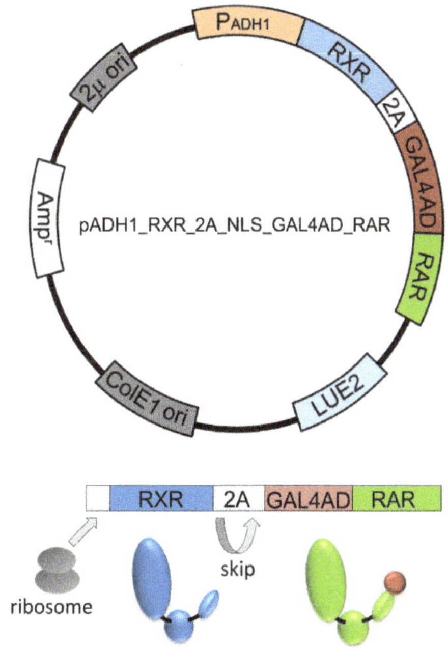

Figure 1. Schematic of bicistronic protein expression from a single expression vector. A bicistronic expression vector contains FMDV 2A peptide sequence (2A) between RXR cDNA and the GAL4 activation domain-RAR fusioncDNA under the control of the ADH1 promoter. The plasmid also contains Col E1 and 2 micro origins for the autonomous replication in *E. coli* and yeast respectively, ampicillin resistance, and a *LEU2* selectable marker.Schematic of RXR and RAR expression via 2A-mediated translational skip mechanism is presented.

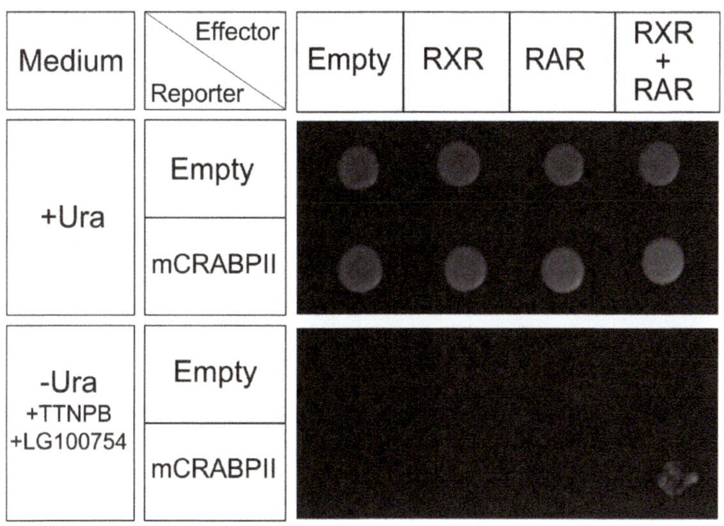

Figure 2. A modified yeast one-hybrid system. Yeast cells transformed with the indicated plasmids. The transformants were grown on synthetic complete media including uracil (*Ura*) or lacking uracil but containing the RAR-specific ligand (*TNPB*) and the antagonist of RXR homodimers (*LG*100754). The plates were photographed after 2 days (+Ura) or 5 days (−Ura + TNPB + LG100754) growth at 30˚C.

ing RAR-specific ligand TTNPB [16] and antagonist of RXR homodimers LG100754 [17] [18]. The transformants expressing either RAR or RXR alone could not grow, whereas the transformants expressing both RAR and RXR could grow in a RARE-dependent manner. These results indicated that RAR/RXR heterodimer could activate the reporter gene via the RARE.

The human genomic library after 5FOA selection [8] was transformed with the human RAR/RXR expression vector (**Figure 1**). More than 1 × 10^7 of the library was selected for uracil prototroph. Human genomic fragments were recovered from the validated colonies by colony-direct PCR and sequenced. Two hundred and eleven unique sequences were obtained from 364 clones.

3.2. Experimental Validation of the RAR/RXR-Binding Sites

RAREs are typically composed of two directrepeats of a core motif, RGKTCA [19]. The classical RARE is a 5 bp-spaced direct repeat (DR5). Furthermore, the heterodimers also bind to direct repeats separated by 1 bp (DR1) or 2 bp (DR2) [20]. To analyze the obtained genomic sequences, AGGTCA motifs were computationally searched. For the stringency of the search, we allowed up to 2-bp mismatches. As an initial test, we examined direct interaction *in vitro* synthesized RAR/RXR and the known RARE (rCRBPII RARE) (**Figure 3**). Incubation of the the cy5-labeled rCRBPII oligonucleotides with the combination of RAR and RXR retarded complexes, but no shifted band was observed with either receptor alone (**Figure 3**, lanes 2-4). The complexes represented a sequence-specific interaction between the rCRBPII probe and the RAR/RXR heterodimer, since the formation of this complex was specifically reduced with molar excess of unlabeled competitors (**Figure 3**, lanes 5-7). Moreover, the addition of anti-RXR or anti-RAR antibody created a slower-migrating complex (**Figure 3(a)**, lanes 7 and 8). No super shifted bands were observed with anti-HNF4 antibodies (**Figure 3**, lane 9). These results indicated that the sequence-specific binding complex contained both RAR and RXR, presumably as a heterodimer.

As a next step, we examined whether the 193 predicted RARE should interact with RAR/RXR heterodimer. At least 160 predicted RAREs in the obtained genomic sequences could directly interact with RAR/RXR heterodimer (**Table 2**). These RAR/RXR-binding sites were located around or in the genes with various functions,

Lane	1	2	3	4	5	7	8	9	10
Protein	-	RXR	RAR	RXR + RAR					
Competitor	-	-	-	-	RARE	Random	-	-	-
Antibody	-	-	-	-	-	-	RXR	RAR	HNF4

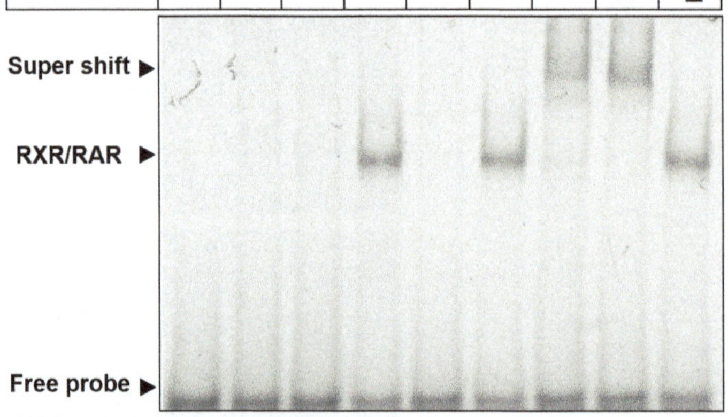

Figure 3. Binding of RAR/RXR heterodimer to RARE by mobility gel shift assays. Cy5-labeled double-stranded rCRBPII RARE was incubated with *in vitro* transcribed/translated human RAR alpha (RAR) and human RXR alpha (RXR). In a competition assay, 100-fold molar excess of the unlabeled oligonucleotides (RARE or Random) were added to the reaction mixture. In a supershift experiment, the indicated antibodies were incubated in the reaction mixture. Binding reactions were resolved by electrophoresis on a 4% acrylamide gel in 0.5 × TBE.

such as cytoskeleton (TUBB6, KIF16B, KRT18 and CTNNA1), extracellular matrix (LARGE), signal transduction (CREB5 and STAT3), transcription (TAF12, POLQ and TRERF1), translation (EIF2B3 and EIF5B) and development (INHBA, ISX, KIAA1715, LHX1, and PBX3). Remarkably, several genes previously known to be regulated by ATRA including FBLN1 [21] and RAR beta [11] [22] were also included. Although the RARE of human RAR betagene was already reported, the exon 9 was a novel binding site. Furthermore, we confirmed that RAR/RXR could directly interact with non-canonical motifs, such as inverted repeats (IRs) and everted repeats (ERs) (**Table 2**).

3.3. Regulatory SNPs in RAR/RXR-Binding Sites

Functional rSNPs in transcription factor-binding sites may predictably lead to differences in gene expression and associate with disease susceptibility. Then, we identify 23SNPs on each RAR/RXR-binding sequence by using the NCBI's SNP database and examined the effect on the RAR/RXR-binding affinity. As a result, 19 functional rSNPs were identified by analyzing the difference in the DNA-binding affinities (**Table 3**).

4. Discussion

We identified 211 of the human genomic fragments containing putative RAREs using the modified yeast one-hybrid. Theses sequences contain canonical RAREs (DR1, DR2, and DR5) but also contain half-site arrangements. At least 160 of putative RAREs in the obtained genomic fragments could directly interact with RAR/RXR heterodimer. Interestingly, some putative RAREs differed from canonical RAREs also directly bound to RAR/RXR heterodimer. Furthermore, 19 functional rSNPs on the RAR/RXR-binding sites were identified by analyzing the difference in DNA-binding affinity.

Most of researches historically focused on promoter regions to find transcription factor-binding sites, but 90% of the identified RAR/RXR-binding sites located in over 10 kb from the transcription start site of genes (**Table 2**). In this study, we could not identify all previously reported canonical RAR/RXR binding sites [11] [23]-[25], although a modified yeast one-hybrid assay is one of the effective methods to identify transcription factor-binding sites as a genome-wide scale. There are some explanations for missed RAR/RXR-DNA interactions. First, the quality of a library will affect the efficiency of identification of protein-DNA interactions. Second, some RAREs may exist adjacent to sequences recognized by yeast transcription factors and may be discarded during the negative selection by 5FOA [8].

RAR/RXR heterodimers contribute to transactivation through widely spaced (up to 150 bp) DR elements [26]. A response element compose of palindromic arrangement of consensus motifs with no spacer nucleotide between the two half-sites (IR0) is known to be bound and activated by RAR/RXR heterodimers [27] [28]. In this study, we confirmed that several types of RAREs including IR0 and DR elements spaced by more than 5 bp directly interact with RAR/RXR heterodimerby EMSA (**Table 2**). These results are consistent with above previous researches.

Genome-scale sequencing has led to the discovery of millions of human SNPs [29]. There are several examples of rSNPs associated with disease susceptibility [30]-[32]. In this study, we identified 19 functionalr SNPs including the 50 kb downstream region of EPHA7 gene. A recent study revealed a strong correlation between expression of EPHA7 and glioblastoma multiforme patient survival [33]. ATRA induces cell differentiation and causes inhibition of cell proliferation in a variety of cancer cell lines including glioblastoma cell lines [34]-[37]. ATRA enhances cytotoxicity of paclitaxel in glioblastoma xenografts and can be therapeutically useful against glioblastoma [38]-[40]. Our results suggest that the direct interaction RAR/RXR with the sequence around EPHA7 gene may possibly affect early stage of neural differentiation and the therapy of glioblastoma. Further studies will be necessary to clarify the relationship between the EPHA7 polymorphism (rs6907105) and the effects of ATRA therapy.

Recently, genome-wide ChIP analyses of RAR/RXR-binding sites were reported using several human cancer cell lines, but such studies are insufficient to get a complete overview of the target genes under its control [41] [42]. As mentioned above, ChIP techniques potentially include indirect transcription factor-DNA interactions [7]. In contrast, our strategy for identification of RAR/RXR sites in human genome took a fundamentally different approach based on the direct interaction between the RAR/RXR heterodimer and human genomic sequences using a yeast genetic selection. Our finding will provide insights into the molecular mechanisms underlying the physiological and pathological actions of RAR/RXR heterodimer.

Acknowledgements

We gratefully acknowledge our laboratory members for their helpful discussion and encouragement.

References

[1] Niederreither, K. and Dolle, P. (2008) Retinoic Acid in Development: Towards an Integrated View. *Nature Reviews Genetics*, **9**, 541-553. http://dx.doi.org/10.1038/nrg2340

[2] Ferrari, N., Pfahl, M. and Levi, G. (1998) Retinoic Acid Receptor Gamma1 (RARgamma1) Levels Control RARbeta2 Expression in SK-N-BE2 (c) Neuroblastoma Cells and Regulate a Differentiation-Apoptosis Switch. *Molecular and Cellular Biology*, **18**, 6482-6492.

[3] Berry, D.C. and Noy, N. (2009) All-Trans-Retinoic Acid Represses Obesity and Insulin Resistance by Activating both Peroxisome Proliferation-Activated Receptor β/δ and Retinoic Acid Receptor. *Molecular and Cellular Biology*, **29**, 3286-3296. http://dx.doi.org/10.1128/MCB.01742-08

[4] Tobita, T., Takeshita, A., Kitamura, K., Ohnishi, K., Yanagi, M., Hiraoka, A., Karasuno, T., Takeuchi, M., Miyawaki, S., Ueda, R., Naoe, T. and Ohno, R. (1997) Treatment with a New Synthetic Retinoid, Am80, of Acute Promyelocytic Leukemia Relapsed from Complete Remission Induced by All-Trans Retinoic Acid. *Blood*, **90**, 967-973.

[5] Fukasawa, H., Kagechika, H. and Shudo, K. (2006) Retinoid Therapy for Autoimmune Diseases. *Nihon Rinsho Meneki Gakkai Kaishi*, **29**, 114-126. http://dx.doi.org/10.2177/jsci.29.114

[6] Farnham, P.J. (2009) Insights from Genomic Profiling of Transcription Factors. *Nature Reviews Genetics*, **10**, 605-616. http://dx.doi.org/10.1038/nrg2636

[7] Wells, J., Yan, P.S., Cechvala, M., Huang, T. and Farnham, P.J. (2003) Identification of Novel pRb Binding Sites Using CpG Microarrays Suggests That E2F Recruits pRb to Specific Genomic Sites during S Phase. *Oncogene*, **22**, 1445-1460. http://dx.doi.org/10.1038/sj.onc.1206264

[8] Taniguchi-Yanai, K., Koike, Y., Hasegawa, T., Furuta, Y., Serizawa, M., Ohshima, N., Kato, N. and Yanai, K. (2010) Identification and Characterization of Glucocorticoid Receptor-Binding Sites in the Human Genome. *Journal of Receptors and Signal Transduction*, **30**, 88-105. http://dx.doi.org/10.3109/10799891003614816

[9] Durand, B., Saunders, M., Leroy, P., Leid, M. and Chambon, P. (1992) All-Trans and 9-cis Retinoic Acid Induction of CRABPII Transcription Is Mediated by RAR-RXR Heterodimers Bound to DR1 and DR2 Repeated Motifs. *Cell*, **71**, 73-85. http://dx.doi.org/10.1016/0092-8674(92)90267-G

[10] Mangelsdorf, D.J., Umesono, K., Kliewer, S.A., Borgmeyer, U. and Ong, E.S. (1991) Evans RMA Direct Repeat in the Cellular Retinol-Binding Protein Type II Gene Confers Differential Regulation by RXR and RAR. *Cell*, **66**, 555-561. http://dx.doi.org/10.1016/0092-8674(81)90018-0

[11] de Thé, H., Vivanco-Ruiz, M.M., Tiollais, P., Stunnenberg, H. and Dejean, A. (1990) Identification of a Retinoic Acid Responsive Element in the Retinoic Acid Receptor Beta Gene. *Nature*, **343**, 177-180. http://dx.doi.org/10.1038/343177a0

[12] de Felipe, P. (2002) Polycistronic Viral Vectors. *Current Gene Therapy*, **2**, 355-378. http://dx.doi.org/10.2174/1566523023347742

[13] Ryan, M.D., King, A.M. and Thomas, G.P. (1991) Cleavage of Foot-and-Mouth Disease Virus Polyprotein Is Mediated by Residues Located within a 19 Amino Acid Sequence. *Journal of General Virology*, **72**, 2727-2732. http://dx.doi.org/10.1099/0022-1317-72-11-2727

[14] Hasegawa, K., Cowan, A.B., Nakatsuji, N. and Suemori, H. (2007) Efficient Multicistronic Expression of a Transgene in Human Embryonic Stem Cells. *Stem Cells*, **25**, 1707-1712. http://dx.doi.org/10.1634/stemcells.2006-0813

[15] Ito-Harashima, S., Kuroha, K., Tatematsu, T. and Inada, T. (2007) Translation of the Poly(A) Tail Plays Crucial Roles in Nonstop mRNA Surveillance via Translation Repression and Protein Destabilization by Proteasome in Yeast. *Genes & Development*, **21**, 519-524. http://dx.doi.org/10.1101/gad.1490207

[16] Aström, A., Pettersson, U., Krust, A., Chambon, P. and Voorhees, J.J. (1990) Retinoic Acid and Synthetic Analogs Differentially Activate Retinoic Acid Receptor Dependent Transcription. *Biochemical and Biophysical Research Communications*, **173**, 339-345. http://dx.doi.org/10.1016/S0006-291X(05)81062-9

[17] Schulman, I.G., Li, C., Schwabe, J.W. and Evans, R.M. (1997) The Phantom Ligand Effect: Allosteric Control of Transcription by the Retinoid X Receptor. *Genes & Development*, **11**, 299-308. http://dx.doi.org/10.1101/gad.11.3.299

[18] Lala, D.S., Mukherjee, R., Schulman, L.G., Canan-Koch, S.S., Dardashti, L.J., Nadzan, A.M., Croston, G.E., Evans, R.M. and Heyman, R.A. (1996) Activation of Specific RXR Heterodimers by an Antagonist of RXR Homodimers. *Nature*, **383**, 450-453. http://dx.doi.org/10.1038/383450a0

[19] Mangelsdorf, D.J. and Evans, R.M. (1995) The RXR Heterodimers and Orphan Receptors. *Cell*, **83**, 841-850.

http://dx.doi.org/10.1016/0092-8674(95)90200-7

[20] Durand, B., Saunders, M., Leroy, P., Leid, M. and Chambon, P. (1992) All-trans and 9-cis Retinoic Acid Induction of CRBPII Transcription Is Mediated by RARA/RXRA Heterodimers Bound to DR1 and DR2 Repeated Motifs. *Cell*, **71**, 73-85. http://dx.doi.org/10.1016/0092-8674(92)90267-G

[21] Li, C., McFadden, S.A., Morgan, I., Cui, D., Hu, J., Wan, W. and Zeng, J. (2010) All-Trans Retinoic Acid Regulates the Expression of the Extracellular Matrix Protein Fibulin-1 in the Guinea Pig Sclera and Human Scleral Fibroblasts. *Molecular Vision*, **16**, 689-697.

[22] de Thé, H., Marchio, A., Tiollais, P. and Dejean, A. (1989) Differential Expression and Ligand Regulation of the Retinoic Acid Receptor Alpha and Beta Genes. *EMBO Journal*, **8**, 429-433.

[23] Loudig, O., Babichuk, C., White, J., Abu-Abed, S., Mueller, C. and Petkovich, M. (2000) Cytochrome P450RAI (CYP26) Promoter: A Distinct Composite Retinoic Acid Response Element Underlies the Complex Regulation of Retinoic Acid Metabolism. *Molecular Endocrinology*, **14**, 1483-1497. http://dx.doi.org/10.1210/mend.14.9.0518

[24] Morrison, A., Moroni, M.C., Ariza-McNaughton, L., Krumlauf, R. and Mavilio, F. (1996) *In Vitro* and Transgenic Analysis of a Human HOXD4 Retinoid-Responsive Enhancer. *Development*, **122**, 1895-1907.

[25] Pikarsky, E., Sharir, H., Ben-Shushan, E. and Bergman, Y. (1994) Retinoic Acid Represses Oct-3/4 Gene Expression through Several Retinoic Acid-Responsive Elements Located in the Promoter-Enhancer Region. *Molecular and Cellular Biology*, **14**, 1026-1038.

[26] Kato, S., Sasaki, H., Suzawa, M., Masushige, S., Tora, L., Chambon, P. and Gronemeyer, H. (1995) Widely Spaced, Directly Repeated PuGGTCA Elements Act as Promiscuous Enhancers for Different Classes of Nuclear Receptors. *Molecular and Cellular Biology*, **15**, 5858-5867.

[27] Zhang, X.K., Lehmann, J., Hoffmann, B., Dawson, M.I., Cameron, J., Graupner, G., Hermann, T., Tran, P. and Pfahl, M. (1992) Homodimer Formation of Retinoid X Receptor Induced by 9-Cis Retinoic Acid. *Nature*, **358**, 587-591. http://dx.doi.org/10.1038/358587a0

[28] Lee, C.H. and Wei, L.N. (1999) Characterization of an Inverted Repeat with a Zero Spacer (IR0)-Type Retinoic Acid Response Element from the Mouse Nuclear Orphan Receptor TR2-11 Gene. *Biochemistry*, **38**, 8820-8825. http://dx.doi.org/10.1021/bi9903547

[29] Sherry, S.T., Ward, M.H., Kholodov, M., Baker, J., Phan, L., Smigielski, E.M. and Sirotkin, K. (2001) dbSNP: The NCBI Database of Genetic Variation. *Nucleic Acids Research*, **29**, 308-311. http://dx.doi.org/10.1093/nar/29.1.308

[30] Yanai, K., Saito, T., Hirota, K., Kobayashi, H., Murakami, K. and Fukamizu, A. (1997) Molecular Variation of the Human Angiotensinogen Core Promoter Element Located between the TATA Box and Transcription Initiation Site Affects Its Transcriptional Activity. *Journal of Biological Chemistry*, **272**, 30558-30562. http://dx.doi.org/10.1074/jbc.272.48.30558

[31] Ponomarenko, J.V., Orlova, G.V., Merkulova, T.I., Gorshkova, E.V., Fokin, O.N., Vasiliev, G.V., Frolov, A.S. and Ponomarenko, M.P. (2002) rSNP Guide: An Integrated Database-Tools System for Studying SNPs and Site-Directed Mutations in Transcription Factor Binding Sites. *Human Mutation*, **20**, 239-248. http://dx.doi.org/10.1002/humu.10116

[32] Ono, S., Ezura, Y., Emi, M., Fujita, Y., Takada, D., Sato, K., Ishigami, T., Umemura, S., Takahashi, K., Kamimura, K., Bujo, H. and Saito, Y. (2003) A Promoter SNP (-1323T>C) in G-Substrate Gene (GSBS) Correlates with Hypercholesterolemia. *Journal of Human Genetics*, **48**, 447-450. http://dx.doi.org/10.1007/s10038-003-0055-x

[33] Wang, L.F., Fokas, E., Juricko, J., You, A., Rose, F., Pagenstecher, A., Engenhart-Cabillic, R. and An, H.X. (2008) Increased Expression of EphA7 Correlates with Adverse Outcome in Primary and Recurrent Glioblastoma Multiforme Patients. *BMC Cancer*, **8**, 79. http://dx.doi.org/10.1186/1471-2407-8-79

[34] Breitman, T.R., Collins, S.J. and Keene, B.R. (1981) Terminal Differentiation of Human Promyelocytic Leukemic Cells in Primary Culture in Response to Retinoic Acid. *Blood*, **57**, 1000-1004.

[35] Reynolds, C.P., Kane, D.J., Einhorn, P.A., Matthay, K.K., Crouse, V.L., Wilbur, J.R., Shurin, S.B. and Seeger, R.C. (1991) Response of Neuroblastoma to Retinoic Acid *in Vitro* and *in Vivo*. *Progress in Clinical and Biological Research*, **366**, 203-211.

[36] Das, A., Banik, N.L. and Ray, S.K. (2007) Differentiation Decreased Telomerase Activity in Rat Glioblastoma C6 Cells and Increased Sensitivity to IFN-Gamma and Taxol for Apoptosis. *Neurochemical Research*, **32**, 2167-2183. http://dx.doi.org/10.1007/s11064-007-9413-y

[37] Haque, A., Das, A., Hajiaghamohseni, L.M., Younger, A., Banik, N.L. and Ray, S.K. (2007) Induction of Apoptosis and Immune Response by All-Trans Retinoic Acid Plus Interferon-Gamma in Human Malignant Glioblastoma T98G and U87MG Cells. *Cancer Immunology, Immunotherapy*, **56**, 615-625. http://dx.doi.org/10.1007/s00262-006-0219-6

[38] Karmakar, S., Banik, N.L., Patel, S.J. and Ray, S.K. (2006) Curcumin Activated Both Receptor-Mediated and Mitochondria-Mediated Proteolytic Pathways for Apoptosis in Human Glioblastoma T98G Cells. *Neuroscience Letters*, **407**, 53-58. http://dx.doi.org/10.1016/j.neulet.2006.08.013

[39] Karmakar, S., Banik, N.L., Patel, S.J. and Ray, S.K. (2007) Combination of All-Trans Retinoic Acid and Taxol Regressed Glioblastoma T98G Xenograft in Nude Mice. *Apoptosis*, **12**, 2077-2087. http://dx.doi.org/10.1007/s10495-007-0116-2

[40] Karmakar, S., Roy Choudhury, S., Banik, N.L. and Ray, S.K. (2010) Activation of Multiple Molecular Mechanisms for Increasing Apoptosis in Human Glioblastoma T98G Xenograft. *Journal of Cancer Science & Therapy*, **2**, 107-113. http://dx.doi.org/10.4172/1948-5956.1000033

[41] Hua, S., Kittler, R. and White, K.P. (2009) Genomic Antagonism between Retinoic Acid and Estrogen Signaling in Breast Cancer. *Cell*, **137**, 1259-1271. http://dx.doi.org/10.1016/j.cell.2009.04.043

[42] Ross-Innes, C.S., Stark, R., Holmes, K.A., Schmidt, D., Spyrou, C., Russell, R., Massie, C.E., Vowler, S.L., Eldridge, M. and Carroll, J.S. (2010) Cooperative Interaction between Retinoic Acid Receptor-α and Estrogen Receptor in Breast Cancer. *Genes & Development*, **24**, 171-182. http://dx.doi.org/10.1101/gad.552910

18

Identification of CAR/RXRα Heterodimer Binding Sites in the Human Genome by a Modified Yeast One-Hybrid Assay

Kenta Hosoda[1], Yuichiro Kanno[2], Masashi Sato[1], Jun Inajima[2], Yoshio Inouye[2], Kazuyuki Yanai[1*]

[1]Department of Biomolecular Science, Faculty of Science, Toho University, Chiba, Japan
[2]Faculty of Pharmaceutical Sciences, Toho University, Chiba, Japan
Email: [*]yanai@biomol.sci.toho-u.ac.jp

Abstract

The constitutive androstane receptor (CAR) is a transcription factor that belongs to the nuclear receptor superfamily. CAR binds as a heterodimer with the retinoid X receptor α (RXRα) to CAR response elements (CAREs) and regulates the expression of various drug metabolizing enzymes and transporters. To identify CAR/RXRα binding sites in the human genome, we performed a modified yeast one-hybrid assay that enables rapid and efficient identification of genomic targets for DNA-binding proteins. DNA fragments were recovered from positive yeast colonies by PCR and sequenced. A motif enrichment analysis revealed that the most frequent motif was a direct repeat (DR) of RGKTCA-like core sequence spaced by 4 bp. Next, we predicted 149 putative CAR/RXRα binding sites from 414 unique clones, by searching for DRs, everted repeats (ERs) and inverted repeats (IRs) of the RGKTCA-like core motif. Based on gel mobility shift assays, the CAR/RXRα heterodimer could directly interact with the 108 predicted sequences, which included not only classical CAREs but also a wide variety of arrangements. Furthermore, we identified 17 regulatory polymorphisms on the CAR/RXRα-binding sites that may influence individual variation in the expression of CAR-regulated genes. These results provide insights into the molecular mechanisms underlying the physiological and pathological actions of CAR/RXRα heterodimers.

Keywords

Constitutive Androstane Receptor, Retinoid X Receptor, Transcription, SNP, Polymorphism, Nuclear Receptor

*Corresponding author.

1. Introduction

Constitutive androstane receptor (CAR), a DNA-binding and ligand-regulated transcription factor that belongs to the superfamily of nuclear receptors [1], is mainly expressed in the liver and acts as a chemical sensor of xenobiotics [2]. In addition, it is expressed in a wide range of organs such as the small intestine, kidney, adrenals, testis, and brain [3]. Generally, it is localized in the cytoplasm in a complex with heat shock protein 90 and the CAR cytoplasmic retention protein and is translocated to the nucleus in response to the stimulation of cells by phenobarbital (PB) and other CAR activators [4] [5]. In the nucleus, CAR binds as a heterodimer with the retinoid X receptor α (RXRα) to CAR response element (CARE) and regulates the expression of various drug metabolizing enzymes and transporters including CYP2B6 [6], CYP3A4 [7], CYP2C9 [8], UDP-glucuronosyltransferase type 1A1 [9], sulfotransferase 2A1 [10], and ABCG2 [11].

Initial studies indicated that CAR/RXRα bound 5 bp-spaced direct repeats (DR5) [2]. Later studies found that CAR/RXRα bound DR4 motif within phenobarbital response enhancer modules (PBREMs) in a proximal promoter region of the human CYP2B6 gene and a variety of DRs [6] [12]. Moreover, CAR can also bind to everted repeat (ER) and inverted repeat (IR) arrangements [7] [10] [13] [14]. Because most known CAREs have been identified within promoter regions, CAREs at distal regions, *i.e.*, more than 10 kb upstream and downstream of transcription start sites (TSS) of target genes, remain unknown. Although DNA microarray experiments have been performed to study the expression profiles of genes regulated specifically by CAR/RXRα, microarray approaches cannot determine whether the regulated genes are primary or secondary target genes [13] [15]-[16]. Recently, genome-wide screening to identify transcription factor-binding sites, using methods such as ChIP-on-chip and ChIP-seq, has been performed against the many nuclear receptors [17]-[23]. Such a whole genome-approach has not been adopted previously for CAR/RXRα.

We previously developed a modified yeast one-hybrid (MY1H) system that enabled rapid and efficient identification of genomic targets for DNA-binding proteins [24]. Here, using this system, we reported functional screening for CAR/RXRα binding sites in the human genome. We demonstrated that 108 human genomic fragments could directly interact with the CAR/RXRα heterodimer by electrophoretic mobility shift assays (EMSAs). Moreover, we identified 17 regulatory single nucleotide polymorphisms (rSNPs) within the identified CAR/RXRα binding sites.

2. Materials and Methods

2.1. Plasmid Constructions

Human RXRα was amplified by the polymerase chain reaction (PCR) from uterus cDNA (PCR Ready-cDNA, Maxim Biotech, Inc., San Francisco, CA, USA). This cDNA fragments were cloned into pGADT7 (CLONTECH, Mountain View, CA, USA) and reamplified by PCR with primers (**Table 1**, RXRα_F and RXRα_R) to generate the restriction sites for subcloning. pGADT7 was cleaved with *Hind*III and ligated with linker DNA to remove the nuclear localization signal (NLS) and GAL4 activation domain (GAL4AD). The resulting plasmid was designated as pADH1. Then, RXRα cDNA was inserted into pADH1 and the resulting plasmid was named pADH1_RXRα. A foot-and-mouth disease virus (FMDV) 2A sequence was amplified by annealing synthetic oligonucleotides (**Table 1**, FMDV 2A_F1, and FMDV 2A_R1) and reamplified by PCR with primers (**Table 1**, FMDV 2A_F2, and FMDV 2A_R2) to generate the restriction sites. FMDV 2A fragment was inserted at the C-terminal end of RXR in pADH1_RXRα and the resulting plasmid was named pADH1_RXRα_2A. The fragment including NLS-GAL4AD was amplified by PCR from pGADT7 with primers (**Table 1**, NLS_GAL4AD_F and NLS_GAL4AD_R) and inserted into pADH1. The resulting plasmid was named pADH1_NLS_GAL4AD. Similarly, three tandem copies of NLS_GAL4AD were inserted at pADH1 and the resulting plasmid was designated pADH1_NLS_GAL4AD × 3. Human CAR cDNA was cloned by PCR with primers (**Table 1**, NLS_GAL4AD_F and NLS_GAL4AD_R) from pcDNA-CAR [25] and reamplified by PCR with primers (**Table 1**, pADH1-hCAR_F and pADH1-hCAR_R) to generate the restriction sites for subcloning. The cDNA was inserted into pADH1_NLS_GAL4AD × 3 and the resulting plasmid was named pADH1_NLS_GAL4AD × 3_ CAR. The NLS_GAL4AD × 3_CAR fragment obtained from pADH1_NLS_GAL4AD × 3_CAR was inserted at the C-terminal end of RXRα_2A in pADH1_RXRα_2A and the resulting plasmid was named pADH1_RXR α_2A_NLS_GAL4AD × 3_CAR. pSUR (GeneBank AB425277) was constructed as previously described [24] and used as a reporter in yeast one-hybrid assays. pSUR-DR4 and pSUR-CYP2B6-NR3 were constructed

Table 1. Primer sequences.

Primer name	Sequence (5'→3')
hRXRα_F	AAAAAGCTTACGCGTGCCGCCACCATGGACACC
hRXRα_R	TTTAAGCTTTCTAGACTACTCGAGAGTCATTTGGTGCGGCGC
FMDV-2A_F1	AAAAGATCTTAAAATTGTCGCTCCTGTCAAACAAACTCTTAACTTTGATTTACTCAAACTGGCTG
FMDV-2A_R1	AAATCTAGAGGATCCTTTACTAGTTGGACCTGGATTGCTTTCTACATCCCCAGCCAGTTTGAGTAAATCA
FMDV-2A_F2	AAAGTCGACAAAATTGTCGCTCCTGTCAA
FMDV-2A_R2	TTTTCTAGAGAATTCCCGCGGCTCGAGACGCGTTGGACCTGGATTGCTTTC
NLS-GAL4AD_F	AAAAGATCTATGGATAAAGCGGAATTAATTCCCGAGC
NLS-GAL4AD_R	TTTCTCGAGTTTGGATCCCTCTTTTTTTGGGTTTGGTGG
pADH1-hCAR_F	CTTGTCGACAGATCTGCCGCCACCATGGCCAGTAGGGAAGATGA
pADH1-hCAR_R	CTTTCTAGACTACTCGAGTCAGCTGCAGATCTCCTGGAGCAGCGG
hCAR_F	AAAGTCGACGGATCCGCCGCCACCATGGCCAGT
hCAR_R	TTTTCTAGACTACTCGAGTCAGCTGCAGATCTCCTGGA
dpSUR_F	AAAAAGTTATCAAGAGACTGC
dpSUR_R	CTAATGCTTCAACTAACTCCA
SPO13-S	CGGCTATTTCTCAATATACTCC
Consensus DR4_F	GATCAGTTCATGGCAGTTCATGGCAGTTCAGATC
Consensus DR4_R	GATCTGAACTGCCATGAACTGCCATGAACTGATC
random_F	CGCGTTGTGTGTGTTTTATTCC
random_R	GGAATAAAACACACACAACGCG

by inserting 3 copies of consensus DR4 [26] and 4 copies of CYP2B6 XREM-NR3 [27] upstream of the SPO13 promoter of pSUR, respectively. For *in vitro* transcription/translation, the cDNAs for human RXRα and CAR were inserted into the *Hind*III/*Xho* I site and the *Sal* I site of pSP64 Poly (A) (Promega, Madison, WI, USA), respectively.

2.2. A modified Yeast One-Hybrid Assays

The human genomic library for a modified yeast one-hybrid assay was generated as previously described [24]. The effector plasmid, pADH1_RXRα_2A_NLS_GAL4AD × 3_CAR, was transformed into the 5FOA-selected yeast containing the human genomic library using polyethylene glycol/lithium acetate. The obtained transformants were grown on synthetic complete media lacking leucine, tryptophan, and uracil but containing 25 μg/ml 6-azauracil for 3 weeks at 30 °C. Human genomic fragments were recovered from the positive colonies by colony-direct PCR with primers corresponding to the vector sequences (**Table 1**, dpSUR_F and dpSUR_R). The PCR fragments were directly sequenced with the primers (**Table 1**, SPO13-S) and used for further experiments.

2.3 EMSAs

The TNT SP6 High Yield system from Promega was used to prepare the human CAR and RXRα proteins. Double-stranded DNA probes (**Table 1**, consensus DR4_F and consensus DR4_R) were used with both ends labeled with Cy5. Proteins were incubated with 200 ng of calf thymus DNA (Invitrogen, Carlsbad, CA, USA) and 1 pmol of the labeled oligonucleotide at 4°C in the presence or absence of the unlabeled oligonucleotides. The binding reaction was carried out in the EMSA binding buffer containing 12 mM HEPES (pH 7.9), 60 mM KCl, 4 mM $MgCl_2$, 1 mM EDTA, 12% glycerol, and 0.5% Nonidet P-40. The reaction mixtures were directly loaded onto 4% nondenaturing polyacrylamide gels made in 0.5 × TBE. In the supershift experiments, CAR and RXR*α* proteins were incubated with antibodies against CAR (Santa Cruz Biotechnology, Santa Cruz, CA, USA, sc-13065) or RXR*α* (Santa Cruz Biotechnology, sc-553X) overnight before incubation with the probe. After electrophoresis was performed at 4˚C, the gels were analyzed using a bio-imaging analyzer (FLA-7000 FUJIFILM). The consensus DR4 and random sequences (**Table 1**, random_F and random_R) were used as positive and negative controls, respectively.

2.4. Bioinformatics

To map the obtained sequences on the human genome assembly (GRCh37), the cloned sequences were analyzed using NCBI's BLAST and the RGKTCA motif was searched using EMBOSS fuzznuc of Galaxy (https://usegalaxy.org/) [28]-[30]. For stringency of the search, we allowed up to 2-bp mismatches in the DR4 motif and 1-bp mismatches in the other motifs. The nearest gene and the distance from the center of the binding site to the transcriptional start site of the gene within 1000 kb were identified with GREAT (http://bejerano.stanford.edu/great/public/html/) [31].

3. Results

3.1. Interaction between the CAR/RXR*α* Heterodimer and CARE in Yeast

To simultaneously express CAR and RXR*α* proteins in yeast, we constructed effector plasmids by placing FMDV 2A peptide between CAR and RXR*α* [32]. Although CAR is expressed as a fusion to the NLS_GAL4AD, RXR*α* was expressed as the native protein to minimize the effect of RXR*α* homodimer on reporter activations (**Figure 1(a)**). To evaluate the function of CAR/RXR*α* heterodimer, yeast cells were transformed with these effectors and the indicated reporters (**Figure 1(b)**). The transformants expressing either CAR or RXR*α* alone were unable to grow, whereas the transformants expressing both CAR and RXR*α* were able to grow in a CARE-dependent manner (**Figure 1(b)**). The known CAR activators, CITCO [14] and PB [33] had no effect on the yeast growth in our assay conditions (data not shown). These results indicated that CAR/RXR*α* heterodimer could activate the reporter gene via CAREs.

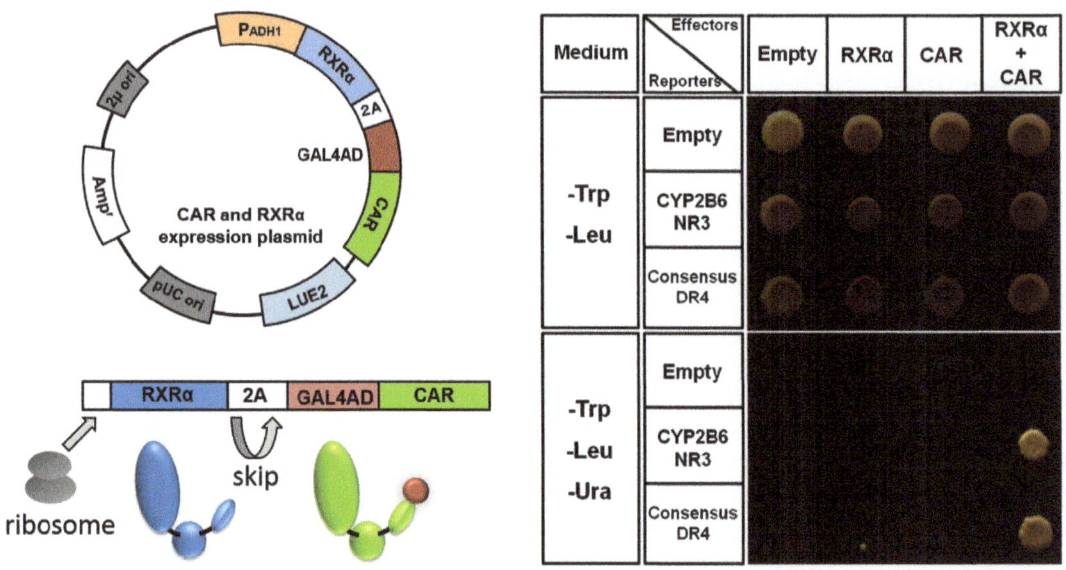

(a) (b)

Figure 1. Yeast one-hybrid assay of the CAR/RXR*α* heterodimer. (a) Schematic diagrams of the effector plasmid. FMDV 2A peptide sequence was placed between CAR and RXRα under the control of the ADH1 promoter. CAR and RAR*α* were expressed via a 2A-mediated translational skip mechanism; (b) An examination of the CAR/RXR*α* heterodimer in yeast cells. Yeast cells were transformed with each effector and reporter plasmids and were grown on synthetic complete media lacking leucine and tryptophan (-Trp -Leu) or and lacking leucine, tryptophan, and uracil (-Trp -Leu -Ura). The plates were photographed after 4 days of growth at 30˚C.

3.2. Identification of CAR/RXRα Heterodimer-Binding Sites in the Human Genome

The human genomic library was constructed by inserting approximately 300 bp fragments in reporter plasmids and treating with 5FOA to efficiently eliminate false-positive interactions in the yeast one-hybrid assay [24]. The CAR/RXRα expression plasmids were transformed into the library and more than 2.8×10^6 were selected on synthetic complete media lacking leucine, tryptophan, and uracil but containing 25 μg/ml 6-azauracil. After 3 weeks, 421 positive colonies were picked from the selection plates and colony-direct PCR was performed to recover human genomic fragments. The PCR fragments were directly sequenced and 414 unique sequences were obtained. As a result of genome mapping, the majority of the 414 unique sequences were located at distal sites, far from the TSS, or in introns. Next, we performed motif enrichment among the 414 unique clones using MEME-ChIP [34]. The most frequently observed motif was a DR4 motif which is known as a classical CARE (**Figure 2**). Then, to obtain putative CAR/RXRα binding sites from the 414 unique clones, we selected these sequences based on the half-site core motif (RGKTCA). As a result, a total of 149 putative CAR/RXRα binding sites were obtained (**Figure 3(b)**). Interestingly, these sequences contained various types of elements such as DRs, ERs and IRs.

Figure 2. Motif enrichment. The most significant motifs found by MEME-ChIP (http://meme.nbcr.net/meme/doc/meme-chip.html) in the clones obtained from the modified yeast one-hybrid assay. E value: 9.7×10^{-34}. R = A or G, K = T or G, W = A or T.

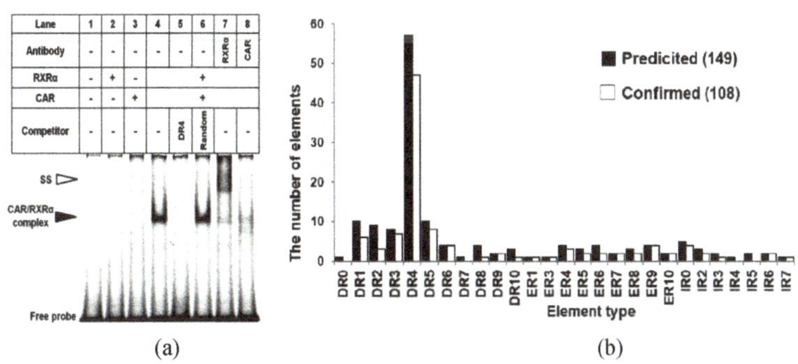

Figure 3. Validation assays for direct interactions between the CAR/RXRα heterodimer and the predicted elements. (a) EMSAs were performed with the *in vitro*-synthesized human CAR and RXRα proteins. The proteins were incubated with 1 pmol of the Cy5-labeled consensus DR4. In a competition assay, 100-fold molar excess of the unlabeled oligonucleotides (DR4 or Random) were added to the reaction mixture. In a supershift experiment, the indicated antibodies were incubated in the reaction mixture. Closed and open arrowheads indicated the CAR/RXRα protein-DNA complexes and the super shifted band, respectively. (b) The number of the predicted CAR/RXRα binding motif and the confirmed. The predicted CAR/RXRα binding sites were obtained by EMBOSS fuzznuc of Galaxy based on RGKTCA motif. For the stringency of the search, motif substitution allowed up to 1-bp mismatches. Motif substitution allowed up to 2-bp mismatches only for the DR4 motif. The direct interaction was confirmed by EMSA competition assays.

3.3. Experimental Validation of the Putative CAR/RXR*α* Binding Sites

To examine the direct interaction of putative CAR/RXR*α* binding sites with the CAR/RXR*α* heterodimer, we performed EMSA. Incubation of the Cy5-labeled consensus DR4 with the combination of CAR and RXR*α*, but not either receptor alone, produced retarded complexes (**Figure 3(a)**, lanes 2 - 4). The complexes represented a sequence-specific interaction between consensus DR4 and the human CAR/RXR*α* proteins, since the formation of this complex was specifically reduced with molar excess of unlabeled competitors (**Figure 3(a)**, lanes 5 and 6). Moreover, the addition of anti-RXR*α* antibody created a slower-migrating complex and the addition of anti-CAR antibody resulted in the disappearance of the band (**Figure 3(a)**, lanes 7 and 8). No supershifted bands were observed with anti-HNF4 antibodies (data not shown). These results indicated that the sequence-specific binding complex contained both CAR and RXR*α*, presumably as a heterodimer.

To examine the direct interaction of CAR/RXR*α* with the 149 putative CAR/RXR*α* binding sites, we performed a semi-quantitative EMSA competition assay. In this method, it is possible to examine binding intensity using a 10- and 100-fold molar excess of unlabeled competitors. Typical examples are shown in **Figure 4**. Sequence I (lanes 15 - 20) was equivalent in binding intensity to the positive control (Consensus DR4, lanes 9 - 14). That is, sequence I had strong binding affinity with the CAR/RXR*α* heterodimer. In the same way, sequence II (lanes 21 - 26) had moderate binding affinity and sequence III (lanes 27 - 32) had low binding affinity. Thus, we evaluated 149 putative CAR/RXR*α* binding sites and confirmed that at least 108 CAR/RXR*α* binding sites could directly interact with the CAR/RXR*α* heterodimer by EMSA (**Table 2**). These sequences contained not only classical CAREs but also a wide variety of additional arrangements.

3.4. rSNPs in the Identified CAR/RXR*α* Binding Sites

rSNPs in transcription factor-binding sites, which alter the ability of a transcription factor to interact with DNA, may lead to predictable differences in gene expression and may be associated with disease susceptibility. Therefore, we tried to identify rSNPs in each of the CAR/RXR*α* binding sequences using the NCBI SNP database. Thirty five rSNPs were identified and the effect on the binding affinity between CAR/RXR*α* and 35 rSNPs was examined by EMSA. As a result, 17 rSNPs were identified based on differences in the DNA-binding affinities (**Table 3**)

4. Discussion

In the present study, we identified CAR/RXR*α*-binding sites in the human genome by a modified yeast one-hybrid assay. Next, we demonstrated that the 108 obtained sequences could directly interact with the CAR/RXR*α* heterodimer by EMSAs. These sequences contained not only classical CAREs but also a wide variety of arrangements (e.g., DRs, ERs, and IRs). Moreover, 17 functional rSNPs on the CAR/RXR*α* binding sites were identified by analyzing differences in DNA-binding affinity.

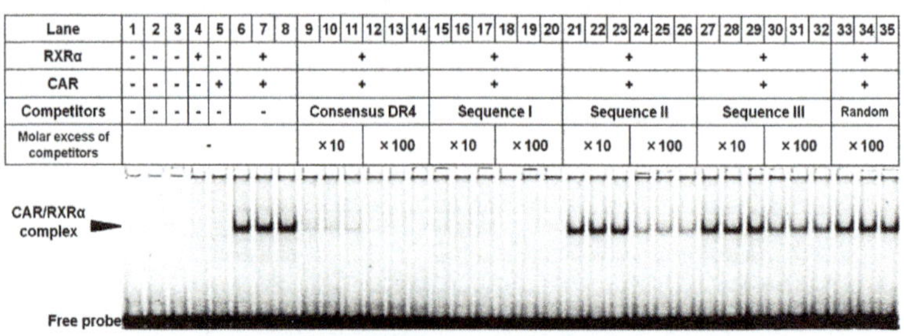

Figure 4. Typical examples of evaluation assays for the putative CAR/RXR*α* binding sites. EMSAs were performed with the *in vitro*-synthesized human CAR and RXR*α* proteins. The proteins were incubated with 1 pmol of the Cy5-labeled consensus DR4. In a competition assay, 10 or 100-fold molar excess of the unlabeled oligonucleotides were added to the reaction mixture. The closed arrowhead indicates the CAR/RXR*α* protein-DNA interaction. Sequence I is a strong binding sequence (**Table 2**, #30). Sequence II is a moderate binding sequence **Table 2**, #79). Sequence III is a week binding sequence (**Table 2**, #89).

Table 2. CAR/RXR*α*-binding sites.

#	Position	Length (bp)	Neighboring Gene (Distance to TSS)	CAR/RXR*α* bindnig sequence	Motif	Competition (%)	
						×10	×100
1	Chr18: 55201286-55201390	105	FECH (+52,631), ONECUT2 (+98,421)	tggAGTTCAaAGGTCAcat	DR1	14.6 ± 1.6	95.2 ± 0.4
2	Chr7: 138292264-138292495	232	SVOPL (+71,410), TRIM24 (+147,301)	cgcAGGTCAaAGGTCAggt	DR1	10.7 ± 2.2	78.7 ± 2.9
3	Chr5: 76841961-76842018	58	OTP (+92,532), PDE8B (+335,284)	cagAGGTCGgAGGTCAagt	DR1	17.9 ± 0.5	43.5 ± 2.1
4	Chr20: 23387626-23387738	113	NAPB (+14,474), GZF1 (+42,682)	caaAGTTCAaAGGTCActg	DR1	52.0 ± 1.3	115.8 ± 0.9
5	Chr13: 59071533-59071644	112	PCDH17 (+865,800)	ttaAGTGCAgGGGTCAgga	DR1	25.2 ± 0.9	44.6 ± 0.8
6	Chr17: 1918186-1918361	176	RTN4RL1 (+9,904), RPA1 (+185,001)	tggAGCTCAgAGTTCAaag	DR1	41.4 ± 1.8	90.3 ± 1.1
7	Chr4: 121700806-121701088	283	MAD2L1 (−712,934), PRDM5 (+143,066)	ttgAGGTCAagAGTTCAagg	DR2	30.2 ± 2.6	101.9 ± 1.4
8	ChrX: 40929970-40930061	92	MED14 (−335,212), USP9X (−14,872)	tgtGGGTGAacAGGTCAatg	DR2	71.5 ± 3.0	102.7 ± 0.3
9	Chr13: 49397697-49397876	180	FNDC3A (−152,261), CYSLTR2 (+116,834)	ggcGGATCAtgAGGTCAaga	DR2	22.2 ± 0.8	50.2 ± 2.1
10	Chr17: 8796437-8796604	168	PIK3R6 (−25,527), PIK3R5 (+19,313)	tggAGGTCAgagAGGTCAagc	DR3	35.3 ± 1.1	110.3 ± 9.6
11	Chr2: 103266921-103267061	141	SLC9A2 (+30,825), MFSD9 (+86,346)	tgaAGGTCAaagAGTTTAaca	DR3	56.8 ± 0.5	102.8 ± 1.1
12	Chr5: 142827845-142828030	186	NR3C1 (−43,893), YIPF5 (+722,340)	cagAGGTCAcagAGGTCAtaa	DR3	60.0 ± 0.6	111.6 ± 2.1
13	Chr7: 33720609-33720751	143	BMPER (−223,843), BBS9 (+551,528)	ccaAGTTCAcatAGTTCAtaa	DR3	65.5 ± 0.4	111.3 ± 1.3
14	ChrX: 45732008-45732304	297	CXorf36 (−672,010), ZNF673 (−574,468)	ctgAGGCCAgggAGTTCAcca	DR3	35.9 ± 2.1	91.5 ± 2.9
15	Chr17: 54747517-54747691	175	NOG (+76,544), C17orf67 (+145,767)	cacAGGTCAggcAGTTGAtgc	DR3	13.1 ± 0.9	46.9 ± 1.9
16	Chr6: 4942951-4943334	384	RPP40 (+61,128), CDYL (+166,463)	cggAGGACAgtgGGGTCAggc	DR3	34.4 ± 5.8	89.6 ± 2.6
17	Chr3: 120575966-120576060	95	STXBP5L (−51,037), GTF2E1 (+114,455)	taaAGGTCAaggaAGTTCAatt	DR4	62.8 ± 1.3	107.8 ± 1.4
18	Chr12: 14672183-14672322	140	PLBD1 (+48,538), ATF7IP (+153,642)	ttaAGGTCAactgAGGTCAtat	DR4	51.5 ± 1.2	107.6 ± 0.7
19	Chr1: 220849174-220849356	183	C1orf115 (−14,363), MARK1 (+147,697)	gcaAGGGCAggggAGGTCAat a	DR4	30.6 ± 1.1	77.2 ± 0.6
20	Chr10: 30521090-30521250	161	SVIL (−597,269), MTPAP (+117,097)	agtGGATCActtgAGGTCAgga	DR4	20.6 ± 0.8	69.0 ± 1.1
21	Chr2: 55562474-55562589	116	MTIF2 (−66,148), CCDC88A (+84,525)	gtgAGGTCAttaaAGTTCAtgt	DR4	70.8 ± 0.8	112.6 ± 0.4
22	Chr8: 1165688-1165800	113	C8orf42 (−670,413), DLGAP2 (−283,825)	atgGGTTCAtttgGGTGCAtca	DR4	50.1 ± 0.7	104.2 ± 0.0
23	Chr13: 91370194-91370452	259	GPC5 (−680,612)	ctgGGGTCAtatgATTTCAtag	DR4	41.2 ± 1.2	93.4 ± 0.8
24	Chr3: 177287827-177287945	119	KCNMB2 (−966,338), TBL1XR1 (−372,838)	ttaGGGTCAgctgAGTTCAtat	DR4	61.5 ± 1.9	100.8 ± 1.3
25	Chr6: 42700810-42701034	225	PRPH2 (−10,564), TBCC (+12,962)	tcaAGGTTAcaagAGTTCAaga	DR4	75.4 ± 0.2	127.9 ± 4.5

Continued

26	ChrX: 66743819-66744135	317	EDA2R (−908,105), AR (−19,897)	gagAGGTCAatgaGGTTCAcag	DR4	49.7 ± 0.8	99.8 ± 0.7
27	Chr9: 13275072-13275351	280	MPDZ (−24,841)	taaAGGTTAaatgAGGTCActa	DR4	53.2 ± 0.8	104.7 ± 1.6
28	Chr19: 5982545-5982685	141	RANBP3 (−4295)	tcaAGTTCAtgaaAGGTCAcac	DR4	61.7 ± 1.3	114.6 ± 0.6
29	Chr5: 168554759-168554923	165	PANK3 (−548,227), SLIT3 (+173,292)	caaAGTTCActtaAGTTCAgag	DR4	57.6 ± 0.7	116.7 ± 0.7
30	Chr8: 4846264-4846562	299	CSMD1 (+5915)	caaAGGTCActtaAGGTCAttt	DR4	71.4 ± 1.5	125.1 ± 1.1
31	Chr1: 88626432-88626556	125	PKN2 (−523,428), LMO4 (+832,343)	tttAGGTCAttgaAGTTCAagt	DR4	55.4 ± 0.7	106.2 ± 3.5
32	Chr8: 95584473-95584707	235	ESRP1 (−68,774), KIAA1429 (−18,844)	aagAGGTCAgttgGGGTCAgat	DR4	45.0 ± 1.3	99.8 ± 1.3
33	Chr7: 15597442-15597713	272	DGKB (−716,503), AGMO (+4062)	caaAGGGCAccagAGTTAAcag	DR4	23.9 ± 1.6	67.6 ± 0.8
34	Chr14: 54345870-54346006	137	DDHD1 (−725,892), BMP4 (+75,332)	ctgAGGTCAagtgAGTTCAcct	DR4	70.2 ± 0.9	118.5 ± 1.4
35	Chr15: 49543989-49544276	288	FGF7 (−171,242), GALK2 (+81,711)	ggtGGATCActtgAGGTCAgga	DR4	26.0 ± 2.4	65.0 ± 0.7
36	Chr13: 55084028-55084114	87	NONE	atgGGGTCAacaaAGTTTAaga	DR4	83.3 ± 4.9	119.2 ± 5.2
37	Chr8: 112300235-112300383	149	NONE	aagAGGTTAcaagAGTTCAtat	DR4	69.2 ± 1.1	112.9 ± 0.3
38	Chr3: 28008615-28008785	171	CMC1 (−274,424), EOMES (−244,915)	gtgGGGTCAcatgAGGGCAcat	DR4	51.9 ± 1.8	105.3 ± 1.0
39	Chr15: 87420670-87420730	61	AGBL1 (+735,458)	aagAGGTCAactgAGTTCAtcc	DR4	69.2 ± 0.6	114.4 ± 0.3
40	Chr9: 81575106-81575356	251	TLE4 (−611,647), PSAT1 (+663,172)	aggAGGTGAtaaaGGTTAAatg	DR4	21.0 ± 1.5	64.6 ± 0.5
41	Chr15: 65168224-65168305	82	PIF1 (−50,427), ANKDD1A (−35,836)	atgGTGTCAgaagAGTTCAtca	DR4	64.7 ± 0.6	107.5 ± 0.2
42	Chr10: 31605586-31605754	169	ZEB1 (−2431)	aaaAGGTGAacagAGTTCAttg	DR4	73.5 ± 1.4	113.6 ± 1.0
43	Chr4: 4003949-4004240	292	OTOP1 (+224,526), ADRA2C (+235,799)	atgGGGTCAgctaGGTTTActt	DR4	43.3 ± 0.8	100.1 ± 2.4
44	Chr5: 154114732-154114879	148	LARP1 (+22,344), C5orf4 (+115,407)	ggcAGGTCAcctgAGGTCAgga	DR4	34.2 ± 1.1	97.2 ± 0.3
45	Chr15: 69230539-69230630	92	NOX5 (−76,449), SPESP1 (+7746)	ggtGGGTCAactaAGGTCAgga	DR4	50.2 ± 3.9	100.7 ± 0.9
46	Chr8: 140124212-140124570	359	COL22A1 (−198,155), KCNK9 (+590,908)	gccAGTGCAgaagAGGTCAcac	DR4	27.7 ± 2.0	78.3 ± 0.9
47	Chr8: 107263662-107263790	129	OXR1 (−196,426), ZFPM2 (+932,579)	gtcGGGTCActtaAGGTCAgga	DR4	40.7 ± 0.4	95.7 ± 0.9
48	Chr15: 78475549-78475775	227	IDH3A (+33,943), ACSBG1 (+51,387)	ggaGGGTCAtttgAGGTCAgga	DR4	39.3 ± 0.8	98.1 ± 2.7
49	Chr14: 65099982-65100056	75	PLEKHG3 (−94,292), HSPA2 (+92,833)	attAGTTAAgatgAGGTCAtac	DR4	19.3 ± 5.4	62.2 ± 5.6
50	Chr6: 31790280-31790663	384	HSPA1B (−5040), HSPA1A (+7181)	caaGGATCAaagaAGTTCAgtg	DR4	57.1 ± 1.0	88.6 ± 0.2
51	Chr13: 90222946-90223063	118	NONE	gtaAGGTCAtaaaAGTTCAggt	DR4	63.1 ± 1.8	110.6 ± 0.4
52	ChrX: 132353555-132353667	113	TFDP3 (−1235)	ttaAGGTTAaatgAGGTCAtaa	DR4	55.1 ± 0.2	97.8 ± 0.4

Continued

53	Chr12: 80739933-80740212	280	PTPRQ (−98,053), OTOGL (+136,840)	caaGGTTGAttagAGTTCAgcc	DR4	48.2 ± 1.3	93.5 ± 0.3
54	Chr6: 133030154-133030300	147	TAAR1 (−63,085), VNN1 (+4967)	aaaAGGTCGtctgAGGGCAtca	DR4	37.7 ± 2.3	89.0 ± 0.3
55	Chr20: 38461220-38461351	132	MAFB (+856,590), DHX35 (+870,305)	acaGGGTCAcaaaAGGCCAata	DR4	34.3 ± 0.2	88.9 ± 1.6
56	Chr18: 609485-609841	357	CLUL1 (−7037), CETN1 (+29,294)	ggtGGGTCAtttgAGGTCAgga	DR4	41.9 ± 1.1	96.9 ± 0.6
57	Chr4: 4081417-4081706	290	OTOP1 (+147,059), ADRA2C (+313,266)	ggcGGATCActtgAGGTCAgga	DR4	17.5 ± 1.0	60.6 ± 0.7
58	Chr4: 27697777-27697856	80	STIM2 (+835,504)	gggAGGTCActcaAGTTCActg	DR4	66.1 ± 1.5	114.3 ± 1.4
59	Chr11: 70447681-70447959	279	CTTN (+203,208), SHANK2 (+488,022)	ggtGGATCAcctgAGGTCAgga	DR4	17.5 ± 0.9	55.7 ± 2.6
60	Chr17: 71023806-71024052	247	SLC39A11 (+64,924), SOX9 (+906,768)	caaGGGTCAaatcAGGGCActg	DR4	40.6 ± 1.4	102.2 ± 0.3
61	Chr7: 87875231-87875351	121	SRI (−25,898), STEAP4 (+60,937)	caaAGGTTAcaagAGTTCTctt	DR4	60.4 ± 1.4	96.3 ± 1.8
62	Chr2: 89376298-89376461	164	RPIA (+385,204)	caaAGGTTAccttAGGTCAtat	DR4	36.9 ± 1.5	92.5 ± 1.1
63	Chr2: 95707549-95707691	143	MAL (+16,141), MRPS5 (+80,134)	gggGGGTTAgcagAGTTGAgct	DR4	33.8 ± 3.4	76.2 ± 5.0
64	Chr4: 88898879-88898955	77	PKD2 (−29,882), SPP1 (+2,115)	cagAGTTCAattccAGTTGAaca	DR5	17.6 ± 2.2	73.4 ± 0.3
65	Chr1: 177083927-177084138	212	ASTN1 (+49,991), PAPPA2 (+651,726)	gctGGGTCAggaaaAGTTCAtgg	DR5	69.6 ± 9.1	98.0 ± 0.2
66	Chr6: 133330363-133330649	287	EYA4 (−231,989), RPS12 (+194,798)	agtGGTTCAatcacAGTTCActg	DR5	49.6 ± 0.2	89.9 ± 1.0
67	Chr14: 35253086-35253208	123	CFL2 (−69,118), BAZ1A (+91,706)	aaaAGTTCAaacacAGTTCAttt	DR5	71.6 ± 1.9	109.9 ± 0.9
68	Chr1: 84080031-84080193	163	TTLL7 (+384,721)	aagAGGTCAagtcaGGTACAaac	DR5	37.2 ± 0.7	86.4 ± 1.0
69	Chr5: 89560627-89560695	69	CETN3 (+144,942)	aggAGGTCAggaggAGTTCActg	DR5	67.0 ± 2.5	102.2 ± 0.8
70	ChrX: 36376152-36376213	62	PRRG1 (−832,345), CXorf59 (+311,130)	aggAGGTCAatatgAGGTCAcag	DR5	39.2 ± 0.3	95.8 ± 1.2
71	Chr1: 48570878-48571087	210	SLC5A9 (−117,374), LOC388630 (−108,421)	gaaAGGTCActcagAGTTCAttt	DR5	68.6 ± 9.7	120.4 ± 0.4
72	Chr10: 16445680-16445852	173	FAM188A (−543,247), PTER (−33,201)	taaAGTTCAtggtaaAGTTCAtgg	DR6	74.0 ± 1.2	119.3 ± 0.6
73	Chr4: 80819846-80819929	84	GK2 (−490,516), ANTXR2 (+174,589)	gaaAGTTCAtattgtAGGTCAttc	DR6	64.7 ± 1.0	112.9 ± 1.4
74	Chr2: 33407228-33407375	148	RASGRP3 (−331,640), LTBP1 (+234,933)	ctaAGTTCAtgatagAGTTCAaag	DR6	65.3 ± 1.3	112.5 ± 1.2
75	Chr6: 23048624-23048748	125	PRL (−745,604)	ttgAGCTCAattcctAGTTCAcca	DR6	27.2 ± 0.8	74.3 ± 0.8
76	Chr2: 135675135-135675350	216	CCNT2 (−1149)	ttgAGGTCAggagttcgAGGGCAagg	DR8	11.7 ± 1.0	54.8 ± 2.6
77	Chr20: 36825053-36825172	120	BPI (−107,439), TGM2 (−31,413)	ctgGGGTCAcagggccagAGGTCActg	DR9	18.8 ± 3.0	70.2 ± 3.6
78	Chr21: 34190796-34190941	146	C21orf62 (−4816)	tagGGGTCAcctcggtagGGGTCAggg	DR9	17.9 ± 0.6	58.7 ± 0.7
79	Chr1: 154209880-154209999	120	HAX1 (−35,099), UBAP2L (+16,615)	tcaGGGTCAaacaatgctgAGTTCAaat	DR10	34.4 ± 1.2	90.0 ± 0.6
80	Chr14: 98635538-98635762	225	NONE	gacTGAACTcTGGTCAttt	ER1	13.0 ± 5.0	58.3 ± 3.2
81	Chr12: 34343448-34343791	344	ALG10 (+168,404)	tggTGCCCTcctAGGTCAtgg	ER3	37.6 ± 0.8	41.6 ± 4.5

Continued

82	Chr12: 10876774-10876871	98	CSDA (−870)	attTGGCCCctagGGTTCAgac	ER4	19.6 ± 0.2	42.6 ± 0.6
83	Chr9: 130011528-130011906	379	ANGPTL2 (−126,673), GARNL3 (−15,039)	tctTGAACTcctgGGCTCAagt	ER4	4.4 ± 2.5	55.2 ± 0.9
84	Chr10: 58646226-58646342	117	ZWINT (−525,250)	ctgTGACCTcttgAGAACTtaa	ER4	19.7 ± 0.5	81.6 ± 1.7
85	Chr20: 7438451-7438525	75	HAO1 (+482,605), BMP2 (+689,743)	acaTGAACTtttggAGGTCAtat	ER5	78.2 ± 0.5	115.0 ± 0.4
86	Chr1: 242533417-242533615	199	MAP1LC3C (−371,131), PLD5 (+154,482)	caaGGAACTaaataAGTTCAttt	ER5	30.9 ± 1.1	75.6 ± 0.9
87	Chr1: 219789689-219789818	130	SLC30A10 (+312,239), LYPLAL1 (+442,562)	atgTGAACTcttctaAGTTCAtgg	ER6	79.0 ± 0.7	111.1 ± 0.7
88	Chr9: 46385533-46385719	187	NONE	gtgTGACCTttttaaAGTTAAatt	ER6	26.9 ± 2.5	87.2 ± 1.4
89	Chr3: 74240550-74240667	118	PDZRN3 (−566,537), CNTN3 (+329,734)	aacTGACCCattttcaGGTTCActg	ER7	30.0 ± 4.9	53.0 ± 2.5
90	Chr8: 41543991-41544261	271	NKX6−3 (−39,251), ANK1 (+210,154)	tgcTGACCTtgctgagAGTTCAtgt	ER7	66.1 ± 0.2	101.5 ± 0.5
91	Chr6: 108964946-108965090	145	ARMC2 (−204,601), FOXO3 (+82,949)	agcTGAACTctgcactaAGGTCAgat	ER8	65.3 ± 1.6	103 ± 0.3
92	Chr14: 91348266-91348370	105	TTC7B (−65,557), RPS6KA5 (+178,675)	gatTAAACTctaatcaaAGGTCAaag	ER8	76.5 ± 1.2	106.7 ± 3.1
93	Chr20: 1565505-1565776	272	SIRPD (−27,298), SIRPB1 (+35,048)	tctTGACCCttttgtcaaAGGTCAatt	ER9	58.9 ± 1.5	101.4 ± 0.6
94	Chr18: 66213526-66213902	377	TMX3 (+168,639)	tttTGAACTctgtctgaaAGGTCActt	ER9	72.2 ± 1.5	105.3 ± 0.5
95	Chr1: 42946546-42946764	219	PPIH (−177,393), PPCS (+24,482)	ggaTGACCTgaggccaggAGTTCAaga	ER9	54.4 ± 0.4	96.9 ± 1.5
96	Chr5: 88641418-88641561	144	MEF2C (−441,568)	accTTAACTctgtctggaAGGTCAaga	ER9	21.9 ± 1.0	79.5 ± 1.9
97	Chr5: 54048617-54048892	276	ESM1 (+232,659), SNX18 (+235,162)	agaTGACCTctgcccgctaAGTTCAagt	ER10	53.4 ± 1.5	111.2 ± 0.8
98	Chr17: 22251179-22251327	149	MTRNR2L1 (+228,816)	accTGAACTttcaaaggaaGGTTCAatt	ER10	31.8 ± 1.0	101.7 ± 1.7
99	Chr21: 42948507-42948610	104	TMPRSS2 (−68,567), RIPK4 (+238,690)	tggAGTTCATGACCTtgt	IR0	50.8 ± 2.9	126.6 ± 2.0
100	Chr22: 51188718-51188943	226	ACR (+12,179), RABL2B (+33,256)	caaAGGACATGAACTctt	IR0	6.4 ± 1.0	52.9 ± 1.1
101	Chr5: 113265138-113265327	190	KCNN2 (−432,783), YTHDC2 (+415,842)	atgAGTTCATGTCCTttg	IR0	19.2 ± 2.8	72.2 ± 3.0
102	Chr19: 27990058-27990456	399	NONE	caaAGGACATGAACTcat	IR0	20.4 ± 2.7	65.6 ± 0.7
103	Chr3: 135695587-135695826	240	PPP2R3A (+11,192), MSL2 (+218,981)	ttaAGTTCAtcTGAAATtta	IR2	15.9 ± 2.3	46.0 ± 1.5
104	Chr16: 27714860-27714998	139	GTF3C1 (−153,678), GSG1L (+359,901)	cttAGTTCAgcTGACCTctg	IR2	48.7 ± 2.4	116.6 ± 1
105	Chr14: 58472137-58472540	404	SLC35F4 (−139,747), C14orf37 (+146,508)	aggAGGTCAaagTGTCCTctg	IR3	8.8 ± 6.8	54.4 ± 6.9
106	Chr7: 152388241-152388469	229	ACTR3B (−68,496), XRCC2 (−15,105)	tggAGTTCAttggcaTGACCTcgg	IR6	12.4 ± 3.7	85.6 ± 1.3
107	Chr5: 162859214-162859417	204	CCNG1 (−5261)	taaAGGTCAttgcctTAACCTagg	IR6	16.1 ± 1.2	70.0 ± 1.1
108	Chr18: 30395764-30395944	181	ASXL3 (−762,687), KLHL14 (−42,880)	aaaAGTTCAacctcccTTAACTatg	IR7	32.9 ± 1.4	95.7 ± 0.8

Table 3. rSNPs in the identified CAR/RXRα binding sites.

dbSNP	Gene (Distance to TSS)	Sequences	Motif	Concentration of competitor	EMSA Competition (%)
rs115089792	NONE	atgGGGTCAacaaAGTTTAaga	DR4	×10	62.8 ± 0.6
		atgGTGTCAacaaAGTTTAaga			36.0 ± 0.4
rs371021620	FAM188A (−543,247), PTER (−33,201)	taaAGTTCAtggtaaAGTTCAtgg	DR6	×10	64.0 ± 1.1
		taaAATTCAtggtaaAGTTCAtgg			19.4 ± 0.8
rs62097972	TMX3 (+168,639)	tttTGAACTctgtctgaaAGGTCActt	ER9	×10	74.4 ± 0.4
		tttTGAATTctgtctgaaAGGTCActt			Not detected
rs181870712	CSMD1 (+5915)	caaAGGTCActtaAGGTCAttt	DR4	×10	62.9 ± 1.1
		caaAAGTCActtaAGGTCAttt			26.0 ± 1.5
rs117662912	CETN3 (+144,942)	aggAGGTCAggaggAGTTCActg	DR5	×10	56.9 ± 0.9
		aggAAGTCAggaggAGTTCActg			17.9 ± 0.6
rs113398776	SLC9A2 (+30,825), MFSD9 (+86,346)	tgaAGGTCAaagAGTTTAaca	DR3	×100	95.5 ± 0.9
		tgaAGGTCAaagAGATTAaca			9.5 ± 1.1
rs190307664	ESM1 (+232,659), SNX18 (+235,162)	agaTGACCTctgcccgctaAGTTCAagt	ER10	×100	101.8 ± 1.5
		agaTGACCTctgcccgctaACTTCAagt			36.5 ± 0.9
rs115070434	MPDZ (−24,841)	taaAGGTTAaatgAGGTCActa	DR4	×100	102.8 ± 1.1
		taaAGGTTAaatgAGGTCCcta			66.3 ± 5.7
rs148735323	NOX5 (−76,449), SPESP1 (+7746)	ggtGGGTCAactaAGGTCAgga	DR4	×100	97.6 ± 0.4
		ggtGGATCAactaAGGTCAgga			71.1 ± 0.8
rs145145243	EYA4 (−231,989), RPS12 (+194,798)	agtGGTTCAatcacAGTTCActg	DR5	×100	94.4 ± 0.9
		agtGGTTCAatcacAGTTCTctg			60.7 ± 1.5
rs369662801	RTN4RL1 (+9904), RPA1 (+185,001)	tggAGCTCAgAGTTCAaag	DR1	×100	82.1 ± 0.5
		tggAGCTCAgCGTTCAaag			22.1 ± 5.5
rs79536015	RPIA (+385,204)	caaAGGTTAccttAGGTCAtat	DR4	×100	96.5 ± 0.2
		caaAGGTTAccttAGGTCGtat			67.5 ± 0.9
rs4497728 rs4303542 rs4307980	MTRNR2L1 (+228,816)	accTGAACTttcaaaggaaGGTTCAatt	ER10	×100	94.5 ± 2.1
		accTGAAATttcaaaggaaGGTTCAatt			20.2 ± 0.5
		accTGAACAttcaaaggaaGGTTCAatt			23.5 ± 0.6
		accTGAACTttcaaaggaaGTTTCAatt			68.6 ± 0.9
rs138716773	PRL (−745,604)	ttgAGCTCAattcctAGTTCAcca	DR6	×100	62.6 ± 2.4
		ttgAGCTTAattcctAGTTCAcca			12.0 ± 3.9
rs11914034	ACR (+12,179), RABL2B (+33,256)	caaAGGACATGAACTctt	IR0	×100	46.9 ± 0.7
		caaAGGACGTGAACTctt			77.6 ± 0.5

It has been reported that CAR/RXRα can bind promiscuously to multiple DNA binding motifs such as DR1 [35], DR3 [7], DR4 [8] [12] [36]-[38], DR5 [2] [11], ER6 [7], ER7 [39], ER8 [40], ER9 [39], ER10 [39], IR2 [10], IR3 [41], and IR6 [42]. In addition to these motif types, our results indicated that CAR/RXRα could bind to DR2, DR6, DR8-10, ER1, ER3, ER4, ER5, IR0, and IR7 (**Figure 3(b)** and **Table 2**). As these motif types overlap with the binding sequences for other nuclear receptors, CAR/RXRα would compete for binding to these elements and cross talk with other nuclear receptors [35] [38] [41]. On the other hand, CAR/RXRα may preferably bind to DR4 relative to the other motifs, because DR4 motif was enriched by motif enrichment among 414 clones (**Figure 2**).

We could not identify the previously reported CAR/RXRα-binding sites in this study. There are some explanations for missed CAR/RXRDNA interactions. First, the quality of a library will affect the efficiency of identification of protein-DNA interactions. Second, some CAREs may exist adjacent to sequences recognized by yeast transcription factors and may be discarded during the negative selection by 5FOA [8]. However, we could identified new CAR/RXRα binding sites located in or in the proximity of genes with various functions such as metabolic process, cell cycle, cell proliferation and apoptosis. In particular, Heat shock 70 kDa protein 1A (HSPA1A) is known as a stress inducible gene that promotes liver tumor cell proliferation [43]. Importantly, the HSPA1A gene was reported as a CAR-dependent gene in a microarray analysis of CAR-knockout mice [15]. Furthermore, we confirmed the CAR-dependent activation of human HSPA1A mRNA expression in human hepatoma HepG2 cells by real-time PCR assays (data not shown). CAR is thought to be essential for liver tumor promotion via the direct regulation of the Mdm2 gene [44] [45]. In addition to this mechanism, our results suggested that CAR is associated with liver tumor promotion via direct regulation of HSPA1A.

ChIP-on-chip or ChIP seq techniques are powerful methods for identifying transcription factor-DNA interactions. However, these methods can also identify potential indirect transcription factor-DNA interactions. In contrast, our strategy for the identification of CAR/RXRα binding sites in the human genome was fundamentally different and was based on the direct interaction of CAR/RXRα with human genomic sequences using yeast genetic selection. We expect that these findings will provide insights into the molecular mechanisms underlying the physiological and pathological actions of CAR/RXRα heterodimers.

5. Conclusion

In this paper, we identified CAR/RXRα-binding sequences in the human genome by a modified yeast one-hybrid assay. The binding sites were located in the proximity of genes with various functions such as metabolic process, cell cycle, cell proliferation and apoptosis. Motif enrichment analysis revealed that the most frequently observed motif was a DR4, although the identified sequences contained not only classical CAREs but also a wide variety of arrangements. Next, we demonstrated that 108 human genomic fragments could directly interact with the CAR/RXRα heterodimer by EMSAs. Furthermore, we identified 17 regulatory polymorphisms on the CAR/RXRα-binding sites which may influence individual variation of expression levels of CAR-regulated genes. We expect that these findings will provide insights into the molecular mechanisms underlying the physiological and pathological actions of CAR/RXRα heterodimers.

Acknowledgements

We would like to thank Mr. Takatsugu Kosugi, Mr. Shogo Wakabayashi, and Mr. Natsuhiro Honda for their helpful discussion and excellent technical assistance.

Declaration of Interest

This work was supported by JSPS KAKENHI Grant Number 30360704. The authors report no conflicts of interest. The authors alone are responsible for the content and writing of this paper.

References

[1] Chawla, A., Repa, J.J., Evans, R.M. and Mangelsdorf, D.J. (2001) Nuclear Receptors and Lipid Physiology: Opening the X-Files. *Science*, **294**, 1866-1870. http://dx.doi.org/10.1126/science.294.5548.1866

[2] Baes, M., Gulick, T., Choi, H.S., Martinoli, M.G., Simha, D. and Moore, D.D. (1994) A New Orphan Member of the Nuclear Hormone Receptor Superfamily That Interacts with a Subset of Retinoic Acid Response Elements. *Molecular and Cellular Biology*, **14**, 1544-1552.

[3] Lamba, J.K., Lamba, V., Yasuda, K., Lin, Y.S., Assem, M., Thompson, E., Strom, S. and Schuetz, E. (2004) Expression of Constitutive Androstane Receptor Splice Variants in Human Tissues and Their Functional Consequences. *Journal of Pharmacology and Experimental Therapeutics*, **311**, 811-821. http://dx.doi.org/10.1124/jpet.104.069310

[4] Kobayashi, K., Sueyoshi, T., Inoue, K., Moore, R. and Negishi, M. (2003) Cytoplasmic Accumulation of the Nuclear Receptor CAR by a Tetratricopeptide Repeat Protein in HepG2 Cells. *Molecular Pharmacology*, **64**, 1069-1075. http://dx.doi.org/10.1124/mol.64.5.1069

[5] Yoshinari, K., Kobayashi, K., Moore, R., Kawamoto, T. and Negishi, M. (2003) Identification of the Nuclear Receptor CAR:HSP90 Complex in Mouse Liver and Recruitment of Protein Phosphatase 2A in Response to Phenobarbital, *FEBS Letters*, **548**, 17-20. http://dx.doi.org/10.1016/S0014-5793(03)00720-8

[6] Goodwin, B., Moore, L.B., Stoltz, C.M., McKee, D.D. and Kliewer, S.A. (2001) Regulation of the Human CYP2B6 Gene by the Nuclear Pregnane X Receptor. *Molecular Pharmacology*, **60**, 427-431.

[7] Goodwin, B., Hodgson, E., D'Costa, D.J., Robertson, G.R. and Liddle, C. (2002) Transcriptional Regulation of the Human CYP3A4 Gene by the Constitutive Androstane Receptor. *Molecular Pharmacology*, **62**, 359-365. http://dx.doi.org/10.1124/mol.62.2.359

[8] Gerbal-Chaloin, S., Daujat, M., Pascussi, J.M., Pichard-Garcia, L., Vilarem, M.J. and Maurel, P. (2002) Transcriptional Regulation of CYP2C9 Gene Role of Glucocorticoid Receptor and Constitutive Androstane Receptor. *The Journal of Biological Chemistry*, **277**, 209-217. http://dx.doi.org/10.1074/jbc.M107228200

[9] Sugatani, J., Kojima, H., Ueda, A., Kakizaki, S., Yoshinari, K., Gong, Q.H., Owens, I.S., Negishi, M. and Sueyoshi, T. (2001) The Phenobarbital Response Enhancer Module in the Human Bilirubin UDP-glucuronosyltransferase UGT1A1 Gene and Regulation by the Nuclear Receptor CAR. *Hepatology*, **33**, 1232-1238. http://dx.doi.org/10.1053/jhep.2001.24172

[10] Chen, X., Zhang, J., Baker, S.M. and Chen, G. (2007) Human Constitutive Androstane Receptor Mediated Methotrexate Induction of Human Dehydroepiandrosterone Sulfotransferase (hSULT2A1). *Toxicology*, **231**, 224-233. http://dx.doi.org/10.1016/j.tox.2006.12.019

[11] Benoki, S., Yoshinari, K., Chikada, T., Imai, J. and Yamazoe, Y. (2012) Transactivation of *ABCG2* through a Novel *cis*-Element in the Distal Promoter by Constitutive Androstane Receptor but Not Pregnane X Receptor in Human Hepatocytes. *Archives of Biochemistry and Biophysics*, **517**, 123-130. http://dx.doi.org/10.1016/j.abb.2011.10.014

[12] Sueyoshi, T., Kawamoto, T., Zelko, I., Honkakoski, P. and Negishi, M. (1999) The Repressed Nuclear Receptor CAR Responds to Phenobarbital in Activating the Human *CYP2B6* Gene. *Journal of Biological Chemistry*, **274**, 6043-6046. http://dx.doi.org/10.1074/jbc.274.10.6043

[13] Ueda, A., Hamadeh, H.K., Webb, H.K., Yamamoto, Y., Sueyoshi, T., Afshari, C.A., Lehmann, J.M. and Negishi, M. (2002) Diverse Roles of the Nuclear Orphan Receptor CAR in Regulating Hepatic Genes in Response to Phenobarbital. *Molecular Pharmacology*, **61**, 1-6. http://dx.doi.org/10.1124/mol.61.1.1

[14] Maglich, J.M., Parks, D.J., Moore, L.B., Collins, J.L., Goodwin, B., Billin, A.N., Stoltz, C.A., Kliewer, S.A., Lambert, M.H., Willson, T.M. and Moore, J.T. (2003) Identification of a Novel Human Constitutive Androstane Receptor (CAR) Agonist and Its Use in the Identification of CAR Target Genes. *Journal of Biological Chemistry*, **278**, 17277-17283. http://dx.doi.org/10.1074/jbc.M300138200

[15] Tojima, H., Kakizaki, S., Yamazaki, Y., Takizawa, D., Horiguchi, N., Sato, K. and Mori, M. (2012) Ligand Dependent Hepatic Gene Expression Profiles of Nuclear Receptors CAR and PXR. *Toxicology Letters*, **212**, 288-297. http://dx.doi.org/10.1016/j.toxlet.2012.06.001

[16] Tamasi, V., Juvan, P., Beer, M., Rozman, D. and Meyer, U.A. (2009) Transcriptional Activation of PPAR Alpha by Phenobarbital in the Absence of CAR and PXR. *Molecular Pharmaceutics*, **6**, 1573-1581. http://dx.doi.org/10.1021/mp9001552

[17] Cui, J.Y., Gunewardena, S.S., Rockwell, C.E. and Klaassen, C.D. (2010) ChIPing the Cistrome of PXR in Mouse Liver. *Nucleic Acids Research*, **38**, 7943-7963. http://dx.doi.org/10.1093/nar/gkq654

[18] Mahony, S., Mazzoni, E.O., McCuine, S., Young, R.A., Wichterle, H. and Gifford, D.K. (2011) Ligand-Dependent Dynamics of Retinoic Acid Receptor Binding during Early Neurogenesis. *Genome Biology*, **12**, R2. http://dx.doi.org/10.1186/gb-2011-12-1-r2

[19] Delacroix, L., Moutier, E., Altobelli, G., Legras, S., Poch, O., Choukrallah, M.A., Bertin, I., Jost, B. and Davidson, I. (2010) Cell-Specific Interaction of Retinoic Acid Receptors with Target Genes in Mouse Embryonic Fibroblasts and Embryonic Stem Cells. *Molecular and Cellular Biology*, **30**, 231-244. http://dx.doi.org/10.1128/MCB.00756-09

[20] Hua, S., Kittler, R. and White, K.P. (2009) Genomic Antagonism between Retinoic Acid and Estrogen Signaling in Breast Cancer. *Cell*, **137**, 1259-1271. http://dx.doi.org/10.1016/j.cell.2009.04.043

[21] Verzi, M.P., Shin, H., San Roman, A.K., Liu, X.S. and Shivdasani, R.A. (2013) Intestinal Master Transcription Factor CDX2 Controls Chromatin Access for Partner Transcription Factor Binding. *Molecular and Cellular Biology*, **33**, 281-292. http://dx.doi.org/10.1128/MCB.01185-12

[22] Reddy, T.E., Pauli, F., Sprouse, R.O., Neff, N.F., Newberry, K.M., Garabedian, M.J. and Myers, R.M. (2009) Genomic Determination of the Glucocorticoid Response Reveals Unexpected Mechanisms of Gene Regulation. *Genome Re-*

search, **19**, 2163-2171. http://dx.doi.org/10.1101/gr.097022.109

[23] Nielsen, R., Pedersen, T.A., Hagenbeek, D., Moulos, P., Siersbaek, R., Megens, E., Denissov, S., Børgesen, M., Francoijs, K.J., Mandrup, S. and Stunnenberg, H.G. (2008) Genome-Wide Profiling of PPARγ: RXR and RNA Polymerase II Occupancy Reveals Temporal Activation of Distinct Metabolic Pathways and Changes in RXR Dimer Composition during Adipogenesis. *Genes & Development*, **22**, 2953-2967. http://dx.doi.org/10.1101/gad.501108

[24] Taniguchi-Yanai, K., Koike, Y., Hasegawa, T., Furuta, Y., Serizawa, M., Ohshima, N., Kato, N. and Yanai, K. (2010) Identification and Characterization of Glucocorticoid Receptor-Binding Sites in the Human Genome. *Journal of Receptors and Signal Transduction*, **30**, 88-105. http://dx.doi.org/10.3109/10799891003614816

[25] Kanno, Y., Aoki, S., Mochizuki, M., Mori, E., Nakahama, T. and Inouye, Y. (2005) Expression of Constitutive Androstane Receptor Splice Variants in Rat Liver and Lung and Their Functional Properties. *Biological and Pharmaceutical Bulletin*, **28**, 2058-2062. http://dx.doi.org/10.1248/bpb.28.2058

[26] Auerbach, S.S., Dekeyser, J.G., Stoner, M.A. and Omiecinski, C.J. (2007) CAR2 Displays Unique Ligand Binding and RXRα Heterodimerization Characteristics. *Drug Metabolism and Disposition*, **35**, 428-439. http://dx.doi.org/10.1124/dmd.106.012641

[27] Wang, H., Faucette, S., Sueyoshi, T., Moore, R., Ferguson, S., Negishi, M. and LeCluyse, E.L. (2003) A Novel Distal Enhancer Module Regulated by Pregnane X Receptor/Constitutive Androstane Receptor Is Essential for the Maximal Induction of CYP2B6 Gene Expression. *Journal of Biological Chemistry*, **278**, 14146-14152. http://dx.doi.org/10.1074/jbc.M212482200

[28] Goecks, J., Nekrutenko, A. and Taylor, J., The Galaxy Team (2010) Galaxy: A Comprehensive Approach for Supporting Accessible, Reproducible, and Transparent Computational Research in the Life Sciences. *Genome Biology*, **11**, R86. http://dx.doi.org/10.1186/gb-2010-11-8-r86

[29] Blankenberg, D., Von Kuster, G., Coraor, N., Ananda, G., Lazarus, R., Mangan, M., Nekrutenko, A. and Taylor, J. (2010) Galaxy: A Web-Based Genome Analysis Tool for Experimentalists. *Current Protocols in Molecular Biology*, **19**, 10.1-10.21.

[30] Giardine, B., Riemer, C., Hardison, R.C., Burhans, R., Elnitski, L., Shah, P., Zhang, Y., Blankenberg, D., Albert, I., Taylor, J., Miller, W., Kent, W.J. and Nekrutenko, A. (2005) Galaxy: A Platform for Interactive Large-Scale Genome Analysis. *Genome Research*, **15**, 1451-1455. http://dx.doi.org/10.1101/gr.4086505

[31] McLean, C.Y., Bristor, D., Hiller, M., Clarke, S.L., Schaar, B.T., Lowe, C.B., Wenger, A.M. and Bejerano, G. (2010) GREAT Improves Functional Interpretation of *cis*-Regulatory Regions. *Nature Biotechnology*, **28**, 495-501. http://dx.doi.org/10.1038/nbt.1630

[32] Hasegawa, K., Cowan, A.B., Nakatsuji, N. and Suemori, H. (2007) Efficient Multicistronic Expression of a Transgene in Human Embryonic Stem Cells. *Stem Cells*, **25**, 1707-1712. http://dx.doi.org/10.1634/stemcells.2006-0813

[33] Kawamoto, T., Sueyoshi, T., Zelko, I., Moore, R., Washburn, K. and Negishi, M. (1999) Phenobarbital-Responsive Nuclear Translocation of the Receptor CAR in Induction of the *CYP2B* Gene. *Molecular and Cellular Biology*, **19**, 6318-6322.

[34] Machanick, P. and Bailey, T.L. (2011) MEME-ChIP: Motif Analysis of Large DNA Datasets. *Bioinformatics*, **27**, 1696-1697. http://dx.doi.org/10.1093/bioinformatics/btr189

[35] Miao, J., Fang, S., Bae, Y. and Kemper, J.K. (2006) Functional Inhibitory Cross-Talk between Constitutive Androstane Receptor and Hepatic Nuclear Factor-4 in Hepatic Lipid/Glucose Metabolism Is Mediated by Competition for Binding to the DR1 Motif and to the Common Coactivators, GRIP-1 and PGC-1α. *Journal of Biological Chemistry*, **281**, 14537-14546. http://dx.doi.org/10.1074/jbc.M510713200

[36] Burk, O., Arnold, K.A., Geick, A., Tegude, H. and Eichelbaum, M. (2005) A Role for Constitutive Androstane Receptor in the Regulation of Human Intestinal MDR1 Expression. *Biological Chemistry*, **386**, 503-513. http://dx.doi.org/10.1515/BC.2005.060

[37] Yoshinari, K., Ohno, H., Benoki, S. and Yamazoe, Y. (2012) Constitutive Androstane Receptor Transactivates the Hepatic Expression of Mouse *Dhcr*24 and Human *DHCR*24 Encoding a Cholesterogenic Enzyme 24-Dehydrocholesterol Reductase. *Toxicology Letters*, **208**, 185-191. http://dx.doi.org/10.1016/j.toxlet.2011.11.003

[38] Chen, Y., Ferguson, S.S., Negishi, M. and Goldstein, J.A. (2003) Identification of Constitutive Androstane Receptor and Glucocorticoid Receptor Binding Sites in the *CYP2C*19 Promoter. *Molecular Pharmacology*, **64**, 316-324. http://dx.doi.org/10.1124/mol.64.2.316

[39] Frank, C., Gonzalez, M.M., Oinonen, C., Dunlop, T.W. and Carlberg, C. (2003) Characterization of DNA Complexes Formed by the Nuclear Receptor Constitutive Androstane Receptor. *Journal of Biological Chemistry*, **278**, 43299-43310. http://dx.doi.org/10.1074/jbc.M305186200

[40] Kast, H.R., Goodwin, B., Tarr, P.T., Jones, S.A., Anisfeld, A.M., Stoltz, C.M., Tontonoz, P., Kliewer, S., Willson, T.M. and Edwards, P.A. (2002) Regulation of Multidrug Resistance-Associated Protein 2 (ABCC2) by the Nuclear Recep-

tors Pregnane X Receptor, Farnesoid X-Activated Receptor, and Constitutive Androstane Receptor. *Journal of Biological Chemistry*, **277**, 2908-2915. http://dx.doi.org/10.1074/jbc.M109326200

[41] Min, G., Kim, H., Bae, Y., Petz, L. and Kemper, J.K. (2002) Inhibitory Cross-Talk between Estrogen Receptor (ER) and Constitutively Activated Androstane Receptor (CAR) CAR Inhibits ER-Mediated Signaling Pathway by Squelching p160 Coactivators. *Journal of Biological Chemistry*, **277**, 34626-34633. http://dx.doi.org/10.1074/jbc.M205239200

[42] Xie, W., Barwick, J.L., Simon, C.M., Pierce, A.M., Safe, S., Blumberg, B., Guzelian, P.S. and Evans, R.M. (2000) Reciprocal Activation of Xenobiotic Response Genes by Nuclear Receptors SXR/PXR and CAR. *Genes & Development*, **14**, 3014-3023. http://dx.doi.org/10.1101/gad.846800

[43] Wu, F.H., Yuan, Y., Li, D., Liao, S.J., Yan, B., Wei, J.J., Zhou, Y.H., Zhu, J.H., Zhang, G.M. and Feng, Z.H. (2012) Extracellular HSPA1A Promotes the Growth of Hepatocarcinoma by Augmenting Tumor Cell Proliferation and Apoptosis-Resistance. *Cancer Letters*, **317**, 157-164. http://dx.doi.org/10.1016/j.canlet.2011.11.020

[44] Huang, W., Zhang, J., Washington, M., Liu, J., Parant, J.M., Lozano, G. and Moore, D.D. (2005) Xenobiotic Stress Induces Hepatomegaly and Liver Tumors via the Nuclear Receptor Constitutive Androstane Receptor. *Molecular Endocrinology*, **19**, 1646-1653. http://dx.doi.org/10.1210/me.2004-0520

[45] Yamamoto, Y., Moore, R., Goldsworthy, T.L., Negishi, M. and Maronpot, R.R. (2004) The Orphan Nuclear Receptor Constitutive Active/Androstane Receptor Is Essential for Liver Tumor Promotion by Phenobarbital in Mice. *Cancer Research*, **64**, 7197-7200. http://dx.doi.org/10.1158/0008-5472.CAN-04-1459

19

Composition of the Putative Prepore Complex of *Bacillus thuringiensis* Cry1Ab Toxin

Manoj S. Nair[1,2*], Donald H. Dean[1]

[1]Department of Chemistry and Biochemistry, The Ohio State University, Columbus, OH, USA
[2]Aaron Diamond AIDS Research Center, Rockefeller University, New York, NY, USA
Email: *mnair@adarc.org

Abstract

Prepore formation is hypothesized to be an obligate step in the insertion of Cry1Ab toxin into insect brush border membrane vesicles. We examined the architecture of the putative prepore when isolated using the published protocols [1] [2]. Our results demonstrate that the putative prepore form of Cry1Ab is a combination of receptor proteins attached to the toxin, when purified. The results also suggest that this prepore form as prepared by the methods published is different from other membrane-extracted oligomeric forms of Cry toxins and prepore of other toxins in general. While most other known prepores are composed of multimers of a single protein, the Cry1Ab prepore, as generated, is a protein-receptor complex oligomer and monomers of Cry toxins.

Keywords

Cry1Ab Protein, Protein-Receptor Interactions, Brush Border Membrane Vesicles, Small Unilamellar Vesicles, Prepore, LC-MS/MS

1. Introduction

Pore-forming toxins represent approximately 30% of bacterial toxins [3]. Highly opportunistic bacterial pathogens produce these toxins to perturb the rigid compartmentalization of plasma membrane by clustering into homo or hetero assemblies within the membrane. Even as pore formation is the main proposed pathogenicity determinant of these toxins, very little is resolved about this crucial step in the mechanism of action of these toxins [4]. This family of toxins has received considerable attention due to their widespread use as insecticides, more

*Corresponding author.

so in transgenic crops and the mechanism of action has been studied for more than two decades [5]-[8].

Most pore forming toxins including the Cry toxins are known to oligomerize during their mechanism of action. It is not clear whether the process begins with monomers of Cry toxins that dimerize and then add on monomers till a definite oligomeric shape is achieved or if monomers interact with preformed higher oligomers to form the channel structure. Bacterial toxins, in general, have been shown to form "prepore" oligomers that allow their insertion into target cell membranes. The term "prepore" refers to an intermediate state of these toxins in which the monomers of the toxin assemble to form a precursor that is competent to insert into these membranes. Many bacterial toxins that form prepore do undergo a substantial conformational change in the process, including a change in the secondary structure of the huge regions involved in the pore formation as in cytolysins [9] [10] and in anthrax toxins [11]. While receptor binding is either obligatory or enhances the formation of the prepore form in these toxins, the oligomers made from these toxins are composed majorly if not exclusively of the actual toxin that forms the pore [12] [13]. Oligomers made of exclusively Cry1Aa, Cry1Ab and Cry1Ac toxins have been isolated from the protease-treated brush border membrane vesicles in several studies [14] [15]. However, none of these studies examine the kinetics of formation of a prepore oligomer form of Cry1A toxins.

Cry1Ab toxin from *Bacillus thuringiensis* has been the model for studying prepore formation and oligomerization of Cry toxins [16]-[19]. In the prepore model, Cry1Ab as represented is a tetramer formed by the conformational change induced in the monomer upon binding cadherin receptor and losing alpha helix 1 that allows the oligomer to bind aminopeptidase N and alkaline phosphatase to mediate insertion of the toxin into the membrane [16] [20]. The conformational change was considered significant enough that the affinity of the cadherin receptor for Cry1Ab drastically reduced from the initial 1 nM affinity even as it was hypothesized to be obligatory for binding APN and alkaline phosphatase with high affinity [16] [21]. Given that the proportion of all the receptors for Cry protein was high on the insect cell surface, it was surprising that the Cry toxin was considered independent of the receptors as it bound or released from any of the receptors during pore formation. Another intriguing factor was that a polyclonal antibody designed to react with denatured Cry1Ab (potentially capable of interacting with multiple epitopes) would not react with the prepore. The aim of this study was to isolate the prepore as published [1] [2] for characterization of the mass of the toxin and to determine the regions of the monomer toxin that were retained in the prepore oligomer.

2. Materials and Methods

2.1. Source for Strains and Antibodies

Bacillus thuringiensis sotto (for Cry1Aa), *Bacillus thuringiensis 4Q7* (acrystalliferous strain for expression of Cry1Ab F371C mutant) and *Bacillus thuringiensis HD*1-19 (for Cry1Ab) strains were obtained from *Bacillus* Genetic Stock Center, The Ohio State University. Antibodies used in this study include anti-Cry1Ab polyclonal antibody generated against denatured Cry1Ab monomer protein and anti-Cry1Ab polyclonal antibody kindly provided by Dr. Alejandra Bravo of Universidad Nacional Autónoma de México (which is referred to in this manuscript as the anti-prepore antibody). Antibody to the cadherin (anti Bt-R1) was kindly provided by Dr. Lee A. Bulla Jr. from University of Texas at Dallas.

2.2. Production of Prepore Oligomers

Prepore was produced by two separate methods that were published [1] [2]. Briefly, in the first method, 1 μg protoxin was mixed with 10 μg insect brush border membrane vesicles (BBMV) in the presence of 50 μl of solubilization buffer (50 mM Na_2CO_3 pH 10.5 + 0.2% *β*-mercaptoethanol) and incubated at room temperature (25˚C - 28˚C) for 15 min. In the second method, 200 ng of protoxin was incubated with scFv73 in a 1:2 mass ratio and 5% midgut juice was added in 100 μl of solubilization buffer containing 60 μM small unilamelar vesicles made of 1,2-Dioleoyl-sn-Glycero-3-phosphocholine (DOPC). The mixture was incubated at 37˚C for 1 hr, followed by precipitating the membrane bound toxin at 400,000× g in a Beckman L8 ultracentrifuge. A control reaction, which lacked any small unilamellar vesicles, was used to show no precipitate formation in the absence of SUV. The pellet was resuspended in 50 μl of buffer in presence of 10% n-octyl-*β*-D-glucopyranoside and clarified by centrifugation, treated with loading dye and boiled for 3 - 5 min. Western blot analysis of the protein was performed using polyclonal anti-prepore antibody (1:50,000; 1hr) and secondary HRP antibody (1:10,000; 1hr) and was detected using chemiluminescence substrate (Bio-Rad).

2.3. Purification of the Prepore Oligomers

Cry1Ab prepore oligomers were prepared and purified using the two published methods [1] [2] using a Superdex 200 HR 16/60 (GE Healthcare) on an AKTA Explorer 100 (GE Healthcare). Several prepore samples were purified and pooled to yield 1 - 5 μg of sample. Sample volume of 2 ml was loaded into the column for every run. The flow rate was 1 ml/min and fractions of 2 ml were collected in a mobile phase containing protease inhibitors and 0.01% n-octyl-β-D-glucopyranoside. Purified proteins were analyzed using Western blots with polyclonal antibody prepared against prepore form of Cry1Ab at dilutions mentioned above. Purified samples isolated by method 1 were compared against buffer control in a Beckman XL-I analytical ultracentrifuge to obtain the closest molecular masses using sedimentation velocity measurements.

2.4. Mass Spectrometric Identification of Prepore Contents

In order to determine the identity of the proteins in the purified prepore complex, SDS-PAGE-resolved Cry1Ab prepore was subjected to LC-MS/MS analysis, performed at the W.M. Keck Facility, New Haven, CT. Briefly, purified prepore was run on SDS-PAGE gel and the entire lane from the gel, which showed cross-reactivity to the prepore antibody, was cut into 1 mm slices and digested with trypsin using in-gel digestion and the resulting samples were desalted in a 100 micron ID C18 column (Waters) in a gradient of 2% - 98% acetonitrile in the presence of 0.01% trifluoroacetic acid. Protein was considered identified if two or more peptides matched to the same protein accession numbers in the database (MASCOT analysis).

2.5. Toxicity Measurements of Isolated Prepore Form to Manduca Sexta Larva

Although the purified form of the prepore had a mixture of oligomer and monomer forms, we nevertheless performed limited toxicity bioassays on them to measure any improvement over the purified monomer toxins as reported. Toxicity levels were determined by estimating the median lethal concentration (LC_{50}) on first instar *M. sexta* larvae using the diet surface contamination assay [22]. Sixteen first instar larvae were used for each concentration of the toxin, and a total of six concentrations of each toxin were used. Mortalities were recorded after 5 days. The LC_{50} for each toxin was calculated by Probit analysis using SoftTox (WindowsChem Software, Inc.).

3. Results

3.1. Formation and Purification of Prepore Complex

Using the published protocols [1] [2] of prepore discovery, we examined oligomer formation of Cry1Aa and Cry1Ab either with BBMV or with scFv73 in the presence of lipid vesicles. Notably, while our Cry1Ab polyclonal antibody raised against denatured monomer toxin did not recognize any oligomer product on western blots (data not shown), the polyclonal prepore antibody we received showed considerable amounts of prepore for both Cry1Aa and Cry1Ab, while also showing equal or higher amounts of monomers of the toxin on these blots (**Figure 1(a)**). Based on the hypothesis that receptor interaction may have modified the protein to form oligomers, we decided to measure the reactivity of the Cry1Ab to anti-Bt-R1 antibody after production of the prepore. We purified the prepore made using the BBMV-based method [1] by Superdex 200 gel filtration column as one single peak (**Figure 2(a)**) and tested the sample on western blot against the cadherin antibody. The blot indicated presence of a 75 kDa predominant band that was not present in the monomer form of Cry1Ab (**Figure 1(b)**). The purified prepore when subjected to sedimentation velocity measurements indicated the presence of multiple peaks suggestive of molecular complexes of different sizes (SedFit analysis) as indicated in **Figure 2(b)**.

3.2. Mass Spectrometry Analysis of Prepore of Cry1Ab

Protein samples verified for formation of the prepore form were purified in order to determine the regions of Cry1Ab present/absent in the prepore. Since SDS-PAGE of purified prepore showed multiple bands, we decided to characterize each band to determine toxin regions present in the prepore form and its variations if any, from the monomer. Proteins were identified from in-gel trypsin digests of the SDS-PAGE bands using LC MS/MS analysis of the digested bands. Proteins were identified from a MASCOT search with two criteria: two or more MS/MS spectra match the same protein in the database and that each of the matched peptides was ensured to be from trypsin digestion.

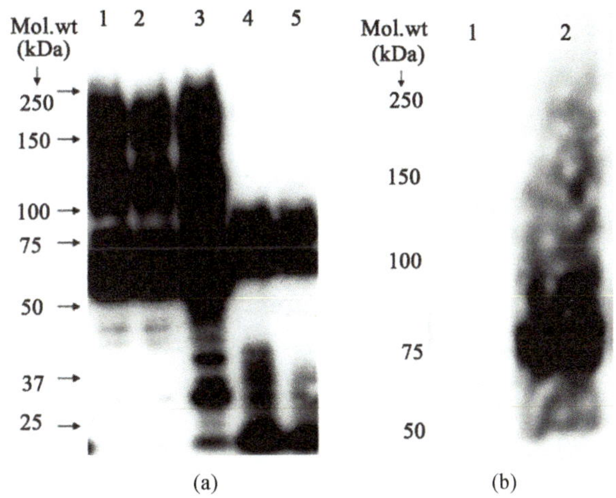

Figure 1. (a) Detection of prepore from Cry toxins using anti polyclonal antisera obtained from Bravo, A, *et al.* [1] at 1:50,000 dilution. Lane 1 = Prepore from Cry1Abwt produced in *Bt*HD1-19. Lane 2 = Prepore from Cry1Ab F371C mutant produced in *Bt*4Q7. Lane 3 = Prepore from Cry1Aa wt produced in *Bt sotto* 4E3. Lane 4 = Cry1Ab produced in *E.coli*. Lane 5 = Cry1Aa produced in *E.coli*; (b) Detection of cadherin association in purified prepore of Cry1Ab. Monomer and prepore obtained were treated with anti Bt-R1 antibody at 1:10,000 dilution of the primary antibody. Lane 1 = Purified Cry1Ab monomer toxin (200 ng), Lane 2 = Purified Cry1Ab prepore toxin (50 ng).

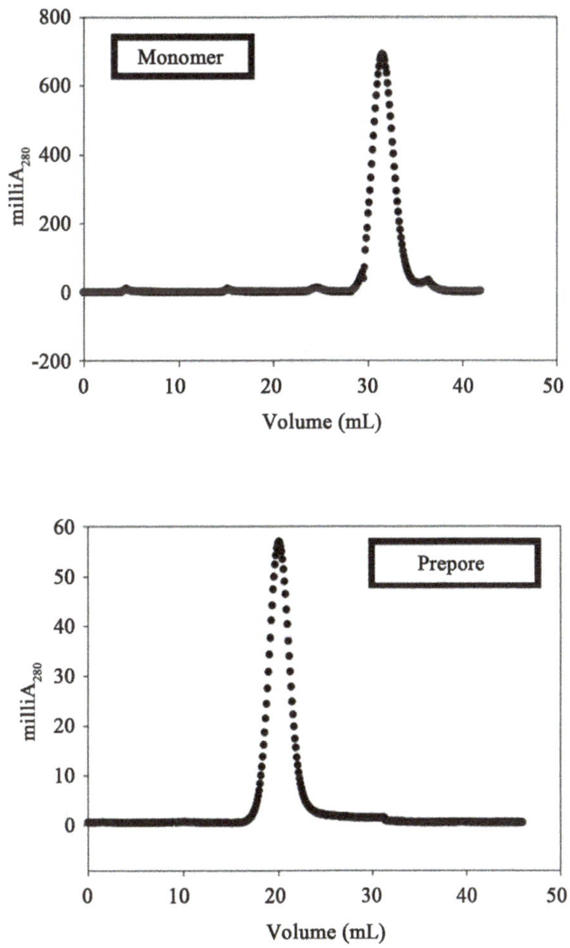

(a)

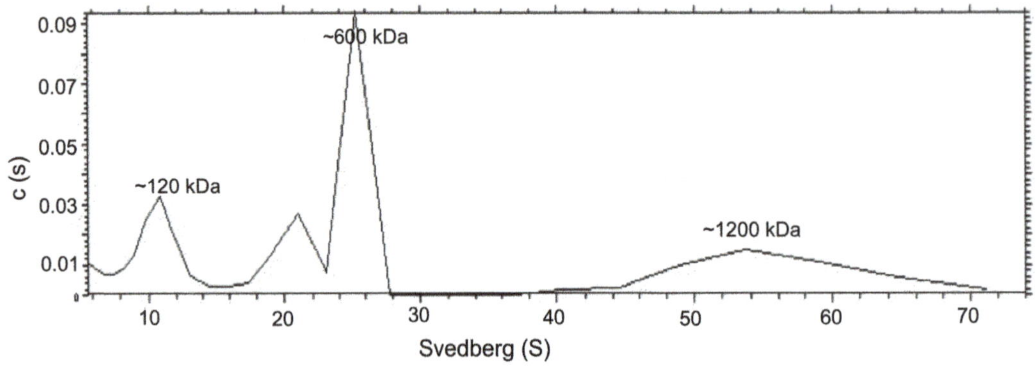

(b)

Figure 2. (a) Superdex 200 HR 16//60 gel filtration column purification of the prepore complex. The two chromatograms show the elution profile of monomer toxin versus the prepore complex as indicated; (b) Sed-Fit analysis of sedimentation velocity measurements in a Beckman XL-I analytical ultracentrifuge of prepore sample purified from Superdex 200 HR column. Molecular weights indicated on each peak were deciphered using a Nomogram to an approximation.

Prepore examined from the two methods showed presence of Cry1Ab toxin. However, the purified prepore in both the methods had additional proteins that were identified by the sensitive MS/MS process. The entire list of pertinent proteins identified is listed in **Table 1**. Prominent candidates include insect cadherin and aminopeptidase in the prepore that was produced in the BBMV-based method [1] and peptides homologous to the single chain antibody (scFv73) in the method that was based on using lipid vesicles [2].

Of interest was the observation [2] [23] that presence of small unilamellar vesicles allowed prepore form of the protein to be inserted into the membrane and that the prepore form was collected by ultracentrifugation of the vesicles. Since models of insertion have predicted alpha helix 4 and 5 of domain I to be involved in membrane insertion, we were curious to determine the regions of prepore embedded in the artificial membrane in the presence of the cadherin-like peptide (scFv73). The prepore of Cry1Ab was extracted out of the lipid vesicles using n-octyl-β-D-glucopyranoside, and the regions of the toxin that were present were identified from the peptides that were detected in the LC-MS/MS. As shown in **Table 2**, alpha helices 5, 6, and 7 of domain I was presented in the membranes. However, more peptides were identified from domain II and domain III of the prepore form of the toxin. The regions of Cry1Ab to which the peptides were mapped have been highlighted (**Figure 3**) with differential coloring for each domain, indicating presence of all three domains of the Cry1Ab toxin in the putative prepore isolated.

3.3. Toxicity Measurements of Monomer and Prepore Forms of Proteins

Even though we could not quantitate the amount of prepore form of Cry1Ab in the reaction mixture due to interference of monomers in all methods, we decided to perform a comparison of toxicity measurements of prepore as isolated by the published procedures to the pure monomer based on the premise that the conformational change would enhance the toxicity of Cry1Ab protein due to the presence of the active intermediate. Our studies using diet surface contamination assays with *Manduca sexta* first instar larvae showed overlapping LC_{50} confidence limits for prepore and monomer as listed in **Table 3**.

4. Discussion

The putative prepore formation of Cry1Ab toxin was reproduced using published protocols [1] [2] and materials provided by the Bravo group who pioneered the serial-receptor binding model. The goal of this study was to examine the resultant form using proteomic and analytical methods. The oligomer form of the toxin was generated in the presence of the cadherin receptor and/or the scFv73 peptide that mimicked the cadherin receptor. Purification of the prepore complex via molecular sieving resulted in a single peak albeit one that eluted at a molecular weight of >250 kDa through a Superdex 200 size exclusion chromatography column (*i.e.* beyond the void volume). This peak reacted to polyclonal Cry1Ab antibody on western blot from a reducing SDS-PAGE gel run as a tetramer form. However, the SDS- PAGE of the single peak generated smaller bands of monomer toxin

Table 1. Major proteins of known function isolated from LC-MS/MS from prepore purified from (a) BBMV-based method or (b) small unilamellar vesicle method of obtaining prepore.

(a)

GI accession number of match	Protein identified	Maximum number of peptides identified
gi|40255	Insecticidal crystal protein	13
gi|143099	Insecticidal crystal protein	13
gi|20465244	Cadherin from *M. sexta*	5
gi|2499901	APN-like protein (Membrane alanyl aminopeptidase precursor Cry1Ac receptor)	32
gi|8488965	Aminopeptidase 2	24
gi|20279109	Aminopeptidase 3	18

(b)

GI accession number of match	Protein identified	Maximum number of peptides identified
gi|40255	Insecticidal crystal protein	21
gi|143099	Insecticidal crystal protein	16
gi|1902832	Single chain (scFv) antibody Mol.wt. 26267	3
gi|106429	Ig heavy chain V region (alpha-phOx15)-human fragment Mol.wt. 13696	2

Table 2. Sequence of major peptides matched to Cry1Ab protein obtained from the LC-MS/MS of prepore sample isolated from detergent solubilization of small unilamellar vesicles of Cry1Ab and purified by gel filtration. The order of the peptides in the table is in decreasing order of the number of hits obtained from MASCOT.

Sequence of Peptides identified	Region of Toxin	Residue positions on toxin
DVSVFGQR	Domain I (alpha 5)	D174-R181
TLSSTLYR	Domain II (beta)	T361-R368
LSHVSMFR	Domain II (beta)	L430-R437
WYNTGLER	Domain I (loop between alpha 6 & 7)	W210-R217
TSPGQISTL	Domain III	T502-R511
GSAQGIEGSIR	Domain II	G282-R292
GPGFTGGDILR	Domain III	G490-R500
VNITAPLSQR	Domain III	V512-R521
IVAQLGQGVYR	Domain II	I350-R360
WGFDAATINSR	Domain I (loop between alpha 5 & 6)	W182-R192
EWEADPTNPALR	Domain I (loop between alpha 3 & 4)	E116-R127
EIYTNPVLENFDGSFR	Domain II	E266-R281
IEFVPAEVTFEAEYDLER	Domain III end (C-term of active toxin)	I602-R619
LEGLSNLYQIYAESFR	Domain I	L100-R115
ELTLTVLDIVSLFPNYDSR	Domain I (alpha 7)	E235-R253
SAEFNNIIPSSQITQIPLTK	Domain III	S458-K477
SPHLMDILNSITIYTDAHR	Domain II	S293-R312

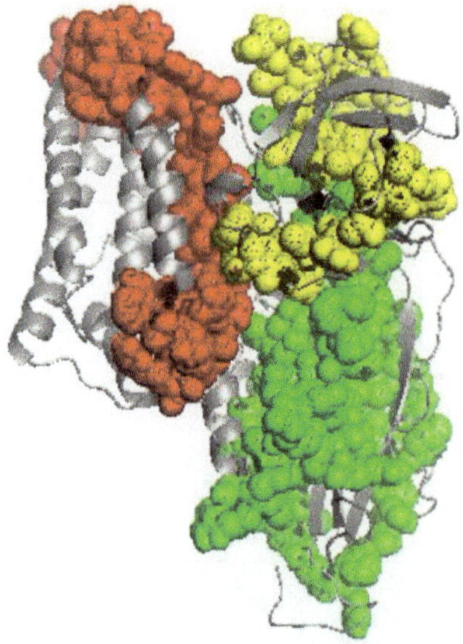

Figure 3. Regions of Cry1Ab that were identified based on peptides matched from LC-MS/MS of digested Cry1Ab prepore samples. Red color indicates peptides matched to Domain I residues, green color indicates peptides matched to Domain II residues and yellow color indicates peptides matched to Domain III residues.

Table 3. Toxicity measurements of monomer and prepore forms of Cry1Ab as measured by LC_{50} of toxin on *Manduca sexta* larvae.

Cry1Ab monomer	20.0 (7.5 - 31.7)
Cry1Ab oligomers formed in solution	28.0 (10.0 - 46.5)
Cry1Ab toxin extracted from membrane**	25.0 (5.0 - 35.0)
Cry1Aa prepore oligomer	17.1 (6.2 - 35.3)
Cry1Ab prepore oligomer	26.2 (3.9 - 40.8)
Cry1Ab F371C prepore oligomer	>2000

**Cry1Ab toxin in this group was obtained by dissolving the BBMV after insertion of toxin using 10 mM HEPES pH 7.5 buffer + 1%β-octyl glucoside detergent.

and other proteins, certainly ones that were associated with the toxin in the putative prepore form as prepared. Sedimentation velocity measurements of the purified oligomer indicated that the native forms of the toxin persisted in many sizes ranging from monomer to dodecamer as deciphered by the limits of nomogram analysis. To test if only a limited portion of the toxin was associated with the membrane in the prepore as predicted by models, we extracted the membrane inserted form of prepore complex obtained by using scFv73 peptide with a detergent. LC-MS analysis of the resulting tryptic digest indicated presence of peptides from all domains of the Cry1Ab in the extracted form. It also indicated presence of scFv73 peptide or the receptors depending on the method of prepore formation that was pursued. Toxicity measurements of the Superdex 200 column purified oligomer did not provide a LC_{50} that was any significantly higher than the monomer form.

We now discuss our observation that the putative prepore does not react against a polyclonal antibody formed against denatured Cry1Ab. The possibility that a conformational change would obviate binding of the polyclonal antibody exists but seems unlikely since the antibody binds to both native and denatured forms of the toxin. We suggest that the Cry1Ab toxin in the putative prepore is masked by the scFv73 peptide or receptors to the extent of blocking it from interaction to the anti-Cry1Ab antibody. In addition, the anti-prepore antibody, which does bind to the putative prepore is observed to bind to purified cadherin (data not shown).

All of these data in combination suggest that while putative prepore forms of Cry1Ab can be successfully generated, the resultant product as examined is not a pure toxin. Since very low amounts of toxin are effective at mediating toxicity, based on the hypothesis of the serial receptor binding model [16], it is likely that a very small fraction of oligomer may mediate most of the toxicity. However, our LC_{50} analysis of the purified product does not show significant variation in overall toxicity between monomer and produced oligomer forms. To be more precise, the contribution of each form to the overall toxicity is not accounted by the procedures used [1] [2] to isolate and purify the putative prepore toxin. Based on the data, the monomer appears to account for most of the toxicity in solution, suggesting a possibility of oligomers of Cry toxin being formed only post-insertion in the presence of the membrane bilayers.

The current data also questions the nature of the membrane bound tetramer as proposed by the serial receptor binding model, which indicates that only two alpha helices of the toxin from each monomer are present within the membrane. The rest of the toxin, if not associated with the membrane, should be easily digested by proteases. Our tryptic digests of membrane extracted prepore isolates presented almost equal number of peptides from all three domains of active Cry1Ab toxin (5 from Domain I, 7 from Domain II and 5 from Domain III) and none from the C-terminal region of the protoxin form used in these prepore formation methods or any peptides matching alpha helix 1 of active toxin. In addition, there are receptor peptides that were bound to the toxin. Our previous studies also highlight an intact 60 kDa monomer toxin that lacks alpha helix 1 as the form that inserts in the membrane [24]. The cumulative results suggest that the prepore form we isolated and purified is a form of the *Bt*-booster of Cry toxins that has been successfully isolated at higher concentrations of receptors and has shown to have higher toxicity than the toxin alone [25]. The verity of the serial receptor binding model has also been challenged by other investigators investigating the mechanism of toxin induced cytotoxicity via a Bt-R1 mediated adenyl cyclase-protein kinase A signaling pathway [26] [27]. In fact, the authors of the serial- receptor binding model have since modified their view of the mechanism [20] indicating the deficiencies of their model as proposed [16].

In summary, we performed a proteomic and biophysical analysis of the putative prepore of Cry1Ab to determine the size and the contents of the complex and have discovered that the toxin does exist as a higher order tetramer as the serial-receptor model hypothesizes but not as a pure toxin protein; it is complexed with receptor proteins and forms several higher order species besides the tetramer. We conclude that the putative prepore of Cry proteins is not a classic prepore as proposed for other protein toxins and should be referred to as a "complex oligomer" of Cry proteins and receptor proteins used to generate it. We did not detect alpha-helix 1 in the proteomic analysis of the complex oligomer, so it remains to directly determine if it is proteolytically removed upon interaction with the receptor or later upon insertion into the membrane. Our efforts toward resolving the structure of the prepore complex will address further details on the exact nature of the complex oligomer. From these studies, we question the exact role of a prepore generated using protocols described [1] [2] as an intermediate in the mode of action of the Cry toxins and suggest technical artifacts that need addressed before referring to the moiety as a typical prepore toxin.

Acknowledgements

We thank Dr. Daniel Zeigler for providing strains of *Bacillus thuringiensis* required for the study, Dr. Lee A. Bulla Jr. for providing us with anti-Bt-R1 antibody that reacts with cadherin and Dr. Alejandra Bravo for providing us with anti-Cry1Ab polyclonal antibody that reacts with the prepore form. We also want to thank Tom Abbott and Kathryn Stone at the Keck Biotechnology Resource Laboratory, Yale School of Medicine for performing the LC-MS experiments and MASCOT data analyses related to this work and Dr. James Lear for running purified prepore samples on the Beckman XL-I for us. This work was funded by NIH R01A29092 to DHD and Michael J. Adang.

References

[1] Gomez, I., Sanchez, J., Miranda, R., Bravo, A. and Soberon, M. (2002) Cadherin-Like Receptor Binding Facilitates Proteolytic Cleavage of Helix Alpha-1 in Domain I and Oligomer Pre-Pore Formation of *Bacillus thuringiensis* Cry1Ab Toxin. *FEBS Letters*, **513**, 242-246. http://dx.doi.org/10.1016/S0014-5793(02)02321-9

[2] Jimenez-Juarez, N., Munoz-Garay, C., Gomez, I., Saab-Rincon, G., Damian-Almazo, J.Y., Gill, S.S., Soberon, M. and Bravo, A. (2007) *Bacillus thuringiensis* Cry1Ab Mutants Affecting Oligomer Formation Are Non-Toxic to *Manduca sexta* Larvae. *The Journal of biological chemistry*, **282**, 21222-21229. http://dx.doi.org/10.1074/jbc.M701314200

[3] Alouf, J.E. and Popoff, M.R. (2006) The Comprehensive Sourcebook of Bacterial Protein Toxins. 3rd Edition, Elsevier, Amsterdam, Boston.

[4] Schnepf, E., Crickmore, N., Van Rie, J., Lereclus, D., Baum, J., Feitelson, J., Zeigler, D.R. and Dean, D.H. (1998) *Bacillus thuringiensis* and Its Pesticidal Crystal Proteins. *Microbiology and Molecular Biology Reviews: MMBR*, **62**, 775-806.

[5] Knowles, B.H. (1994) Mechanism of Action of *Bacillus thuringiensis* Insecticidal δ-Endotoxins. *Advances in Insect Physiology*, **24**, 275-308. http://dx.doi.org/10.1016/s0065-2806(08)60085-5

[6] Dean, D.H., Rajamohan, F., Lee, M.K., Wu, S.J., Chen, X.J., Alcantara, E. and Hussain, S.R. (1996) Probing the Mechanism of Action of *Bacillus thuringiensis* Insecticidal Proteins by Site-Directed Mutagenesis—A Minireview. *Gene*, **179**, 111-117. http://dx.doi.org/10.1016/S0378-1119(96)00442-8

[7] Chattopadhyay, A., Bhatnagar, N.B. and Bhatnagar, R. (2004) Bacterial Insecticidal Toxins. *Critical Reviews in Microbiology*, **30**, 33-54.

[8] Soberon, M., Pardo, L., Munoz-Garay, C., Sanchez, J., Gomez, I., Porta, H. and Bravo, A. (2010) Pore Formation by Cry Toxins. *Advances in Experimental Medicine and Biology*, **677**, 127-142. http://dx.doi.org/10.1007/978-1-4419-6327-7_11

[9] Mueller, M., Grauschopf, U., Maier, T., Glockshuber, R. and Ban, N. (2009) The Structure of a Cytolytic Alpha-Helical Toxin Pore Reveals Its Assembly Mechanism. *Nature*, **459**, 726-730. http://dx.doi.org/10.1038/nature08026

[10] Shepard, L.A., Heuck, A.P., Hamman, B.D., Rossjohn, J., Parker, M.W., Ryan, K.R., Johnson, A.E. and Tweten, R.K. (1998) Identification of a Membrane-Spanning Domain of the Thiol-Activated Pore-Forming Toxin *Clostridium perfringens* Perfringolysin O: An Alpha-Helical to Beta-Sheet Transition Identified by Fluorescence Spectroscopy. *Biochemistry*, **37**, 14563-14574. http://dx.doi.org/10.1021/bi981452f

[11] Collier, R.J. and Young, J.A.T. (2003) Anthrax Toxin. *Annual Review of Cell and Developmental Biology*, **19**, 45-70. http://dx.doi.org/10.1146/annurev.cellbio.19.111301.140655

[12] Young, J.A. and Collier, R.J. (2007) Anthrax Toxin: Receptor Binding, Internalization, Pore Formation, and Translocation. *Annual Review of Biochemistry*, **76**, 243-265. http://dx.doi.org/10.1146/annurev.biochem.75.103004.142728

[13] Tilley, S.J., Orlova, E.V., Gilbert, R.J., Andrew, P.W. and Saibil, H.R. (2005) Structural Basis of Pore Formation by the Bacterial Toxin Pneumolysin. *Cell*, **121**, 247-256. http://dx.doi.org/10.1016/j.cell.2005.02.033

[14] Aronson, A.I., Geng, C. and Wu, L. (1999) Aggregation of *Bacillus thuringiensis* Cry1A Toxins upon Binding to Target Insect Larval Midgut Vesicles. *Applied and Environmental Microbiology*, **65**, 2503-2507.

[15] Tomimoto, K., Hayakawa, T. and Hori, H. (2006) Pronase Digestion of Brush Border Membrane-Bound Cry1Aa Shows That Almost the Whole Activated Cry1Aa Molecule Penetrates into the Membrane. *Comparative Biochemistry and Physiology Part B: Biochemistry & Molecular Biology*, **144**, 413-422. http://dx.doi.org/10.1016/j.cbpb.2006.04.013

[16] Bravo, A., Gomez, I., Conde, J., Munoz-Garay, C., Sanchez, J., Miranda, R., Zhuang, M., Gill, S.S. and Soberon, M. (2004) Oligomerization Triggers Binding of a *Bacillus thuringiensis* Cry1Ab Pore-Forming Toxin to Aminopeptidase N Receptor Leading to Insertion into Membrane Microdomains. *Biochimica et Biophysica Acta*, **1667**, 38-46. http://dx.doi.org/10.1016/j.bbamem.2004.08.013

[17] Rausell, C., Muñoz-Garay, C., Miranda-CassoLuengo, R., Gomez, I., Rudiño-Piñera, E., Soberón, M. and Bravo, A. (2004) Tryptophan Spectroscopy Studies and Black Lipid Bilayer Analysis Indicate That the Oligomeric Structure of Cry1Ab Toxin from *Bacillus thuringiensis* Is the Membrane-Insertion Intermediate. *Biochemistry*, **43**, 166-174. http://dx.doi.org/10.1021/bi035527d

[18] Gomez, I., Dean, D.H., Bravo, A. and Soberon, M. (2003) Molecular Basis for *Bacillus thuringiensis* Cry1Ab Toxin Specificity: Two Structural Determinants in the *Manduca sexta* Bt-R_1 Receptor Interact with Loops Alpha-8 and 2 in Domain II of Cy1Ab Toxin. *Biochemistry*, **42**, 10482-10489. http://dx.doi.org/10.1021/bi034440p

[19] Gomez, I., Miranda-Rios, J., Rudino-Pinera, E., Oltean, D.I., Gill, S.S., Bravo, A. and Soberon, M. (2002) Hydropathic Complementarity Determines Interaction of Epitope ^{869}HITDTNNK^{876} in *Manduca sexta* Bt-R_1 Receptor with Loop 2 of Domain II of *Bacillus thuringiensis* Cry1A Toxins. *The Journal of Biological Chemistry*, **277**, 30137-30143. http://dx.doi.org/10.1074/jbc.M203121200

[20] Pacheco, S., Gomez, I., Arenas, I., Saab-Rincon, G., Rodriguez-Almazan, C., Gill, S.S., Bravo, A. and Soberon, M. (2009) Domain II Loop 3 of *Bacillus thuringiensis* Cry1Ab Toxin Is Involved in a "Ping Pong" Binding Mechanism with *Manduca sexta* Aminopeptidase-N and Cadherin Receptors. *The Journal of Biological Chemistry*, **284**, 32750-32757. http://dx.doi.org/10.1074/jbc.M109.024968

[21] Arenas, I., Bravo, A., Soberon, M. and Gomez, I. (2010) Role of Alkaline Phosphatase from *Manduca sexta* in the Mechanism of Action of *Bacillus thuringiensis* Cry1Ab Toxin. *The Journal of Biological Chemistry*, **285**, 12497-12503. http://dx.doi.org/10.1074/jbc.M109.085266

[22] Lee, M.K., Milne, R.E., Ge, A.Z. and Dean, D.H. (1992) Location of a *Bombyx mori* Receptor Binding Region on a

Bacillus thuringiensis Delta-Endotoxin. *The Journal of Biological Chemistry*, **267**, 3115-3121.

[23] Jimenez-Juarez, N., Munoz-Garay, C., Gomez, I., Gill, S.S., Soberon, M. and Bravo, A. (2008) The Pre-Pore from *Bacillus thuringiensis* Cry1Ab Toxin Is Necessary to Induce Insect Death in *Manduca sexta*. *Peptides*, **29**, 318-323. http://dx.doi.org/10.1016/j.peptides.2007.09.026

[24] Nair, M.S. and Dean, D.H. (2008) All Domains of Cry1A Toxins Insert into Insect Brush Border Membranes. *The Journal of Biological Chemistry*, **283**, 26324-26331. http://dx.doi.org/10.1074/jbc.M802895200

[25] Chen, J., Hua, G., Jurat-Fuentes, J.L., Abdullah, M.A. and Adang, M.J. (2007) Synergism of *Bacillus thuringiensis* Toxins by a Fragment of a Toxin-Binding Cadherin. *Proceedings of the National Academy of Sciences of the United States of America*, **104**, 13901-13906. http://dx.doi.org/10.1073/pnas.0706011104

[26] Zhang, X., Candas, M., Griko, N.B., Taussig, R. and Bulla Jr., L.A. (2006) A Mechanism of Cell Death Involving an Adenylyl Cyclase/PKA Signaling Pathway Is Induced by the Cry1Ab Toxin of *Bacillus thuringiensis*. *Proceedings of the National Academy of Sciences of the United States of America*, **103**, 9897-9902. http://dx.doi.org/10.1073/pnas.0604017103

[27] Griko, N., Zhang, X., Ibrahim, M., Midboe, E.G. and Bulla Jr., L.A. (2008) Susceptibility of *Manduca sexta* to Cry1Ab Toxin of *Bacillus thuringiensis* Correlates Directly to Developmental Expression of the Cadherin Receptor BT-R_1. *Comparative Biochemistry and Physiology Part B*: *Biochemistry & Molecular Biology*, **151**, 59-63. http://dx.doi.org/10.1016/j.cbpb.2008.05.016

20

Influence of Cell Confluency on the Expression of the α4 Integrin Subunit of Retinal Pigment Epithelial Cells

Jean-Michel Bourget[1], Mohib Morcos[2,3], Karine Zaniolo[1], Sylvain L. Guérin[1,4], Stéphanie Proulx[1,4*]

[1]Centre de recherche du CHU de Québec and Centre de recherche en organogénèse expérimentale de l'Université Laval/LOEX, axe médecine régénératrice, Hôpital du Saint-Sacrement, Québec, Canada
[2]Département d'anatomie et de cytologie pathologiques, CHU de Québec, Hôpital du Saint-Sacrement, Québec, Canada
[3]Département de biologie moléculaire, biochimie médicale et pathologie, Faculté de médecine, Université Laval, Québec, Canada
[4]Département d'Ophtalmologie, Faculté de médecine, Université Laval, Québec, Canada
Email: [*]stephanie.proulx@fmed.ulaval.ca

Abstract

Integrins are a family of transmembrane glycoproteins that mediate cell-cell and cell-extracellular matrix interactions. The integrin α4 subunit is widely expressed by cells from the immune system and its expression by non-hematopoietic cells is scarce. In the present study, gene and protein expression of this integrin subunit was characterized in proliferating and quiescent human RPE cells. Immunofluorescent studies confirm that the α4 subunit is expressed *in vitro* by RPE cells, a result that has been validated by immunofluorescence and FACS analyses. The accumulation of the α4 integrin at cell-cell junctions in post-confluent RPE cell cultures negatively correlated with the level of expression of the mRNA transcript. Accordingly, transient transfection analyses reveal that the α4 promoter activity is considerably reduced when RPE cells form a confluent monolayer. Moreover, transfection of recombinant constructs bearing 5'-deletions of the α4 promoter segment allows the localization of strong negative regulatory elements on the −76 to −300 region of the α4 gene suggesting that its expression is intimately linked to the proliferative state of primary cultured RPE cells.

Keywords

Retinal Pigment Epithelium, Integrin Alpha 4 Subunit, Cell Culture, Confluency, Promoter

[*]Corresponding author.

1. Introduction

The retinal pigment epithelium (RPE) is a monolayer of polarized cells that separates the retina from the underlying vasculature. As part of the blood-ocular barrier, the RPE is responsible for the transport of nutrients from the choroidal blood vessels to the photoreceptors and, conversely, for the transport of waste products from the photoreceptors to the choroidal blood vessels [1]. In response to a traumatic retinal injury, RPE cells can become proliferative, leading to retinal disorders and vision loss [2]. A change in the expression of integrins has been observed when RPE cells become proliferative [3] [4].

Integrins form a family of transmembrane glycoproteins that regulate cell-cell and cell-extracellular matrix (ECM) interactions [5] [6]. They control many cellular processes, including adhesion, migration, proliferation, survival and differentiation [7] [8]. Integrins are heterodimeric molecules made-up of an α subunit, which confers most of the ECM ligand specificity, and a β subunit, that interacts with the intracellular cytoskeleton via adaptor proteins and determines the broad class of the receptor [9] [10]. To date, 8 β- and 18 α-subunits have been identified that can associate with each other to produce the 24 integrin heterodimers reported to date [11]-[15].

The $\alpha4$ subunit can associate with either the $\beta1$ or the $\beta7$ subunit [8] [13] [14] [16]. Integrins $\alpha4\beta1$, also known as very late antigen-4 (VLA-4) [17], and $\alpha4\beta7$ mediate interaction with the immunoglobulin vascular cell adhesion molecule-1 (VCAM-1) as well as with the alternatively spliced CS-1 region of the ECM protein fibronectin (FN) [8] [13] [14] [16]. However, $\alpha4\beta7$ binds these two ligands with a lesser affinity than $\alpha4\beta1$ [18].

The integrin $\alpha4\beta1$ mediates both cell-cell and cell-ECM interactions and plays an important role in development and cell differentiation. This integrin is expressed by a variety of hematopoietic cells including lymphocytes, monocytes and eosinophils [19]-[22]. It plays an important role in the regulation of the immune response by promoting the recruitment of lymphocytes at inflammation sites [23]-[25] by its interaction with VCAM-1 on activated vascular endothelial cells, allowing for diapedesis [26].

Apart from the immune cells, the $\alpha4\beta1$ integrin has also been identified on dermal fibroblasts [27], in the epicardium [28] [29], on developing myocytes [30] [31], on neural crest cells [32] [33], on rhabdomyosarcomas [17], on melanoma [34]-[36] and on corneal epithelial cells [37]. Moreover, this integrin also plays an important role in development and cell differentiation. Indeed, interaction between $\alpha4\beta1$ and VCAM-1 is important in myogenic differentiation and $\alpha4$ expression is induced during differentiation of C2C12 cells (a mouse myoblast cell line) and during muscle differentiation [30]. *In vivo*, antibodies that block either $\alpha4$ or VCAM-1 inhibit the development of the sympathetic innervation of the heart [38]. Knockout mice lacking $\alpha4$ die at embryonic day 11 due to cardiac and placental formation failure [29] [39].

In the embryonic retina, $\alpha4\beta1$ is expressed by retinal ganglion cells and undifferentiated retinal neuroblasts. It is proposed to function as a mediator for neurite outgrowth by its interaction with VCAM-1, FN and osteopontin in this developing tissue [40] and it is required for cell survival in the developing retina [41]. In the mature human retina, $\alpha4$ expression is observed on cells from the outer nuclear layer, including rods and cones, cells from the inner nuclear layer, ganglion cells and RPE cells [3] [42] [43]. The present study was undertaken in order to evaluate whether expression of the integrin $\alpha4$ subunit changes with increasing cell densities that replicate the proliferative states of primary cultured RPE cells (from sub-confluence (proliferating) to post-confluence (quiescent).

2. Materials and Methods

This study was conducted in accordance with our institution's guidelines and the Declaration of Helsinki. The protocols were also approved by our institution's ethics committee for the protection of human subjects.

2.1. Cell Culture

Primary cultures of RPE cells were established from human eyes rejected for transplantation and obtained from the Centre Universitaire d'Ophtalmologie (CUO) Eyebank Inc. (Québec, Qc, Canada) 24 h postmortem. Dissection was done according to a previously published method with some modifications [44]. Briefly, pieces of dissected RPE/choroid were incubated 45 minutes at 37˚C in 2.4% dispase (Roche Diagnostics, Laval, Qc, Canada). RPE sheets were gently detached by spraying growth medium against the tissue. RPE sheets were then cultured in Keratinocyte-SFM medium (Life Technologies, Burlington, Ont., Canada) supplemented with 5% bovine calf serum (Hyclone, Logan, UT) and various vitamins and proteins, as described [45]. For all experiments, cultures were used at passage 2 and cells were plated at different densities as follows: sub-confluent cells: 2.5×10^4 cells/cm^2, 2 days in culture (which corresponded to approximately 80% coverage of the culture plate at the time the cells were harvested); confluent cells: 1.5×10^5 cells/cm^2, 2 days in culture (which corresponded to 100% coverage of the culture plate at the time the cells were harvested); post-confluent cells: 1.5×10^5 cells/cm^2, 2 weeks in culture (which corresponded to maintaining the cells at full confluency for 12 days prior to cell harvesting).

2.2. Immunostaining

The localization of the integrin α4 subunit at the cell membrane was monitored by immunohistochemistry on paraffin-embedded sections of human retina-choroid tissues and by immunofluorescence on *in vitro* human primary RPE cells cultured on glass coverslips.

For immunohistochemical *in situ* staining, human retina-choroid sections were fixed in 4% paraformaldehyde (Electron microscopy Sciences, Hatfield, PA), embedded in paraffin, and cut in 4 µm sections. After deparaffinisation and rehydration, samples were processed in a PT-Link (Dako, Mississauga, Ont., Canada) for antigen retrieval (Tris-EDTA, pH 9, 85°C, 20 min). Immunohistochemical staining was performed manually using the EnVision Flex, High pH Kit (Dako) according to the manufacturer's protocol and using a rabbit monoclonal antibody against integrin α4 (clone EPR1355Y, Millipore, Billerica, MA). Nuclei were counterstained using Harris' hematoxylin. For negative controls, the primary antibody was replaced with normal rabbit serum.

Immunofluorescence was performed on cultured RPE cells grown on glass coverslips. A monoclonal antibody directed against the α4 integrin subunit (mouse anti-human integrin α4, clone HP2/1, Cedarlane, Burlington, ON, Canada) and the secondary antibody Alexa Fluor 488 goat anti-mouse IgG (Life Technologies) were used to identify the presence of this integrin subunit on RPE cells *in vitro*, as described previously [46]. For negative controls, the primary antibody was omitted.

2.3. FACS Analyses

RPE cells grown at different confluencies were harvested in PBS/EDTA 2 mM and fixed in 90% ethanol. Cells (2×10^5) were incubated for one hour with a mouse monoclonal antibody directed against the α4 integrin subunit (Cedarlane). As a negative control, this antibody was replaced by a mouse primary antibody isotype control (Dako). Cells were then incubated for 45 min with a goat anti-mouse phycoerythrin-conjugated secondary antibody (Cedarlane) and then analyzed by flow cytometry (Facscalibur, BD Biosciences, Mississauga, Ont., Canada).

2.4. Semi-Quantitative Reverse Transcription-Polymerase Chain Reaction

Whole RNA extracts from RPE cells grown at various cell densities (sub-confluency, confluency and post-confluency) were reverse transcribed and PCR-amplified as described [45]. PCR products were in the linear range between 25 and 37 cycles for sub-confluent and confluent cells, and between 27 and 39 for post-confluent cells. The total number of cycles used for semi-quantitative RT-PCR was 32 and the experiment was done in triplicate using different populations of RPE cells derived from the retina of three different donors. The DNA sequence of the primers used for the amplification of the human α4 transcript were: 5'-TGGCGTGGTACAACTTGACTG-3' (forward primer), and 5'-CATGCGCAACATTTCATCCT-3' (reverse primer; 772-bp PCR product) and cycle parameters were 94°C for 30 sec, 55°C for 30 sec and 72°C for 30 sec. The 18S primers (Ambion; Life Technologies) gave a PCR product of 489 bp. Bands were analyzed with the GelDoc2000 gel documentation system (Bio-Rad Laboratories, Mississauga, Ont., Canada) and the Quantity One 1-D image analysis software (Bio-Rad Laboratories).

2.5. Transient Transfections and CAT Assays

Plasmids containing the chloramphenicol acetyltransferase (CAT) reporter gene from the plasmid pSKCAT fused to various 5' deletions of the α4 gene (−1000α4CAT, −800α4CAT, −600α4CAT, −400α4CAT, −300α4CAT, −200α4CAT, −76α4CAT, −41α4CAT) were obtained from Dr. Glenn D. Rosen (Stanford University Medical Center, Stanford, CA). RPE cells were transfected using the calcium phosphate precipitation method as described [45]. Cells were harvested 48 h following transfection and CAT activities determined [47] and normalized to the amount of human growth hormone (hGH) encoded by the co-transfected plasmid pXGH5 and assayed using a kit for quantitative measurement of hGH (Immunocorp, Montréal, Qc, Canada) as previously described [47]. The value presented for each individual test plasmid transfected corresponds to the mean of at least three separate transfections done with RPE cells from three different donors, each in triplicate. Student's t-test was performed for comparison of the groups. Differences were considered to be statistically significant at $P < 0.05$. All data are expressed as mean ± SD.

3. Results

3.1. The $\alpha 4$ Integrin Subunit Is Expressed on RPE Cells *in Situ* and *in Vitro*

The retinal pigment epithelium has a natural brown pigmentation, as seen in the *in situ* sections (**Figure 1(B)**, **Figure 1(D)**). Nevertheless, expression of the $\alpha 4$ protein was readily detected *in situ* in the RPE monolayer over their natural brown pigmentation (**Figure 1(A)**, **Figure 1(C)**), especially in the sections that were cut obliquely (**Figure 1(C)**). Meanwhile, other cells from the choroid also stained positive for this integrin. Expression of the $\alpha 4$ subunit was also monitored *in vitro* on primary human RPE cells cultured at varying cell densities. As shown on **Figure 2(A)**, a weak $\alpha 4$ signal could be observed on sub-confluent RPE cells, with a more intense staining at the cell edges. Cells that reached confluency had a stronger $\alpha 4$ staining at cell-cell contacts (**Figure 2(B)**). RPE cells that were left for 2 weeks at confluency (post-confluent cells) were tightly packed and also showed a positive $\alpha 4$ staining at cell-cell contacts (**Figure 2(C)**). The expression of the $\alpha 4$ integrin subunit by RPE cells has been further demonstrated by FACS analyses that clearly confirm the expression of this protein *in vitro* (**Figure 3**). The number of $\alpha 4$ positive cells reduced from 97.7% to 88.2% between the sub- and post-confluent states ($p = 0.007$; **Figure 3(a)**). The mean fluorescence intensity of $\alpha 4$ positive cells was similar in sub-confluent, confluent and post-confluent cultures (**Figure 3(b)**).

3.2. Transcription of the $\alpha 4$ Gene in RPE Cell Cultures

In order to evaluate the amount of $\alpha 4$ mRNA transcript present at different cell densities, total RNA from cul-

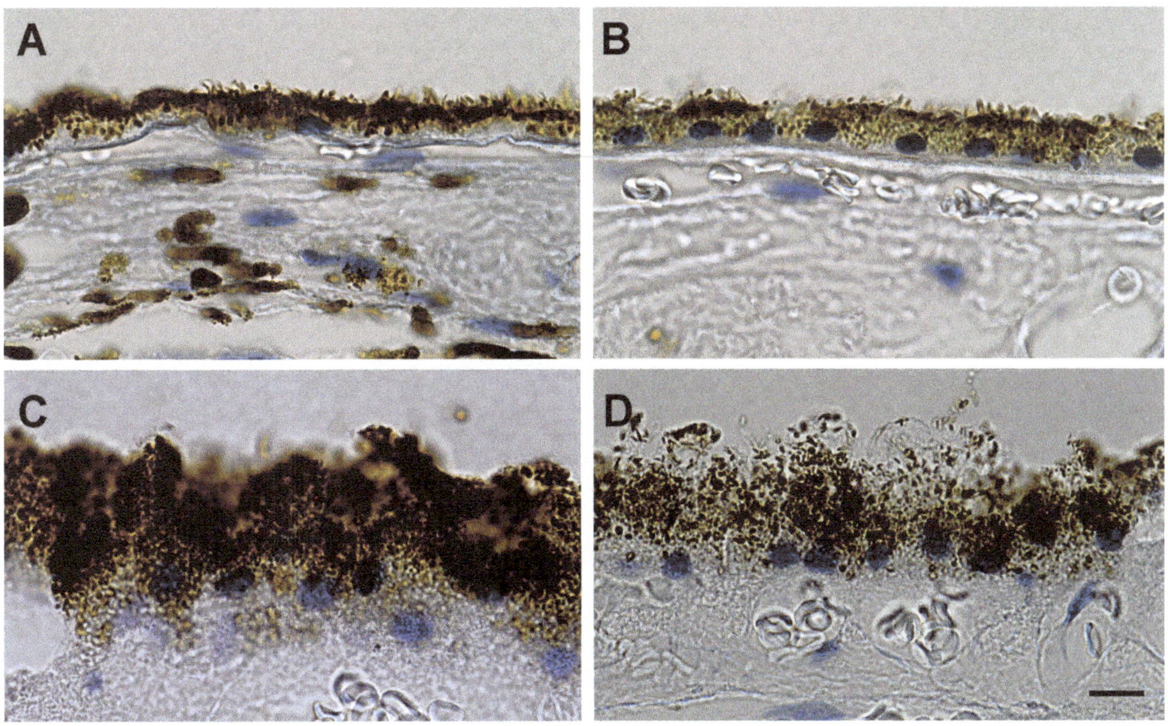

Figure 1. Immunohistochemical analysis of the $\alpha 4$ integrin subunit on paraffin sections of a human retina/choroid tissue. (A) (C) The $\alpha 4$ integrin subunit is detected diffusely throughout the RPE surface. (A) cross section, (C) oblique section; (B) (D) Negative controls using normal rabbit serum (B) cross section; (D) oblique section. Note that the retinal pigment epithelium possesses melanin granules that give a natural brown pigmentation to this monolayer. Scale bar represents 10 μm.

Figure 2. Immunofluorescence analysis of the $\alpha4$ integrin subunit on cultured human RPE cells. (A) Sub-confluent cells are positively labeled for $\alpha4$ at cell borders (arrow). The $\alpha4$ staining is also present on confluent (B) and post-confluent (C) cells. Scale bar represents 10 μm and all pictures were taken at the same magnification.

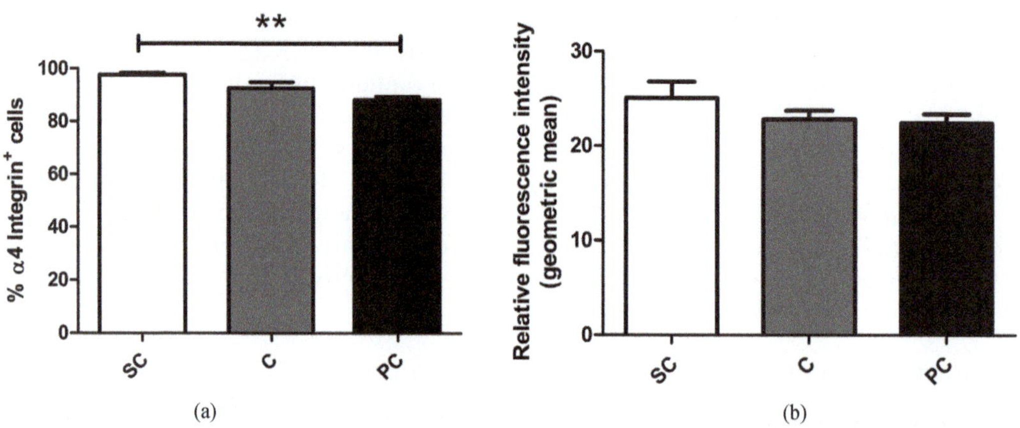

Figure 3. Expression of the integrin subunit $\alpha4$ was monitored by flow cytometry in primary cultures of human RPE cells grown to sub-confluency (SC), confluency (C) and post-confluency (PC). (a) Percentage of $\alpha4$ positive cells; (b) Relative fluorescence intensity, geometric means of the $\alpha4$ positive cells. Mean ± SEM, n = 4, one-way ANOVA with a subsequent Bonferroni post-hoc test, $^{**}P < 0.01$.

tured RPE cells was isolated and used for semi-quantitative RT-PCR measurements. When normalized to the amount of transcripts encoded by the ribosomal 18S gene, a significant decrease (2.4 times) in the intensity of the PCR product corresponding to the $\alpha4$ gene was observed between sub-confluent and confluent RPE cells (**Figure 4**). Transcription of $\alpha4$ decreased further down (8.1 times) to a near undetectable level as RPE cells progressed from confluency to post-confluency. It is noteworthy that the intensity of the 18S PCR product remained almost unchanged (**Figure 4**).

Given that the level of $\alpha4$ integrin subunit mRNA was influenced by cell confluency, we next determined whether the transcriptional activity normally driven by the $\alpha4$ gene promoter was also similarly regulated upon transient transfection of $\alpha4$ promoter/CAT recombinant constructs into sub-confluent, confluent and post-confluent RPE cells. As shown in **Figure 5(a)**, an 8-fold decrease in promoter activity was observed between sub-confluent and confluent cells when the −200$\alpha4$CAT plasmid was used. The $\alpha4$ promoter activity decreased further to a level 17-fold lower than that measured in sub-confluent cells when the −200$\alpha4$CAT construct was transfected in 2 weeks post-confluent RPE cells. Therefore, the activity of the $\alpha4$ promoter is repressed when RPE cells reach confluency. Furthermore, the −1000$\alpha4$CAT plasmid, which bears the longest $\alpha4$ promoter segment and is thus the most representative of the native gene, is highly repressed in cultured RPE cells, regardless of the state of confluency reached by the cells (**Figure 5(a)**). These results suggest that strong negative regulatory elements that repress $\alpha4$ gene transcription when RPE cells reach confluency must be located on the 800 bp segment from the $\alpha4$ promoter comprised between positions −200 and −1000.

To more precisely delineate the position of these negative regulatory elements along the $\alpha4$ promoter, CAT constructs bearing various deletions of the $\alpha4$ promoter, ranging from −41 bp to −1000 bp relative to the $\alpha4$ mRNA start site were transfected into sub-confluent cultures of RPE cells. The minimal $\alpha4$ promoter sequences that directed maximal expression in RPE cells were found to be contained on the −76$\alpha4$CAT construct (**Figure 5(b)**). Deleting further the $\alpha4$ promoter down to position −41 resulted in a drastic 20-fold reduction in basal

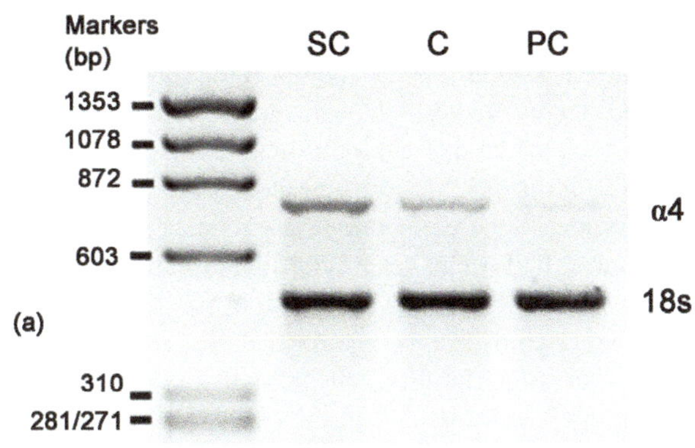

(b)

	SC	C	PC
α4	193	80	10
18S	841	865	893

Figure 4. Expression of the *α*4 transcripts in primary cultured human RPE cells. (a) RT-PCR amplification products corresponding to the *α*4 integrin mRNA transcript were obtained from sub-confluent (SC), confluent (C) and post-confluent (PC) cultures of RPE cells and normalized to the 18S PCR product for semi-quantitative evaluation. The position of the PCR product corresponding to *α*4 (772 bp) is shown along with that corresponding to the 18S rRNA (489 bp); (b) Band density was calculated by the QuantityOne Image analysis software and reported as unit of intensity per millimeters.

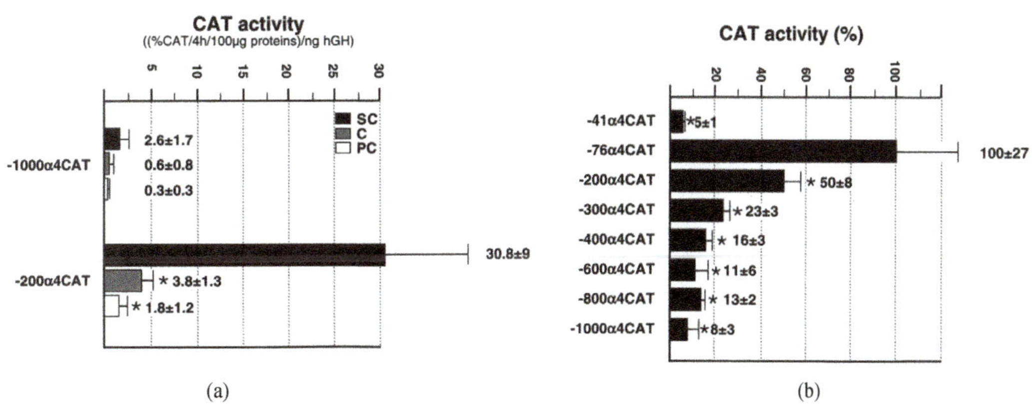

Figure 5. *α*4 promoter activity in primary cultured RPE cells. (a) Sub-confluent (black box), confluent (grey box) and post-confluent (white box) cultures of RPE cells (n = 6) were transfected with plasmids bearing fragments from the human *α*4 gene promoter extending either to 5' position −200 (in −200*α*4CAT) or −1000 (in −1000*α*4CAT) and fused to the CAT reporter gene. Cells were harvested 2 days following transfection and CAT activities determined and normalized to the amount of hGH secreted into the culture media. Asterisks (*) indicate CAT activities at both confluency and post-confluency that are statistically different from those measured in sub-confluent cells ($P < 0.05$; paired samples, t-test). S.D. is also provided; (b) *α*4/CAT recombinant plasmids bearing 5'-deletions of the *α*4 promoter (5' end-point ranging between positions −1000 and −41 relative to the *α*4 mRNA start site) were transfected into sub-confluent cultures of RPE cells (n = 6). Cells were harvested and CAT activities determined and normalized as in (a). *: indicates CAT activities from transfected RPE cells that are statistically different from those measured with the −76*α*4CAT plasmid ($P < 0.05$; paired samples, t-test). S.D. is also provided.

promoter function. Extending further the 5'-end of the *α*4 promoter resulted in a progressive reduction of *α*4 promoter activity that reached a level 13-times lower with plasmid −1000*α*4CAT relative to the activity directed by the −76*α*4CAT construct. However, most of this negative regulatory influence is primarily directed by the *α*4 promoter segment comprised between positions −76 to −300 as it accounted for nearly 80% of the repressive influence observed in RPE cells.

4. Discussion

When cultured RPE cells are left at post-confluency, they become quiescent and acquire morphological characteristics similar to those observed *in vivo*. Using RPE cells grown to the same cell densities as those reported in the present study, we previously established that sub-confluent and confluent RPE cells cultured *in vitro* are actively proliferating whereas post-confluent cells are quiescent [45]. In the present paper, we reported that a marked reduction in the transcription of the endogenous $\alpha4$ gene as well as in the activity directed by the $\alpha4$ promoter is observed when cultured RPE cells reach confluency, suggesting that $\alpha4$ gene expression is intimately related to the proliferative state of RPE cells. In addition, we demonstrated that the quantity of $\alpha4$ protein is similar between sub-confluent and confluent cells, and that the protein accumulates at the membrane level once cells reach confluency. This suggests that an immature form of the protein was present within sub-confluent cells, which increasingly translocated to the cell membrane during the maturation period of cell culture. The localisation at the cell junction upon reaching confluency is not unique to the $\alpha4$ integrin subunit. Indeed, other cell adhesion molecules, such as N-Cadherin [48], have been shown to localise at the cell periphery when maintaining cultured cells at post-confluency for several weeks.

The stronger $\alpha4$ staining at cell-cell junctions observed with confluent cells (as compared to sub-confluent cells) could seem contradictory with regards to the down-regulation of the mRNA observed in **Figure 4**. However, this apparent discrepancy can be explained by the accumulation of the protein at cell-cell contact leading to a negative feedback loop. Moreover, giving that the $\alpha4\beta1$ integrin has been reported to bind to itself [36] [49], the increase in cell-cell contacts, which are typical of confluent cultures, may facilitate detection of this integrin by promoting its clustering at the cell edges and could therefore explain the high fluorescence staining at this particular location in confluent cells.

Expression of the $\alpha4$ integrin subunit in native RPE cells has been previously documented [42]. However, as RPE cells are mitotically inactive *in situ*, $\alpha4$ expression in RPE cells is intriguing given its well-known role in the regulation of the immune response [24] [25], as well as in development, cell differentiation and migration [29] [30] [38] [39] [50]. In diseases such as proliferative vitreous retinopathy, the normally quiescent RPE cells dedifferentiate, proliferate and secrete extracellular matrix molecules that form fibrocellular membranes [51]. Fibronectin, a ligand of the $\alpha4\beta1$ integrin, is a major component of these epiretinal membranes [52]. The presence of a constitutive $\alpha4$ expression in the native cells might facilitate RPE dedifferentiation, migration and proliferation and contribute to the progression of the disease. To that regard, blocking antibodies directed against the $\alpha4$ subunit might prove a promising approach for the treatment of this fibrocellular disease.

The $\alpha4$ promoter is more efficiently transcribed into sub-confluent cells and becomes heavily down-regulated as cells reach confluency. This pattern of expression is consistent with the data from the RT-PCR analyses (**Figure 4**) but yet differs from that previously observed with the $\alpha5$ integrin promoter in RPE cells [45], as maximal level of $\alpha5$ promoter activity was seen at confluency, and then decreased when cells reached post-confluency. Although constructs of similar lengths were used for the transfection studies, both the $\alpha4$ and $\alpha5$ promoters bear clearly different regulatory regions, suggesting that these genes are controlled by different sets of transcription factors. Most interestingly, our results also demonstrate the presence of one (or many) negative regulatory element(s) located between position −76 and −200 relative to the $\alpha4$ mRNA start site. Of particular interest is the sequence −76/−200 which functions as a powerful transcriptional repressor in RPE cells [53]-[55]. This reduced promoter activity must be mediated by regulatory proteins other than the ZEB transcription factor since ZEB sites have only been located at positions −361 and −399 [56].

5. Conclusion

This study establishes that cell confluency generates a downregulation of the $\alpha4$ gene expression that is dependent upon strong negative regulatory elements located on the −76 to −300 region of the promoter. Given that the expression of the $\alpha4$ integrin subunit is scarce in non-haematopoietic cells and that it binds an alternatively spliced segment of FN that is normally expressed only during wound healing [57]-[59], its expression by RPE cells remains intriguing and deserves more studies.

Acknowledgements

The authors would like to thank the CUO Eyebank (Québec, Qc, Canada) for providing the human globes. This

Eyebank is partly supported by an infrastructure from the Vision Health Research Network of the "Fonds de recherche du Québec-Santé" (FRQS). We also would like to thank Marcelle Giasson for her help with the CAT assays and Sébastien Larochelle for help with flow cytometry analysis. This study was supported by grants from the Natural Sciences and Engineering Research Council of Canada (NSERC) to S.P. (grant #418598-2012) and S.L.G. (grant #138624-2012).

References

[1] Rizzolo, L.J., Peng, S., Luo, Y. and Xiao, W. (2011) Integration of Tight Junctions and Claudins with the Barrier Functions of the Retinal Pigment Epithelium. *Progress in Retinal and Eye Research*, **30**, 296-323. http://dx.doi.org/10.1016/j.preteyeres.2011.06.002

[2] Chiba, C. (2014) The Retinal Pigment Epithelium: An Important Player of Retinal Disorders and Regeneration. *Experimental Eye Research*, **123**, 107-114. http://dx.doi.org/10.1016/j.exer.2013.07.009

[3] Robbins, S.G., Brem, R.B., Wilson, D.J., O'Rourke, L.M., Robertson, J.E., Westra, I., *et al.* (1994) Immunolocalization of Integrins in Proliferative Retinal Membranes. *Investigative Ophthalmology & Visual Science*, **35**, 3475-3485.

[4] Hoffmann, S., He, S., Jin, M., Ehren, M., Wiedemann, P., Ryan, S.J., *et al.* (2005) A Selective Cyclic Integrin Antagonist Blocks the Integrin Receptors Alphavbeta3 and Alphavbeta5 and Inhibits Retinal Pigment Epithelium Cell Attachment, Migration and Invasion. *BMC ophthalmology*, **5**, 16. http://dx.doi.org/10.1186/1471-2415-5-16

[5] Hynes, R.O. (2002) Integrins: Bidirectional, Allosteric Signaling Machines. *Cell*, **110**, 673-687. http://dx.doi.org/10.1016/S0092-8674(02)00971-6

[6] Vigneault, F., Zaniolo, K., Gaudreault, M., Gingras, M.E. and Guerin, S.L. (2007) Control of Integrin Genes Expression in the Eye. *Progress in Retinal and Eye Research*, **26**, 99-161. http://dx.doi.org/10.1016/j.preteyeres.2006.10.004

[7] Takada, Y., Ye, X. and Simon, S. (2007) The Integrins. *Genome Biology*, **8**, 215. http://dx.doi.org/10.1186/gb-2007-8-5-215

[8] Elner, S.G. and Elner, V.M. (1996) The Integrin Superfamily and the Eye. *Investigative Ophthalmology & Visual Science*, **37**, 696-701.

[9] Legate, K.R. and Fassler, R. (2009) Mechanisms That Regulate Adaptor Binding to Beta-Integrin Cytoplasmic Tails. *Journal of Cell Science*, **122**, 187-198. http://dx.doi.org/10.1242/jcs.041624

[10] Wiesner, S., Legate, K.R. and Fassler, R. (2005) Integrin-Actin Interactions. *Cellular and Molecular Life Sciences: CMLS*, **62**, 1081-1099. http://dx.doi.org/10.1007/s00018-005-4522-8

[11] Shattil, S.J., Kim, C. and Ginsberg, M.H. (2010) The Final Steps of Integrin Activation: The End Game. *Nature Reviews: Molecular Cell Biology*, **11**, 288-300. http://dx.doi.org/10.1038/nrm2871

[12] Cordes, N. and Park, C.C. (2007) Beta1 Integrin as a Molecular Therapeutic Target. *International Journal of Radiation Biology*, **83**, 753-760. http://dx.doi.org/10.1080/09553000701639694

[13] Aota, S. and Yamada, K.M. (1995) Fibronectin and Cell Adhesion: Specificity of Integrin-Ligand Interaction. *Advances in Enzymology and Related Areas of Molecular Biology*, **70**, 1-21.

[14] Wehrle-Haller, B. and Imhof, B.A. (2003) Integrin-Dependent Pathologies. *Journal of Pathology*, **200**, 481-487. http://dx.doi.org/10.1002/path.1399

[15] Lowell, C.A. and Mayadas, T.N. (2012) Overview: Studying Integrins *in Vivo*. *Methods in Molecular Biology*, **757**, 369-397. http://dx.doi.org/10.1007/978-1-61779-166-6_22

[16] Hemler, M.E., Elices, M.J., Parker, C. and Takada, Y. (1990) Structure of the Integrin Vla-4 and Its Cell-Cell and Cell-Matrix Adhesion Functions. *Immunological Reviews*, **114**, 45-65. http://dx.doi.org/10.1111/j.1600-065X.1990.tb00561.x

[17] Hemler, M.E., Huang, C. and Schwarz, L. (1987) The Vla Protein Family. Characterization of Five Distinct Cell Surface Heterodimers Each with a Common 130,000 Molecular Weight Beta Subunit. *The Journal of Biological Chemistry*, **262**, 3300-3309.

[18] Chan, B.M., Elices, M.J., Murphy, E. and Hemler, M.E. (1992) Adhesion to Vascular Cell Adhesion Molecule 1 and Fibronectin. Comparison of $\alpha_4\beta_1$ (Vla-4) and $\alpha_4\beta_7$ on the Human B Cell Line JY. *The Journal of Biological Chemistry*, **267**, 8366-8370.

[19] Zhang, Y. and Wang, H. (2012) Integrin Signalling and Function in Immune Cells. *Immunology*, **135**, 268-275. http://dx.doi.org/10.1111/j.1365-2567.2011.03549.x

[20] Ferguson, T.A., Mizutani, H. and Kupper, T.S. (1991) Two Integrin-Binding Peptides Abrogate T Cell-Mediated Immune Responses *in Vivo*. *Proceedings of the National Academy of Sciences of the United States of America*, **88**, 8072-8076. http://dx.doi.org/10.1073/pnas.88.18.8072

[21] Williams, D.A., Rios, M., Stephens, C. and Patel, V.P. (1991) Fibronectin and Vla-4 in Haematopoietic Stem Cell-Microenvironment Interactions. *Nature*, **352**, 438-441. http://dx.doi.org/10.1038/352438a0

[22] Crisa, L., Cirulli, V., Ellisman, M.H., Ishii, J.K., Elices, M.J. and Salomon, D.R. (1996) Cell Adhesion and Migration Are Regulated at Distinct Stages of Thymic T Cell Development: The Roles of Fibronectin, VLA4, and VLA5. *Journal of Experimental Medicine*, **184**, 215-228. http://dx.doi.org/10.1084/jem.184.1.215

[23] Lim, Y.C., Wakelin, M.W., Henault, L., Goetz, D.J., Yednock, T., Cabanas, C., *et al.* (2000) $\alpha_4\beta_1$-Integrin Activation Is Necessary for High-Efficiency T-Cell Subset Interactions with Vcam-1 under Flow. *Microcirculation*, **7**, 201-214. http://dx.doi.org/10.1111/j.1549-8719.2000.tb00121.x

[24] Osborn, L., Hession, C., Tizard, R., Vassallo, C., Luhowskyj, S., Chi-Rosso, G., *et al.* (1989) Direct Expression Cloning of Vascular Cell Adhesion Molecule 1, a Cytokine-Induced Endothelial Protein That Binds to Lymphocytes. *Cell*, **59**, 1203-1211. http://dx.doi.org/10.1016/0092-8674(89)90775-7

[25] Elices, M.J., Osborn, L., Takada, Y., Crouse, C., Luhowskyj, S., Hemler, M.E., *et al.* (1990) Vcam-1 on Activated Endothelium Interacts with the Leukocyte Integrin Vla-4 at a Site Distinct from the Vla-4/Fibronectin Binding Site. *Cell*, **60**, 577-584. http://dx.doi.org/10.1016/0092-8674(90)90661-W

[26] Bonder, C.S., Norman, M.U., Swain, M.G., Zbytnuik, L.D., Yamanouchi, J., Santamaria, P., *et al.* (2005) Rules of Recruitment for Th1 and Th2 Lymphocytes in Inflamed Liver: A Role for Alpha-4 Integrin and Vascular Adhesion Protein-1. *Immunity*, **23**, 153-163. http://dx.doi.org/10.1016/j.immuni.2005.06.007

[27] Gailit, J., Pierschbacher, M. and Clark, R.A. (1993) Expression of Functional $\alpha_4\beta_1$ Integrin by Human Dermal Fibroblasts. *Journal of Investigative Dermatology*, **100**, 323-328. http://dx.doi.org/10.1111/1523-1747.ep12470011

[28] Kwee, L., Baldwin, H.S., Shen, H.M., Stewart, C.L., Buck, C., Buck, C.A., *et al.* (1995) Defective Development of the Embryonic and Extraembryonic Circulatory Systems in Vascular Cell Adhesion Molecule (Vcam-1) Deficient Mice. *Development*, **121**, 489-503.

[29] Yang, J.T., Rayburn, H. and Hynes, R.O. (1995) Cell Adhesion Events Mediated by Alpha 4 Integrins Are Essential in Placental and Cardiac Development. *Development*, **121**, 549-560.

[30] Rosen, G.D., Sanes, J.R., LaChance, R., Cunningham, J.M., Roman, J. and Dean, D.C. (1992) Roles for the Integrin Vla-4 and Its Counter Receptor Vcam-1 in Myogenesis. *Cell*, **69**, 1107-1119. http://dx.doi.org/10.1016/0092-8674(92)90633-N

[31] Sheppard, A.M., Onken, M.D., Rosen, G.D., Noakes, P.G. and Dean, D.C. (1994) Expanding Roles for Alpha 4 Integrin and Its Ligands in Development. *Cell Communication and Adhesion*, **2**, 27-43. http://dx.doi.org/10.3109/15419069409014200

[32] Stepp, M.A., Urry, L.A. and Hynes, R.O. (1994) Expression of Alpha 4 Integrin Mrna and Protein and Fibronectin in the Early Chicken Embryo. *Cell Communication and Adhesion*, **2**, 359-375. http://dx.doi.org/10.3109/15419069409014210

[33] Kil, S.H., Krull, C.E., Cann, G., Clegg, D. and Bronner-Fraser, M. (1998) The α_4Subunit of Integrin Is Important for Neural Crest Cell Migration. *Developmental Biology*, **202**, 29-42. http://dx.doi.org/10.1006/dbio.1998.8985

[34] Garofalo, A., Chirivi, R.G., Foglieni, C., Pigott, R., Mortarini, R., Martin-Padura, I., *et al.* (1995) Involvement of the Very Late Antigen 4 Integrin on Melanoma in Interleukin 1-Augmented Experimental Metastases. *Cancer Research*, **55**, 414-419.

[35] Saini, A., Seller, Z., Davies, D., Marshall, J.F. and Hart, I.R. (1997) Activation Status and Function of the Vla-4 ($\alpha_4\beta_1$) Integrin Expressed on Human Melanoma Cell Lines. *International Journal of Cancer*, **73**, 264-270. http://dx.doi.org/10.1002/(SICI)1097-0215(19971009)73:2<264::AID-IJC17>3.0.CO;2-F

[36] Qian, F., Vaux, D.L. and Weissman, I.L. (1994) Expression of the Integrin $\alpha_4\beta_1$ on Melanoma Cells Can Inhibit the Invasive Stage of Metastasis Formation. *Cell*, **77**, 335-347. http://dx.doi.org/10.1016/0092-8674(94)90149-X

[37] Lauweryns, B., van den Oord, J.J., Volpes, R., Foets, B. and Missotten, L. (1991) Distribution of Very Late Activation Integrins in the Human Cornea. An Immunohistochemical Study Using Monoclonal Antibodies. *Investigative Ophthalmology & Visual Science*, **32**, 2079-2085.

[38] Wingerd, K.L., Goodman, N.L., Tresser, J.W., Smail, M.M., Leu, S.T., Rohan, S.J., *et al.* (2002) Alpha 4 Integrins and Vascular Cell Adhesion Molecule-1 Play a Role in Sympathetic Innervation of the Heart. *The Journal of Neuroscience: The Official Journal of the Society for Neuroscience*, **22**, 10772-10780.

[39] Hynes, R.O. (1996) Targeted Mutations in Cell Adhesion Genes: What Have We Learned from Them? *Developmental Biology*, **180**, 402-412. http://dx.doi.org/10.1006/dbio.1996.0314

[40] Hikita, S.T., Cann, G.M., Wingerd, K.L., Mullick, L.H., Wayne, W.C., Webb, S.W., *et al.* (2003) Integrin $\alpha_4\beta_1$ (Vla-4) Expression and Activity in Retinal and Peripheral Neurons. *Molecular and Cellular Neuroscience*, **23**, 427-439. http://dx.doi.org/10.1016/S1044-7431(03)00065-4

[41] Leu, S.T., Jacques, S.A., Wingerd, K.L., Hikita, S.T., Tolhurst, E.C., Pring, J.L., *et al.* (2004) Integrin $\alpha_4\beta_1$ Function Is Required for Cell Survival in Developing Retina. *Developmental Biology*, **276**, 416-430. http://dx.doi.org/10.1016/j.ydbio.2004.09.003

[42] Brem, R.B., Robbins, S.G., Wilson, D.J., O'Rourke, L.M., Mixon, R.N., Robertson, J.E., *et al.* (1994) Immunolocalization of Integrins in the Human Retina. *Investigative Ophthalmology & Visual Science*, **35**, 3466-3474.

[43] Gullapalli, V.K., Sugino, I.K. and Zarbin, M.A. (2008) Culture-Induced Increase in Alpha Integrin Subunit Expression in Retinal Pigment Epithelium Is Important for Improved Resurfacing of Aged Human Bruch's Membrane. *Experimental Eye Research*, **86**, 189-200. http://dx.doi.org/10.1016/j.exer.2007.10.009

[44] Pfeffer, B.A. (1991) Improved Methodology for Cell Culture of Human and Monkey Retinal Pigment Epithelium. In: Osborne, N. and Chader, J., Eds., *Progress in Retinal Research*, Pergamon Press, Oxford, 251-291. http://dx.doi.org/10.1016/0278-4327(91)90015-T

[45] Proulx, S., Guérin, S.L. and Salesse, C. (2003) Effect of Quiescence on Integrin $\alpha_5\beta_1$ Expression in Human Retinal Pigment Epithelium. *Molecular Vision*, **9**, 473-481.

[46] Proulx, S., Landreville, S., Guerin, S.L. and Salesse, C. (2004) Integrin Alpha5 Expression by the Arpe-19 Cell Line: Comparison with Primary Rpe Cultures and Effect of Growth Medium on the Alpha5 Gene Promoter Strength. *Experimental Eye Research*, **79**, 157-165. http://dx.doi.org/10.1016/j.exer.2004.03.004

[47] Pothier, F., Ouellet, M., Julien, J.P. and Guerin, S.L. (1992) An Improved Cat Assay for Promoter Analysis in Either Transgenic Mice or Tissue Culture Cells. *DNA and Cell Biology*, **11**, 83-90. http://dx.doi.org/10.1089/dna.1992.11.83

[48] Kaida, M., Cao, F., Skumatz, C.M., Irving, P.E. and Burke, J.M. (2000) Time at Confluence for Human Rpe Cells: Effects on the Adherens Junction and *in Vitro* Wound Closure. *Investigative Ophthalmology & Visual Science*, **41**, 3215-3224.

[49] Altevogt, P., Hubbe, M., Ruppert, M., Lohr, J., von Hoegen, P., Sammar, M., *et al.* (1995) The Alpha 4 Integrin Chain Is a Ligand for $\alpha_4\beta_7$ and $\alpha_4\beta_1$. *The Journal of Experimental Medicine*, **182**, 345-355. http://dx.doi.org/10.1084/jem.182.2.345

[50] Sengbusch, J.K., He, W., Pinco, K.A. and Yang, J.T. (2002) Dual Functions of $\alpha_4\beta_1$ Integrin in Epicardial Development: Initial Migration and Long-Term Attachment. *The Journal of Cell Biology*, **157**, 873-882. http://dx.doi.org/10.1083/jcb.200203075

[51] Pastor, J.C. (1998) Proliferative Vitreoretinopathy: An Overview. *Survey of Ophthalmology*, **43**, 3-18. http://dx.doi.org/10.1016/S0039-6257(98)00023-X

[52] Casaroli Marano, R.P. and Vilaro, S. (1994) The Role of Fibronectin, Laminin, Vitronectin and Their Receptors on Cellular Adhesion in Proliferative Vitreoretinopathy. *Investigative Ophthalmology & Visual Science*, **35**, 2791-2803.

[53] Audet, J.F., Masson, J.Y., Rosen, G.D., Salesse, C. and Guerin, S.L. (1994) Multiple Regulatory Elements Control the Basal Promoter Activity of the Human Alpha 4 Integrin Gene. *DNA and Cell Biology*, **13**, 1071-1085. http://dx.doi.org/10.1089/dna.1994.13.1071

[54] Rosen, G.D., Barks, J.L., Iademarco, M.F., Fisher, R.J. and Dean, D.C. (1994) An Intricate Arrangement of Binding Sites for the Ets Family of Transcription Factors Regulates Activity of the Alpha 4 Integrin Gene Promoter. *Journal of Biological Chemistry*, **269**, 15652-15660.

[55] Postigo, A.A., Sheppard, A.M., Mucenski, M.L. and Dean, D.C. (1997) C-Myb and Ets Proteins Synergize to Overcome Transcriptional Repression by Zeb. *The EMBO Journal*, **16**, 3924-3934. http://dx.doi.org/10.1093/emboj/16.13.3924

[56] Postigo, A.A. and Dean, D.C. (1997) Zeb, a Vertebrate Homolog of Drosophila Zfh-1, Is a Negative Regulator of Muscle Differentiation. *The EMBO Journal*, **16**, 3935-3943. http://dx.doi.org/10.1093/emboj/16.13.3935

[57] Ffrench-Constant, C. and Hynes, R.O. (1989) Alternative Splicing of Fibronectin Is Temporally and Spatially Regulated in the Chicken Embryo. *Development*, **106**, 375-388.

[58] Ffrench-Constant, C., Van de Water, L., Dvorak, H.F. and Hynes, R.O. (1989) Reappearance of an Embryonic Pattern of Fibronectin Splicing during Wound Healing in the Adult Rat. *The Journal of Cell Biology*, **109**, 903-914. http://dx.doi.org/10.1083/jcb.109.2.903

[59] Ffrench-Constant, C. (1995) Alternative Splicing of Fibronectin—Many Different Proteins but Few Different Functions. *Experimental Cell Research*, **221**, 261-271. http://dx.doi.org/10.1006/excr.1995.1374

21

Hsp90 and ECM29 Are Important to Maintain the Integrity of Mammalian 26S Proteasome

Jean-Rene Q. Acquah[1], Kousuke Haratake[1], Randeep Rakwal[2], Heiichiro Udono[3], Tomoki Chiba[1*]

[1]Graduate School of Life and Environmental Sciences, University of Tsukuba, Ibaraki, Japan
[2]Organization for Educational Initiatives, University of Tsukuba, Ibaraki, Japan
[3]Graduate School of Medicine, Dentistry and Pharmaceutical Sciences, Okayama University, Okayama, Japan
Email: *tchiba@biol.tsukuba.ac.jp

Abstract

The proteasome is a protease complex composed of a core particle (CP) and regulatory particles (RPs) that bind to both ends of the CP. ECM29 is a protein that associates with the proteasome and is involved in the maintenance and regulation of the proteasome assembly. However, ECM29 deficient mice can form functional proteasome. In this paper we sought to identify the mechanisms and/or proteins that help and allow the maintenance of the proteasome activity in the absence of ECM29. We analyzed the proteasome components of ECM29-deficient mice and identified Hsp90 as a protein associated with the proteasome. Furthermore, the inhibition of Hsp90 led to a partial disassembly of the proteasome only in ECM29-deficient cells. Those findings attest to the importance of Hsp90 for the maintenance of the proteasome integrity in the absence of ECM29.

Keywords

Mammalian Proteasome, ECM29, Hsp90, Proteasome Structure

1. Introduction

In eukaryote cells, the 26S proteasome is the main machinery for the degradation of short-lived proteins [1] [2]. The 26S proteasome is the result of the association of the 20S core particule (CP) with a regulatory particle (RP) PA700 (19S) at each of its extremity [3] [4]. The CP, site of the enzymatic activity, is a hollow cylindrical structure formed by four stacked α and β rings, arranged in a sequence $\alpha\beta\beta\alpha$. Each of these rings is composed of 7 subunits ($\alpha 1$ - $\alpha 7$, $\beta 1$ - $\beta 7$) [5] [6]. The PA700 consists of a base subcomplex and a lid subcomplex. It pos-

*Corresponding author.

sesses the ability to bind, deubiquitylate and unfold targeted proteins prior to their degradation by the CP [7]-[11]. The CP can also bind to other RPs such as PA200, PA28α/β hetero-heptamer and PA28γ homo-heptamer. This will result in the formation of hybrid proteasomes with specific functions. For example, PA28α/β hetero-heptamer can associate with the CP and is involved in the immune response [12]-[15]. The absence of PA28γ homo-heptamer leads to a retardation in cell proliferation [16] [17]. PA200 binding to the proteasome is necessary for a normal spermatogenesis [18]-[21]. Finally ECM29, a conserved HEAT-like repeat protein has been reported to bind to the CP and/or PA700 to maintain the 26S proteasome structure [22]-[25]. In addition to those particles, many other proteins are known to bind and possibly assist the function of the proteasome [26]-[28].

The molecular chaperone Hsp90, one of the most abundant intracellular proteins (1% - 2% of total cellular proteins), helps the final maturation of specific client proteins [29]. Its importance has been shown in various cellular processes such as signal transduction, intracellular transport and the response to oxidative stress [30]-[34]. In mammals, Hsp90α and Hsp90β are the two main isoforms of Hsp90. Hsp90α is inducible under stress conditions, while Hsp90β is constitutively expressed. Recent studies mainly in yeast also implied that Hsp90 is involved in the maintenance of the 26S structure of the proteasome [35]-[37].

In an attempt to identify new proteasome components in the absence ofECM29, we surprisingly found a relationship between ECM29 and Hsp90 in the maintenance of the proteasome structure. To purify the proteasome, we used a slightly modified affinity method used by Legget *et al.* [23] in yeast. Here, we applied this method for the first time directly on the liver of WT and ECM29 deficient mice. After purification and analysis by Mass Spectrometry (MS), we identified Hsp90, and found that the amount of Hsp90α bound to the proteasome was higher in the ECM29 deficient mice. Furthermore, the inhibition of Hsp90 led to a partial dissociation of the proteasome only in ECM29 deficient cells. We also observed that the ECM29 KO cells were more resistant to the treatment by geldanamycin (GA), an Hsp90 inhibitor. Our data clearly show that, Hsp90 associates with the proteasome and helps to maintain the structure of the proteasome in mammalian cells. Moreover, those results suggest a cooperation between Hsp90 and ECM29 to maintain the structure of the proteasome.

2. Materials and Methods

2.1. Knock Out and Transgenic Mice

Kiaa0368 KO (ECM29 deficient) mice were generated, and its details will be published elsewhere. Transgenic mice that ubiquitously express HTBT-tagged Rpn11 (subunit of PA700 lid) were a gift from Professor Heiichiro Udono (Okayama University). The KO mice with the HTBT-tagged Rpn11 were obtained after several cycles of breeding. The mice genotypes were checked by PCR. The primers used for ECM29 were 5'-CCAGTGTAGCAG GAGTTCTTTCAGG-3', 5'-CATATTTGGTTTTAGATCAGTCCAG-3' and 5'-GATCAGGATGATCTGGACG AAGAG-3'. They amplified a WT band of 2.2 Kbp and a KO band of 600 bp. The primers used to check the presence of the HTBT-tagged Rpn11 were5'-TTGTGGATCCCATTCAGAGTGTAA-3' and 5'-TATCGTAGTC ATGATGGTGGTGAT-3'. They amplified a product length of 600 bp.

All animal experiments were performed in accordance with the guidelines of the University of Tsukuba's Regulation of Animal Experiments Committee.

2.2. Preparation of MEFs and Cell Culture

Mouse embryonic fibroblast (MEF) cells were isolated from 13.5 days post coitus embryos and cultured in Dulbecco's modified Eagle's medium (D-MEM, high glucose) (Wako) supplemented with 10% fetal bovine serum and 1% penicillin/streptomycin (Invitrogen) at 37˚C, 5% CO_2. MEFs were immortalized by transfecting SV40 large T antigen (kindly provided from Dr. N. Mizushima [The University of Tokyo, Japan]) using Lipofectamine 2000 (Invitrogen) according to the manufacturer's instruction.

2.3. Cell Viability Assay

The MEF cells (WT and ECM29 deficient) were seeded into a 96-well plate at 5000 cells/well and grown for 24 h. The growing medium was replaced by Geldanamycin (GA) (250 nM or 750 nM) or 0.5% DMSO (control) in D-MEM and the cells were incubated for 48 h. The cell proliferation assay was then conducted according to the manufacturer's instructions (Cell Counting Kit 8, Dojindo). The results are expressed as the relative viability of cells compared to their respective control.

2.4. Cells Extract Preparation and Glycerol Density Gradient Centrifugation

The cells were lysed in 500 µl buffer (50 mM TrisHCl (pH 7.5), 5 mM $MgCl_2$, 0.5% NP-40, 2 mM ATP and 1mM DTT) for 30 min on ice. The lysate was cleared by centrifugation for 30 min at 15,000 g and the supernatant was collected.

600 µl of cell lysates were separated by 10% - 40% glycerol density gradient centrifugation (GDGC) for 22 h at 25,000 rpm at 4°C (Beckman Coulter Optima-LE 80k Ultracentrifuge, SW41 rotor). The gradients were fractionated using liquid layer injector fractionator (ADVANTEC) equipped with fraction collector (ATTO).

2.5. Purification of the Proteasome

The mouse livers were dissected and homogenized in 5 volumes per liver's weight of buffer A (25 mM TrisHCl (pH 7.5), 10% Glycerol, 5 mM $MgCl_2$, 1 mM PMSF, 5 mM ATP). The lysate was cleared by centrifugation for 20 min at 10,000 rpm. The supernatant was kept and mixed with 250 µl Streptavidine-Sepharose's high performance beads (GE healthcare) overnight in cold room. The beads were collected, washed twice in buffer A and mixed with 2% (v/v) of AcTEV (Novex) at 30°C for 2 h. The mix was centrifuged at 10,000 rpm for 1 min and the eluates were collected. The elution step was repeated 6 times.

2.6. Mass Spectrometry Analysis

The MS analyses were performed as previously described [38]. In brief, the excised spots from the gel were rinsed with 100 µl (50 mM ammonium bicarbonate (NH_4HCO_3) solution in 50% acetonitrile) and vacuum-dried. For the in-gel digestion, we added 2 µl of trypsin solution (10 mM NH_4HCO_3 and 10 ng/µl of trypsin), and incubated at 37°C for 4 h to overnight. The peptides were extracted with 10 µl (75% ACN containing 0.025% trifluoroacetic acid) by ultrasonication. The spot on the MALDI sample plate was loaded with 0.5 µl of the matrix (α-Cyano-4-hydroxycinnamic acid (4-CHCA, Shimadzu Biotech, Kyoto, Japan) and 1 µl of the peptide sample solution. The samples were analyzed by the AXIMA Performance MALDI-TOF-MS (Shimadzu Biotech, Kyoto, Japan) and identified by Peptide Mass Fingerprint (PMF) analysis.

2.7. Electrophoresis and Western Blot

Native-PAGE was performed using 3% - 10% gradient gels (Wako Pure Chemical Industries) in electrophoresis buffer (180 mM Tris borate (pH 8.3). Protein samples were mixed with xylene cyanol, run at 100 V until the xylene cyanol dye exited the gel and transferred to PVDF membrane.

Purified proteasome, cleared lysate or GDGC fractions were mixed with 2 × SDS sample buffer, separated by SDS-PAGE and transferred to PVDF membrane.

The membrane was probed with primary antibodies, HRP-conjugated secondary antibodies, and then detected with chemiluminescence reagent (Amersham ECL plus Western Blotting Detection Reagents, GE Healthcare).

2.8. Antibodies

Anti-PA28γ monoclonal (BD Transduction Laboratories; 611,180), anti-β5 rabbit polyclonal (Abcam; ab3330), anti-Hsp86 rabbit polyclonal (thermo scientific), anti-PA200 rabbit polyclonal (Santa Cruz Biotechnology) and Peroxidase-conjugated anti-rabbit or anti-mouse (Jackson Immuno Research) antibodies were purchased. Anti-ECM29 and anti-β7 antibodies were raised from rabbit using the recombinant his-tagged β7 and the recombinant his-tagged ECM29 (100 C-terminal amino acid). Anti-PA28α and anti-Rpt1 antibodies are described previously [39] [40].

2.9. Proteasome Activity

2 µl of purified proteasome were mixed with 98 µl of buffer (100 mM TrisHCl (pH 7.5) and 0.01 mM of fluorescent substrate (N-suc-leu-leu-val-tyr-AMC) (Peptide Institute)) and incubated at 37°C. The fluorescence was read on the 1420 Multilabel counter (Perkin Elmer ex355/em410). For each purified proteasome, the concentration was measured by Bradford assay and the protease activity is as the N-suc-LLVY-AMC hydrolysis/ng/min.

10 µl of GDGC fraction was mixed with 90 µl of buffer (100 mM TrisHCl (pH 8.0), 50 mM KCl, 10 mM $MgCl_2$, 0.056 mM of fluorescent substrate (N-suc-leu-leu-val-tyr-AMC), 1.12 mM ATP and 39 µl of water). The

fluorescence was read on the 1420 Multilabel counter (Perkin Elmer ex355/em410).

2.10. Statistical Analysis

Student's *t*-tests were performed for statistical analyses. Data in all graphs are presented as means ± S.D.

3. Results

3.1. Purification of Functional Proteasomes from Mice Liver

The purification method for yeast proteasome by HTBH-tagged Rpn11 was described previously by Leggett *et al.* [23] [41]. We followed the same method using WT and ECM29 deficient mice with HTBH-tagged Rpn11 (**Figure 1(a)**). The purified samples (**Figure 1(b)**) had an SDS-PAGE profile similar to the ones obtained by previous studies [23] [41]. It has been shown that the presence of 100 - 500 mM of NaCl is sufficient to dissociate proteins associated to the proteasome [7] [23] [42]. During the purification process on WT mice, we washed the proteasome bound to the streptavidin beads with buffer A containing an increasing concentration of NaCl (50, 100, 150, 200, 250 and 300 mM). The Eluate of each of the washing step was then analyzed by west-

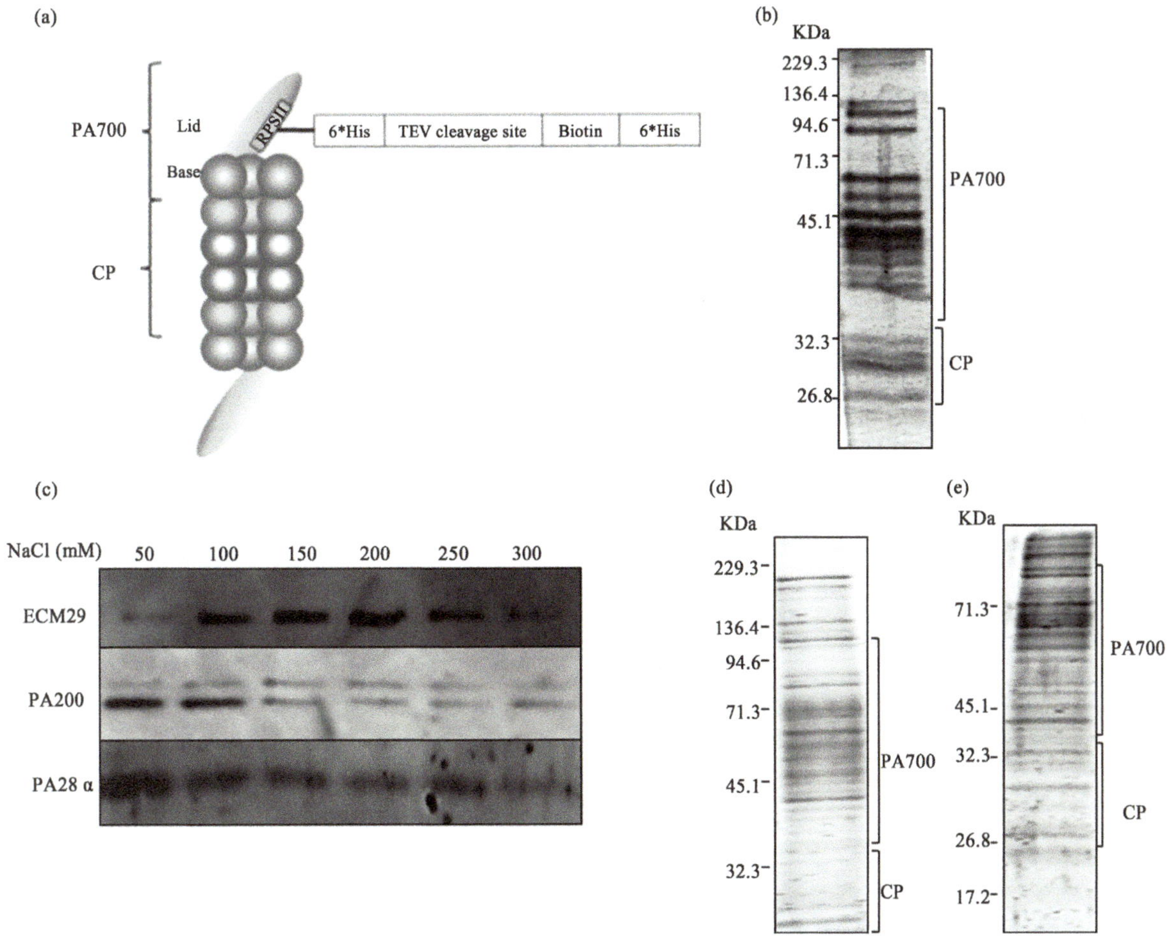

Figure 1. Profile of purified Proteasome. (a) Schematic representation of the proteasome with HTBH tag fused to the C-terminus of Rpn11. The HTBH tag consists of two hexahistidine tags, a TEV cleavage site and a biotin site. (b) The proteasome purified from WT mice liver with buffer A containing 300 mM of NaCl was separated by SDS-PAGE (10% gel) and stained by coomassie blue. From 130 KDa to 40 KDa are the proteins that compose PA700. From 35 KDa to 25 KDa are the proteins that compose CP. (c) The proteasome bound to the streptavidin beads was washed with buffer A containing 50, 100, 150, 200, 250 and 300 mM of NaCl. The eluates were analyzed by WB for the respective proteins. (d) (e) The proteasome was purified as in A with Buffer A (0 mM NaCl). The samples were separated by SDS-PAGE and stained by Sypro-Ruby (D. 7.5% gel, E. 12% gel).

ern blot (WB) of proteasome associated proteins (ECM29, PA200, PA28α) (**Figure 1(c)**). We noticed that lower concentration of NaCl (from 50 mM) also causes the dissociation of these proteins associated to the proteasome (**Figure 1(c)**). To avoid losing proteins associated to the proteasome, we excluded the NaCl and the surfactant from the purification buffer in following analyses. The SDS-PAGE profile of the proteasome purified by the modified buffer contained more bands (**Figure 1(d)** and **Figure 1(e)**), showing that more proteins were co-purified with the proteasome.

To assert that we purified functional proteasomes, we first checked the peptidase activity of the purified samples from WT and ECM29 deficient mice. A significant protease activity was detected for the purified samples compared to the control (mice with no tag) (**Figure 2(a)**). This result was consistent with findings of a previous study showing that ECM29 inhibits the peptidase activity of the proteasome [43]. It also proved that there is no unspecific binding of the proteasome to the streptavidin beads.

Finally, we confirmed by WB the presence of the CP subunit, the PA700, PA200, PA28α and PA28γ. According to the results, our method retained the CP, the PA700 and other known associated proteins such as PA200, PA28α, PA28γ and ECM29 (**Figure 2(b)**). The WB of the control also showed that the purification of these proteasome components is specific. Taken together, these data confirmed that our modified method could purify an active proteasome including hybrid proteasomes directly from the liver.

3.2. More Hsp90 Bind to the Proteasome in the Absence of ECM29

Next, we analyzed the components of the purified proteasome by MS and identified Hsp90α and Hsp90β as candidate proteins. The presence of Hsp90α and the specificity of its association with the proteasome was further confirmed by WB of the purified samples from WT, ECM29 deficient and control mice (**Figure 3(a)**). Interestingly, we found that the quantity of Hsp90α binding to the proteasome was higher in the case of the ECM29 deficient mice (**Figure 3(a)** and **Figure 3(b)**). Hsp90 is a stress-induced chaperone protein in mammalian cells. The variation observed might be a result of the increase of the total cytosolic Hsp90 in ECM29 KO condition. To eliminate this possibility, we checked the amount of Hsp90 in the lysate of liver from WT and ECM29 deficient mice (**Figure 3(c)** and **Figure 3(d)**). No noticeable variation was observed. Taken together our data imply a link between Hsp90 function and the absence of ECM29.

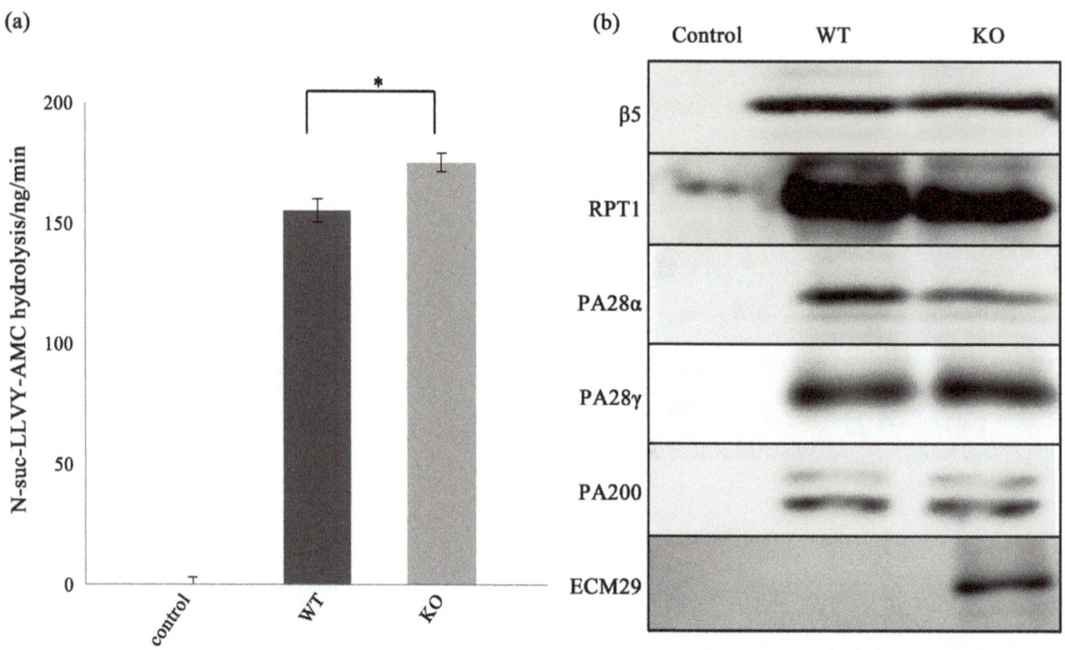

Figure 2. The purified proteasomes contain the 26S subunit and the well known proteasome associated proteins. (a) The purified samples from WT, ECM29 deficient and control (without the tag) liver were incubated with fluorescent substrate for 30 min. The results represent the mean ± S.D (n = 3). $^*p < 0.005$. The protease activity is expressed by N-suc-LLVY-AMC hydrolysis/ng/min. (b) Purified samples from WT, Ecm29 KO, and control mice were immunoblotted with indicated antibodies.$\beta 5$ (CP subunit), RPT1 (PA700 subunit), PA28α, PA28γ, PA200, ECM29 (proteasome associated proteins).

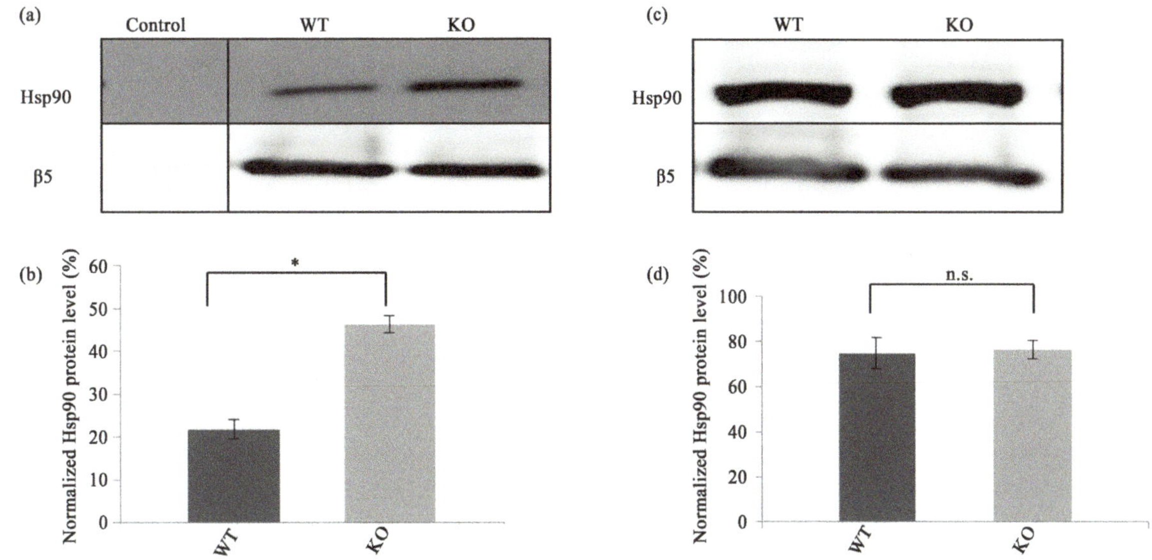

Figure 3. Hsp90 binds more abundantly to the proteasome in the absence of ECM29. (a) The purified samples from WT, ECM29 deficient and control mice were analyzed by WB against Hsp90 and $\beta 5$. (b) The intensity of Hsp90 signal from A was measured by ImageJ and normalized with the signal intensity of $\beta 5$, subunit of CP. (c) Liver lysate were analyzed by WB against Hsp90 and $\beta 5$ from WT and ECM29. (d) The intensity of Hsp90 signal from (c) was measured by ImageJ and normalized with the signal intensity of $\beta 5$.The results represent the mean ± S.D of 3 independent experiments. $^{*}p < 0.005$. n.s. not significant.

3.3. Hsp90 Helps to Maintain the Proteasome Integrity

As ECM29 has been reported to stabilize the 26S proteasome complex in yeast, we focused on the proteasome structure. To elucidate the functional link between Hsp90 and ECM29 on the structure of the proteasome, we ran a glycerol density gradient centrifugation experiment (GDGC) with WT and ECM29 deficient cell lysates following treatment with Geldanamycin (GA), an inhibitor of Hsp90. We then checked the proteasome activity of each fraction. GA treatment induced no differences for the WT cells (**Figure 4(a)**), but the peak of the proteasome activity shifted from the 20th fraction to the lighter 19th fraction in ECM29 deficient cells (**Figure 4(b)**). This shift suggests a partial disassembly of the proteasome, and these results imply that the presence of ECM29 stabilizes the proteasome structure when Hsp90 is absent. In ECM29 deficient cells, the shift correlated with the decrease of Hsp90α present in the fraction of maximum activity upon GA treatment (**Figure 4(b)**). However, no noticeable decrease of Hsp90α was observed for the WT lysate (**Figure 4(a)**).This may suggest a preferential binding of Hsp90 to the proteasome in the absence of ECM29.

To confirm the direct effect of Hsp90 on the proteasome and avoid any interference by the components of the lysates, we treated the purified proteasome with GA. Native-PAGE followed by WB showed that GA treatment of the proteasome lacking ECM29 led to an increase in intermediate varieties of proteasome and lower molecular weight sub-complex containing RPT1, a subunit of PA700 (**Figure 5(a)** and **Figure 5(d)**). Likewise, in the proteasome lacking ECM29, the GA treatment led also to the increase in intermediate varieties of proteasome containing $\beta 5$, a subunit of the CP (**Figure 5(b)** and **Figure 5(e)**). Western blots against PA200, PA28α or PA28γ did not show any differences, suggesting that Hsp90 is not involved in the association or dissociation of these RPs even in the absence of ECM29 (**Figure 5(c)**, **Figure 5(f)** and data not shown). Taken together, these results suggest that in the absence of ECM29, the inhibition of Hsp90 causes a dissociation of the PA700 from the CP proteasome.

3.4. ECM29 Deficient Cells are More Resistant to GA Treatment

Finally we checked the impact of the GA treatment on the survival of WT and ECM29-deficient cells. We present the cell viability by the percentage of viable cells compared to the control cells (WT and ECM29 deficient cells treated with DMSO only) to account for a possible difference of growth rates. Surprisingly, we found that ECM29 deficient cells were more resistant to the GA treatment (**Figure 6**).

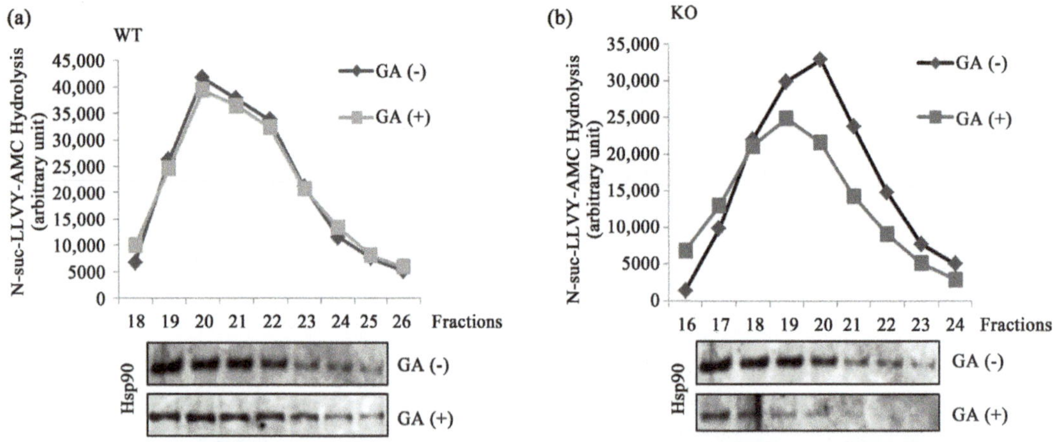

Figure 4. Inhibition of Hsp90 leads to the disassembly of the proteasome in ECM29 deficient cells. (a) WT cell lysates treated without GA (−) or with GA (+) (60 μM) were subjected to GDGC. The proteasome activities (top) and WB of Hsp90 (bottom) in the indicated fractions are shown. (b) ECM29 deficient cell lysates were analyzed as in (a).

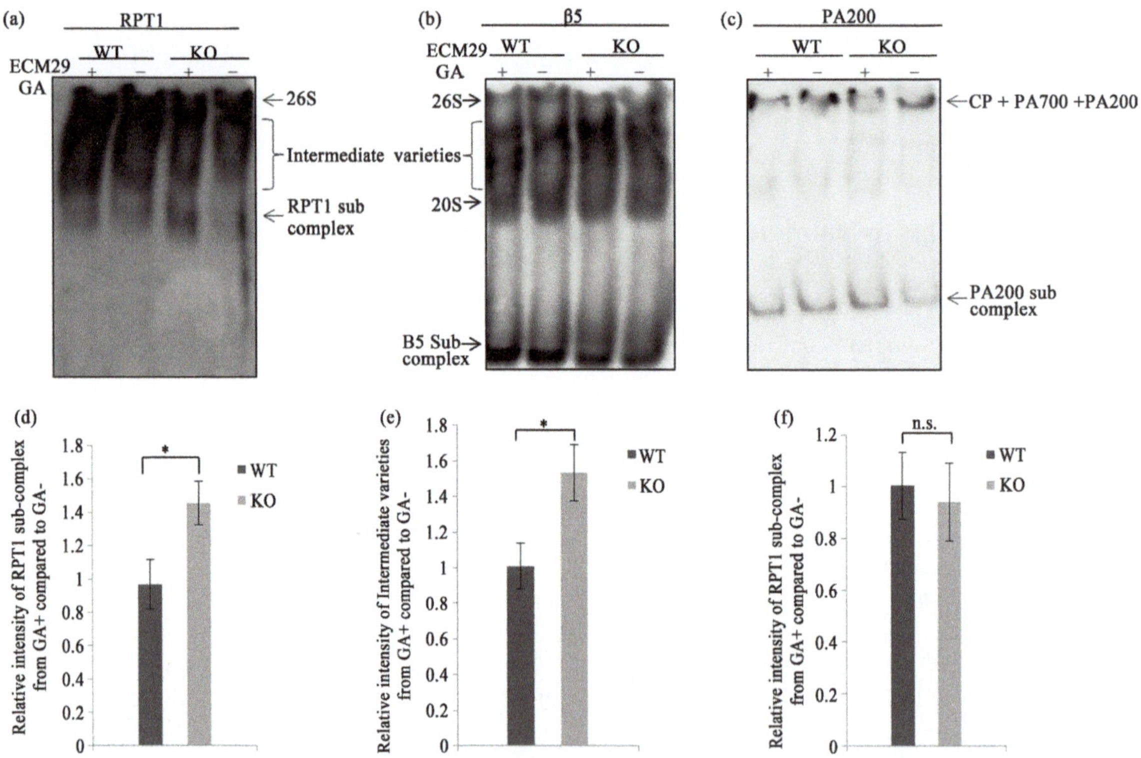

Figure 5. GA treatment of purified proteasome leads to a disassembly of PA700 from the CP. (a)-(c) The purified proteasome from WT mice and ECM29 deficient mice were treated with GA (150 μM), separated by a native-PAGE and subjected to WB against the indicated proteins (RPT1, $\beta 5$ and PA200). (d)-(f) The intensity of the RPT1 sub complex (a), the intermediate forms (b) and PA200 sub complex (c) were measured (WT and KO) with or without GA treatment and the relative intensity of GA+ compared to GA− was calculated. The results are shown in (d), (e), and (f), respectively. The results represent the mean ± S.D. $^{*}p < 0.005$. n.s. not significant.

4. Discussion

The proteasome is involved in numerous cellular processes and its dysfunction leads to serious diseases. A major challenge to understand this complex is to identify the proteins that associate with the proteasome and to elucidate the effect of those proteins on the proteasome. To achieve this goal we based our strategy on the purification of the proteasome by a very efficient and gentle technique using an HTBH-tag. For the first time, we ap-

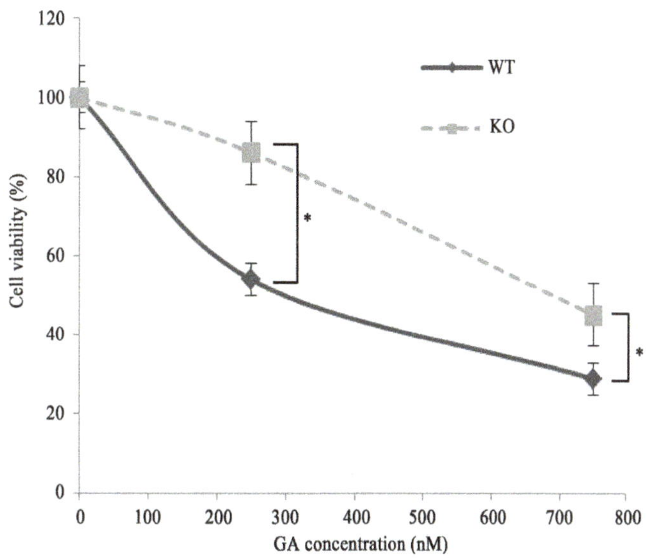

Figure 6. ECM29 deficient cells are more resistant to GA treatment. WT and ECM29 deficient cells were treated with GA (250 nM or 750 nM) for 48 h. The cell viability was then measured by WST8 assay and presented as the relative viability to their respective untreated control (0.5% DMSO). The results represent the mean ± S.D (n = 5) $^*p < 0.005$.

plied this method directly on WT and ECM29 deficient mice liver and identified Hsp90 by MS. In mammalian cells, only one study discussed the involvement of Hsp90 in the proteasome assembly *in vitro* [37]. In yeast, previous studies [35] [36] revealed an interaction between the proteasome and Hsp90. Imai *et al.* [35] reported that Hsp90 plays a role in the maintenance of the proteasome structure, since the proteasome of the yeast Hsp90 mutant was disassembled at non-permissive temperature. They also reported that the GA-treatment of wild-type yeast proteasome didn't lead to the disassembly of the 26S structure.

In the present study, we confirm by MS and WB that Hsp90 is associated with the proteasome in mammalian cells. We find that the amount of Hsp90 associates to the proteasome is higher in the absence of ECM29. We also shed more light on the function of Hsp90 and ECM29 which helps to maintain the 26S structure of the proteasome. Indeed, we reveal a link between the Hsp90 and ECM29 functions since GA treatment led to the partial dissociation of the proteasome only in the absence of ECM29. Our data suggest that ECM29 stabilizes the proteasome structure after Hsp90 has completed its function. This explains why the inhibition of Hsp90 doesn't lead to the dissociation of the proteasome in the WT lines. Accordingly, in the absence of ECM29, the inhibition of Hsp90 leads to the disassembly of the 26S proteasome.

Curiously, in the GDGC experiment, the GA treatment of the WT lysates didn't lead to a marked shift of Hsp90 around the fraction of maximum proteasome activity as observed in the ECM29 deficient cell lysates (**Figure 4(a)**, **Figure 4(b)**). As explained by Schneider *et al.* [44], Hsp90 can still remain bound to its protein client while inhibited. In this case, the decrease of Hsp90 observed in the ECM29 deficient cells (**Figure 4(b)**) would be the consequence of the dissociation of PA700 from CP and couldn't be observed in WT cells since there was no dissociation. Also, we detected Hsp90 in the heaviest fractions of WT cells but not of ECM29 deficient cells (**Figure 4(a)** and **Figure 4(b)**). Since Hsp90 stabilizes the structure of the proteasome in the absence of ECM29, Hsp90 would be mostly bound to the proteasome. In the WT cells, Hsp90 is free from this role by the presence of ECM29 and bind to other client proteins present in the heaviest fractions. This may suggest a preferential binding of Hsp90 to the proteasome in the absence of ECM29.

GA treatment of cells causes a disruption of Hsp90 protein clients and an increase in reactive oxygen species (ROS) level leading to the cell death [45]. Previous studies imply that an increase in CP and hybrid proteasomes helps the cells to eliminate the ROS more efficiently [46]-[48]. We have shown that, upon GA treatment, a partial dissociation of the 26S occurs in the absence of ECM29. This means that in ECM29 deficient cells, the formation of CP and hybrid proteasomes is facilitated, ensuring a better elimination of ROS and thus a better viability (**Figure 6**).

Taken together, our data provide a better understanding of how the structure of the proteasome is maintained and show a functional link between ECM29 and Hsp90 on the proteasome structure. However, the precise mechanism of this relation under normal and stress conditions must be thoroughly investigated in the future.

Acknowledgements

We thank the members of the Chiba Laboratory for engaging with us in helpful discussion. This work was supported by Grants-in-Aid from the Ministry of Education, Culture, Sports, Science, and Technology (MEXT) of Japan (T.C.). J.Q.A was supported by the MEXT scholarship for foreign student.

Conflict of Interest

The authors have declared that no conflict of interest exists.

References

[1] Rock, K.L., Gramm, C., Rothstein, L., Clark, K., Stein, R., Dick, L., Hwang, D. and Goldberg, A.L. (1994) Inhibitors of the Proteasome Block the Degradation of Most Cell Proteins and the Generation of Peptides Presented on MHC Class I Molecules. *Cell*, **78**, 761-771. http://dx.doi.org/10.1016/S0092-8674(94)90462-6

[2] Lee, D.H. and Goldberg, A.L. (1998) Proteasome Inhibitors: Valuable New Tools for Cell Biologists. *Trends in Cell Biology*, **8**, 397-403. http://dx.doi.org/10.1016/S0962-8924(98)01346-4

[3] Hoffman, L., Pratt, G. and Rechsteiner, M. (1992) Multiple Forms of the 20S Multicatalytic and the 26 S Ubiquitin/ATP-Dependent Proteases from Rabbit Reticulocyte Lysate. *The Journal of Biological Chemistry*, **267**, 22362-22368.

[4] Da Fonseca, P.C.A. and Morris, E.P. (2008) Structure of the Human 26S Proteasome: Subunit Radial Displacements Open the Gate into the Proteolytic Core. *The Journal of Biological Chemistry*, **283**, 23305-23314. http://dx.doi.org/10.1074/jbc.M802716200

[5] Löwe, J., Stock, D., Jap, B., Zwickl, P., Baumeister, W. and Huber, R. (1995) Crystal Structure of the 20S Proteasome from the Archaeon T. Acidophilum at 3.4 A Resolution. *Science*, **268**, 533-539. http://dx.doi.org/10.1126/science.7725097

[6] Groll, M., Ditzel, L., Löwe, J., Stock, D., Bochtler, M., Bartunik, H.D. and Huber R. (1997) Structure of 20S Proteasome from Yeast at 2.4 Å Resolution. *Nature*, **386**, 463-471. http://dx.doi.org/10.1038/386463a0

[7] Glickman, M.H., Rubin, D.M., Fried, V.A. and Finley, D. (1998) The Regulatory Particle of the *Saccharomyces cerevisiae* Proteasome. *Molecular and Cellular Biology*, **18**, 3149-3162. http://dx.doi.org/10.1128/MCB.18.6.3149

[8] Glickman, M.H., Rubin, D.M., Coux, O., Wefes, I., Pfeifer, G., Cjeka, Z., Baumeister, W., Fried, V.A. and Finley, D. (1998) A Subcomplex of the Proteasome Regulatory Particle Required for Ubiquitin-Conjugate Degradation and Related to the COP9-Signalosome and eIF3. *Cell*, **94**, 615-623. http://dx.doi.org/10.1016/S0092-8674(00)81603-7

[9] Larsen, C.N. and Finley, D. (1997) Protein Translocation Channels in the Proteasome and Other Proteases. *Cell*, **91**, 431-434. http://dx.doi.org/10.1016/S0092-8674(00)80427-4

[10] Baumeister, W., Walz, J., Zuhl, F. and Seemuller, E. (1998) The Proteasome: Paradigm of a Self-Compartmentalizing Protease. *Cell*, **92**, 367-380. http://dx.doi.org/10.1016/S0092-8674(00)80929-0

[11] Pickart, C. (1997) Targeting of Substrates to the 26S Proteasome. *The FASEB Journal*, **11**, 1055-1066.

[12] Ahn, K., Erlander, M., Leturcq, D., Peterson, P.A., Früh, K. and Yang, Y. (1996) *In Vivo* Characterization of the Proteasome Regulator PA28. *The Journal of Biological Chemistry*, **271**, 18237-18242. http://dx.doi.org/10.1074/jbc.271.30.18237

[13] Kuehn, L. and Dahlmann, B. (1997) Structural and Functional Properties of Proteasome Activator PA28. *Molecular Biology Reports*, **24**, 89-93. http://dx.doi.org/10.1023/A:1006897801858

[14] Preckel, T., Fung-Leung, W.P., Cai, Z., Vitiello, A., Salter-Cid, L., Winqvist, O., Wolfe, T.G., Von Herrath, M., Angulo, A., Ghazal, P., Lee, J.D., Fourie, A.M., Wu, Y., Pang, J., Ngo, K., Peterson, P.A., Früh, K. and Yang, Y. (1999) Impaired Immunoproteasome Assembly and Immune Responses in $PA28^{-/-}$ Mice. *Science*, **286**, 2162-2165. http://dx.doi.org/10.1126/science.286.5447.2162

[15] Cascio, P., Call, M., Petre, B.M., Walz, T. and Goldberg, A.L. (2002) Properties of the Hybrid Form of the 26S Proteasome Containing both 19S and PA28 Complexes. *The EMBO Journal*, **21**, 2636-2645. http://dx.doi.org/10.1093/emboj/21.11.2636

[16] Zhang, Z. and Zhang, R. (2008) Proteasome Activator PA28γ Regulates p53 by Enhancing Its MDM2-Mediated Degradation. *The EMBO Journal*, **27**, 852-864. http://dx.doi.org/10.1038/emboj.2008.25

[17] Murata, S., Kawahara, H., Tohma, S., Yamamoto, K., Kasahara, M., Nabeshima, Y.I., Tanaka, K. and Chiba, T. (1999) Growth Retardation in Mice Lacking the Proteasome Activator PA28γ. *The Journal of Biological Chemistry*, **274**,

38211-38215. http://dx.doi.org/10.1074/jbc.274.53.38211

[18] Ortega, J., Heymann, J.B., Kajava, A.V., Ustrell, V., Rechsteiner, M. and Steven, A.C. (2005) The Axial Channel of the 20S Proteasome Opens upon Binding of the PA200 Activator. *Journal of Molecular Biology*, **346**, 1221-1227. http://dx.doi.org/10.1016/j.jmb.2004.12.049

[19] Iwanczyk, J., Sadre-Bazzaz, K., Ferrell, K., Kondrashkina, E., Formosa, T., Hill, C.P. and Ortega1, J. (2006) Structure of the Blm10-20S Proteasome Complex by Cryo-Electron Microscopy, Insights into the Mechanism of Activation of Mature Yeast Proteasomes. *Journal of Molecular Biology*, **363**, 648-659. http://dx.doi.org/10.1016/j.jmb.2006.08.010

[20] Blickwedehl, J., Olejniczak, S., Cummings, R., Sarvaiya, N., Mantilla, A., Chanan-Khan, A., Pandita, T.K., Schmidt, M., Thompson, C.B. and Bangia, N. (2012) The Proteasome Activator PA200 Regulates Tumor Cell Responsiveness to Glutamine and Resistance to Ionizing Radiation. *Molecular Cancer Research*, **10**, 937-944. http://dx.doi.org/10.1158/1541-7786.MCR-11-0493-T

[21] Khor, B., Bredemeyer, A.L., Huang, C.Y., Turnbull, I.R., Evans, R., Maggi, L.B., White, J.M., Walker, L.M., Carnes, K., Hess, R.A. and Sleckman, B.P. (2006) Proteasome Activator PA200 Is Required for Normal Spermatogenesis. *Molecular and Cellular Biology*, **26**, 2999-3007. http://dx.doi.org/10.1128/MCB.26.8.2999-3007.2006

[22] Gorbea, C., Goellner, G.M., Teter, K., Holmes, R.K. and Rechsteiner, M. (2004) Characterization of Mammalian Ecm29, a 26S Proteasome-Associated Protein That Localizes to the Nucleus and Membrane Vesicles. *The Journal of Biological Chemistry*, **279**, 54849-54861. http://dx.doi.org/10.1074/jbc.M410444200

[23] Leggett, D.S., Hanna, J., Borodovsky, A., Crosas, B., Schmidt, M., Baker, R.T., Walz, T., Ploegh, H. and Finley, D. (2002) Multiple Associated Proteins Regulate Proteasome Structure and Function. *Molecular Cell*, **10**, 495-507. http://dx.doi.org/10.1016/S1097-2765(02)00638-X

[24] Lehmann, A., Niewienda, A., Jechow, K., Janek, K. and Enenkel, C. (2010) Ecm29 Fulfils Quality Control Functions in Proteasome Assembly. *Molecular Cell*, **38**, 879-888. http://dx.doi.org/10.1016/j.molcel.2010.06.016

[25] Kajavaa, A.V., Gorbeab, C., Ortegac, J., Rechsteinerb, M. and Steven, A.C. (2004) New HEAT-Like Repeat Motifs in Proteins Regulating Proteasome Structure and Function. *Journal of Structural Biology*, **146**, 425-430. http://dx.doi.org/10.1016/j.jsb.2004.01.013

[26] Elsasser, S., Chandler-Militello, D., Müller, B., Hanna, J. and Finley, D. (2004) Rad23 and Rpn10 Serve as Alternative Ubiquitin Receptors for the Proteasome. *The Journal of Biological Chemistry*, **279**, 26817-26822. http://dx.doi.org/10.1074/jbc.M404020200

[27] Schmidt, M., Hanna, J., Elsasser, S. and Finley, D. (2005) Proteasome-Associated Proteins: Regulation of a Proteolytic Machine. *The Journal of Biological Chemistry*, **386**, 725-737. http://dx.doi.org/10.1515/BC.2005.085

[28] Guerrero, C., Tagwerker, C., Kaiser, P. and Huang, L. (2006) An Integrated Mass Spectrometry-Based Proteomic Approach: Quantitative Analysis of Tandem Affinity-Purified *in Vivo* Cross-Linked Protein Complexes (QTAX) to Decipher the 26S Proteasome-Interacting Network. *Molecular and Cellular Proteomics*, **5**, 366-378. http://dx.doi.org/10.1074/mcp.M500303-MCP200

[29] Taipale, M., Jarosz, D.F. and Lindquist, S. (2010) Hsp90 at the Hub of Protein Homeostasis: Emerging Mechanistic Insights. *Nature Reviews Molecular Cell Biology*, **11**, 515-528. http://dx.doi.org/10.1038/nrm2918

[30] Nathan, D.F., Vos, M.H. and Lindquist, S. (1997) *In Vivo* Functions of the *Saccharomyces cerevisiae* Hsp90 Chaperone. *Proceedings of the National Academy of Sciences of the United States of America*, **94**, 12949-12956. http://dx.doi.org/10.1073/pnas.94.24.12949

[31] Zhao, R., Davey, M., Hsu, Y.C., Kaplanek, P., Tong, A., Parsons, A.B., Krogan, N., Cagney, G., Mai, D., Greenblatt, J., Boone, C., Emili, A. and Houry, W.A. (2005) Navigating the Chaperone Network: An Integrative Map of Physical and Genetic Interactions Mediated by the Hsp90 Chaperone. *Cell*, **120**, 715-727. http://dx.doi.org/10.1016/j.cell.2004.12.024

[32] Picard, D. (2002) Heat-Shock Protein 90, a Chaperone for Folding and Regulation. *Cellular and Molecular Life Sciences*, **59**, 1640-1648. http://dx.doi.org/10.1007/PL00012491

[33] Smith, D.F. (1993) Dynamics of Heat-Shock Protein 90-Progesterone Receptor Binding and the Disactivation Loop Model for Steroid Receptor Complexes. *Molecular Endocrinology*, **7**, 1418-1429. http://www.ncbi.nlm.nih.gov/pubmed/7906860

[34] Pratt, W.B., Morishima, Y., Peng, H.M. and Osawa, Y. (2010) Proposal for a Role of the Hsp90/Hsp70-Based Chaperone Machinery in Making Triage Decisions When Proteins Undergo Oxidative and Toxic Damage. *Experimental Biology and Medicine*, **235**, 278-289. http://dx.doi.org/10.1258/ebm.2009.009250

[35] Imai, J., Maruya, M., Yashiroda, H., Yahara, I. and Tanaka, K. (2003) The Molecular Chaperone Hsp90 Plays a Role in the Assembly and Maintenance of the 26S Proteasome. *The EMBO Journal*, **22**, 3557-3567. http://dx.doi.org/10.1093/emboj/cdg349

[36] Verma, R., Chen, S., Feldman, R., Schieltz, D., Yates, J., Dohmen, J. and Deshaies, R.J. (2000) Proteasomal Proteo-

mics: Identification of Nucleotide-Sensitive Proteasome-Interacting Proteins by Mass Spectrometric Analysis of Affinity-Purified Proteasomes. *Molecular Biology of the Cell*, **11**, 3425-3439. http://dx.doi.org/10.1091/mbc.11.10.3425

[37] Yamano, T., Mizukami, S., Murata, S., Chiba, T., Tanaka, K. and Udono, H. (2008) Hsp90-Mediated Assembly of the 26S Proteasome Is Involved in Major Histocompatibility Complex Class I Antigen Processing. *The Journal of Biological Chemistry*, **283**, 28060-28065. http://dx.doi.org/10.1074/jbc.M803077200

[38] Hori, M., Nakamachi, T., Rakwal, R., Shibato, J., Ogawa, T., Aiuchi, T., Tsuruyama, T., Tamaki, K. and Shioda, S. (2012) Transcriptomics and Proteomics Analyses of the PACAP38 Influenced Ischemic Brain in Permanent Middle Cerebral Artery Occlusion Model Mice. *Journal of Neuroinflammation*, **9**, 256-274. http://www.jneuroinflammation.com/content/pdf/1742-2094-9-256.pdf http://dx.doi.org/10.1186/1742-2094-9-256

[39] Tanahashi, N., Yokota, K., Ahn, J.Y., Chung, C.H., Fujiwara, T., Takahashi, E., DeMartino G.N., Slaughter, C.A., Toyonaga, T., Yamamura, K., Shimbara, N. and Tanaka, K. (1997) Molecular Properties of the Proteasome Activator PA28 Family Proteins and Gamma-Interferon Regulation. *Genes to Cells*, **2**, 195-211. http://onlinelibrary.wiley.com/doi/10.1046/j.1365-2443.1997.d01-308.x/epdf http://dx.doi.org/10.1046/j.1365-2443.1997.d01-308.x

[40] Tanahashi, N., Suzuki, M., Fujiwara, T., Takahashi, E., Shimbara, N., Chung, C.H. and Tanaka, K. (1998) Chromosomal Localization and Immunological Analysis of a Family of Human 26S Proteasomal ATPases. *Biochemical and Biophysical Research Communications*, **243**, 229-232. http://dx.doi.org/10.1006/bbrc.1997.7892

[41] Leggett, D.S., Glickman, M.H. and Finley, D. (2005) Purification of Proteasomes, Proteasome Subcomplexes, and Proteasome-Associated Proteins from Budding Yeast. *Methods in Molecular Biology*, **301**, 57-70. http://dx.doi.org/10.1385/1-59259-895-1:057

[42] Besche, H.C., Haas, W., Gygi, S.P. and Goldberg, A.L. (2009) Isolation of Mammalian 26S Proteasomes and p97/VCP Complexes Using the Ubiquitin-Like Domain from HHR23B Reveals Novel Proteasome-Associated Proteins. *Biochemistry*, **48**, 2538-2549. http://dx.doi.org/10.1021/bi802198q

[43] De La Mota-Peynado, Lee, S.Y.C., Pierce, B.M., Wani, P., Singh, C.R. and Roelofs, J. (2013) The Proteasome-Associated Protein Ecm29 Inhibits Proteasomal ATPase Activity and *in Vivo* Protein Degradation by the Proteasome. *The Journal of Biological Chemistry*, **288**, 29467-29481. http://dx.doi.org/10.1074/jbc.M113.491662

[44] Schneider, C., Sepp-Lorenzino, L., Nimmesgern, E., Ouerfelli, O., Danishefsky, S., Rosen, N. and Hartl, F.U. (1996) Pharmacologic Shifting of a Balance between Protein Refolding and Degradation Mediated by Hsp90. *Proceedings of the National Academy of Sciences of the United States of America*, **93**, 14536-14541. http://dx.doi.org/10.1073/pnas.93.25.14536

[45] Aparajita, D., and Arthur, I.C. (2006) Geldanamycin, an Inhibitor of Hsp90, Potentiates Cytochrome P4502E1-Mediated Toxicity in HepG2 Cells. *Journal of Pharmacology and Experimental Therapeutics*, **317**, 1391-1399. http://dx.doi.org/10.1124/jpet.106.101808

[46] Pickering, A.M., Koop, A.L., Teoh, C.Y., Ermak, G., Grune T. and Davies K. J. (2010) The Immunoproteasome, the 20S Proteasome, and the PA28$\alpha\beta$ Proteasome Regulator Are Oxidative-Stress-Adaptive Proteolytic Complexes. *Biochemical Journal*, **432**, 585-594. http://dx.doi.org/10.1042/BJ20100878

[47] Seifert, U., Bialy, L.P., Ebstein, F., Bech-Otschir, D., Voigt, A., Schröter, F., Prozorovski, T., Lange, N., Steffen, J., Rieger, M., Kuckelkorn, U., Aktas, O., Kloetzel, P.M. and Krüger, E. (2010) Immunoproteasomes Preserve Protein Homeostasis upon Interferon-Induced Oxidative Stress. *Cell*, **142**, 613-624. http://dx.doi.org/10.1016/j.cell.2010.07.036

[48] Charity, T.A., Robyn, M.K., Xiaorong, W. and Lan, H. (2011) Oxidative Stress-Mediated Regulation of Proteasome Complexes. *Molecular and Cellular Proteomics*, **10**, Article ID: 006924.

22

Rad7 E3 Ubiquitin Ligase Attenuates Polyubiquitylation of Rpn10 and Dsk2 Following DNA Damage in *Saccharomyces cerevisiae*

Joseph M. Benoun, Danielle Lalimar-Cortez, Analila Valencia, Adriana Granda, Destaye M. Moore, Eric P. Kelson, Paula L. Fischhaber*

Department of Chemistry and Biochemistry, California State University Northridge, Northridge, CA, USA
Email: *paula.fischhaber@csun.edu

Abstract

During Nucleotide Excision Repair (NER) in the yeast *S. cerevisiae*, ubiquitylation of Rad4 is carried out by the E3 ubiquitin ligase that includes Rad7-Elc1-Cul3 and is critical to optimal NER. Rad7 E3 activity targets Rad4 for degradation by the proteaseome but, in principle, could also trigger other DNA damage responses. We observed increased nuclear ubiquitin foci (Ub-RFP) formation in *S. cerevisiae* containing a Rad7 E3 ligase mutant (*rad7SOCS*) in response to DNA damage by benzo[a]pyrenediolepoxide (BPDE) in dividing cells. Immunoblots reveal that ubiquitin conjugates of Rpn10 and Dsk2 accumulate in greater abundance in *rad7SOCS* compared to *RAD7* in dividing cells in response to BPDE which makes Rpn10 and Dsk2 candidates for being the ubiquitylated species observed in our microscopy experiments. Microscopy analysis with strains containing Dsk2-GFP shows that Dsk2-GFP and Dsk2-GFP/Ub-RFP colocalized in nuclear foci form to an increased extent in a *rad7SOCS* mutant background in dividing cells than in a *RAD7* wild-type strain. Further, Dsk2-GFP in the *rad7SOCS* strain formed more foci at the plasma membrane following BPDE treatment in dividing cells relative to strains containing *RAD7* or a *rad7Δ* deletion mutant. In response to a different agent, UV irradiation, levels of ubiquitylated proteins were increased in *rad7SOCS* relative to *RAD7*, and the proteasomal deubiquitylase subunit, Rpn11 was even monoubiquitylated in the absence of damaging agents. Together these data show that Rad7 E3 activity attenuates ubiquitylation of proteins regulating the shuttling of polyubiquitylated proteins to the proteasome (Dsk2 and Rpn10) and removal of ubiquitin chains just prior to degradation (Rpn11). Since Rad7 E3 ligase activity has been shown to increase ubiquitylation of its target proteins, yet our results show increased ubiquitylation in the absence of Rad7 E3, we suggest that

*Corresponding author.

Rad7 E3 action regulates ubiquitin ligase and deubiquitylase (DUB) activities that act on Rpn10, Dsk2 and Rpn11.

Keywords

Ubiquitin, Rad7, Rpn10, Rpn11, Dsk2, DNA Damage Response

1. Introduction

Nucleotide Excision Repair (NER) in eukaryotes repairs DNA damage via a multistep pathway functioning in two modes with differing DNA damage recognition steps: global genome repair (GGR) and transcription-coupled repair (TCR). In GGR in *S. cerevisiae*, the DNA lesion is recognized by the Rad4-Rad23 complex. Damage recognition is followed by unwinding of the DNA surrounding the damage and damage verification by multiprotein complex TFIIH, followed by assembly of the NER complex by Rad14 and RPA. Complex assembly is followed by 5' and 3' incisions of the damaged strand by the Rad1-Rad10 complex and Rad2 which is interweaved with gap-filling by DNA polymerase [1] [2]. The NER pathway is largely conserved in eukaryotes, making *S. cerevisiae* a valuable model system for the regulation of NER activity.

Rad4 is ubiquitylated in response to DNA damage by the E3 ubiquitin ligase complex that includes Rad7 (Rad7-Elc1-Cul3) targeting Rad4 for degradation by the proteasome [3]. It is currently unknown if ubiquitylation of Rad4 may serve other roles. In humans, for example, the Rad4 ortholog, XPC becomes ubiquitylated during repair, but ubiquitylation increases DNA binding activity, especially on UV-damaged DNA, rather than targeting XPC for degradation [4]. Further, the Rad7-Rad16 complex along with ABF1 plays roles in histone modification and creates the torsional strain in DNA required for NER [5]-[7]. Prior reports have shown that the Rad23 ubiquitin like domain (UBL) binds the proteasome and influences NER, by mechanisms that are not yet clear [3] [8]. Numerous proteins are ubiquitylated in response to DNA damage, including several proteasome receptors and shuttle factors including Rpn10, Dsk2 and Rad23, which become ubiquitylated in order to regulate the recognition and shuttling of polyubiquitylated proteins destined for degradation by the proteasome [9]-[11].

Dsk2, Rad23, and Ddi1 are three UBL/UBA-containing shuttle factors responsible for recruiting K48-linked ubiquitylated proteins to the proteasome for destruction [12] [13]. Dsk2 and Rad23 also function in cell cycle regulation, enabling spindle pole body (SPB) duplication at the G1/S boundary and working with the Anaphase Promoting Complex to enable progression through anaphase [14] [15]. *S. cerevisiae* cells deleted of *DSK2, RAD23* and *DDI1* exhibit a delay in cell cycle progression in M phase indicating an essential role in cell cycle regulation and some genetic redundancy in their functions [15]. A *dsk2Δ rad23Δ* double mutant exhibits a G2/M arrest phenotype while a *dsk2Δ rad23Δ ddi1Δ* triple mutant additionally exhibits duplicated SPB's that have not separated [15] [16]. Hence, Dsk2, Rad23 and Ddi1 may promote progression through M phase by aiding removal of SPB bridge proteins so that spindle poles can separate. Importantly, Dsk2 and Rad23 are ubiquitylated in response to DNA damage [17] [18]. Ubiquitylation of Rpn10 alters its binding to other polyubiquitylated proteins being recruited to the proteaseome for degradation [8] [10]. Monoubiquitylation of Rpn10 promotes its dissociation from the proteasome, which, in turn, facilitates binding of Dsk2 [19]. Interestingly *rad23Δ rpn10Δ* double mutants exhibit sensitivity to cellular stress, cell cycle delays and defective SPB duplication while Dsk2 overexpression blocks SPB duplication [20] [21]. Overexpression of either Rad23 or Rpn10 will rescue sensitivity to cellular stress, but not in the case of a Rad23 mutant that lacks two phosphorylation sites in the UBL domain [22]. Therefore, Rad23, Dsk2 and Rpn10 play key roles in cell cycle progression, which is regulated in part by their own states of ubiquitylation.

Roles for NER factors in regulating proteolytic activity and cell-cycle regulation have also been shown. Several studies showed that the proteasome negatively regulates NER through mechanisms involving Rad4 and Rad23 [3] [23]-[26]. In a screen for mutants giving rise to yeast spindle morphogenesis phenotypes, *bim1Δ rad4Δ* and *bim1Δ rad7Δ* double mutants exhibited increased delayed anaphase over *bim1Δ* single mutants [27]. The *bim1Δ rad7Δ* double mutant also exhibited delayed nuclear positioning and misoriented spindles, while a *rad4Δ* single mutant exhibited delayed nuclear positioning [27]. Bim1 is a component of the cortical microtubule capture site that physically interacts with the APC and is required for nuclear migration and the assembly and disassembly of spindle poles [28] [29]. These results indicate NER factors Rad4 and Rad7 play an ill-defined role in spindle checkpoint regulation. Additionally, mutants of the proteasome regulatory particle deubiquitylase (DUB), Rpn11 exhibit UV-sensitive phenotypes and cell cycle-arrest in M phase, suggesting that Rpn11 plays multiple roles in the UV damage response. Mutants in the metalloprotease motif of Rpn11 have been shown to exhibit not only UV sensitivity, but also defects in tubular organization and cell cycle-arrest [30]. Recent reports show that Rpn11 and Ubp6, another DUB, are situated near each other in the proteasome lid and work together to trim ubiquitin chains prior to proteasomal degradation [31] [32].

Given prior reports showing *RAD7* involvement in cell cycle and its known role as the E3 ubiquitin ligase ubiquitylating Rad4 in NER, it is of interest to determine whether Rad7 might exert a DNA damage response that regulates the proteasome in a manner that might reinforce cell cycle checkpoints following DNA damage. The present study demonstrates that a *RAD7* mutant lacking only the E3 ligase function exhibits several phenotypes consistent with a role in regulation of the ubiquitylation states of Dsk2, Rpn10 and Rpn11. These phenotypes were observed in the context of damaging agents giving rise to DNA damage only repaired by NER, strongly suggesting an NER-dependent DNA damage response. Connections between NER and checkpoint activation have been reported in the literature, but thus far indicate primarily a role for NER in recruitment of checkpoint activation kinases and the 9-1-1 clamp to DNA damage sites [33]. Our observations suggest that *RAD7* plays other roles that have not been previously characterized.

2. Materials and Methods

2.1. Preparation of Yeast Strains *Ub-RFP*, and *rad4Δ*

The *S. cerevisiae UB14* gene was tagged at the N-terminus with the gene for monomeric red fluorescent protein (mRFP) from *Discoma* [34] to prepare a strain containing fluorescently-labeled ubiquitin (Ub-RFP) in the W303-1A genetic background by adaptamer-mediated PCR [35]. All strains used in this study are detailed in **Table 1**. The resulting strain was crossed to produce strains PF038-1D and PF040-3A used in microscopy experiments. The presence of the RFP tag in frame with no mutations was confirmed by PCR, fluorescence microscopy and sequencing. Functional Ub-RFP was confirmed by immunoblotting of yeast WCEs using standard methods with a ubiquitin antibody (Ubiquitin (P4D1) Mouse mAb, Cat No 3936, Cell Signalling, Danvers, MA) and *α-β*-actin (Abcam ab8224, Cambridge, MA).

A yeast strain containing a Rad7 E3 ligase mutant (*rad7SOCS*) was prepared by cloning and site-directed mutatgenesis of *RAD7* by mutating amino acids L168 and C172 to alanines. Plasmids were constructed via adaptamer-mediated PCR and mutagenized using the Quickchange II XL kit (Agilent) and manufacturer's instructions. Yeast transformation and backselection were carried out as referenced above. Successful backselected clones were sequenced confirming the presence of the *rad7SOCS* mutations and no additional mutations. The resulting strain was crossed to prepare strains PF084-7A and PF090-1D.

A strain containing a *RAD4* disruption was prepared by standard targeted integration of the *rad4::URA3* cassette from plasmid pNF416 [36]. A *rad4::URA3*$^{+}$ transformant was crossed to produce strain PF097-2C. UV survival experiments confirmed identical UV sensitivity to previously published *rad4Δ* strains.

2.2. Preparation of Diploid Strains Containing *Dsk2-GFP Ub-RFP* and *Rpn10-GFP Ub-RFP*

Diploid strains were prepared by mating strains YMR276W and YHR200W each with PF039-2D, PF087-2C and PF125-23A to yield a panel of diploid yeast strains containing *Dsk2-GFP Ub-RFP RAD7* (PF165), *Dsk2-GFP Ub-RFP rad7SOCS* (PF166), *Dsk2-GFP Ub-RFP rad7Δ* (PF167), *Rpn10-GFP Ub-RFP RAD7* (PF162), *Rpn10-GFP Ub-RFP rad7SOCS* (PF163), and *Rpn10-GFP Ub-RFP rad7Δ* (PF164) Diploids were selected on SC agar lacking methionine and lysine and diploid status confirmed by the presence of the GFP and RFP chromophores.

2.3. General Microscopy

Microscopy was conducted on a Zeiss AxioImager M1 microscope with a Plan—Apochromat 100×, 1.4 numerical aperture (NA) objective oil immersion lens as previously reported [37]. Except for timelapse experiments, Ub-RFP exposure times were 800 ms. Unless indicated, images were acquired as Z-stacks of 11 images with a focal plane offset of 0.3 μm per slice. Foci were counted by manually inspecting images for regions of bright,

Table 1. Strains used in this study.

Strain Name	Genotype	Where Published
MGSC104[a]	*MAT**α** rad7Δ::LEU2 ade2-1 trp1-1 can1-100 leu2-3,112 his3-11,15 ura3-1*	[47]
PF038-1D	*MAT**a** ADE2 lys2Δ trp1-1 can1-100 his 3-11,15 leu2-3,112 ura3-1 RAD10-YFP Ub-RFP*	This manuscript
PF040-3A	*MAT**a** ADE2 lys2Δ trp1-1 can1-100 his 3-11,15 leu2-3,112 ura3-1 Rad14-CFP Ub-RFP*	This manuscript
PF084-7A	*MAT**a** ADE2 lys2Δ trp1-1 can1-100 his 3-11,15 leu2-3,112 ura3-1 RAD10-YFP Ub-RFP rad7SOCS(L168A, C172A)*	This manuscript
PF090-1D	*MAT**a** ADE2 lys2Δ TRP can1-100 his 3-11,15 leu2-3,112 ura3-1 Rad14-CFP Ub-RFP rad7SOCS(L168A, C172A)*	This manuscript
PF097-2C	*MAT**a** ADE2 lys2Δ trp1-1 can1-100 his3-11,15 leu2-3,112 ura3-1 RAD10-YFP Ub-RFP rad4::URA3*	This manuscript
PF162	*MAT**a**/MATα LYS2/lys2Δ TRP1/trp1-1 CAN1/can1-100 his 3-11,15/his3Δ1 leu2-3,112/leu2Δ0 MET15/met15Δ0 ura3-1/ura3Δ0 RPN10/Rpn10-GFP RAD10/Rad10-YFP UB14/Ub-RFP*	This manuscript
PF163	*MAT**a**/MATα LYS2/lys2Δ CAN1/can1-100 his 3-11,15/his3Δ1 leu2-3,112/leu2Δ0 MET15/met15Δ0 ura3-1/ura3Δ0 RPN10//Rpn10-GFP RAD10/Rad10-YFP UB14/Ub-RFP RAD7/rad7SOCS(L168A, C172A)*	This manuscript
PF164	*MAT**a**/MATα LYS2/lys2Δ CAN1/can1-100 his 3-11,15/his3Δ1 leu2-3,112/leu2Δ0 MET15/met15Δ0 ura3-1/ura3Δ0 RPN10/Rpn10-GFP RAD10/Rad10-YFP UB14/Ub-RFP RAD7/rad7Δ::LEU2*	This manuscript
PF165	*MAT**a**/MATα LYS2/lys2Δ TRP1/trp1-1 CAN1/can1-100 his 3-11,15/his3Δ1 leu2-3,112/leu2Δ0 MET15/met15Δ0 ura3-1/ura3Δ0 DSK2/Dsk2-GFP RAD10/Rad10-YFP UB14/Ub-RFP*	This manuscript
PF166	*MAT**a**/MATα LYS2/lys2Δ CAN1/can1-100 his 3-11,15/his3Δ1 leu2-3,112/leu2Δ0 MET15/met15Δ0 ura3-1/ura3Δ0 DSK2/Dsk2-GFP RAD10/Rad10-YFP UB14/Ub-RFP RAD7/rad7SOCS(L168A, C172A)*	This manuscript
PF167	*MAT**a**/MATα LYS2/lys2Δ CAN1/can1-100 his 3-11,15/his3Δ1 leu2-3,112/leu2Δ0 MET15/met15Δ0 ura3-1/ura3Δ0 DSK2/Dsk2-GFP RAD10/Rad10-YFP UB14/Ub-RFP RAD7/rad7Δ::LEU2*	This manuscript
PF168	*MAT**a** his3Δ1 leu2Δ0 met15Δ0 ura3Δ0 Rpn11-TAP rad7SOCS(L168A, C172A)*	This manuscript
W1588-4C	*MAT**a** ade2-1 lys2Δ trp1-1 can1-100 his 3-11,15 leu2-3,112 ura3-1*	[48]
W303-1A	*MAT**a** ade2-1 lys2Δ trp1-1 can1-100 his 3-11,15 leu2-3,112 ura3-1 rad5-535*	[48]
YFR004W	*MAT**a** his3Δ1 leu2Δ0 met15Δ0 ura3Δ0 Rpn11-TAP*	Thermo Scientific TAP-tag library
YHR200W	*MAT**a** his3Δ1 leu2Δ0 met15Δ0 ura3Δ0 Rpn10-GFP*	Invitrogen Yeast GFP library
YMR276W	*MAT**a** his3Δ1 leu2Δ0 met15Δ0 ura3Δ0 Dsk2-GFP*	Invitrogen Yeast GFP library

[a]All strains are haploid and derivatives of W303-1A and W303-1B [48] except for TAP-tagged strains which are of the BY4741 background and PF162, PF163, PF164, PF165, PF166 and PF167 which are diploid crosses of the W303-1B and BY4741 strain backgrounds. Additionally, all strains are wild-type for the *RAD5* gene unless otherwise noted.

punctate fluorescence. Images were contrast-enhanced to make the fluorescent foci visible. Data represent numbers of foci observed per cell from at least 100 cells per condition unless otherwise stated.

2.4. Induction of Foci with N-acetoxy-2-acetylaminofluorene (AAAF) or Benzo[a]pyrene-r-7, t-8-dihydrodiol-t-9,10-epoxide(±), (anti) (BPDE)

Cultures for AAAF or BPDE pulse-chase experiments were propagated in SC medium supplemented with 200 μg/mL Adenine (SC + ade) at 23°C. Overnight cultures were freshly diluted, incubated (3 h) induced with AAAF or BPDE (10 μM final) or mock induced with ethanol (AAAF control) or tetrahydrofuran (BPDE control). Aliquots of cultures were processed for microscopy as previously described [37].

2.5. Timelapse Pulse-Chase Experiments

To manage photobleaching in timelapse experiments, cells were imaged as Z-stacks containing only three slices (as opposed to 11 as in other experiments). Foci were counted in the second slice, using the first and third to verify the presence of the focus in the second. Accordingly, foci counts for these experiments represent only about 10% of each cell thickness. YFP images were acquired using 400 ms for the 0 - 30 min timepoints and 800 ms for the 35 and 40 min time points. RFP exposure times were 200 ms for 0 - 30 min, then 400 ms for 35 and 40 min time points. 600 cells per condition were analyzed.

2.6. DAPI Stained Images

4',6-diamidino-2-phenylindole (DAPI) stained images were prepared as previously reported [37].

2.7. Immunoblots of Cell Cycle-Arrested WCEs Treated with BPDE

Log phase cultures of strains PF038-1D, PF084-7A, PF097-2C and MGSC104 were synchronized by incubation with nocodazole, (10 μg/mL, 1 hr, 23°C), released back into cell cycle by filtering, washing and incubation in fresh SC + ade medium (5 min, 23°C). Released cells were DNA damage-induced (or not) by addition of BPDE to 10 μM and swirling (3 min). Cells were filtered, washed, resuspended in freshly prepared native lysis buffer (25 mM Tris-HCl, pH 8.0, 250 mM NaCl, 5 mM Na_2EDTA, 10% Glycerol, 1 μM Phenylmethylsulfonyl fluoride, 2.5 μM Benzamidine, 0.7 μM Leupeptin, 1.45 μM Pepstatin, 5 μM β-Mercaptoethanol) and bead-beaten to prepare WCEs, which were quickly centrifuged to remove beads and cellular debris and snap-frozen. Aliquots normalized for total protein content were analyzed by standard SDS-PAGE and immunoblotting. Primary antibodies were obtained commercially [AbCam (Rpn10, Dsk2, PCNA [PC10], Rpn1, Rpt1, RNAPII-CTD, H2B, H3K79), Active Motif (H2A), Cell Signaling (ubiquitin), Santa Cruz (β-actin)] or were gifts of the laboratory of Errol Friedberg (Rad4, Rad23, Rad7 and Rad16).

2.8. Immunoblots of Cell Cycle-Arrested WCEs Prepared under Denaturing Conditions (for Ni-NTA Affinity Capture Experiments)

Cells transformed with plasmids YEplac195 or YEplac195 CUP1::His7-Ub were cultured in Synthetic Complete medium lacking uracil and supplemented with 200 mg/L adenine (SC-ura+ade) + 10 μM $CuSO_4$ (12 h), cell cycle-arrested with nocodazole, released briefly back into cell cycle by filtering, washing and incubation in fresh SC-ura + ade medium (5 min, 30°C) and DNA damage-induced with BPDE as described above. Cells were centrifuged, resuspended in denaturing lysis buffer (20% TCA w/v), bead-beaten, centrifuged and protein pellets resuspended in urea binding buffer (20 mM sodium phosphate, 500 mM NaCl, 8 M urea, pH 7.8), vortexed and filtered. WCEs were subjected to standard affinity capture under denaturing conditions on Ni-NTA agarose (Invitrogen) using manufacturer's instructions. Eluates were analyzed by SDS-PAGE and immunoblotting. Primary antibodies were obtained commercially [AbCam (Rpn10, Dsk2) and Genscript (α-TAP)].

2.9. Immunoblots of Cell Cycle-Arrested, UV-Treated WCEs Prepared under Nondenaturing Conditions

Strains PF038-1D, PF084-7A and MGSC104 were cultured in YPD, cell cycle-arrested with nocodazole, released briefly back into cell cycle by filtering, washing and incubation in fresh YPD (5 min, 30°C), centrifuged to remove medium, resuspended in a small volume of water (15 mL), transferred to a sterile petri dish and DNA damaged with UV-C (100 J/m^2). Damaged cells were centrifuged and resuspended in native lysis buffer prior to cell disruption and processing as described above in section 2.7. Aliquots normalized for total protein content were analyzed by SDS-PAGE and immunoblotting with an ubiquitin primary antibody (Cell Signaling).

2.10. Immunoblots of Cell Cycle-Arrested WCEs Prepared under Nondenaturing Conditions

Strains YFR004W and PF168 were transformed with plasmids YEplac195 or YEplac195 CUP1::His7-Ub, cultured in Synthetic Complete medium lacking uracil and supplemented with 200 mg/L adenine (SC-ura + ade) + 10 μM $CuSO_4$ (12 h), cell cycle-arrested with nocodazole, released briefly back into cell cycle by filtering, washing and incubation in fresh SC-ura + ade medium (5 min, 30°C), centrifuged to remove medium, resuspended in denaturing lysis buffer (20% TCA w/v), bead-beaten, centrifuged and protein pellets resuspended in urea binding buffer (20 mM sodium phosphate, 500 mM NaCl, 8 M urea, pH 7.8), vortexed and filtered. WCEs were subjected to affinity capture under denaturing conditions on Ni-NTA agarose (Invitrogen) using manufacturer's instructions. Eluates were analyzed by SDS-PAGE and immunoblotting with an antibody to TAP (Genscript).

3. Results

3.1. Ubiquitin Foci Form in the Nucleus in Response to AAAF and BPDE Treatment

To investigate the role of Rad7 in the DNA damage response, the *Saccharomyces cerevisiae* ubiquitin gene (*UBI4*) was N-terminally-tagged with monomeric red fluorescent protein (Ub-RFP). Following treatment of Ub-RFP cells with Benzo[a]pyrenediolepoxide (BPDE) or N-Acetoxy-2-acetylaminofluorene (AAAF), RFP foci were formed in the nucleus that colocalized with the fluorescent signal from DAPI (**Figure 1(a)**, orange arrows). This strain also exhibits large cytosolic RFP foci with or without DNA damage and their brightness and numbers did not appear to change significantly following DNA damage, but this was not investigated rigorously by us (**Figure 1(a)**, and **Figure 1(b)** green arrows). However, a prior RFP screen indicated that an analogous Ub-RFP strain exhibited both pan-nuclear and pancytosolic localization of Ub-RFP, which contrasts with our findings [38]. Based on the classification system used in the prior report, our cytosolic foci could correspond to ubiquitin localized any of a number of cytosolic features including Golgi, peroxisomes, endosomes or actin but this was not examined by us. To confirm the functionality of the Ub-RFP strain we analyzed WCEs for high molecular weight (HMW) ubiquitylated proteins and found similar band intensities when comparing wild-type *UBI4* and *Ub-RFP* strain extracts (**Figure 1(c)**, upper panel). A β-actin loading control shows similar loading between *UBI4* and *Ub-RFP* (**Figure 1(c)**, lower panel).

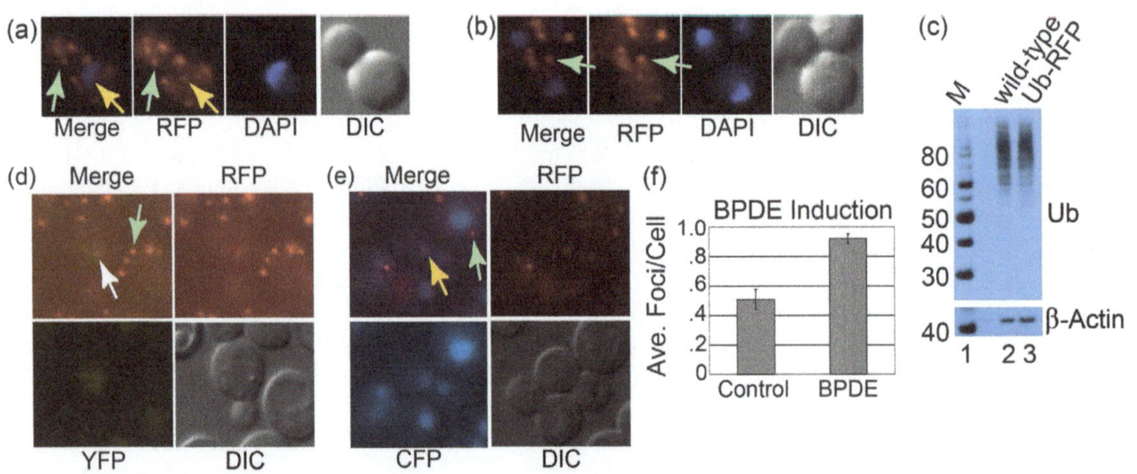

Figure 1. (a) Merged RFP and DAPI fluorescent images are shown (left) of a Ub-RFP cell 15 min following induction with UV-C (20 J/m^2) followed by images containing only RFP and DAPI channels (middle) and a differential interference contrast (DIC) image (right). Bright cytosolic RFP foci are visible throughout the cell (green arrow indicates one focus). A nuclear-localized RFP focus is indicated (orange arrow); (b) Same as (a) except cell shown does not contain a nuclear-localized RFP focus. Bright cytosolic RFP foci are visible throughout the cell (green arrow indicates one focus); (c) Immunoblot of WCEs from the indicated yeast strains probed with an antibody to ubiquitin (Ub). Lower panel shows the same blot reprobed with an antibody to β-Actin. Lane 1: MagicMark™ protein standard, lane 2: wild-type (W1588-4C), lane 3 *Ub-RFP* PF038-1D); (d) The *Ub-RFP Rad10-YFP* strain (PF038-1D) was incubated with BPDE (10 μM, 1 h) and imaged as a single focal plane. Merged, RFP, YFP and DIC images are shown. A white arrow indicates a nuclear-localized RFP focus; (e) Same as panel (c), except the strain was *Ub-RFP Rad14-CFP* (PF040-3A) and the DNA damaging agent was AAAF. A nuclear-localized RFP focus is indicated with an orange arrow. In both (d) and (e), green arrows indicate examples of RFP foci that are not nuclear or colocalized; (f) Graph of nuclear Ub-RFP foci counts from 11-slice Z-stacks (rather than single focal plane images) acquired during experiments described/depicted in panel (d). Data are shown for *Ub-RFP Rad10-YFP* strain (PF038-1D) samples induced with BPDE (10 μM, 1 h) or controls mock-induced with tetrahydrofuran (THF). Error bars represent standard error.

We investigated DNA damage responses using strains containing *Ub-RFP* and either the *Rad10-YFP* or *Rad14-CFP*. Strains bearing these fluorescently labeled NER genes exhibit pan-nuclear YFP or CFP signal in the absence of DNA damage allowing identification of the location of nucleus within the cell [37] [39]. Following DNA damage, YFP and CFP signal is still pan-nuclear but bright, punctate YFP or CFP foci are induced within the nucleus following a temporal delay, thereby enabling the YFP and CFP signal to reveal the location of the nucleus within the cell but also indicating sites of NER within the nucleus [37] [39]. The *Ub-RFP Rad10-YFP* strain was treated with BPDE and imaged, exhibiting Ub-RFP foci localized within the periphery of the pan-nuclear Rad10-YFP signal (**Figure 1(d)**). Analogous experiments with the *Ub-RFP Rad14-CFP* strain and AAAF as the DNA damaging agent similarly showed Ub-RFP foci localized within the nuclear periphery (**Figure 1(e)**). Images of both damaged and undamaged cells showed the aforementioned, large, bright cytosolic Ub-RFP foci, which, as stated, did not change appreciably following BPDE and AAAF treatment (**Figure 1(d)** and **Figure 1(e)**, green arrows). Images from *Ub-RFP Rad10-YFP* BPDE experiments were analyzed by counting the nuclear-localized Ub-RFP foci and which exhibited approximately 1.8-fold induction (**Figure 1(f)**). Attempts at Ub-RFP induction in response to UV were not successful owing to bleaching of the Rad10-YFP and Rad14-CFP fluorophores (data not shown). Together these data show that nuclear ubiquitylated species form in response to BPDE and AAAF, two agents giving rise to damage only repaired by NER.

3.2. Nuclear Ubiquitin Foci Form Rapidly in Response to BPDE Treatment in a *RAD7* E3 Ubiquitin Ligase Mutant

The *Ub-RFP Rad10-YFP* strain was monitored as a function of time to determine the timescale for maximum induction and duration of focus persistence. Nuclear Ub-RFP foci were observed appearing and disappearing in successive images of the same fields of cells during 5 minute intervals acquired immediately following BPDE treatment (**Figure 2(a)**, upper panels). To examine whether the nuclear Ub-RFP foci observed were NER foci containing ubiquitylated Rad4, we prepared a *RAD7* mutant containing two amino acid substitutions abrogating the Rad7 E3 ubiquitin ligase activity that would normally ubiquitylate Rad4 (*rad7SOCS*) [3]. BPDE induction experiments with *rad7SOCS* similarly exhibited nuclear localized Ub-RFP foci (**Figure 2(a)**, lower panels). When foci counts of S, G2 and M cells from *RAD7* and *rad7SOCS* strains were compared, nuclear Ub-RFP foci were in greater abundance at the earliest time point in *rad7SOCS* (**Figure 2(b)**). We observed approximately 0.023 nuclear foci per cell in *rad7SOCS* at zero min with the signal becoming more diminished in successive

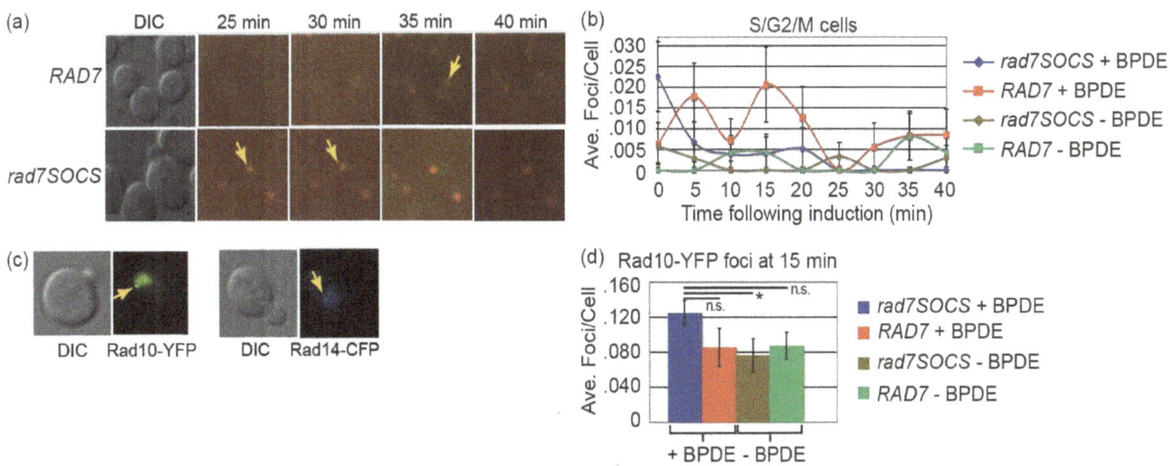

Figure 2. Timelapse experiments with the *Ub-RFP Rad10-YFP* strains following BPDE treatment and imaging at 5 minute intervals. Representative DIC and merged RFP/YFP images are shown from strains containing either the *RAD7* (PF038-1D, upper images) or *rad7SOCS* (PF084-7A, lower images). Gold arrows indicate nuclear-localized Ub-RFP foci. Note in the lower panels a Ub-RFP focus is shown which persisted through two timepoints. (b) Quantification of timelapse experiments. (c) Representative images from *RAD10-YFP* and *RAD14-CFP* strains showing pan-nuclear YFP or CFP signal and a Rad10-YFP or Rad14-CFP focus. Gold arrows indicate nuclear-localized foci. (d) Quantification of Rad10-YFP foci observed at 15 min in timelapse experiments. In (b) and (d), the plotted values represent means (λ's) from ideal Poisson distributions least squares fit to the observed distributions of foci per cell. The least squares deviations were propagated into the overall errors, which are depicted as the error bars in the graphs.

time points while *RAD7* cells exhibited 0.005 foci at zero minutes and peaked at approximately 0.020 nuclear foci per cell at 15 min post-induction (**Figure 2(b)**). The "zero minutes" timepoint reflects when cells were harvested following BPDE treatment, but processing for microscopy takes approximately three minutes. The Ub-RFP foci observed in these experiments are extremely rare events, even at peak focus induction, in part because only a fraction (~1/3) of each cell is being imaged. Hence, to ensure meaningful signal-to-noise, we analyzed a minimum of 600 cells for each condition at each time point.

Our observation of increased abundance of Ub-RFP foci in *rad7SOCS* relative to *RAD7* at zero minutes suggested that the foci were not NER foci since, in principle, the inability to ubiquitylate Rad4 would ostensibly result in *fewer* ubiquitylated NER protein complexes in *rad7SOCS*, relative to *RAD7*. To further cull the data for possible NER foci, YFP and CFP images acquired in parallel with Ub-RFP following BPDE and AAAF damage were inspected for colocalized Ub-RFP/Rad10-YFP and Ub-RFP/Rad14-CFP foci and none were observed (data not shown). For reference, examples of typical Rad10-YFP and Rad14-CFP foci are shown (**Figure 2(c)**, gold arrows). Since DNA damaging agents have been shown to induce Rad10-YFP foci following a temporal delay [37] [39], as a control, we also quantitatively analyzed images from Rad10-YFP/Ub-RFP/BPDE experiments at 15 minutes for Rad10-YFP foci. Numbers of Rad10-YFP foci were not dramatically altered in response to BPDE treatment, with the BPDE-treated *rad7SOCS* cells exhibiting a slight induction (~1.5-fold) of Rad10-YFP foci relative to uninduced controls while no induction was observed in *RAD7* (**Figure 2(d)**). In summary, only limited Rad10-YFP focus induction was observed after 15 minutes and the temporal pattern differed from that of Ub-RFP.

3.3. Polyubiquitylated Rpn10 and Dsk2 Are Present at Increased Levels Following BPDE Treatment in a *RAD7* E3 Ubiquitin Ligase Mutant

Both the pattern of Ub-RFP focus induction and the lack of induction of Ub-RFP/Rad14-CFP or Ub-RFP/Rad10-YFP colocalized foci suggested the Ub-RFP foci were not NER foci, so we performed experiments to determine the molecular basis for the nuclear ubiquitin-RFP foci. We hypothesized that nuclear Ub-RFP foci may derive from ubiquitylated histones, PCNA, or components of the proteasome [10] [40]-[44].

To test these hypotheses we prepared yeast WCEs from isogenic *RAD7* wild-type and *rad7SOCS* strains cell cycle-arrested at G2/M, released briefly into cell cycle and DNA damaged with BPDE using conditions similar to microscopy experiments. Immunoblots with antibodies to Rpn10, Dsk2, PCNA, Rad16 and Rpt1 exhibited HMW species in WCEs from *rad7SOCS* in greater abundance than *RAD7*, or uninduced controls (bands indicated with arrows in **Figure 3(a)**). Further, the HMW bands were not observed in a *rad7Δ* mutant strain that contained a wild-type ubiquitin gene instead of Ub-RFP (lanes 5 and 6, **Figure 3(a)**) indicating that their formation in higher abundance is specific to the lack of the E3 ubiquitin ligase function of Rad7 but also dependent on the presence of the of Rad7 polypeptide itself. Notably, native (unubiquitylated) Dsk2 protein abundance was also significantly increased in *rad7SOCS* following BPDE treatment (lane 4, **Figure 3(a)**). The observed pattern of HMW band induction aligns with results from the zero time point in our timelapse microscopy experiments (**Figure 2(b)**), making these HMW species candidates for being the ubiquitylated proteins in microscopy since all are nuclear-localized in dividing cells.

Membranes were immunoblotted for several other candidates including histones H2A and H2B, Rad23, Rad51, H3K79me3 and RNAPII, none of which exhibited a greater abundance of HMW bands in *rad7SOCS* following BPDE over *RAD7* or uninduced controls (**Supplemental Figure S1**). Notably, we observed a slight increase in H3K79me3 signal in *rad7SOCS* following BPDE treatment and a distinct signal increase in *rad7Δ* in the absence of DNA damage (**Supplemental Figure S1**), which could be indicative of altered checkpoint activation in *rad7SOCS* and *rad7Δ* [40]. Control representative blots were also hybridized with antibodies to Rad7, Rad4 and Rpn1, and included extracts from an isogenic *rad4Δ* strain (**Figure 3(b)**). Results from the Rad7 immunoblot indicate that *rad7SOCS* protein is present in lower abundance in the absence of DNA damage but, following BPDE treatment, increases to approximately the level observed in *RAD7* (**Figure 3(b)**). As expected, Rad4 levels are higher in the *rad7SOCS* strain following BPDE treatment since Rad4 expression is induced following DNA damage and degraded more slowly in the absence of polyubiquitylation by Rad7 (**Figure 3(b)**). Rpn1 levels were elevated slightly in *rad7SOCS* following BPDE and in *rad4Δ* without BPDE treatment (**Figure 3(b)**). Rpn1 levels were diminished in *rad7Δ* with or without BPDE treatment and highly diminished in *rad4Δ* following BPDE treatment, but we did not investigate whether the diminishment reflects degradation or loss by some other means (**Figure 3(b)**).

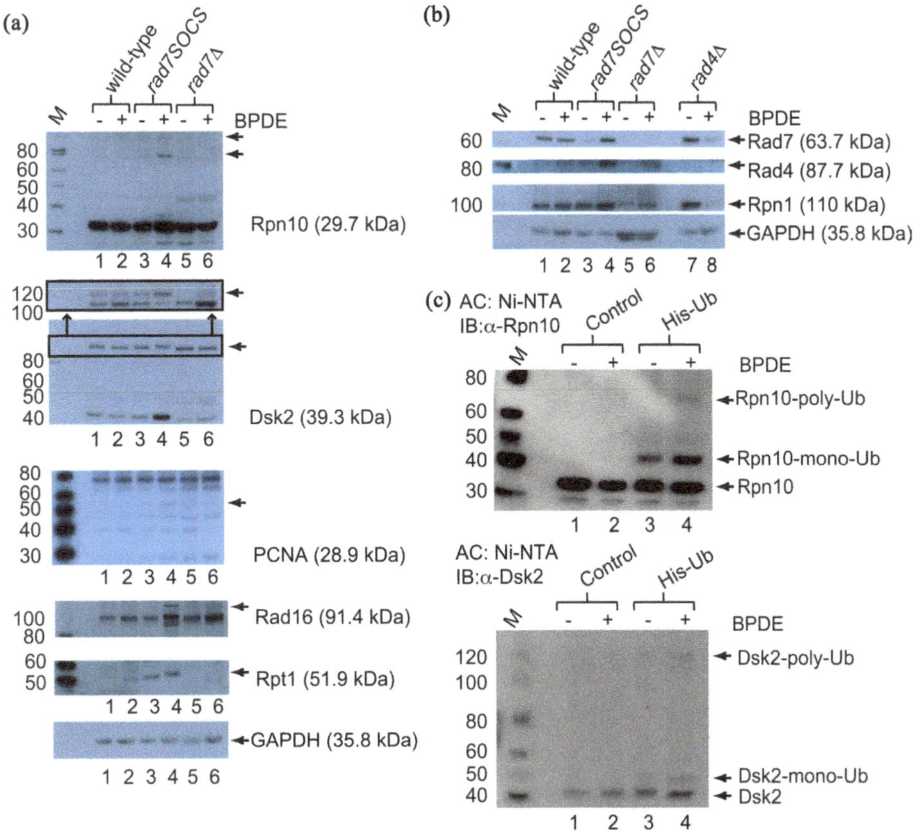

Figure 3. Cells from strains PF038-1D, ("wild-type", lanes 1 and 2), PF084-7A, ("*rad7SOCS*", lanes 3 and 4), MGSC104, ("*rad7Δ*", lanes 5 and 6) or PF097-2C, ("*rad4Δ*", lanes 7 and 8, panel B only) were cell cycle-arrested at the G2/M boundary, released briefly back into cell cycle, treated with 10 μM BPDE (lanes 2, 4, 6 or 8 "+") for 3 minutes or mock treated (lanes 1, 3, 5 and 7 "-") and disrupted. Normalized protein quantities were analyzed by SDS-PAGE and immunoblotted with the indicated antibodies. "M" indicates MagicMark™ molecular weight marker. The expected native protein size is shown in parentheses next to the antibody name. (a) Protein candidates for which we observed HMW species that appeared in greater abundance in *rad7SOCS*, BPDE-treated extracts as compared to controls. Arrows indicate the observed HMW species. The second image panel from the top shows a portion of a Dsk2 immunoblot from a 3% - 8% Tris-Acetate gel which gave better resolution of the corresponding HMW bands than the corresponding 4% - 20% Tris-glycine gel shown immediately below. Glyceraldehyde phosphate dehydrogenase (GAPDH) is a loading control. (b) Control immunoblots showing: (top panel) the *rad7SOCS* polypeptide in the *rad7SOCS* strain, (second panel) the accumulation of Rad4 in the absence of polyubiquitylation and degradation of Rad4 in *rad7SOCS* (compare lane 4 with lanes 1 - 3), (third panel) Rpn1 and (bottom panel) GAPDH as a loading control. (c) Ni-NTA affinity capture experiments from cells transformed either with the empty vector (YEplac 195, "Control", lanes 1 and 2) or with YEplac195 CUP1::His7-Ub ("His-Ub", lanes 3 and 4), cell cycle-arrested, BPDE-treated and harvested under denaturing conditions prior to Ni-NTA affinity capture. Immunoblots for Rpn10 (top panel), or Dsk2 (bottom panel) are shown.

To determine whether the Rpn10 and Dsk2 HMW bands observed in BPDE-treated extracts represented ubiquitylated proteins we performed Nickel Nitriloacetic acid (Ni-NTA) affinity chromatography pull down experiments using extracts from cells transformed with a plasmid containing a His-tagged ubiquitin gene, cell cycle-arrested at the G2/M boundary, released briefly from arrest and DNA damaged with BPDE prior to cell disruption under denaturing conditions. Immunoblots with antibodies to Rpn10 and Dsk2 (**Figure 3(c)**) gave specific HMW bands of similar sizes as had been observed in non-His-tagged WCEs (**Figure 3(a)**). The stoichiometry of ubiquitin conjugates were tentatively assigned as mono- and polyubiquitylated forms based on SDS PAGE mobility. It is important to note that in experiments analyzing Ub-RFP WCE's (**Figure 3(a)**), the Ub-RFP conjugates would be expected to increase the molecular weights of the native proteins in ~38 kDa increments (8 kDa Ub + 30 kDa RFP) while in **Figure 3(c)** the addition of the 6x His tag adds only ~8.8 kDa per His-tagged ubiquitin conjugate (8 kDa Ub + 0.84 kDa 6x His). Any polyubiquitin conjugates pulled down would be expected to have at least one 6x His-containing 8.8 kDa conjugate, but, if polyubiquitylated, could have varying mixtures of additions of 8.8 kDa plus ~38 kDa units, perhaps explaining why HMW bands in the two experiments (**Figure 3(a)** and **Figure 3(c)**) do not have exactly the same electrophoretic mobilities. Native Rpn10 and Dsk2 were pulled down in both His-tagged ubiquitin samples and untagged control experiments, possibly indicating pull down of a background of histidine-rich proteins that are constantly undergoing proteasomal degradation that bind tightly to both Ni-NTA and either Rpn10 or Dsk2 during pull down experiments, even under the denaturing conditions employed.

3.4. Dsk2-GFP Foci Are Formed at Increased Levels in a *RAD7* E3 Ubiquitin Ligase Mutant

To further investigate whether ubiquitylated forms of Dsk2 and Rpn10 were nuclear localized in response to BPDE damage, we prepared yeast diploids containing *DSK2/Dsk2-GFP UB14/Ub-RFP RAD10/Rad10-YFP* or *RPN10/Rpn10-GFP UB14/Ub-RFP RAD10/Rad10-YFP* in either a *RAD7/RAD7*, *rad7SOCS/rad7SOCS* or *rad7Δ/rad7Δ* strain background and carried out BPDE damage induction experiments on cells synchronized at the G2/M boundary and release briefly back into cell cycle.

In microscopy experiments with the *DSK2/Dsk2-GFP UB14/Ub-RFP RAD10/Rad10-YFP* strain panel we made several noteworthy observations. First, we observed nuclear-colocalized Dsk2-GFP/UbRFP foci (**Figure 4(a)**). Some Dsk2-GFP/Ub-RFP foci observed were not perfectly colocalized, but within one pixel of each other

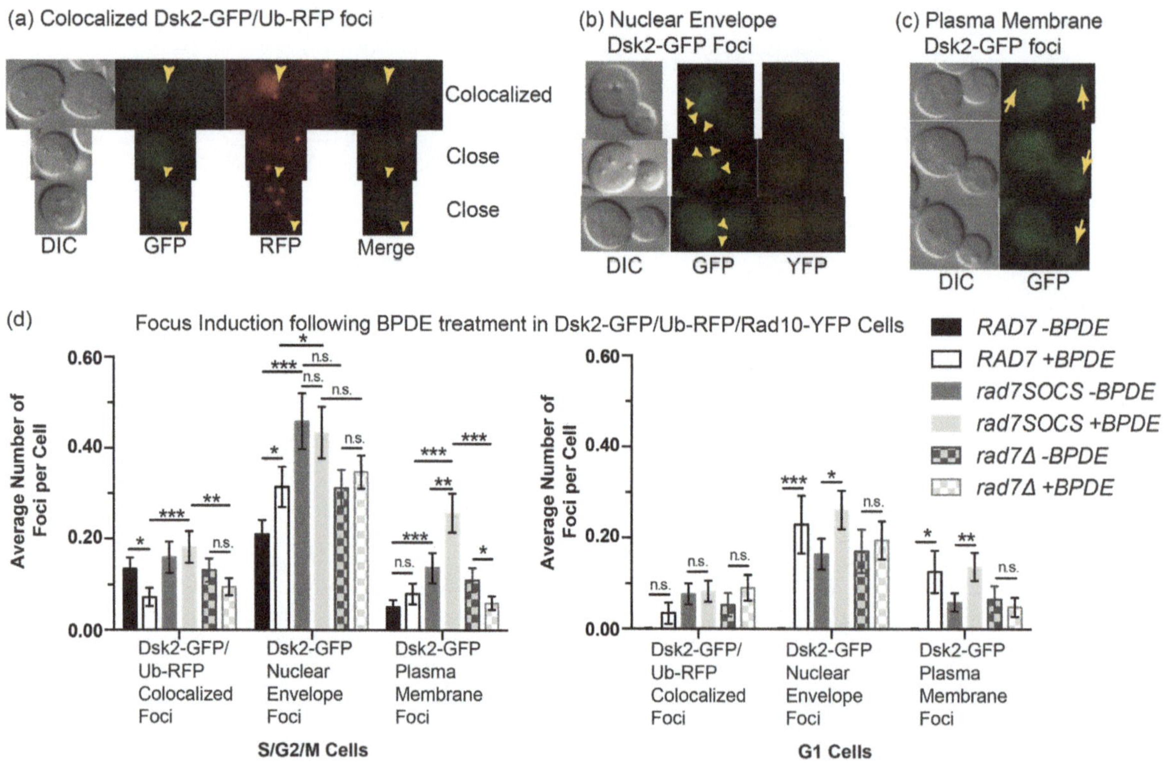

Figure 4. Fluorescence microscopy in Dsk2-GFP/Ub-RFP/Rad10-YFP diploid strains [PF165 (*RAD7*), PF166 (*rad7SOCS*), PF167 (*rad7Δ*)], were cell cycle arrested and treated with BPDE. (a) Examples of nuclear colocalized and "close" Dsk2-GFP/Ub-RFP foci, which were grouped and tallied together as simply "colocalized" for the purposes of our data analysis. DIC, GFP and RFP images are shown as well as merged GFP/RFP images. (b) Examples of nuclear envelope signal in which Dsk2-GFP signal circling portions of the edge of the nuclear envelope was observed. Rad10-YFP pan-nuclear fluorescence shows the approximate location of the nucleus. (c) Examples of plasma membrane Dsk2-GFP foci. Yellow arrows indicate the locations of the foci. (d) Graph of foci counts observed in S, G2 and M phase cells in 3-slice Z-stacks acquired during experiments depicted in panels (a)-(c) (left graph) and corresponding foci counts in G1 phase (right graph). The plotted values represent means (λ's) from ideal Poisson distributions least squares fit to the observed distributions of foci per cell. The least squares deviations were propagated into the overall errors, which are depicted as the error bars in the graphs. Statistical comparisons were carried out via a paired t_{calc} according to the NIST/SEMATECH e-handbook of Statistical Methods and then calculating p from integration of the single tail area beyond the paired t_{calc} in a Gaussian distribution. "***" indicates $p < 0.001$, "**" indicates $0.001 < p < 0.01$, "*" indicates $0.01 < p < 0.05$, "n.s." indicates $0.05 < p$.

and were classified along with the others as "colocalized" (**Figure 4(a)**). Such foci could presumably belong to a large object such as the proteasome to which both Ub-RFP and Dsk2-GFP had localized. Second, in many cells we observed a distinct "ring" of Dsk2-GFP signal that appeared to localize to the nuclear envelope (**Figure 4(b)**). Third, we observed bright Dsk2-GFP foci that appeared to localize to the plasma membrane, frequently at the opposing poles that aligned with the axis of the mitotic spindle, ostensibly at the microtubule organizing center (MTOC) (**Figure 4(c)**). We quantified all these phenomena and found that all were observed in higher abundances in *rad7SOCS* than in either *RAD7* or *rad7Δ* in dividing cells (left graph, **Figure 4(d)**). In the case of Dsk2-GFP plasma membrane foci, abundances were even higher in BPDE-treated dividing cells than in uninduced controls (left graph, **Figure 4(d)**). G1 cells exhibited lower overall abundances of these foci, but we still observed some statistically significant differences between BPDE-treated cells over uninduced controls (right graph, **Figure 4(d)**). Together these data indicate that Dsk2 localization and state of ubiquitylation is influenced by Rad7 E3 activity. Further, our data suggest that *RAD7* could play more than one role in governing Dsk2 localization/ubiquitylation since different phenotypes were observed between *rad7SOCS* and *rad7Δ*. We did not observe significant differences in the appearances of images recorded for the *RPN10/Rpn10-GFP UB14/Ub-RFP RAD10/Rad10-YFP* strain panel (data not shown).

3.5. Ubiquitylated Proteins Form at Increased Levels in a *RAD7* E3 Ubiquitin Ligase Mutant Following UV and Monoubiquitylated Rpn11 Forms Even in the Absence of UV

Next we examined whether ubiquitylated proteins form in greater abundance in *rad7SOCS* in response to DNA damaging agents other than BPDE. WCEs were prepared from cultures synchronized at the G2/M boundary, released briefly back into cell cycle and UV-irradiated. Immunoblots of these WCEs with the ubiquitin antibody showed increased levels of ubiquitylated species in *rad7SOCS* as compared to *RAD*7 and especially as compared to *rad7Δ*, mirroring the patterns observed in microscopy experiments and immunoblots with BPDE-treated extracts (**Figure 5(a)**). When HMW proteins from these extracts were resolved more fully using a 3% - 8% Tris-acetate gradient gel, it became even more clear that HMW ubiquitylated proteins were observed in greater abundance in *rad7SOCS* following UV (**Figure 5(b)**), compared to *rad7Δ* extracts.

Since prior reports identified the existence of UV-sensitive mutants of proteasome DUB, Rpn11 [30], we prepared an Rpn11-TAP strain panel containing either *RAD7* or *rad7SOCS*, transformed with the His-Ub plasmid and subjected protein extracts to Ni-NTA affinity capture experiments under denaturing conditions. In addition to isolating native Rpn11-TAP in both *RAD7* and *rad7SOCS* extracts, we also observed a slightly larger molecular weight band in *rad7SOCS* but not *RAD7* (**Figure 5(c)**, lanes 4 & 6). Capture of native Rpn11-TAP, as in **Figure 3(c)**, likely represents pull down of ubiquitylated species bound to proteasomes that had not fully denatured under the conditions in our experiment and was similar between strains. However, capture of the ~8 kDa larger Rpn11-TAP species observed in *rad7SOCS* could represent monoubiquitylated Rpn11-TAP. The presence of this band in the *rad7SOCS* but not *RAD7* extract indicates that Rad7 E3 function influences the state of ubiquitylation of Rpn11 and normally attenuates accumulation of mono-ubiquitylated Rpn11. We also analyzed analogous extracts that had been UV-treated (**Figure 5(c)**, lanes 5 & 7). These UV-treated extracts were devoid of Rpn11-TAP-positive bands in both *RAD7* and *rad7SOCS* extracts but we do not know whether this reflects degradation of Rpn11 or its depletion through other means in our assay system. These data show that the Rpn11 status changed as a function of *RAD7* genetic status.

4. Discussion

Our data reveal that the E3 ligase function of *RAD7* attenuates ubiquitylation of proteasomal proteins Dsk2, Rpn10 and Rpn11 and alters localization of Dsk2 in response to DNA damaging agents. Intriguingly, we also observed changes in abundances and gel mobilities of several other proteasome regulatory particle subunits that were a function of *RAD7* or *RAD4* gene status (Rpt1 and Rpn1) providing further evidence that NER damage recognition factors interact with the proteasome following DNA damage in a complicated manner. The direct target of Rad7 E3 ligase activity responsible for all these observations remains to be identified, but our data are consistent with it being another E3 or E4 ubiquitin ligase, which is stimulated by the action of Rad7 E3, or, alternatively, a DUB which is inhibited by Rad7 E3 action (**Figure 6**). In principle, there could be more than one target, including a combination of E3 and E4 ligases and DUBs, and targeting could either be direct or part of a

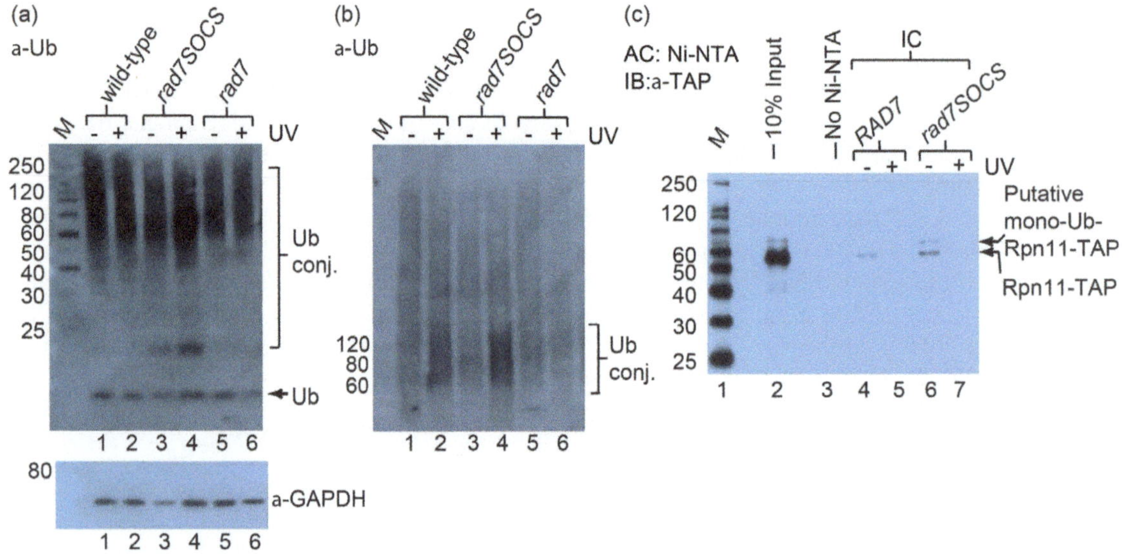

Figure 5. Increased levels of ubiquitylated proteins are observed in *rad7SOCS* cells following UV and increased monoubiquitylated Rpn11 is observed in *rad7SOCS* without damage. (a) Cells from strains PF038-1D, ("wild-type", lanes 1 and 2), PF084-7A, ("*rad7SOCS*", lanes 3 and 4) and MGSC104, ("*rad7Δ*", lanes 5 and 6), were cell cycle arrested at the G2/M boundary, released briefly back into cell cycle, treated with 100 J/m^2 UV-C (lanes 2, 4 or 6, "+") for 3 minutes or mock treated (lanes 1, 3 or 5, "-") and disrupted. Normalized quantities of protein from WCEs were analyzed on a NuPAGE Novex 4% - 12% Bis-Tris gel with MES SDS running buffer SDS-PAGE and immunoblotted with antibodies to Ub (upper panel) or GAPDH (lower panel) as a loading control. "M" indicates MagicMark™ molecular weight marker. (b) Same as upper panel of (a) except that extracts were analyzed on a NuPAGE Novex 3% - 8% Tris-Acetate Gel with Tris-Acetate running buffer. Arrows indicate ubiquitin; brackets indicate putative ubiquitin-conjugates. (c) The *Rpn11-TAP* strain (YFR004W) and an isogenic *Rpn11-TAP/rad7SOCS* derivative (PF168) were transformed with YEplac195 CUP1::His7-Ub, cell cycle arrested, released briefly back into cell cycle and disrupted. Denatured protein extracts were subjected to Ni-NTA affinity capture experiments. Lane 1, MagicMark™ molecular weight marker; lane 2, 10% input; lane 3, affinity capture in which Ni-NTA beads were omitted; lane 4, affinity capture from *Rpn11-TAP*, no UV; lane 5 affinity capture from *Rpn11-TAP* with UV; lane 6 affinity capture from *Rpn11-TAP/rad7SOCS*, no UV; lane 7 affinity capture from *Rpn11-TAP* with UV.

multistep cascade of which Rad7 E3 action is an upstream step.

Cell cycle checkpoints are regulated by proteasomal degradation of key proteins. There is literature precedent for participation of both Rpn10 and Dsk2 in the degradation of cyclins [10] [45]. Dsk2 and Rpn10 are genetically linked with SPB duplication and spindle pole bridge separation, two key events in cell division in *S. cerevisiae*. Dsk2 normally shuttles polyubiquitylated proteins to the proteasome, while Rpn10 provides selective access of polyubiqutylated proteins to the proteasome core for destruction. It is therefore tempting to speculate that the role of Rad7 is to regulate polyubiquitilation of Dsk2 and Rpn10, monoubiquitylation of Rpn11, and ultimately the SPB and spindle pole bridge proteins in a manner that fine tunes checkpoint activation in the face of DNA damage, perhaps depending on whether repair is proceeding effectively. In support of this possibility, we observed alterations in H3K79me3 signal in *rad7SOCS* following BPDE damage and in *rad7Δ* in the absence of BPDE (**Supplemental Figure S1**).

Our microscopy data support this model; *Dsk2-GFP Ub-RFP* diploids revealed slightly higher levels of Dsk2-GFP localized to the nuclear envelope in dividing cells (the location of proteasomes during cell division) and higher Dsk2-GFP/Ub-RFP nuclear colocalization was observed in *rad7SOCS* cells (**Figure 4**). We also observed an increase in abundances of plasma membrane foci in *rad7SOCS* in dividing cells following BPDE treatment, indicating a greater abundance of Dsk2 associated with a structure present at the plasma membrane during M phase and which appeared to be along the mitotic spindle axis, possibly localized to MTOCs (**Figure 4**). This result augurs well with the known requirement for Dsk2 in SPB separation and progression through anaphase [15] [46]. Rad7 influence on Dsk2 localization may also explain prior observations in which delayed nuclear positioning, delayed anaphase and misoriented spindles were observed in *bim1Δ rad7Δ* double mutants [27]. Taken together our results show Rad7 E3 activity influences the states of ubiquitylation of several proteasome-associated proteins.

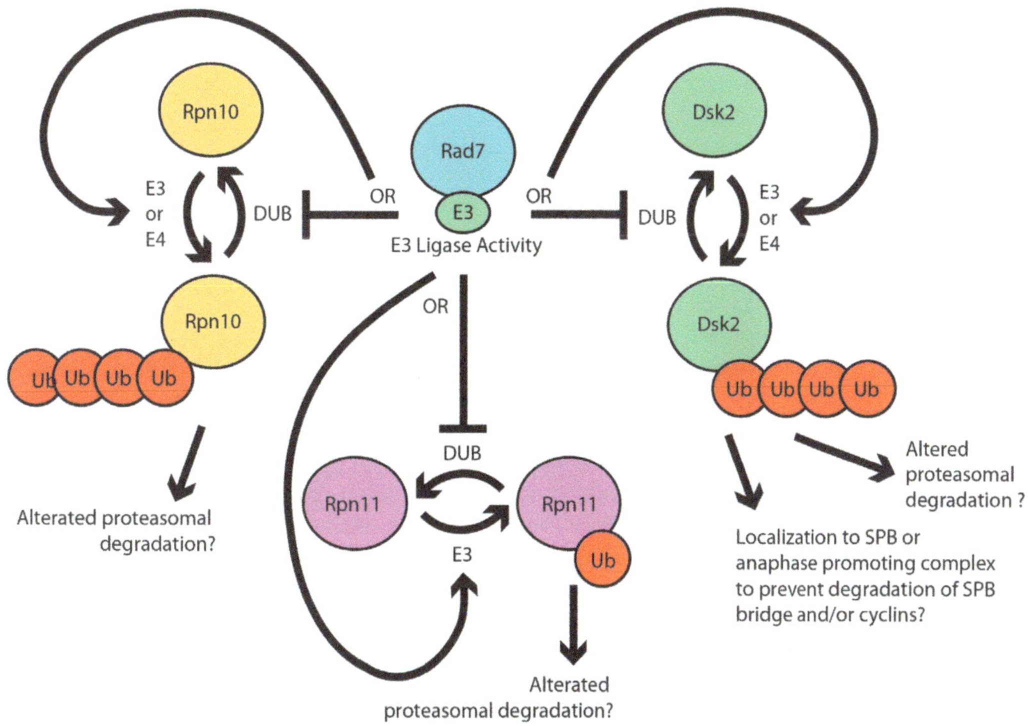

Figure 6. Model for regulation of proteasomal factors Rpn10, Dsk2, and Rpn11. Ubiquitylation of Rpn10, Dsk2 and Rpn11 is augmented in a *rad7SOCS* background indicating that Rad7 E3 normally attenuates ubiquitylation. We propose that Rad7 E3 ligase function may alternately either stimulate ubiquitylation or inhibit deubiquitylation of all three proteins. Such alteration may influence the amount and types of proteins degraded by the proteasome. In the case of Dsk2, alteration may also influence localization to spindle pole bodies or the APC.

Acknowledgements

We thank Dr. Errol C. Friedberg for providing several yeast strains and antibodies and Dr. Helle Ulrich for His-UB plasmids. We thank Dr. Justin Karlin, Dr. Robert Haughton and Ms. Rowen Jane Odango for technical assistance and Dr. Orlando Schärer for critical reading of the manuscript. This work was supported by the NIGMS (SC2GM081155 and SC3GM093858 to PF, S06GM48680 to EK and 5T34GM008395 to DL and AV) and Research Corporation (Cottrell College Science Awards #7146/7267 to PF).

Conflict of Interest

The authors declare that there is no conflict of interest regarding the publication of this paper.

References

[1] Friedberg, E.C., Walker, G.C., Siede, W., Wood, R.D., Schultz, R. A., *et al.* (2005) DNA Repair and Mutagenesis. 2nd Edition, ASM Press, Washington DC.

[2] Staresincic, L., Fagbemi, A.F., Enzlin, J.H., Gourdin, A.M., Wijgers, N., *et al.* (2009) Coordination of Dual Incision and Repair Synthesis in Human Nucleotide Excision Repair. *The EMBO Journal*, **28**, 1111-1120. http://dx.doi.org/10.1038/emboj.2009.49

[3] Gillette, T.G., Yu, S., Zhou, Z., Waters, R., Johnston, S.A., *et al.* (2006) Distinct Functions of the Ubiquitin-Proteasome Pathway Influence Nucleotide Excision Repair. *The EMBO Journal*, **25**, 2529-2538. http://dx.doi.org/10.1038/sj.emboj.7601120

[4] Sugasawa, K., Okuda, Y., Saijo, M., Nishi, R., Matsuda, N., *et al.* (2005) UV-Induced Ubiquitylation of XPC Protein Mediated by UV-DDB-Ubiquitin Ligase Complex. *Cell*, **121**, 387-400. http://dx.doi.org/10.1016/j.cell.2005.02.035

[5] Teng, Y., Liu, H., Gill, H.W., Yu, Y., Waters, R., *et al.* (2008) *Saccharomyces cerevisiae* Rad16 Mediates Ultraviolet-Dependent Histone H3 Acetylation Required for Efficient Global Genome Nucleotide-Excision Repair. *EMBO Reports*, **9**, 97-102. http://dx.doi.org/10.1038/sj.embor.7401112

[6] Yu, S., Owen-Hughes, T., Friedberg, E.C., Waters, R. and Reed, S.H. (2004) The Yeast Rad7/Rad16/Abf1 Complex Generates Superhelical Torsion in DNA That Is Required for Nucleotide Excision Repair. *DNA Repair*, **3**, 277-287. http://dx.doi.org/10.1016/j.dnarep.2003.11.004

[7] Yu, S., Smirnova, J.B., Friedberg, E.C., Stillman, B., Akiyama, M., *et al.* (2009) ABF1-Binding Sites Promote Efficient Global Genome Nucleotide Excision Repair. *The Journal of Biological Chemistry*, **284**, 966-973. http://dx.doi.org/10.1074/jbc.M806830200

[8] Madura, K. (2004) Rad23 and Rpn10: Perennial Wallflowers Join the Melee. *Trends in Biochemical Sciences*, **29**, 637-640. http://dx.doi.org/10.1016/j.tibs.2004.10.008

[9] Crosas, B., Hanna, J., Kirkpatrick, D.S., Zhang, D.P., Tone, Y., *et al.* (2006) Ubiquitin Chains Are Remodeled at the Proteasome by Opposing Ubiquitin Ligase and Deubiquitinating Activities. *Cell*, **127**, 1401-1413. http://dx.doi.org/10.1016/j.cell.2006.09.051

[10] Isasa, M., Katz, E.J., Kim, W., Yugo, V., Gonzalez, S., *et al.* (2010) Monoubiquitination of RPN10 Regulates Substrate Recruitment to the Proteasome. *Molecular Cell*, **38**, 733-745. http://dx.doi.org/10.1016/j.molcel.2010.05.001

[11] Mao, P. and Smerdon, M.J. (2010) Yeast Deubiquitinase Ubp3 Interacts with the 26 S Proteasome to Facilitate Rad4 Degradation. *Journal of Biological Chemistry*, **285**, 37542-37550. http://dx.doi.org/10.1074/jbc.M110.170175

[12] Liu, C., van Dyk, D., Li, Y., Andrews, B. and Rao, H. (2009) A Genome-Wide Synthetic Dosage Lethality Screen Reveals Multiple Pathways That Require the Functioning of Ubiquitin-Binding Proteins Rad23 and Dsk2. *BMC Biology*, **7**, 75. http://dx.doi.org/10.1186/1741-7007-7-75

[13] Finley, D. (2009) Recognition and Processing of Ubiquitin-Protein Conjugates by the Proteasome. *Annual Review of Biochemistry*, **78**, 477-513. http://dx.doi.org/10.1146/annurev.biochem.78.081507.101607

[14] Jaspersen, S.L. and Winey, M. (2004) The Budding Yeast Spindle Pole Body: Structure, Duplication, and Function. *Annual Review of Cell and Developmental Biology*, **20**, 1-28. http://dx.doi.org/10.1146/annurev.cellbio.20.022003.114106

[15] Diaz-Martinez, L.A., Kang, Y., Walters, K.J. and Clarke, D.J. (2006) Yeast UBL-UBA Proteins Have Partially Redundant Functions in Cell Cycle Control. *Cell Division*, **1**, 28. http://dx.doi.org/10.1186/1747-1028-1-28

[16] Kaye, F.J., Modi, S., Ivanovska, I., Koonin, E.V., Thress, K., *et al.* (2000) A Family of Ubiquitin-Like Proteins Binds the ATPase Domain of Hsp70-Like Stch. *FEBS Letters*, **467**, 348-355. http://dx.doi.org/10.1016/S0014-5793(00)01135-2

[17] Heessen, S., Masucci, M.G. and Dantuma, N.P. (2005) The UBA2 Domain Functions as an Intrinsic Stabilization Signal That Protects Rad23 from Proteasomal Degradation. *Molecular Cell*, **18**, 225-235. http://dx.doi.org/10.1016/j.molcel.2005.03.015

[18] Sekiguchi, T., Sasaki, T., Funakoshi, M., Ishii, T., Saitoh, Y.H., *et al.* (2011) Ubiquitin Chains in the Dsk2 UBL Domain Mediate Dsk2 Stability and Protein Degradation in Yeast. *Biochemical and Biophysical Research Communications*, **411**, 555-561. http://dx.doi.org/10.1016/j.bbrc.2011.06.183

[19] Zuin, A., Bichmann, A., Isasa, M., Puig-Sarries, P., Diaz, L.M., *et al.* (2015) Rpn10 Monoubiquitination Orchestrates the Association of the Ubiquilin-Type DSK2 Receptor with the Proteasome. *Biochemical Journal*, In Press. http://dx.doi.org/10.1042/BJ20150609

[20] Lambertson, D., Chen, L. and Madura, K. (1999) Pleiotropic Defects Caused by Loss of the Proteasome-Interacting Factors Rad23 and Rpn10 of *Saccharomyces cerevisiae*. *Genetics*, **153**, 69-79.

[21] Biggins, S., Ivanovska, I. and Rose, M.D. (1996) Yeast Ubiquitin-Like Genes Are Involved in Duplication of the Microtubule Organizing Center. *Journal of Cell Biology*, **133**, 1331-1346. http://dx.doi.org/10.1083/jcb.133.6.1331

[22] Liang, R.Y., Chen, L., Ko, B.T., Shen, Y.H., Li, Y.T., *et al.* (2014) Rad23 Interaction with the Proteasome Is Regulated by Phosphorylation of Its Ubiquitin-Like (UbL) Domain. *Journal of Molecular Biology*, **426**, 4049-4060. http://dx.doi.org/10.1016/j.jmb.2014.10.004

[23] Gillette, T.G., Huang, W., Russell, S.J., Reed, S.H., Johnston, S.A., *et al.* (2001) The 19S Complex of the Proteasome Regulates Nucleotide Excision Repair in Yeast. *Genes & Development*, **15**, 1528-1539. http://dx.doi.org/10.1101/gad.869601

[24] Li, Y., Yan, J., Kim, I., Liu, C., Huo, K., *et al.* (2010) Rad4 Regulates Protein Turnover at a Postubiquitylation Step. *Molecular Biology of the Cell*, **21**, 177-185. http://dx.doi.org/10.1091/mbc.E09-04-0305

[25] Lommel, L., Chen, L., Madura, K. and Sweder, K. (2000) The 26S Proteasome Negatively Regulates the Level of Overall Genomic Nucleotide Excision Repair. *Nucleic Acids Research*, **28**, 4839-4845. http://dx.doi.org/10.1093/nar/28.24.4839

[26] Lommel, L., Ortolan, T., Chen, L., Madura, K. and Sweder, K.S. (2002) Proteolysis of a Nucleotide Excision Repair Protein by the 26 S Proteasome. *Current Genetics*, **42**, 9-20. http://dx.doi.org/10.1007/s00294-002-0332-9

[27] Vizeacoumar, F.J., van Dyk, N., Vizeacoumar, F.S., Cheung, V., Li, J., *et al.* (2010) Integrating High-Throughput Genetic Interaction Mapping and High-Content Screening to Explore Yeast Spindle Morphogenesis. *Journal of Cell Biology*, **188**, 69-81. http://dx.doi.org/10.1083/jcb.200909013

[28] Miller, R.K., Cheng, S.C. and Rose, M.D. (2000) Bim1p/Yeb1p Mediates the Kar9p-Dependent Cortical Attachment of Cytoplasmic Microtubules. *Molecular Biology of the Cell*, **11**, 2949-2959. http://dx.doi.org/10.1091/mbc.11.9.2949

[29] Su, L.K., Burrell, M., Hill, D.E., Gyuris, J., Brent, R., *et al.* (1995) APC Binds to the Novel Protein EB1. *Cancer Research*, **55**, 2972-2977.

[30] Rinaldi, T., Pick, E., Gambadoro, A., Zilli, S., Maytal-Kivity, V., *et al.* (2004) Participation of the Proteasomal Lid Subunit Rpn11 in Mitochondrial Morphology and Function Is Mapped to a Distinct C-Terminal Domain. *Biochemical Journal*, **381**, 275-285. http://dx.doi.org/10.1042/BJ20040008

[31] Bashore, C., Dambacher, C.M., Goodall, E.A., Matyskiela, M.E., Lander, G.C., *et al.* (2015) Ubp6 Deubiquitinase Controls Conformational Dynamics and Substrate Degradation of the 26S Proteasome. *Nature Structural & Molecular Biology*, **22**, 712-719. http://dx.doi.org/10.1038/nsmb.3075

[32] Aufderheide, A., Beck, F., Stengel, F., Hartwig, M., Schweitzer, A., *et al.* (2015) Structural Characterization of the Interaction of Ubp6 with the 26S Proteasome. *Proceedings of the National Academy of Sciences of the United States of America*, **112**, 8626-8631. http://dx.doi.org/10.1073/pnas.1510449112

[33] Novarina, D., Amara, F., Lazzaro, F., Plevani, P. and Muzi-Falconi, M. (2011) Mind the Gap: Keeping UV Lesions in Check. *DNA Repair*, **10**, 751-759. http://dx.doi.org/10.1016/j.dnarep.2011.04.030

[34] Campbell, R.E., Tour, O., Palmer, A.E., Steinbach, P.A., Baird, G.S., *et al.* (2002) A Monomeric Red Fluorescent Protein. *Proceedings of the National Academy of Sciences of the United States of America*, **99**, 7877-7882. http://dx.doi.org/10.1073/pnas.082243699

[35] Reid, R.J., Lisby, M. and Rothstein, R. (2002) Cloning-Free Genome Alterations in *Saccharomyces cerevisiae* Using Adaptamer-Mediated PCR. *Methods in Enzymology*, **350**, 258-277. http://dx.doi.org/10.1016/S0076-6879(02)50968-X

[36] Fleer, R., Nicolet, C.M., Pure, G.A. and Friedberg, E.C. (1987) RAD4 Gene of *Saccharomyces cerevisiae*: Molecular Cloning and Partial Characterization of a Gene That Is Inactivated in *Escherichia coli*. *Molecular and Cellular Biology*, **7**, 1180-1192.

[37] Moore, D.M., Karlin, J., Gonzalez-Barrera, S., Mardiros, A., Lisby, M., *et al.* (2009) Rad10 Exhibits Lesion-Dependent Genetic Requirements for Recruitment to DNA Double-Strand Breaks in *Saccharomyces cerevisiae*. *Nucleic Acids Research*, **37**, 6429-6438. http://dx.doi.org/10.1093/nar/gkp709

[38] Huh, W.K., Falvo, J.V., Gerke, L.C., Carroll, A.S., Howson, R.W., *et al.* (2003) Global Analysis of Protein Localization in Budding Yeast. *Nature*, **425**, 686-691.

[39] Mardiros, A., Benoun, J.M., Haughton, R., Baxter, K., Kelson, E.P., *et al.* (2011) Rad10-YFP Focus Induction in Response to UV Depends on RAD14 in Yeast. *Acta Histochemica*, **113**, 409-415. http://dx.doi.org/10.1016/j.acthis.2010.03.005

[40] Game, J.C. and Chernikova, S.B. (2009) The Role of RAD6 in Recombinational Repair, Checkpoints and Meiosis via Histone Modification. *DNA Repair*, **8**, 470-482. http://dx.doi.org/10.1016/j.dnarep.2009.01.007

[41] Weake, V.M. and Workman, J.L. (2008) Histone Ubiquitination: Triggering Gene Activity. *Molecular Cell*, **29**, 653-663. http://dx.doi.org/10.1016/j.molcel.2008.02.014

[42] Ulrich, H.D. (2005) Mutual Interactions between the SUMO and Ubiquitin Systems: A Plea of No Contest. *Trends in Cell Biology*, **15**, 525-532. http://dx.doi.org/10.1016/j.tcb.2005.08.002

[43] Ulrich, H.D. (2009) Regulating Post-Translational Modifications of the Eukaryotic Replication Clamp PCNA. *DNA Repair*, **8**, 461-469. http://dx.doi.org/10.1016/j.dnarep.2009.01.006

[44] Laporte, D., Salin, B., Daignan-Fornier, B. and Sagot, I. (2008) Reversible Cytoplasmic Localization of the Proteasome in Quiescent Yeast Cells. *Journal of Cell Biology*, **181**, 737-745. http://dx.doi.org/10.1083/jcb.200711154

[45] Funakoshi, M., Tomko Jr., R.J., Kobayashi, H. and Hochstrasser, M. (2009) Multiple Assembly Chaperones Govern Biogenesis of the Proteasome Regulatory Particle Base. *Cell*, **137**, 887-899. http://dx.doi.org/10.1016/j.cell.2009.04.061

[46] Seeger, M., Hartmann-Petersen, R., Wilkinson, C.R., Wallace, M., Samejima, I., *et al.* (2003) Interaction of the Anaphase-Promoting Complex/Cyclosome and Proteasome Protein Complexes with Multiubiquitin Chain-Binding Proteins. *Journal of Biological Chemistry*, **278**, 16791-16796. http://dx.doi.org/10.1074/jbc.M208281200

[47] Verhage, R.A., Van de Putte, P. and Brouwer, J. (1996) Repair of rDNA in *Saccharomyces cerevisiae*: RAD4-Independent Strand-Specific Nucleotide Excision Repair of RNA Polymerase I Transcribed Genes. *Nucleic Acids Research*, **24**, 1020-1025. http://dx.doi.org/5w0229

[48] Thomas, B.J. and Rothstein, R. (1989) Elevated Recombination Rates in Transcriptionally Active DNA. *Cell*, **56**, 619-630. http://dx.doi.org/10.1016/0092-8674(89)90584-9

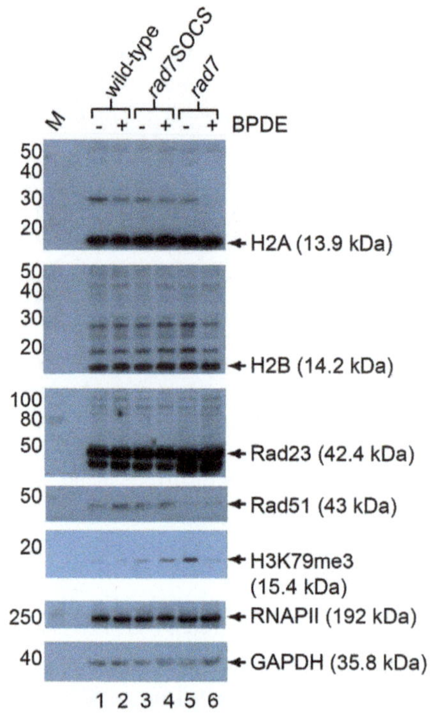

Figure S1. Additional candidates for ubiquitylation that were tested and found not exhibit HMW species that differed between wild-type and *rad7SOCS*. Immunoblots are from the same experiments described for **Figure 3**. Cells from strains PF038-1D, ("wild-type", lanes 1 and 2), PF084-7A, ("*rad7SOCS*", lanes 3 and 4), MGSC104, ("*rad7Δ*", lanes 5 and 6) were cell cycle-arrested at the G2/M boundary, released briefly back into cell cycle, treated with 10 μM BPDE (lanes 2, 4 or 6,"+") for 3 minutes or mock treated (lanes 1, 3 or 5, "-") and disrupted. Normalized protein quantities were analyzed by SDS-PAGE and immunoblotted with the indicated antibodies. "M" indicates MagicMark™ molecular weight marker. The expected native protein size is shown in parentheses next to the antibody name. H2A = Histone 2A, H2B = Histone 2B, Rad23 = Rad23, Rad51 = Rad51, H3K79me3 = Histone 3 K79 trimethylated, RNAPII = RNA Polymerase II. GAPDH is a loading control.

23

Physico-Chemical and Microbiological Analysis of Potable Water in Jericho and Molete Areas of Ibadan Metropolis

Olusola A. Ladokun, Sarah O. Oni

Lead City University, Ibadan, Nigeria
Email: solaJP@gmail.com

Abstract

This study was carried out to determine the quality of some water supplies in Jericho and Molete areas in Ibadan, Oyo State, Nigeria. A total of 6 water samples (Jericho and Molete) from well, borehole and public tap were randomly selected and analyzed in triplicate for some physico-chemical and microbial parameters. The pH ranged from 5.88 ± 0.01 to 6.88 ± 0.06 while total hardness ranged from 2.60 ± 0.04 ppm to 44.00 ± 1.20 ppm. The temperature ranged from 15.00˚C ± 0.07˚C to 18.00˚C ± 0.21˚C while conductivity ranged from 230 ± 1.15 to 460 ± 1.26 mho/cm. Alkalinity, Dissolved Oxygen (DO) and Biochemical Oxygen Demand (BOD) ranged from 4.03 ± 0.02 mg/l to 14.88 ± 0.15 mg/l, 0.4 ± 0.02 mg/l to 4.6 ± 0.02 mg/l and 12.80 ± 0.22 mg/l to 26.40 ± 0.72 mg/l respectively. Microbial analysis revealed that all of these water samples were not free from pathogens and thereby not suitable for drinking.

Keywords

Water Alkalinity, BOD, Physico-Chemical Analysis

1. Introduction

Water is the most abundant substance found on the surface of the earth, covering roughly three quarters of the earth's surface and it plays a very vital role in existence as well as maintenance of life of organisms. Water related diseases continue to be one of the major health problems globally due to consumption of contaminated water. The high prevalence of diarrhea among children and infants can be traced to the consumption of unsafe water [1].

The most dangerous form of water pollution occurs when fecal contaminant like *Escherichia coli* enters the water supply. Also, the fecal-oral routes of transmission, in which pathogens are shed in human or animal faces. Containmants put into water supply which may perpetuate many diseases examples of such pathogen are *Salmonella spp*, *Shigella spp*, *Vibrio choierae*, *Escherichia coli* [2].

Typhoid fever remains a great socio-economic problem in developing countries, Nigeria inclusive. Perforation of intestines is associated with high mortality with wound infection occurring in 50% - 75% of survivals (Archimpong, 1976). Controlling wound sepsis or wound infection with various complications also affects mortality [3] [4] and unsafe drinking water has been the major source of this infection.

This paper is designed to determine the level of pollution or contamination of water in Molete and Jericho areas of Ibadan through the physico-chemical and microbial analysis of the three main water sources in each area.

The values of the physico-chemical and microbiological parameters were determined through laboratory water quality analysis to know if a particular source of water was suitable for drinking. The water quality of potable water in our towns and cities in Nigeria are often not known. Deviations from water quality standard of the World Health Organization (WHO) and European Economic Community (E.E.C) are contributory to health hazards and problems of most Nigerians. The National Agency for Food and Drug Administration and Control (NAFDAC) is established to regulate and maintain standards of food and drugs (including water) to promote the health of Nigerians. However, the physico-chemical properties of water in a given area need to be determined and established.

2. Materials and Methods

2.1. Sample Collection

A total of six (in triplicates) water samples (well, borehole and public tap) from two different areas in Ibadan were collected to determine the quality of drinking water supply in the metropolis as shown in **Table 1**.

Before water samples were collected, cotton wool soaked in 70% ethanol was used to sterilize the nozzle of the tap-outlet and the water was allowed to run for three minutes. The water samples were then collected into 2 L plastic container after thorough washing and rinsing with the respective water samples. The 2 L plastic container was filled to the brim to avoid air bubbles; the cap was then used to seal it firmly. The bottle was labeled and was kept in a cooler box containing ice, before being sent to the laboratory for analyses. Samples for microbial analysis were collected in a sterilized and phosphate free bottle.

2.2. Method of Analysis

The water quality parameters selected for the study were: pH, temperature, dissolved oxygen, biochemical oxygen demand, conductivity, total hardness and alkalinity. pH, temperature and DO parameters were measured *in-situ*, with an ATI-Orion pH meter, thermometer probe and meter respectively. All the parameters were determined using Standard Methods of Analysis of Water and Wastewater [5].

2.3. Microbiological Assessment

The method for enumeration of total coli form bacteria as recommended by [6] was employed to determine the bacteriological quality of water samples.

Table 1. Sample location.

Sample	Location
W1 (public tap water)	Molete, Ibadan.
W2 (bore hole water)	Molete, Ibadan.
W3 (well water)	Molete, Ibadan.
W4 (public tap water)	Jericho, Ibadan.
W5 (borehole water)	Jericho, Ibadan.
W6 (well water)	Jericho, Ibadan.

3. Result and Discussion

The temperature for the two neighborhoods ranged from 15.00˚C ± 0.07˚C to 18.00˚C ± 0.21˚C. The samples were taken in the morning and the temperature obtained for the samples were quite low as compared to the room temperature. The lowest temperature of 15.00˚C ± 0.07˚C was recorded for borehole water at Molete while the highest value of 18.00˚C ± 0.21˚C was recorded for borehole and well water at Jericho, Ibadan as shown in **Figure 1** below.

High temperature is known to increase the toxicity of some substances such as ammonia, reduce the concentration of dissolved oxygen, increase water acidity and influence the activities of some microorganisms.

The DO values range from 0.4 ± 0.02 mg/l to 4.6 ± 0.02 mg/l, Jericho borehole water recorded the highest value and Molete public tap water had the lowest value as shown in **Figure 2** and in **Table 2** below.

All the water samples in these neighborhoods had values below the 5 mg/l WHO threshold [7] [8] as revealed in **Table 2**.

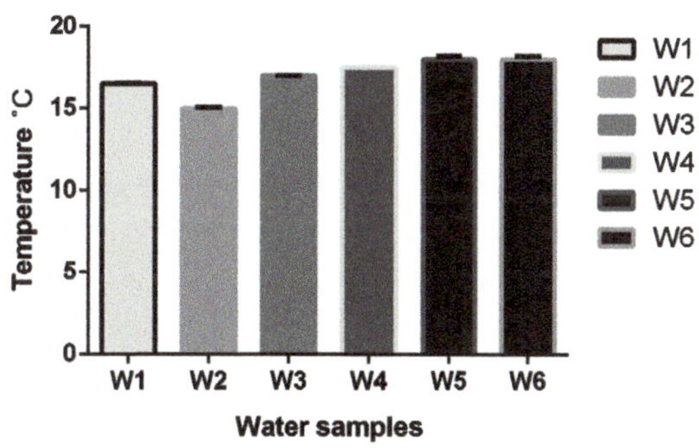

Figure 1. Temperature of the water samples.

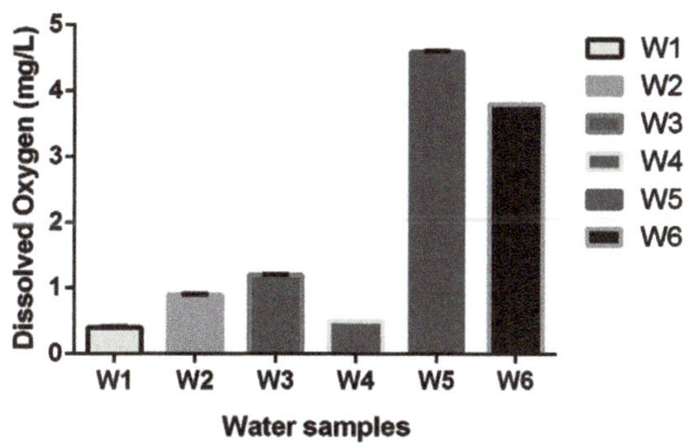

Figure 2. Dissolved oxygen of water samples.

Table 2. Physico-chemical and biological parameters of water samples from molete and jericho, ibadan.

Sample	BOD mg/l	DO mg/l	pH	TEMP ˚C	Hardness mg/ml	Alkalinity (mg/l)	Conductivity mho/cm	Coliform count (Cfu/ml)
W1	24.30 ± 0.08	0.4 ± 0.02	6.60 ± 0.00	16.50 ± 0.02	44.00 ± 1.20	14.88 ± 0.15	460 ± 1.26	$8.5 \times 10^6 \pm 5.23$
W2	22.40 ± 0.06	0.9 ± 0.01	6.41 ± 0.01	15.00 ± 0.07	2.60 ± 0.04	9.25 ± 0.03	290 ± 1.51	$8.0 \times 10^6 \pm 11.14$
W3	18.20 ± 0.35	1.2 ± 0.01	6.87 ± 0.02	17.00 ± 0.01	42.00 ± 0.13	14.64 ± 0.07	320 ± 2.34	$5.4 \times 10^6 \pm 9.67$
W4	26.40 ± 0.72	0.5 ± 0.00	6.84 ± 0.05	17.50 ± 0.00	20.80 ± 0.04	4.03 ± 0.02	230 ± 1.15	$9.8 \times 10^6 \pm 17.81$
W5	12.80 ± 0.22	4.6 ± 0.02	6.88 ± 0.06	18.00 ± 0.21	11.20 ± 0.11	4.03 ± 0.11	290 ± 2.77	$3.0 \times 10^6 \pm 20.21$
W6	15.89 ± 0.13	3.8 ± 0.00	5.88 ± 0.01	18.00 ± 0.20	8.30 ± 0.00	4.26 ± 0.02	300 ± 2.19	$4.2 \times 10^6 \pm 22.10$
WHO	3.0	5	6.5 - 8.5	Ambient	150	---------	1000	0

Results were presented in mean ± s.d for triplicate. BOD: Biochemical Oxygen Demand; DO: Dissolved Oxygen; WHO: World Health Organization.

Adequate dissolved oxygen is necessary for good water quality and the lack of oxygen in body tissues creates a defect of red blood cells. Drinking water with low dissolved oxygen concentration further exacerbates the condition by constricting blood vessels in the lungs. Dissolved oxygen present in drinking water adds taste and it is a highly fluctuating factor in water. Dissolved Oxygen (DO) is a parameter of aeration and inverse pollution. Well aerated water has reduced pollution than a poorly aerated water body. When the DO is high the pollution is low when the DO is low, the pollution level is high. The borehole water samples have high DO than the tap water samples. The Average DO for public tap water at Molete and Jericho was 0.4 ± 0.02 mg/l and 0.5 ± 0.00 mg/l respectively while the average DO for borehole at Jericho was 4.85. It is notable that the borehole at Molete shows rather low DO, similar deviation was observed in the BOD of the Molete well water. This could be due toa nearby septic tank or leaky seepage of industrial effluents.

Water treatment or purification can only be effectively carried out when the water quality is known. The results of BOD and DO of water samples from Molete and Jericho areas of Ibadan have shown that borehole water has lower level of pollution than water from the well and public tap as shown in **Figure 3** below.

The water from the well at Molete has higher pollution level than the well water at Jericho. However the public tap water in Molete has lower level of pollution than that of Jericho. One can trace the higher pollution level of public tap water in Jericho to discharge of industrial wastes, and sewage into the water. Results of BOD analysis shown in **Table 2** and **Figure 3** indicates that BOD is high because any value above 3.00 mg/l is considered high. The borehole water sample is expectedly lower in BOD value than the public tap water and well water because of the depths of the boreholes. However high value of BOD of Borehole water in Molete may be due to poor storage in unclean tank .The public tap water sample from both Molete and Jericho are high because they are drawn from surface (such as stream or rivers) water sources exposed to waste dump and discharge of effluent from the industries which may affect the pH of the water bodies.

The pH of a water body is very important because it may affect the solubility and toxicity of metals in the aquatic system, which may have adverse effects on human health. pH indicates the acidity/alkalinity of the medium. The Ph ranged from 5.88 ± 0.01 to 6.88 ± 0.06 as shown in the **Figure 4** below.

Water with pH lower than 5 may cause problems due to corrosion because many metals become more soluble in low pH waters [9]. The results showed that sample W2, W3, and W6 were slightly acidic while sample W1,

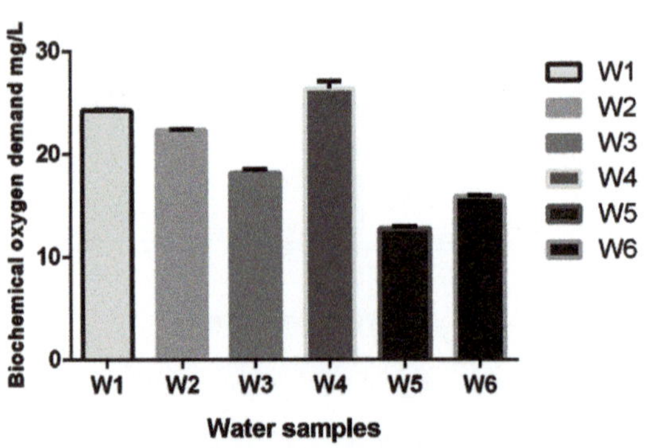

Figure 3. Biochemical oxygen demand of water samples.

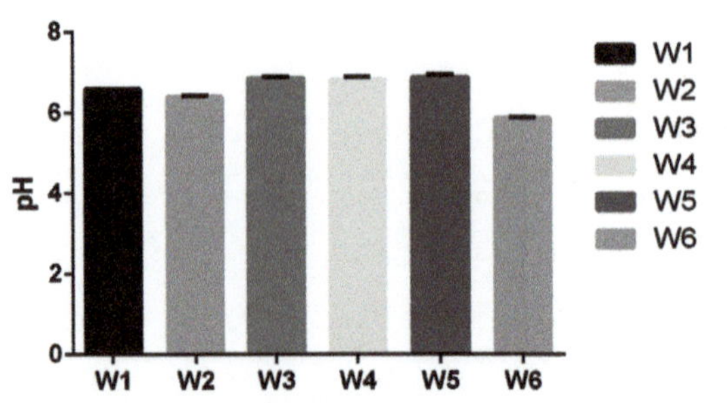

Figure 4. pH of water samples.

W4 and W5 were neutral. A pH value of higher than 8.5 indicates that a significant amount of sodium bicarbonate may be present in the water which contributes to the hardness of the water.

Water hardness occurs as a result of the quantity of calcium, magnesium and carbonates present in the water sample. Generally the hardness value of both Molete and Jericho water samples can be classified as soft based on **Table 2** and **Figure 5**.

The water in these areas are good for drinking and laundry based on the hardness values as lather formation with soap is not hindered or prevented.

The coliform count recorded in this study falls within the slightly polluted limits [9] on the pollution of drinking water. Detection of coliform shows the danger of faecal pollution and the consequent hazard of contracting disease through pathogenic (**Figure 6**).

Organisms disease causing organisms transmitted via drinking water are predominantly of faecal origin, but the coliform count would not constitute much concern without the detection of *E. coli* in the water samples. It has been reported that typical enteropathogenic *Escherichia coli* is a leading cause of the infantile diarrhoea in developing countries, whereas this is rare in industrialized country [10].

4. Recommendation

All the water samples are recommended for both micro-filtration and filtration through activated carbon surface to remove dissolved substances and obnoxious smell originating from decomposition of organic wastes respectively.

5. Conclusion

Water from both Molete and Jericho requires some varying degrees of treatment or purification to make it fit for human consumption. High BOD indicates bacteriological activity demanding oxygen for biodegradation of waste or effluents in water. Therefore treatment requirement is targeted toward reduction of bacteriological activity demanding oxygen for biodegradation of waste or effluents in water. The treatment requirement is targeted toward reduction of bacteria (*E. coli* or Coliform) count.

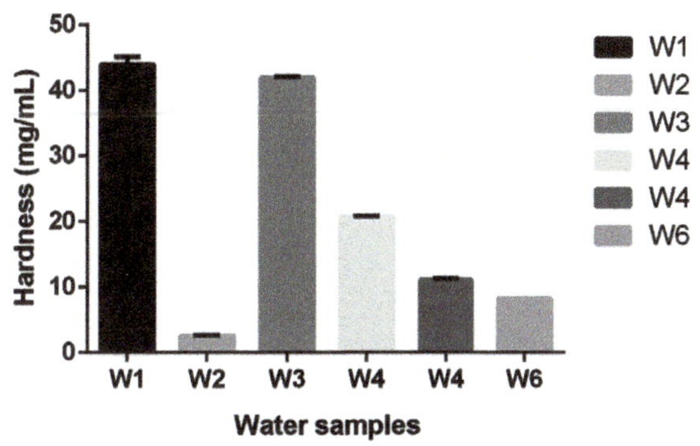

Figure 5. Water Hardness of water samples.

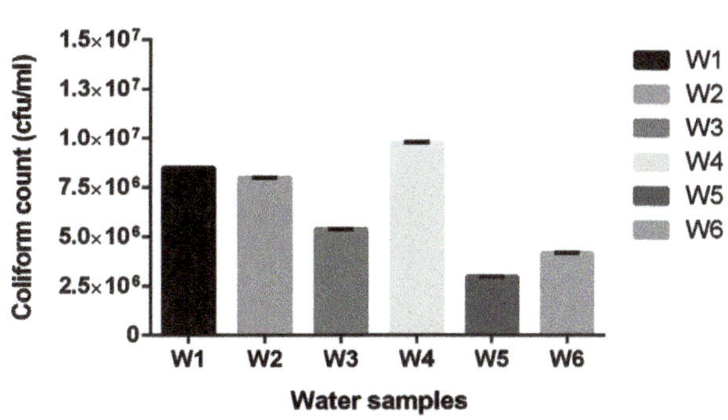

Figure 6. Coliform countof water samples.

The level of bacterial concentration should be reduced by passing the water through ultra-violet (UV) radiations before consumption and also chlorination.

References

[1] Tortora, G.J., Funke, R.B. and Case, L.C. (2002) Microbiology: An Introduction. Media update of 7th Edition, Including Bibliography and Index Publisher Daryl Fox, 258-260.

[2] McLellan, S.L., Daniels, A.D. and Salmore, A.K. (2001) Clonal Populatioms of Thermotolerant *Enterobacteriaceae* in Recreational Water and their Potential Interference with Fecal *Escherichia coli* Count. *Applied and Environmental Microbiology*, **67**, 4934-4938. http://dx.doi.org/10.1128/AEM.67.10.4934-4938.2001

[3] Ajao, O.G. (1982) Typhoid Perforation: Factors Affecting Mortality and Morbidity. *International Surgery*, **67**, 317-319.

[4] Badejo, O.A. and Arigbabu, A.O. (1980) Operative Treatment of Typhoid Perforation and Peritoneal Irrigation: A Comparative Study. *Gut*, **21**, 141-145.

[5] American Public Health Association (APHA) (1999) Standard Methods for the Examination of Water and Wastewater, APHA, Washington DC.

[6] American Public Health Association (APHA) (1998) Standard Methods for the Examination of Water and Wastewater, 20th Edition, APHA, Washington DC.

[7] WHO (2008) Guidelines for drinking-Water Quality [Electronic Resource]: Incorporating 1st and 2nd addenda, Vol.1, Recommendations. 3rd Edition, WHO, Geneva.

[8] UNESCO/WHO/UNEP (1996) Water Quality Assessments—A Guide to Use of Biota, Sediments and Water inEnvironmental Monitoring—Second Edition.

[9] Ajibade, L.T. and Ibrahim, S.I. Assessment of Water Quality and Domestic Uses in Medium-Sized Towns of Niger state, Nigeria (Unpublished).

[10] Trabuisi, L.R., Keller, R. and Tardelli-Gromes, T.A. (2002). Typical and Atypical Enteropathogenic *Escherichia coli*. Emerging *Infectious Diseases*, **8**, 508-513. http://dx.doi.org/10.3201/eid0805.010385

24

Monolayers Langmuir-Blodgett Films of Synthetic Artificial Mimic Molecules That Resemble the Following Tetraether Lipids on Silicon Wafers

Sri Vidawati[1], Ulrich Rothe[2]

[1]Department of Physics, Faculty of Natural Science and Mathematics, University Indonesia, Depok, Indonesia
[2]Institute of Physiological Chemistry, Faculty of Medicine, University Halle-Wittenberg, Hollystr, Halle/Saale, Germany
Email: sri.vidawati@ui.ac.id

Abstract

This study investigated the behavior and molecular organization of synthetic artificial mimic molecules that resemble the following tetraether lipids: di-O-hexadecyl-glycero-3-phosphatidyl-glycerol (DHGPG) and bis-4-dodecylphenyl-12-phosphate. These molecules were analyzed using Langmuir film balance, ellipsometry and atomic force microscopy. The monolayer Langmuir-Blodgett films of DHGPG and bis-4-dodecylphenyl-12-phosphate were stable on the solid surface silicon wafers. The ellipsometry and AFM results showed that monolayers Langmuir-Blodgett films of DHGPG and bis-4-dodecylphenyl-12-phosphate were present, and the thickness of the observed films varied from 1.2 - 5.0 nm.

Keywords

Di-O-Hexadecyl-Glycero-3-Phosphatidyl-Glycerol (DHGPG), Bis-4-Dodecylphenyl-12-Phosphate, Langmuir-Blodgett Film, Ellipsometry, Atomic Force Microscopy (AFM)

1. Introduction

The molecular organization of Langmuir-Blodgett films on the solid surface has applications in a variety of fields in nanotechnology [1], materials science [2] and biochemistry [3]. Monolayer/bilayer of bolaamphiphiles have attracted people's interest for several years because these films provide a key model for understanding the molecular orientation and packing mode of bipolar molecules at an air-water interface. These natural bolaamphiphiles and resembling synthetic analogues can made lipid membranes that are stable under some conditions. The extremely stable biomolecules manufactured by organisms from extreme environments have significant scientific and engineering potential for applications as novel lubricants and monolayer/bilayer lipid sensor devices.

The preparation of arranged films using Langmuir-Blodgett technology is established on the organic molecules, such as lipids or phospholipids, to form a monolayer at the air-water interfaces. The stability and homogeneity of monolayer film is a precondition for the formation of organized lipid Langmuir-Blodgett films with excellent structural integrity. Langmuir-Blodgett film methods have significant advantages in the study of the air-water and as air-solid interfacial behavior of thin solid films. Irwing Langmuir established the first systematic studies of floating surfactant monolayer. However, the first detailed description of sequential monolayer transfer was developed several years later by Katherine Blodgett. These built-up thin films are therefore known as Langmuir-Blodgett films. The most important characteristic of the monolayer properties of bolaamphiphiles is performed by measuring the surface pressure as a function of the area of the water surface, which is performed at constant temperature and is known as the surface pressure-area isotherm.

Compressed and expanded films formed from bolaamphiphiles at an air-water interface have been previously investigated by Langmuir-Blodgett film equipment. The dimensions of lipids are sufficient to form amonolayer/bilayer with a thickness similar to that of common bilayer membranes [4]. Synthetic single-chain bolaamphiphiles are relatively flexible and thus preferentially assume a horseshoe configuration at an air-water interface, at which both hydrophilic head groups are hydrated in the water phase [5]. These configurations (horseshoe and upright standing configurations) were confirmed by computerized calculations by Gabriel *et al.*

This paper presents, we investigated the behavior and properties of monolayer Langmuir-Blodgett films of the synthetic artificial mimic molecules that resemble the tetraether lipids di-O-hexadecyl-glycero-3-phospha-tidyl-glycerol and bis-4-dodecylphenyl-12-phosphate using AFM and ellipsometry. Di-O-hexadecyl-glycero-3-phosphatidyl-glycerol is a molecule containing small-length tetraether lipids; the other side of this molecule contains half tetraether lipids because the molecule consists of only diether lipids. Bis-4-dodecylphenyl-12-phosphate is a molecule of the same length and stiffness, with both ends of a polar head group acting as a natural tetraether lipid; the other side of this molecule contains only half of the tetraether lipid molecule because these molecules contain only a single-chain hydrocarbon. The chemical structures of di-O-hexadecyl-glycero-3-phosphatidyl-glycerol and bis-4-dodecylphenyl-12-phosphate are shown in **Figure 1**.

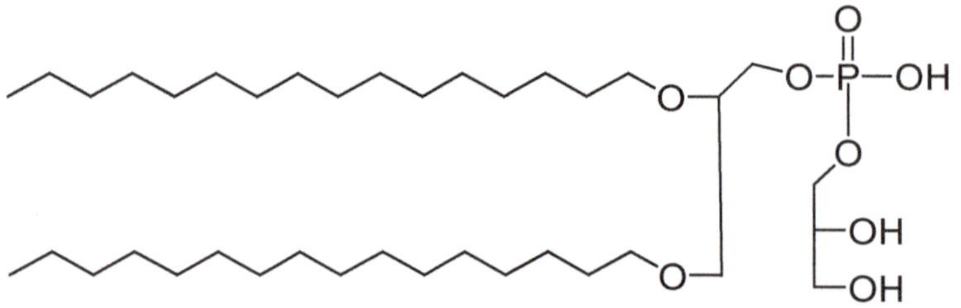

di-O-hexadecyl-glycero-3-phosphatidyl-glycerol (DHGPG)

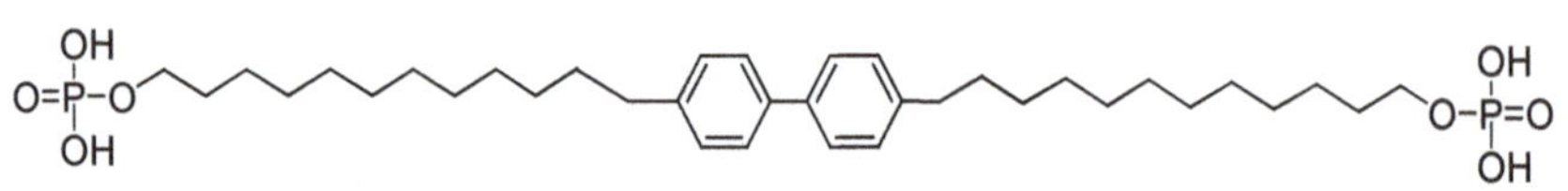

bis-4-dodecylphenyl-12-phosphate

Figure 1. Chemical structure of the synthetic artificial mimic molecules that resemble the following tetraether lipids: di-O-hexadecyl-glycero-3-phosphatidyl-glycerol (DHGPG) and bis-4-dodecylphenyl-12-phosphate.

2. Materials and Methods

2.1. Materials

For all of these experiments, lipids from the archaea *Sulfolobus acidocaldarius* were used. The lipids were prepared from dried bacteria in accordance with the method described by Du Plesis, *et al.* (2008) [6]. The Langmuir-Blodgett film experiments were performed on di-O-hexadecylglycero-3-phosphatidyl-glycerol and bis-4-dodecylphenyl-12-phosphate of the highest purity. All other chemicals were purchased from Merck (Damstadt) or Sigma (Deisenhofen).

2.2. Langmuir-Blodgett Film Methods

The Langmuir film investigations were conducted using a commercial film balance (R & K GmbH, Mainz, Germany) equipped with a rectangular thermostated Teflon trough (area of 418 cm^2, depth of 1 cm). The surface pressure was measured using a filter paper Wilhelmy balance. For the film balance experiments, the lipids were dissolved in 2:1 (v: v) chloroform/methanol at a concentration of 1.0 mmol/L. The subphase was Milli Q pure water (18.2 MΩ, pH 5.6). All of the experiments were carried out at a temperature of 19˚C ± 1˚C. The lipid solution in chloroform was spread onto the subphase.

A 50 µl of 1 m MDHGPG and 70 µl of 1 mM bis-4-dodecylphenyl-12-phosphate were used for each isotherm experiment. After spreading, the lipid films were equilibrated without surface pressure before starting the measurements. All of the films were compressed with a constant velocity of 0.045 $cm^2 \cdot s^{-1}$ (2.5×10^{-2} $nm^2 \cdot s^{-1}$ per molecule). Y-type transfer (on the upward stroke) can be used on either hydrophilic or aminosilanized substrates with a reversed head-tail arrangement, whereas X-type transfer (on the downward stroke) is better on hydrophobic substrates.

2.3. Silicon Wafer

Silicon wafers were used as the substrate for all of the experiments. Silicon wafers were cut to dimensions of 1 cm × 1 cm. We used the Piranha method to clean the silicon wafer substrates. Piranha solution is a mixture of concentrated sulfuric acid and hydrogen peroxide. The silicon wafers were cleaned by immersion in 7:3 concentrated H_2SO_4: H_2O_2 for 20 min and then rinsed with pure water. Thereafter, the silicon wafers were dried in a N_2 gas flow.

Hydrophilic surfaces on the silicon wafers were prepared using a UV lamp (Xeradex, Lampenwerk Wipperfürth, Germany) for 5 minutes to activate the silicon wafer surface. From this method, the water contact angle changes from 48˚ to <20˚ (unmodified silicon wafers).

Hydrophobic surfaces on the silicon wafers were achieved by hydrophobic silanization. The silicon wafer surfaces were activated using hydrophobization with 20 µl of dimethyl C18-chlorosilane, which was placed on the siliconwafer, and the siliconwafer was then covered with a second siliconwafer that was flipped upside down. These siliconwafers were maintained at 70˚C overnight and then washed with chloroform in an ultrasonic bath. By this method, the layer thickness was approximately 1.5 nm, and the contact angle changed to approximately 90˚.

Amino-silanized surfaces on the silicon wafers were prepared using a UV lamp (Xeradex, Lampenwerk Wipperfürth, Germany) for 5 minutes, and the silicon wafers were then submerged in 1% aminopropyldimethylethoxysilane in chloroform at 70˚C overnight and washed with chloroform in an ultrasonic bath several times. The amino-silanized substrates had a layer thickness of approximately 1 nm, and the contact angle changed to approximately 70˚.

2.4. Ellipsometry Measurements

The thicknesses of all of the monolayer Langmuir-Blodgett films were characterized by ellipsometry. The lipids were analyzed after transfer on the various modified silicon wafer. The ellipsometry measurements were performed with a mapping single wavelength ellipsometer SE-400 (Sentech Instruments, Berlin, Germany). Such the light source, a He/Ne laser with a wavelength of l = 623.8 nm was used. The angle of incidence was 60˚. For the measurements, silicon wafers (1 cm × 1 cm) with a natural oxide layer (thickness 2.6 - 2.9 nm) and optical parameters of ns = 3.828 and ks = −i0.200 were used. The refractive index of the tetraether lipids was established to be nu = 1.58. As a special scanning technique, the line scan method was used for the present measurements. The surface was scanned in a single line from the pure silicon oxide layer to the lipid layer. The distance between the measurement points was 200 µm, and the measured area of one detected point was 250 μm^2. For the correct determination of the mean film thickness, an area of 1 mm^2 in the middle of the lipid-covered silicon wafer was scanned by mapping ellipsometry. The ellipsometry data yielded the mean film thickness over a largespot area of 250 μm^2 per scanned point. The presented values were computed according to the statistical standard deviation method.

2.5. Atomic Force Microscopy Measurements

The surface morphology of the monolayer film was analyzed using Atomic force microscopy. Atomic force microscopy was accomplished using a NanoWizard apparatus (JPK instruments, Berlin, Germany) under atmospheric conditions. Commercial pyramidal Si_3N_4 tips (NSC16AlBS, Micromasch, Estonia) mounted on a cantilever with a length of 230 µm, a resonance frequency of approximately 170 kHz and a nominal force constant of approximately 40 N/m were used.

The measurements were accomplished intermittently to protect sample damage. The scan size and scan frequency were using from 0.5 to 1.5 Hz. AFM image characterization was accomplished by showing the signal height in the retrace direction (512 × 512 pixels).

3. Results

3.1. Langmuir Films

The phenomena of the compression-decompression cycles of a Langmuir film at an air-water interface provide detailed information on film stability, the thermodynamics and the kinetics of film formation [7]. The pressure vs. area/molecule (π/A) isotherms for the compression-decompression cycles of the monolayer Langmuir films of the synthetic artificial mimic molecules that resemble tetraether lipids DHGPG are shown in (**Figure 2**). The dynamic compression-decompression cycles of the monolayer Langmuir film of DHGPG were measured of concentrations 50 µl at 1 mM.

As shown in **Figure 2**, the first compression-decompression cycle of the monolayer Langmuir film of DHGPG (50 µl of 1 mM) was compressed at a surface pressure coming the collapse pressure, and then decompressed to a zero surface pressure, obtaining a hysteresis-curve. After the first cycle, the lipid films were equilibrated without a surface pressure for 1 h before a second cycle was started. The hysteresis-curve of the monolayer Langmuir films (50 µl of 1 mM DHGPG) were accomplished after the second cycles, and the resulting hysteresis-curves showed information for the thermodynamic and the stability of the monolayer DHGPG films [7]-[9].

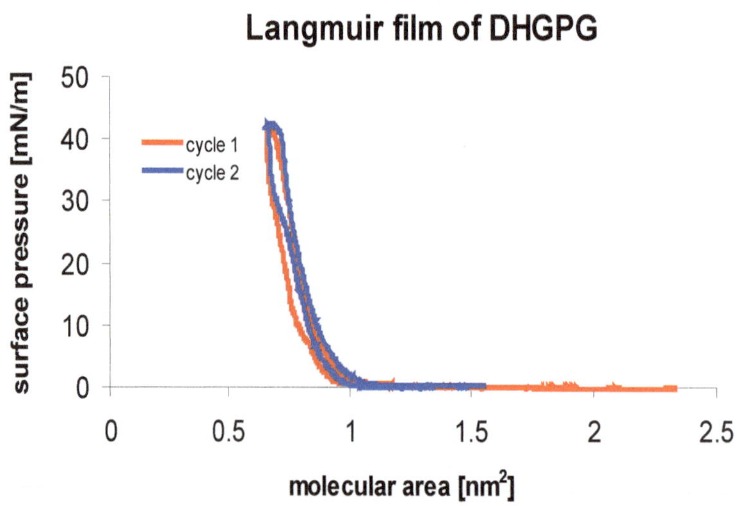

Figure 2. Representative surface pressure-area isotherms for the compression decompression cycles 1 and 2 of the monolayer Langmuir film of DHGPG at 19°C and pH 5.5.

The stability of the monolayer Langmuir film of bis-4-dodecylphenyl-12-phosphate was low (data not shown). It was not possible to compress and decompress the film. However, we attempted to transfer the monolayer Langmuir film of bis-4-dodecylphenyl-12-phosphate (70 µl of 1 mM) to a solid substrate on modified silicon wafers.

Surface isotherms are usually used to describe the strength and stability of Langmuir-Blodgett films. All of the transfer experiments (Y-type and X-type) of the Langmuir-Blodgett films were accomplished immediately before the collapse pressure was achieved. The X-type film transfers method was used for hydrophobically modified surfaces. Y-type film transfers method was used for hydrophilically modified and amino-silanized modified surfaces.

3.2. Ellipsometry

The thicknesses of all of the monolayer Langmuir-Blodgett films were measured by ellipsometry. Ellipsometry data gives the mean film thickness over a large spot area of 250 μm^2 and therefore cannot separate between films with different morphological structures. The Langmuir-Blodgett films were observed on various modified silicon wafer, and different thickness values were produced.

The thickness of the monolayer Langmuir-Blodgett films of DHGPG on the hydrophilic modified silicon wafer and on the hydrophobic modified silicon wafer showed thickness values were only 1.2 nm and 3.0 nm, respectively [8]-[10].

The thickness of the monolayer Langmuir-Blodgett films of bis-4-dodecylphenyl-12-phosphate gave values between 3.5 nm for the amino-silanized modified silicon wafer and 3.5 nm for the hydrophobic modified silicon wafer [8]-[10].

3.3. Atomic Force Microscopy

AFM is used for studied and analyzed the surface morphology and thickness of the monolayer Langmuir-Blodgett films of DHGPG and bis-4-dodecylphenyl-12-phosphate on the molecular scale. Typical AFM images of monolayer DHGPG films after transfer on the hydrophobic and hydrophilic modified silicon wafers are displayed in **Figure 3(a)** and **Figure 3(b)**. From the AFM images, the thicknesses of the monolayer Langmuir-Blodgett films of DHGPG on the hydrophilic of modified silicon wafer (Y-type transfer) and hydrophobic of modified silicon wafer (X-type transfer) were found to be approximately 4.0 - 5.0 nm [8] [9].

The monolayer Langmuir-Blodgett films of bis-4-dodecylphenyl-12-phosphate on hydrophobic of modified silicon wafer (X-type transfer) and amino-silanised of modified silicon wafer (Y-type transfer) were evaluated by AFM (see **Figure 4(a)** and **Figure 4(b)**). The thicknesses of the monolayer Langmuir-Blodgett films of bis-4-dodecylphenyl-12-phosphate were 4.0 - 5.0 nm [8] [9]. All of the monolayer Langmuir-Blodgett film information was for an upright standing configuration [5] [11].

4. Discussion

The study of monolayer/bilayer formation is an interesting branch of physical chemistry that has applications in many fields of science and technology. A thorough understanding of monolayer/bilayer behavior is important for utilizing the Langmuir-Blodgett method, which implicates the position of a film on a water surface with following transfer onto a solid substrate.

The surface isotherms are usually used to describe the strength and stability of Langmuir-Blodgett films. When compression-decompression cycles where measured; statements about the film stability and the kinetics of the film formation are available.

The Langmuir film stability of DHGPG at the air-water interface is shown on a hysteresis-curve from the surface pressure-area (π-A) isotherms (**Figure 2**). A hysteresis loop was performed for 50 µl of 1 mM concentration of lipid solutions. The concentration of DHGPG is also an important parameter for obtaining the stability of the Langmuir films. The dynamic compression-decompression isotherm cycles of the synthetic artificial mimic molecules that resemble the following tetraether lipids: DHGPG Langmuir films provide complete information on the formation, molecular area, stability, collapse behavior, phase transitions, thermodynamics, and kinetics of film formation [7] [9].

The pressure versus area/molecule (π/A) isotherms were recorded at 19˚C ± 1˚C and pH 5.5 for 50 µl of 1 mM concentration of DHGPG; the results are shown in **Figure 2**. DHGPG (50 µl of 1 mM) had a collapse film pressure of approximately 41 mN/m and a molecular area from 0.60 nm^2 to 0.75 nm^2.

The compression and decompression cycles can be essential to reach consistent isotherms stability of the Langmuir films. Highly stable Langmuir films are necessary for finding out the quality of the Langmuir-Blodgett monolayer. **Figure 2** shows the Langmuir films of 50 µl of 1 mM DHGPG. The first cycle had the highest surface pressure at approximately 41 mN/m and a molecular area from 0.60 nm^2 to 0.75 nm^2. The second cycle also displayed the most stable surface pressure of the Langmuir film of DHGPG of 41 mN/m and a molecular area that was almost equal to that observed in the first cycle (between 0.60 nm^2 and 0.75 nm^2). As shown in **Figure 2**, the curves show the thermodynamic features and the excellent stability of the Langmuir film of DHGPG at an air-water interface. The floating monolayer stability is a significant parameter for obtaining very-high-quality Langmuir-Blodgett films. Some efforts have been done to produce high-quality Langmuir-Blodgett films of bolaamphiphiles [8]-[10] [12]-[16].

The data characterized by ellipsometry for 50 µl of 1 mM DHGPG had a high collapse pressure of 41mN/m and a molecular area approximately 0.60 nm^2. The mean thickness of the film was 3.5 nm after hydrophobic transfer (X-type), which suggested an upright standing conformation, whereas a mean thickness of the film of only 1.2 nm was measured after hydrophilic transfer (Y-type), which suggested a horse shoe conformation. One could hypothesize that the existence of the films on the solid substrate would be expected after hydrophobic and hydrophilic transfer.

The thickness of the monolayer Langmuir-Blodgett films of 70 µl of 1 mM bis-4-dodecylphenyl-12-phosphate was measured by ellipsometry. The ellipsometry measurements after transfer Langmuir monolayer film on the hydrophobic surface (X-type transfer) yielded a film height of approximately 3.5 nm, which suggested an upright standing conformation. The thickness of the Langmuir monolayer film transferred onto amino-silanized surface (Y-type transfer) was approximately 3.5 nm, which also indicative an upright standing conformation

Ellipsometry, as a mapping technique, will cover a spot area of approximately 250 μm^2. The ellipsometry data for the monolayer Langmuir-Blodgett films of bis-4-dodecylphenyl-12-phosphate transferred onto hydrophobic and amino-silanized of modified silicon wafers were appropriate with an upright standing confirmation, and the monolayer Langmuir-Blodgett films of DHGPG transferred onto hydrophobic and hydrophilic of modified silicon wafers varied in thickness between 1.2 nm and 3.5 nm.

The AFM results (**Figure 3(a)** and **Figure 3(b)**) for the monolayer Langmuir-Blodgett film of 50 µl of 1 m MDHGPG revealed a relatively homogeneous lipid layer. On the hydrophilic (Y-type transfer) surface of modified silicon wafer, the film had a thickness of approximately 4.0 - 5.0 nm, which suggested an upright stand- ing conformation, whereas 50 µl of 1 mM DHGPG on a hydrophobic surface (X-type transfer) looked like a collection of small flat islands with a thickness of approximately 4.0 - 5.0 nm, which also suggested an upright standing conformation. These results suggest that the monolayer Langmuir-Blodgett films of 50 µl of 1 mM DHGPG on the hydrophilic and hydrophobic surfaces of modified silicon wafers were stable.

Based on the AFM characterization (**Figure 4(a)** and **Figure 4(b)**), the monolayer Langmuir-Blodgett film of 70 µl of 1 mM bis-4-dodecylphenyl-12-phosphate on the amino-silanized surface of modified silicon wafer (Y type transfer) was stable with a thickness of approximately 4.0 - 5.0 nm. The AFM images of the monolayer Langmuir-Blodgett films show some islands and domains. On the hydrophobic surface of modified silicon wafer (X-type transfer), the film of 70 µl of 1 mM bis-4-dodecylphenyl-12-phosphate had a thickness of approximately 3.0 - 4.0 nm, indicating an upright standing configuration.

5. Conclusions

Monolayer Langmuir-Blodgett film studies of di-O-hexadecyl-glycero-3-phosphatidyl-glycerol and bis-4-dodecylphenyl-12-phosphate on silicon wafers are presented in this manuscript. The results show that monolayer films of DHGPG and bis-4-dodecylphenyl-12-phosphate are present, and the thicknesses of the observed films vary between 1.2 and 5.0 nm.

By summarizing our data, we show that the relatively stiff synthetic artificial mimic molecules that resemble the tetraether lipids di-O-hexadecyl-glycero-3-phosphatidyl-glycerol and bis-4-dodecylphenyl-12-phosphate can adopt several conformational supra molecular arrangements. All of the monolayer Langmuir-Blodgett films of di-O-hexadecyl-glycero-3-phosphatidyl-glycerol and bis-4-dodecylphenyl-12-phosphate are stable on a solid surface. Further study is necessary to enable the optimization of monolayer films for biomedical applications.

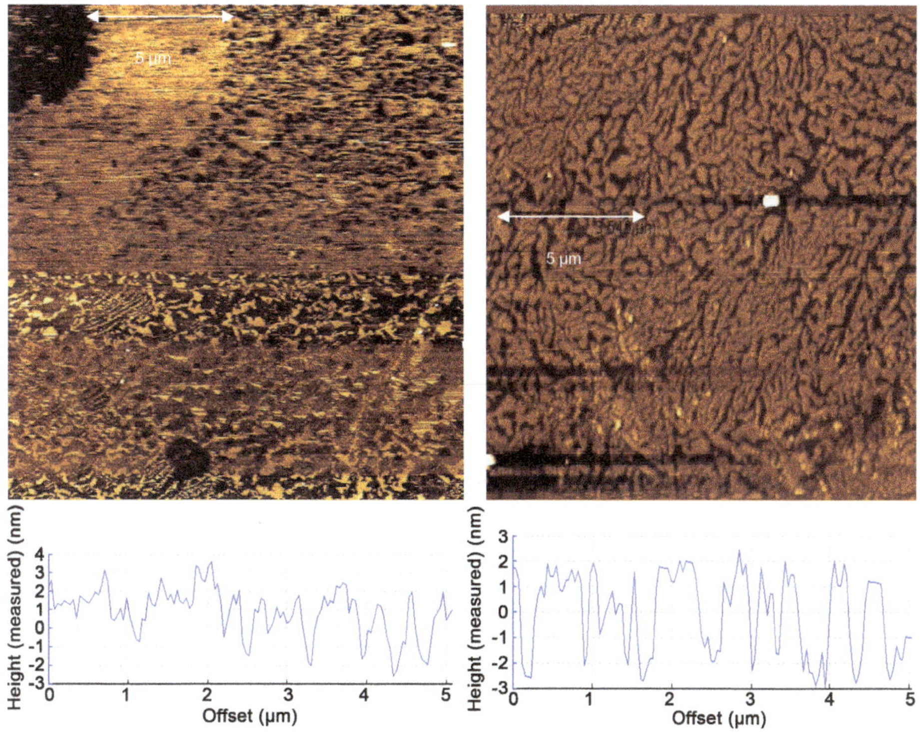

(a) (b)

Figure 3. AFM images of the monolayer Langmuir-Blodgett films of DHGPG. (a) DHGPG hydrophilic surface (Y-type transfer); (b) DHGPG hydrophobic surface (X-type transfer).

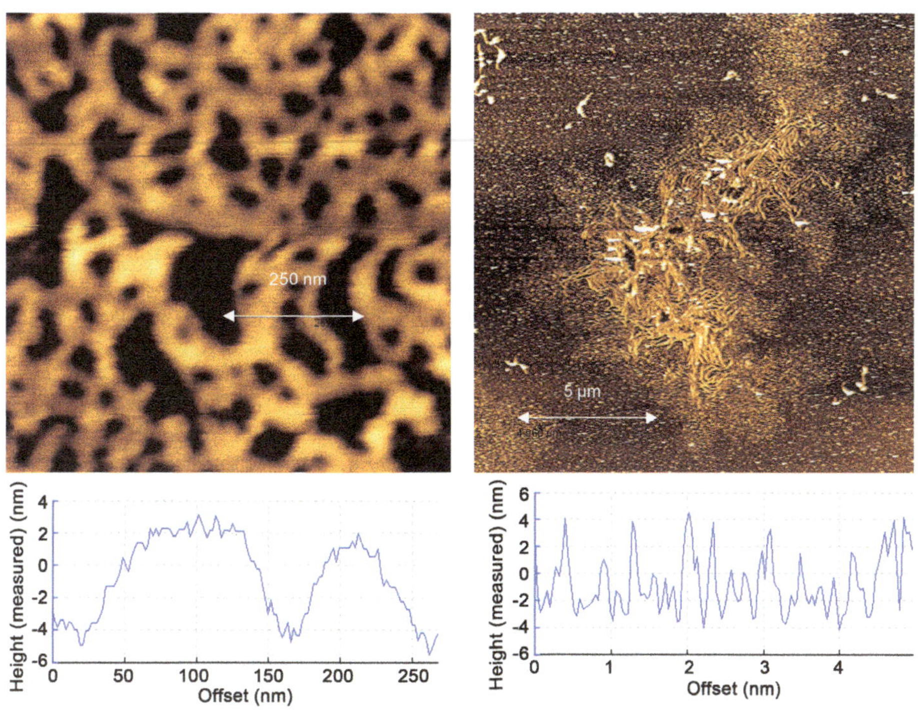

(a) (b)

Figure 4. AFM images of the monolayer Langmuir-Blodgett films of bis-4-dodecylphenyl-12-phosphate: (a) bis-4-dodecylphenyl-12-phosphate amino-silanized surface (Y-type transfer) and (b) bis-4-dodecylphenyl-12-phosphate hydrophobic surface (X-type transfer).

Acknowledgments

This study was supported by the Deutscher Akademischer Austauschdienst (DAAD, SV).

References

[1] Weissbuch, I., Baxter, P.N.W., Cohen, S., Cohen, H., Kjaer, K., Howes, P.B., Als-Nielsen, J., Hanan, G.S., Schubert, U.S., Lehn, J.-M., Leiserowitz, L. and Lahav, M.J. (1998) Self-Assembly at the Air-Water Interface. *In-Situ* Preparation of Thin Films of Metal Ion Grid Architectures. *Journal of the American Chemistry Society*, **120**, 4850-4860. http://dx.doi.org/10.1021/ja980205w

[2] Swift, J.A., Pivovar, A.M., Reynolds, A.M. and Ward, M.D. (1998) Template-Directed Architectural Isomerism of Open Molecular Frameworks: Engineering of Crystalline Clathrates. *Journal of the American Chemistry Society*, **120**, 5887-5894. http://dx.doi.org/10.1021/ja980793d

[3] Sengupta, K., Schilling, J., Marx, S., Fischer, M., Bacher, A. and Sackmann, E. (2003) Mimicking Tissue Surfaces by Supported Membrane Coupled Ultrathin Layer of Hyaluronic Acid. *Langmuir*, **19**, 1775-1781. http://dx.doi.org/10.1021/la026146o

[4] Stern, J., Freisleben, H.J., Janku, S. and Ring, K. (1992) Black Lipid Membranes of Tetraether Lipids from *Thermoplasma acidophilum*. *Biochimica et Biophysica Acta*, **1128**, 227-236. http://dx.doi.org/10.1016/0005-2760(92)90312-J

[5] Fuhrhop, J., David, H.H., Mathieu, J., Liman, U., Winter, H.J. and Boekema, E.J. (1986) Bolaamphiphiles and Monolayer Lipid Membranes Made from 1,6,19,24-Tetraoxa-3,21-cyclohexatriacontadiene-2,5,20,23-tetrone. *Journal of the American Chemistry Society*, **108**, 1785-1791. http://dx.doi.org/10.1021/ja00268a013

[6] Bode, M.L., Buddoo, S.R., Minnaar, S.H. and Du Plessis, C.A. (2008) Extraction, Isolation and NMR Data of the Tetraether Lipid Calditoglycerocaldarchaeol (GDNT) from *Sulfolobus metallicus* Harvested from a Bioleaching Reactor. *Chemistry and Physics of Lipids*, **154**, 94-104. http://dx.doi.org/10.1016/j.chemphyslip.2008.02.005

[7] Vidawati, S., Rothe, U. and Bakowsky, U. (2012) Langmuir Films Stability Phenomenon of Glycerol Dialkyl Nonitol Tetraether at the Air-Water Interface for Variations Spreading Time. *Advances in Biological Chemistry*, **2**, 233-237. http://dx.doi.org/10.4236/abc.2012.23028

[8] Vidawati, S., Sitterberg, J., Bakowsky, U. and Rothe, U. (2010) AFM and Ellipsometric Studies on LB Films of Natural Asymmetric and Symmetric Bolaamphiphilic Archaebacterial Tetraether Lipids on Silicon Wafers. *Colloids and Surfaces B*: *Biointerfaces*, **78**, 303-309. http://dx.doi.org/10.1016/j.colsurfb.2010.03.015

[9] Vidawati, S., Sitterberg, J., Rothe, U. and Bakowsky, U. (2011) Stability of Monomolecular Films of Archaebacterial Tetraether Lipids on Silicon Wafers: A Comparison of Physisorbed and Chemisorbed Monolayers. *Colloids and Surfaces B*: *Biointerfaces*, **87**, 209-216. http://dx.doi.org/10.1016/j.colsurfb.2011.05.005

[10] Bakowsky, U., Rothe, U., Antonopoulos, E., Martini, T., Henkel, L. and Freisleben, H.J. (2000) Monomolecular Organization of the Main Tetraether Lipid from *Thermoplasma acidophilum* at the Water-Air Interface. *Chemistry and Physics of Lipids*, **105**, 31-42. http://dx.doi.org/10.1016/S0009-3084(99)00131-0

[11] Gabriel, J.L., Chong, P.L.G. (2000) Molecular Modeling of Archaebacterial Bipolar Tetraether Lipid Membranes. *Chemistry and Physics of Lipids*, **105**, 193-200. http://dx.doi.org/10.1016/S0009-3084(00)00126-2

[12] Rolandi, R., Schindler, H., De Rosa, M. and Gambacorta, A. (1986) Monolayers of Ether Lipids from Archaebacteria, *European Biophysics Journal*, **14**, 19-27. http://dx.doi.org/10.1007/BF00260399

[13] Dote, J.L., Barger, W.R., Behroozi, F., Chang, E.L., Lo, S.L., Montague, C.E. and Nagumo, M. (1990) Monomolecular Film Behavior of Tetraether Lipids from a Thermoacidophilic Archaebacterium at the Air/Water Interface. *Langmuir*, **6**, 1017-1023. http://dx.doi.org/10.1021/la00095a023

[14] Elferinck, M.G.L., De Wit, J.G., Demel, R., Driessen, A.J.M. and Konings, W.N.J. (1992) Functional Reconstitution of Membrane Proteins in Monolayer Liposomes from Bipolar Lipids of *Sulfolobus acidocaldarius*. *Journal of Biological Chemistry*, **267**, 1375-1381.

[15] De Rosa, M. (1996) Archaeal Lipids: Structural Features and Supramolecular Organization. *Thin Solid Films*, **284-285**, 13-17. http://dx.doi.org/10.1016/S0040-6090(96)08832-3

[16] Dante, S., De Rosa, M., Maccioni, B., Morana, A., Nicolini, C., Rustichelli, F., Troitsky, V.I. and Yang, B. (1995) Thermal, stability of bipolar lipid Langmuir Blodgett films by X-Ray diffraction. *Molecular Crystals and Liquid Crystals Science and Technology*, **262**, 191-207. http://dx.doi.org/10.1080/10587259508033525

The Profile of Secondary Metabolites of Sponge *Clathria reinwardtii* Extract as a Result of Fe Accumulation in Spermonde Archipelago

Lydia Melawaty[1], Kristiana Pasau[2]

[1]Chemical Engineering Department, Paul Christian University of Indonesia, Makassar, Indonesia
[2]Mechanical Engineering Department, Paul Christian University of Indonesia, Makassar, Indonesia
Email: lylamety@yahoo.com, kristianapasau@yahoo.co.id

Abstract

Study on the effect of Fe metal concentrations in Spermonde Archipelago on secondary metabolites production from sponge was conducted. The percent of peak area and *m/z* from GC-MS at t_R = 13.555 were affected by accumulated Fe in *Chlathria reinwardtii*, where at high Fe concentration a lower percent peak area was produced compared to *Clathria reinwardtii* extract at low Fe concentration. Chromatogram CR-LL-DCM produced 2 peaks at t_R = 21.401 and 25.174 which were not observed in CR-BC-DCM chromatogram. This indicated that *Clathria reinwardtii* in Lae-Lae island produced metabolite secondary compounds that were not produced by *Clahtria reinwardtii* in Barrang Caddi island. FT-IR analysis indicated that *Clathria reinwardtii* accumulating Fe with highest and lowest concentration produced strong peaks at wavenumber of about 3400 cm^{-1} for O-H and N-H stretches. O-H pliable and C-O stretch peaks at wavenumber of about 1200 cm^{-1} had changes from medium intensity at low Fe concentration (CR-BC-DCM) to weak at high Fe concentration (CR-LL-DCM).

Keywords

***Clathria reinwardtii*, Fe, GC-MS, FTIR**

1. Introduction

The water environment of Spermonde Archipelago is an abundant source of bioactive compounds. Bioactive compounds from marine environment are primarily in the form of secondary metabolite compounds with great potential to be developed as medicine materials. Sponges produce various bioactive secondary metabolite chemical compounds. Specific chemical compounds are produced to defend themselves from predators, to bind their soft body, and settled. Secondary metabolite compounds are natural chemical substances source that can be found in nature and can be made as reference for medicines development and to support various industrial interests. These substances will never be depleted and continuously produced with interconverted molecular structure from time to time.

Ecologically, sponges occur at various habitat conditions. This poses a question regarding the adaptation mechanism of sponges, which are simple organism, toward their habitat environment conditions. Studies suggest that sponges have a chemical self-defense (secondary metabolite). These chemical compounds are useful to defend themselves from competitor pressure, antagonism reaction, infection or predation by other marine organisms [1]. Sponges produce two types of metabolites during their growth and development, namely primary metabolites and secondary metabolites. Primary metabolites are metabolites formed during growth period and used in metabolism processes that are essential to organism. Whereas, secondary metabolites are compound components produced when primary metabolite requirements have been met and used in evolutionary mechanism of the species or adaptation strategy to environment [2]. The characteristics of secondary metabolites include the formation of secondary metabolites that is highly influenced by the environment conditions of the organism and each of the secondary metabolites are produced by only some certain organisms [3] [4].

Secondary metabolite compounds play important roles in important biological processes including reproduction, competition, and protection against predators [5] [6]. Sponges have highly diverse compounds in their tissues and this has an effect to predators, competitors, community structure and dynamic, and evolution [7] [8]. Understanding the causes of this variation is a valuable information not only about factors affecting the secondary metabolite production but also conditions where sponges invest more on chemical protection.

Essential metals in biota cells are required not only for growth but also for self-protection from predators. Fe is required by sponges to synthesize secondary metabolites. The influence of heavy metal pollution can increase or decrease the secondary metabolites production in sponges.

Water environment conditions in Spermonde Archipelago due to low or high Fe concentration will influence the formation of secondary metabolites in the sponge. Heavy metal Fe and biota cells are needed not only for growth but also for self-defense against predator using the produced metabolites. The effect of Fe pollution can increase or reduce the secondary metabolite production in sponge at Spermonde Archipelago region.

2. Method

Clathria reinwardtii samples were chopped, and then dried under indirect sunlight for 3 days and crushed with blender. Sponge powder was macerated by methanol (Merck) solvent for 1×24 hours three times. Maceration result was separated from residu by decantation and then filtered by Whatman 41 paper. Organic macerates were taken and the solvent was evaporated by using rotavapour Buchi.

Methanol extract was then partitioned with mixture of dichloromethane (Merck) solvent and water at 1:1 ratio. Extraction was performed until the mixture was clear. The obtained dichloromethane extract was collected and then analyzed by FT-IR Shimadzu model IR Prestige-21 and GC-MS Agilent Technologies 7890A.

3. Results and Discussion

GC-MS result of dichloromethane extract from sponge *Clathria reinwardtii* provided chromatograms as can be seen in **Figure 1**. GC chromatograms shows retention time difference and percentage of peak area of compound components in each of dichloromethane extract from *Clathria reinwardtii* with highest and lowest Fe concentration. The same is true for compound fragmentation indicated by different Mass Spectrophotometry spectrum for each of the dichloromethane extract from the same sponge species but different in accumulated Fe concentration.

Both of these chromatograms show a peak with 99% similarity to α-limonen (compound from data center) and have a very high percentage of peak area at retention time (t_R) = 13.555, namely 62.82% with m/z 136.1 (CR-LL-DCM) and 90.32% with m/z = 136.1 (CR-BC-DCM) (**Figure 2**). Data on percentage of peak area and m/z at t_R = 13.555 was influenced by Fe accumulated in *Clathria reinwardtii* where at high Fe concentration the percentage of peak area was lower than *Clathria reinwardtii* extract at lower Fe concentration.

CR-LL-DCM chromatogram produced 2 peaks at t_R = 21.401 and 25.174 which were not found in chromate-gram CR-BC-DCM. This indicated that *Clathria reinwardtii* in Lae-Lae islands produced secondary metabolite compounds that were not produced by *Clathria reinwardtii* in Barrang Caddi islands as a mechanism to protect themselves against hi pollutant Fe concentration either in the *Clathria reinwardtii* theirself or in the sediment.

Compound produced by CR-LL-DCM at t_R = 21.401 (**Figure 3**) had 38% similarity to compound 1,1-dimethyl thiourea with percentage of peak area = 13.37 and m/z value = 405. Similarity level was very low due to different fragmentation pattern and m/z value of the compound from data center was 1,1-dimethyl thiourea = 104.

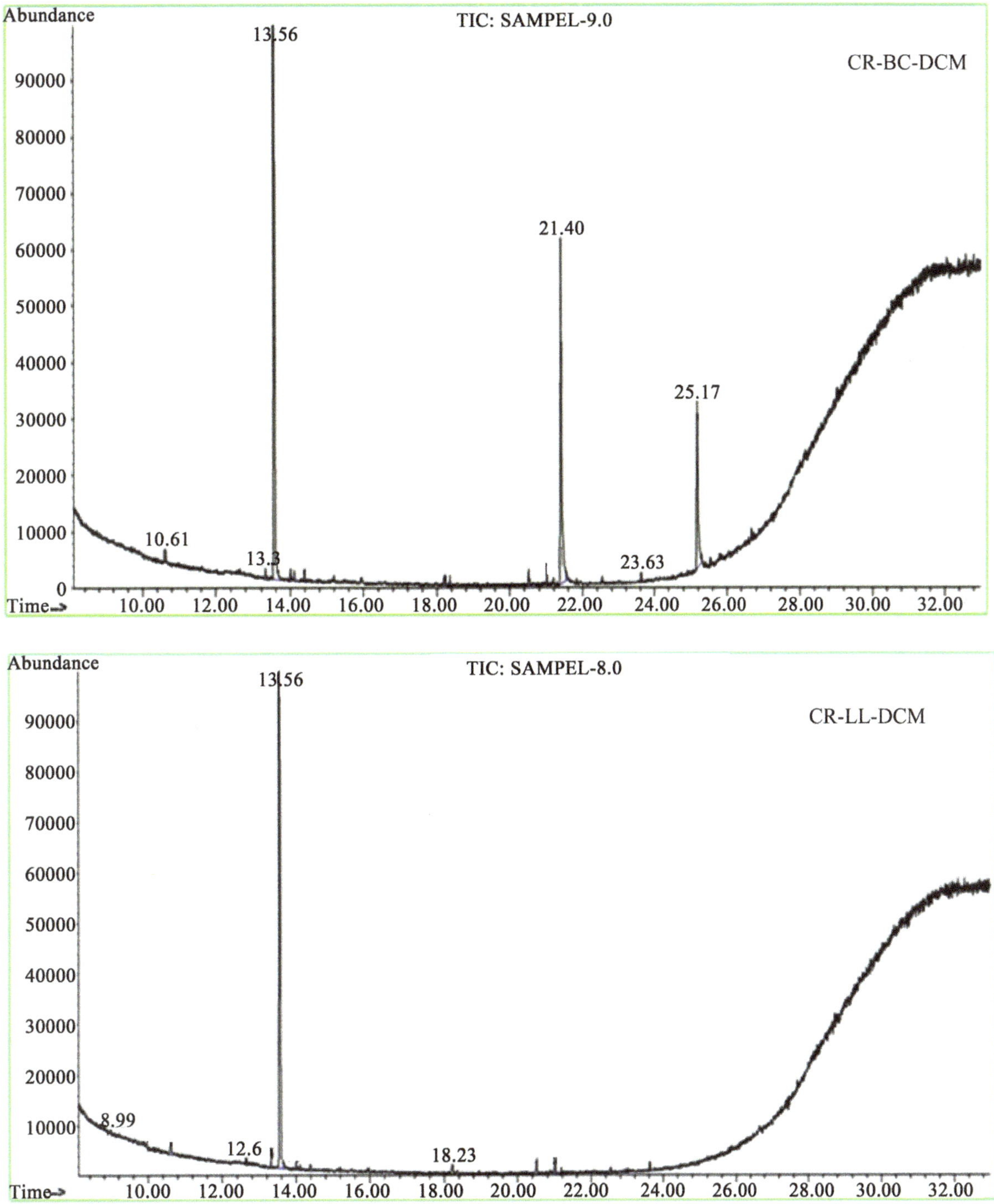

Figure 1. GC-MS chromatograms of dichloromethane extract of *Clathria reinwardtii.*

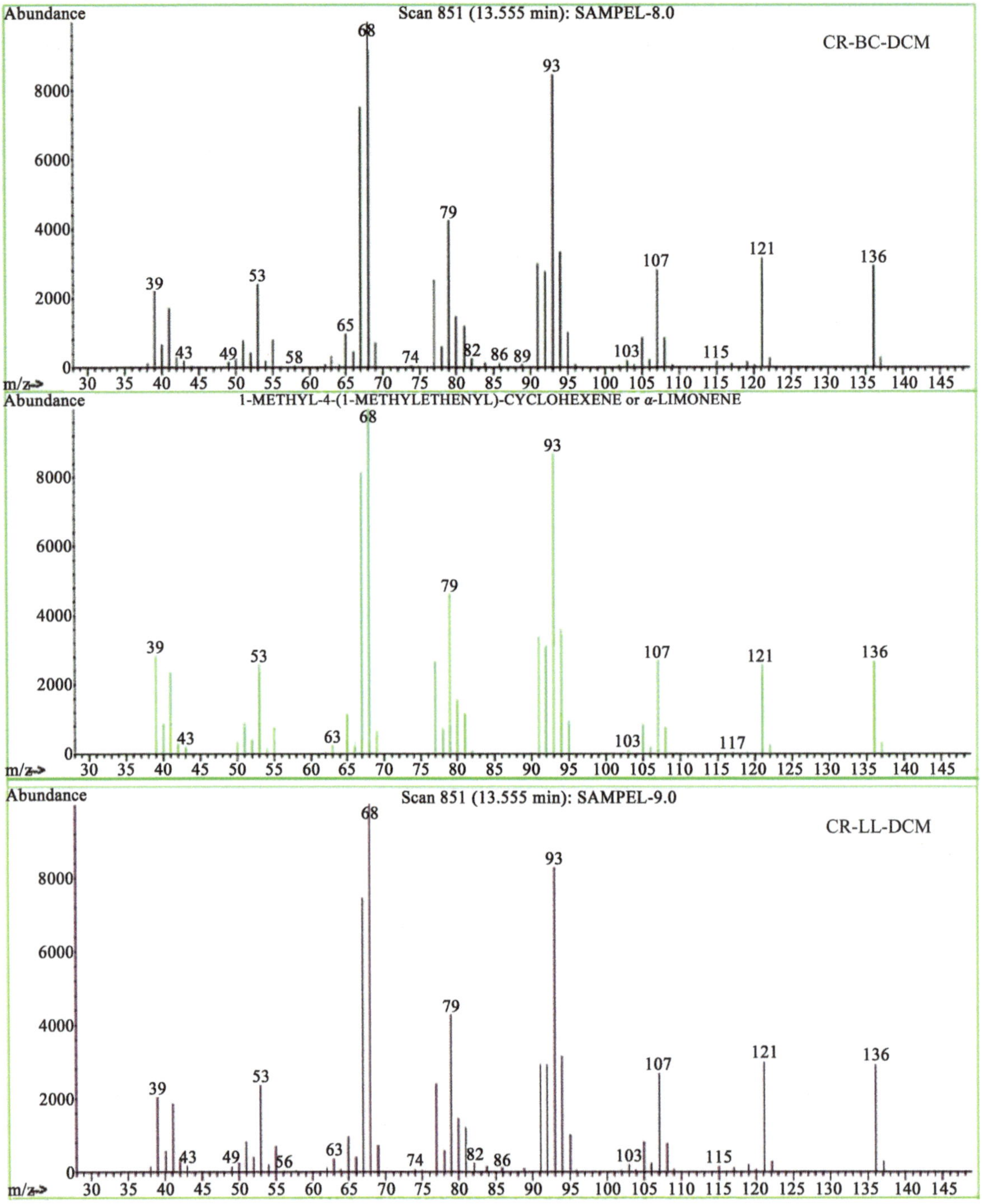

Figure 2. Mass spectrophotometry spectra at t_R = 13,555 of *Clathria reinwardtii.*

The compound at t_R = 25.174 (**Figure 4**) had 83% similarity to the compound (1-methyl ethyl)-2-[(aminocarbonil) oxylmethyl]-2-methyl penthyl ester carbamate with percentage of peak area = 12.04 and *m/z* value = 281. The similarity level was only 83% because the fragmentation pattern is not fully equal and the *m/z* value from the central data was (1-methyl ethyl)-2-[(aminocarbonil) oxylmethyl]-2-methyl penthyl ester carbamate = 260.

FT-IR analysis results as shown in **Figure 5** indicates that *Clathria reinwardtii* accumulating Fe at highest and lowest concentration produced strong peak at wave number of about 3400 cm^{-1} for O-H and N-H stretch. Flexible peak O-H and C-O stretch at wave number of about 1200 cm^{-1} subjected to change from medium intensity at low Fe concentration (CR-BC-DCM) to weak intensity at high Fe concentration (CR-LL-DCM).

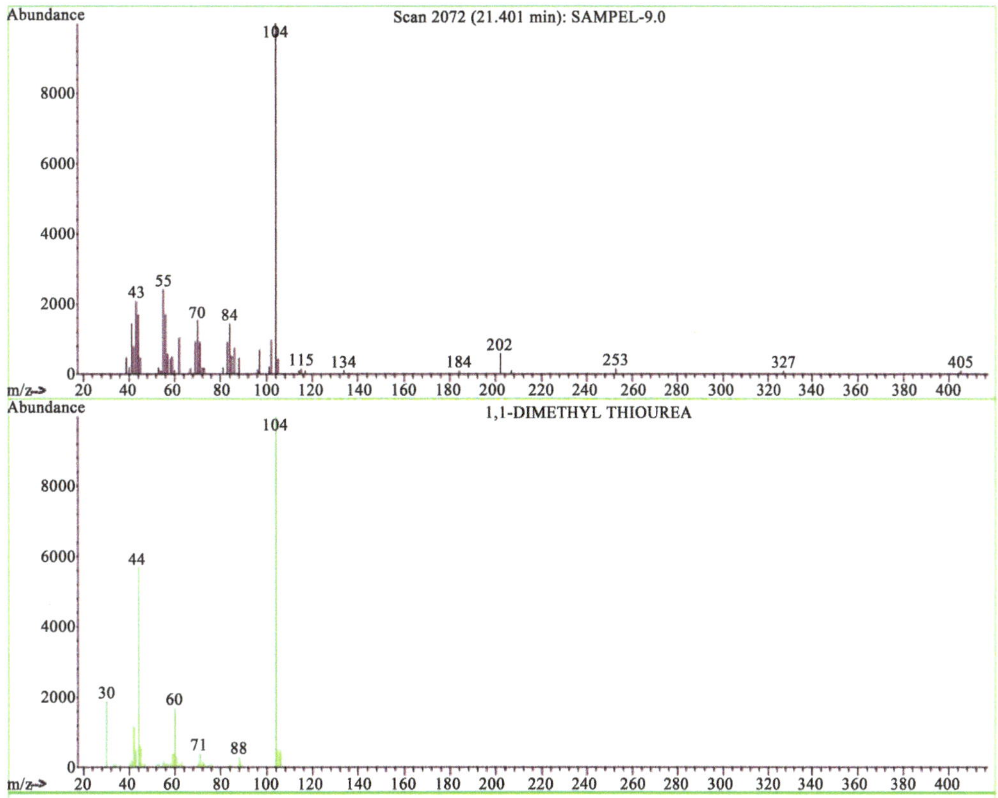

Figure 3. Mass spectrophotometry spectra at t_R = 21,401 of *Clathria reinwardtii* (CR-LL-DCM).

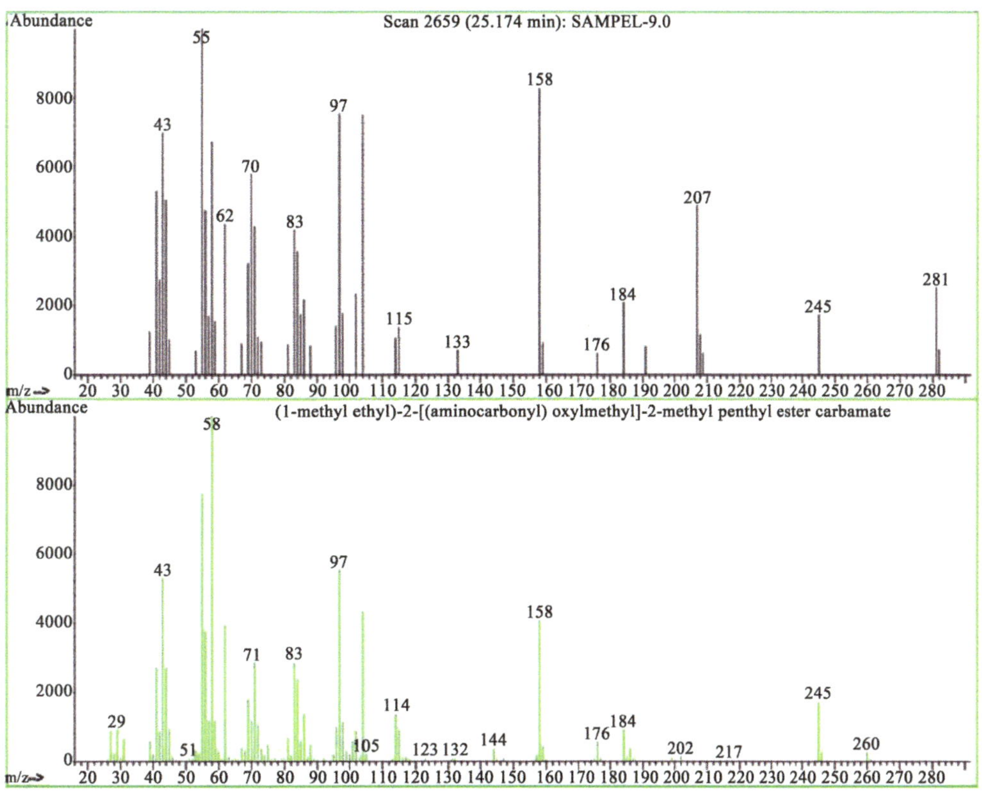

Figure 4. Mass spectrophotometry spectra at t_R = 25,174 from *Clathria reinwardtii* (CR-LL-DCM).

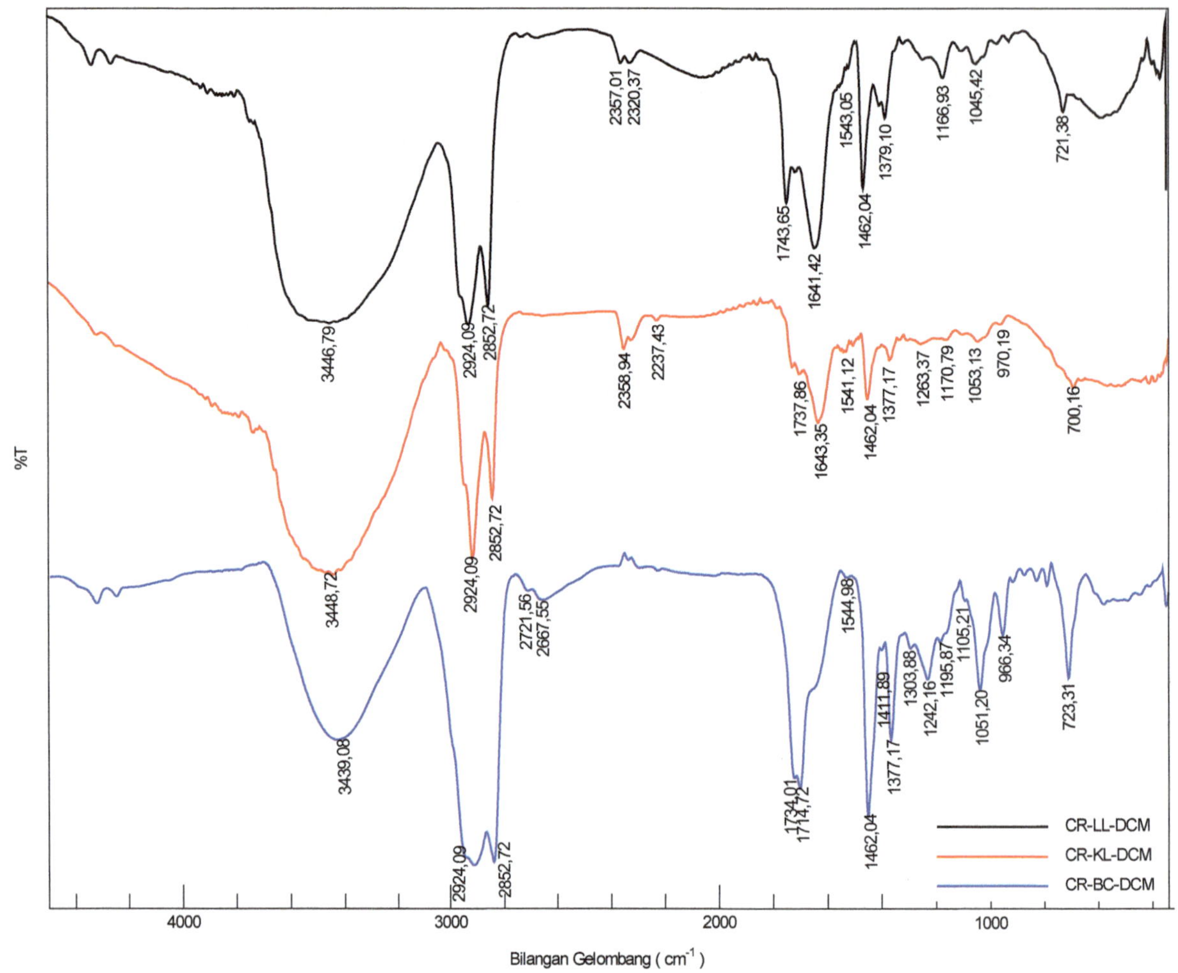

Figure 5. FT-IR spectrum of *Clathria reinwardtii* secondary metabolites.

The change in intensity of O-H flexible peak and C-O stretch from medium to weak indicated the influence of Fe at high concentration resulting the weakening of O-H flexible and C-O stretch peaks. These peaks underwent shift of 21 cm^{-1} to bigger wave number direction or lower wavelength (blue shift). The wave number shift can be caused by induction, mesomery effect, and field effect [9]. Peak at 1240.23 cm^{-1} (CR-LL-DCM) underwent only change to low intensity. The change in absorption intensity from strong to medium or from medium to weak indicated that the secondary metabolite concentration decreased in *Clathria reinwardtii* extract due to the effect of high Fe concentration.

The data obtained from measurements results of dichloromethane extract of *Clathria reinwardtii* using GC-MS and FT-IR indicated that the influence of Fe accumulated in high concentration on secondary metabolites that can be produced. The obtained molecular weight from Mass Spectrophotometry for dichloromethane extract of *Clathria reinwardtii* with high Fe concentration was generally greater than extract with low Fe concentration.

4. Conclusions

Chromatograms indicated peaks with 99% similarity to α-limonen with percentage of peak area at retention time (t_R) = 13.555 where at high Fe concentration produced percentage of peak area 62.82% and at low Fe concentration 90.32%.

Chromatograms indicated that *Clathria reinwardtii* at t_R = 21.401 and 25.174 in Lae-Lae island produced secondary metabolite compounds that were not produced by *Clathria reinwardtii* in Barrang Caddi island as a mechanism of self defense against high Fe concentration in the *Clathria reinwardtii* itself or in sediment.

The change in intensity of O-H flexible and C-O stretch peak from medium to weak indicated the influence of Fe at high concentration that weakened the O-H flexible and C-O stretch capability. These peaks underwent shift of 21 cm^{-1} to the direction higher wave number or lower wavelength (blue shift).

References

[1] Thoms, C. and Schupp, P.J. (2007) Chemical Defense Strategies in Sponges: A Review. *Porifera Research Biodiversity, Innovation, and Sustainability*, Rio de Janeiro, 627-637.

[2] Torssel, K.B.G. (1983) Natural Product Chemistry. A Mechanisitc and Biosynthetic Approach to Secondary Metabolism. John Wiley and Sons Ltd., New York.

[3] Crueger, W. and Crueger, A. (1982) Biotechnology: A Textbook of Industrial Microbiology. Science Tech. Inc., Madison.

[4] Madigan, M.T., Martinko, J.M., Parker, J. and Brock, T.D. (2000) Biology of Microorganisms. Prentice-Hall Inc., Upper Saddle River.

[5] McClintock, J.B and Baker, B.J. (2001) Marine Chemical Ecology. CRC Press, Boca Raton.

[6] Paul, V.J and Puglisi, M.P. (2004) Chemical Mediation of Interactions among Marine Organisms. *Natural Product Reports*, **21**, 189-209. http://dx.doi.org/10.1039/b302334f

[7] Cimino, G. and Ghiselin, M.T. (2001) Marine Natural Products Chemistry as an Evolutionary Narrative. In: McClintock, J.B. and Baker, B.J., Eds., *Marine Chemical Ecology*, CRC Press, Boca Raton, 115-154.

[8] Paul, V.J., Cruz-Rivera, E. and Thacker, R.W. (2001) Chemical Mediation of Macroalgal-Herbivore Interactions: Ecological and Evolutionary Perspectives. In: McClintock, J.B. and Baker, B.J., Eds., *Marine Chemical Ecology*, CRC Press, Boca Raton, 227-265.

[9] Supratman, U. (2010) Elucidation of Organic Compound Structures. Spectoscopy Method for Organic Compound Structure Determination. Widya Padjadjaran.

Permissions

All chapters in this book were first published in ABC, by Scientific Research Publishing; hereby published with permission under the Creative Commons Attribution License or equivalent. Every chapter published in this book has been scrutinized by our experts. Their significance has been extensively debated. The topics covered herein carry significant findings which will fuel the growth of the discipline. They may even be implemented as practical applications or may be referred to as a beginning point for another development.

The contributors of this book come from diverse backgrounds, making this book a truly international effort. This book will bring forth new frontiers with its revolutionizing research information and detailed analysis of the nascent developments around the world.

We would like to thank all the contributing authors for lending their expertise to make the book truly unique. They have played a crucial role in the development of this book. Without their invaluable contributions this book wouldn't have been possible. They have made vital efforts to compile up to date information on the varied aspects of this subject to make this book a valuable addition to the collection of many professionals and students.

This book was conceptualized with the vision of imparting up-to-date information and advanced data in this field. To ensure the same, a matchless editorial board was set up. Every individual on the board went through rigorous rounds of assessment to prove their worth. After which they invested a large part of their time researching and compiling the most relevant data for our readers.

The editorial board has been involved in producing this book since its inception. They have spent rigorous hours researching and exploring the diverse topics which have resulted in the successful publishing of this book. They have passed on their knowledge of decades through this book. To expedite this challenging task, the publisher supported the team at every step. A small team of assistant editors was also appointed to further simplify the editing procedure and attain best results for the readers.

Apart from the editorial board, the designing team has also invested a significant amount of their time in understanding the subject and creating the most relevant covers. They scrutinized every image to scout for the most suitable representation of the subject and create an appropriate cover for the book.

The publishing team has been an ardent support to the editorial, designing and production team. Their endless efforts to recruit the best for this project, has resulted in the accomplishment of this book. They are a veteran in the field of academics and their pool of knowledge is as vast as their experience in printing. Their expertise and guidance has proved useful at every step. Their uncompromising quality standards have made this book an exceptional effort. Their encouragement from time to time has been an inspiration for everyone.

The publisher and the editorial board hope that this book will prove to be a valuable piece of knowledge for researchers, students, practitioners and scholars across the globe.

List of Contributors

Xiaoyu Tang, Laura M. C. Ades, Dorothy M. Morré and D. James Morré
MorNuCo, Inc., Purdue Research Park, West Lafayette, IN, USA

Haruko Inose, Kotomi Mukai, Misa Ito and Seiji Masuda
Division of Integrated Life Science, Graduate School of Biostudies, Kyoto University, Kyoto, Japan

Abha J. Chalpe, Chad D. Law, Jennifer N. Dumdie and Kathleen M. Eyster
Division of Basic Biomedical Sciences, Sanford School of Medicine, The University of South Dakota, Vermillion, SD, USA

Keith A. Hansen and Kathleen M. Eyster
Department of Obstetrics and Gynecology, Sanford School of Medicine, The University of South Dakota, Sioux Falls, SD, USA

JoDi Lynn Osborn and Susanna F. Greer
Department of Biology, Georgia State University, Atlanta, USA

Norah Basopo and Thamsanqa Ngabaza
Department of Applied Biology and Biochemistry, National University of Science and Technology, Bulawayo, Zimbabwe

D. S. Jairajpuri
Department of Medical Biochemistry, College of Medicine and Medical Sciences, Arabian Gulf University, Manama, Kingdom of Bahrain

S. Fatima
Department of Biochemistry, Faculty of Life Science, Aligarh Muslim University, Aligarh, India

Z. S. Jairajpuri
Department of Pathology, Hamdard Institute of Medical Sciences and Research, Hamdard University,New Delhi, India

Lidia Syrtsova, Natalia Sanina, Boris Psikha, Ildar Tukhvatullin, Natal'ja Shkondina, Olesia Pokidova and Alexander Kotelnikov
Institute of Problems of Chemical Physics of the Russian Academy of Sciences, Moscow Region,Russian Federation

Kozo Nagano
Nagano Research Institute of Molecular Biology, Yokohama, Japan

Valerija Dunkić, Ivana Radovanović, Nada Bezić and Elma Vuko
Department of Biology, Faculty of Science, University of Split, Split, Croatia

Wataru Sakamoto
Serotec Laboratory, Ebetsu, Japan

Takashi Kanehira and Hirohisa Hongou
Division of Preventive Dentistry, Graduate School of Dental Medicine, Hokkaido University, Sapporo, Japan

Kozo Asano
Division of Applied Bioscience, Research Group of Molecular Bioscience, Graduate School of Agriculture, Hokkaido University, Sapporo, Japan

Pietro Volpe
Department of Biology, University of Rome "Tor Vergata", Rome, Italy
National Institute of Biostructures and Biosystems, Rome, Italy

Humberto Rafael Bravo-D and Alain Cruz-Nolasco
Divisi6n de Estudios de Posgrado, Universidad del Papaloapan, Circuito Central 200, Parque Industrial, wwTuxtepec, Oaxaca, México

Luis Raúl Gutiérrez-Lucas and Ana Karin Navarro-Mtz
Instituto de Biotecnología, Universidad del Papaloapan, Circuito Central 200, Parque Industrial, Tuxtepec, Oaxaca, México

Gad Degani
School of Science and Technology, Tel-Hai Academic College, Tel Hai, Israel MIGAL—Galilee Technology Center, Kiryat Shmona, Israel

Hedi Mighri, Khawla Sabri, Hajer Eljeni, Mohamed Neffati and Ahmed Akrout
Range Ecology Laboratory, Arid Lands Institute, University of Gabès, Medenine, Tunisia

Eric Romney
Arizona College of Osteopathic Medicine, Midwestern University, Glendale, Arizona

Vinay J. Nagaraj
Department of Biochemistry, Midwestern University, Glendale, Arizona

E. Leila Jerome Clay
Departments of Pediatrics and Internal Medicine/ Division of Hematology and Oncology, Georgia Regents University, Augusta, GA, USA

Alison Motsinger-Reif
Department of Statistics, Bioinformatics Research Center, North Carolina State University, Raleigh, NC, USA

Janelle Hoskins
Gentris Corporation, Morrisville, NC, USA

Lindsay Veit
Orthopedic Surgery Resident, University of Toledo Medical Center, Toledo, OH, USA

Ali Calikoglu
Department of Pediatrics/Division of Endocrinology, University of North Carolina School of Medicine, Chapel Hill, NC, USA

Rupa Redding-Lallinger
Departments of Pediatrics and Internal Medicine/ Division of Hematology and Oncology, University of North Carolina School of Medicine, Chapel Hill, NC, USA

Kenta Hosoda, Masashi Sato and Kazuyuki Yanai
Department of Biomolecular Science, Faculty of Science, Toho University, Chiba, Japan

Kenta Hosoda, Masashi Sato and Kazuyuki Yanai
Department of Biomolecular Science, Faculty of Science, Toho University, Chiba, Japan

Jun Inajima, Yoshio Inouye and Yuichiro Kanno
Faculty of Pharmaceutical Sciences, Toho University, Chiba, Japan

Manoj S. Nair and Donald H. Dean
Department of Chemistry and Biochemistry, The Ohio State University, Columbus, OH, USA

Manoj S. Nair
Aaron Diamond AIDS Research Center, Rockefeller University, New York, NY, USA

Jean-Michel Bourget, Karine Zaniolo, Sylvain L. Guérin and Stéphanie Proulx
Centre de recherche du CHU de Québec and Centre de recherche en organogénèse expérimentale del'Université Laval/LOEX, axe médecine régénératrice, Hôpital du Saint-Sacrement, Québec, Canada

Mohib Morcos
Départment d'anatomie et de cytologie pathologiques, CHU de Québec, Hôpital du Saint-Sacrement, Québec, Canada

Département de biologie moléculaire, biochimie médicale et pathologie, Faculté de médecine, Université Laval, Québec, Canada

Sylvain L. Guérin and Stéphanie Proulx
Département d'Ophtalmologie, Faculté de médecine, Université Laval, Québec, Canada

Jean-Rene Q. Acquah, Kousuke Haratake and Tomoki Chiba
Graduate School of Life and Environmental Sciences, University of Tsukuba, Ibaraki, Japan

Randeep Rakwal
Organization for Educational Initiatives, University of Tsukuba, Ibaraki, Japan

Heiichiro Udono
Graduate School of Medicine, Dentistry and Pharmaceutical Sciences, Okayama University, Okayama, Japan

Joseph M. Benoun, Danielle Lalimar-Cortez, Analila Valencia, Adriana Granda, Destaye M. Moore, Eric P. Kelson and Paula L. Fischhaber
Department of Chemistry and Biochemistry, California State University Northridge, Northridge, CA, USA

Olusola A. Ladokun and Sarah O. Oni
Lead City University, Ibadan, Nigeria

Sri Vidawati
Department of Physics, Faculty of Natural Science and Mathematics, University Indonesia, Depok, Indonesia

Ulrich Rothe
Institute of Physiological Chemistry, Faculty of Medicine, University Halle-Wittenberg, Hollystr, Halle/Saale, Germany

Lydia Melawaty
Chemical Engineering Department, Paul Christian University of Indonesia, Makassar, Indonesia

Kristiana Pasau
Mechanical Engineering Department, Paul Christian University of Indonesia, Makassar, Indonesia